Y0-CBI-356

HANDBOOK OF ENGINEERING MANAGEMENT

HANDBOOK OF ENGINEERING MANAGEMENT

JOHN E. ULLMANN
Editor

DONALD A. CHRISTMAN
BERT HOLTJE
Associate Editors

A JAMES PETER BOOK

James Peter Associates, Inc.

A Wiley-Interscience Publication

John Wiley & Sons

New York · Chichester · Brisbane · Toronto · Singapore

Library of Congress Cataloging in Publication Data:

Main entry under title:
Handbook of engineering management.
   "A James Peter book."
   "A Wiley-Interscience publication."
   Includes index.
   1. Engineering—Management.   I. Ullmann, John E.
II. Christman, Donald A.   III. Holtje, Bert.
TA190.H36   1986   620'.0068   85-22805
ISBN 0-471-87828-6

Printed in the United States of America

10 9 8 7 6 5 4 3 2 1

CONTRIBUTORS

ROBERT M. ANDERSON, JR.
General Electric Company
Bridgeport, Connecticut

MICHAEL K. BADAWY
Virginia Polytechnic Institute
Falls Church, Virginia

SUZANNE DUMONT BAYS
Bank of Montreal
Toronto, Ontario

WILLIAM V. BELLANDO
Deloitte, Haskins & Sells
New York, New York

GEORGE BLACK
Executive Vice President
Bozell & Jacobs
Union, New Jersey

PAUL V. BONFIGLIO
West Hempstead, New York

WILLIAM C. BYHAM
Development Dimensions
International
Pittsburgh, Pennsylvania

MAHESH CHANDRA
Hofstra University
Hempstead, New York

BERNARD E. DELURY
Sea-Land Corporation
Iselin, New Jersey

JOHN DIGAETANI
Hofstra University
Hempstead, New York

RALPH DIPIETRO
Montclair, State College
Montclair, New Jersey

MARC DORIO
McCooe Associates
Ridgewood, New Jersey

MICHAEL FRISCH
New York, New York

SAMUEL E. GLUCK
New York, New York

ROBERT R. GODFREY
OMRC Management Consultants
Yorktown Heights, New York

GUS W. GRAMMAS
Columbia University
New York, New York

EARL HITCHCOCK
Midland Park, New Jersey

GERRE L. JONES
Gerre Jones Associates
Washington, DC

ARTHUR D. KELLNER
Roseland, New Jersey

FRANCINE KENEBORUS
Sea-Land Corporation
Iselin, New Jersey

MICHAEL KLAHR
Barry University
Miami Shores, Florida

HAROLD LAZARUS
Hofstra University
Hempstead, New York

GREG LEWIN
Neuberger and Berman
New York, New York

MILTON F. LUNCH
National Society of Professional
Engineers
Alexandria, Virginia

JOHN McCOOE
McCooe Associates
Ridgewood, New Jersey

PATRICK MONTANA
Hofstra University
Hempstead, New York

RONALD B. MORGAN
Burroughs Corporation
Detroit, Michigan

PETER V. NORDEN
IBM Corporation
White Plains, New York

ADOLPH M. ORLANDO
Pelham Manor, New York

JOEL N. ORR
Orr Associates
Danbury, Connecticut

ADA PRESSMAN
Monterey Park, California

KENNETH H. RECKNAGEL
Leadership Development
Associates, Inc.
Westwood, New Jersey

ERCOLE ROSA
IBM Corporation
New York, New York

GEORGE R. ROUKIS
Hofstra University
Hempstead, New York

LEONARD J. SMITH
Leonard J. Smith Associates
Rutherford, New Jersey

EDWARD A. TOMESKI
Barry University
Miami Shores, Florida

ROY UDOLF
Hofstra University
Hempstead, New York

ROY WALTERS
Roy W. Walters & Associates
Mahwah, New Jersey

DONALD G. WEINERT
National Society of Professional Engineers
Alexandria, Virginia

JACK L. YURKIEWICZ
Pace University
New York, New York

PREFACE

This handbook addresses itself to managers in the two kinds of organizations that furnish engineering services. The first group is comprised of the engineering departments of manufacturing firms and other enterprises with technological requirements, such as utilities and government agencies. The second group consists of engineering firms that furnish a diverse range of engineering services to others. The book provides a wealth of information and suggestions on how to run such operations successfully and with proper regard to the available resources and their deployment.

The book is divided into four parts. The first part defines the engineering profession, and discusses its structure and major interrelationships with other businesses as well as its legal regulations. Part II covers the principles of organization: the nature of engineering work, resources required, and managerial tools that are essential for effective monitoring of the jobs.

The third part deals with a wide range of interpersonal skills that are involved in the utilization of human resources in the engineering organization. Finally, the fourth part offers advice to individual engineers on how to function in an engineering organization.

JOHN E. ULLMANN
DONALD A. CHRISTMAN
BERT HOLTJE

Hempstead, New York
Wyckoff, New Jersey
Tenafly, New Jersey
March 1986

ACKNOWLEDGMENTS

I am very much aware of the difficulties in getting busy professionals in the various subject areas to take time out from their demanding schedules to prepare their chapters. My first thanks go to the contributors for their excellent cooperation and interest. I am also most grateful to my friends and colleagues at Hofstra University, Dr. Harold Lazarus and Dr. George S. Roukis, who, in addition to being able contributors, were very helpful in suggesting possible authors of other chapters.

J.E.U.

CONTENTS

III. THE ENGINEER AS MANAGER

IV. MANAGING YOUR CAREER

THE ENGINEER
AS MANAGER

1

ENGINEERING AS A FUNCTION

John E. Ullmann

JOHN E. ULLMANN is professor of Management and Quantitative Methods at Hofstra University. He has a B.S. in civil and mechanical engineering from the University of London and an M.S. and Ph.D. in industrial engineering from Columbia University, and he is a registered professional engineer in New York State. Prior to his teaching career, which included faculty positions at Stevens Institute of Technology, Columbia, and New York universities, Dr. Ullmann was in engineering practice for 14 years. He specializes in product and market analysis, quantitative analysis, and related planning and policy formulation in several areas of business and the public sector. He has been a consultant and expert witness to business and government in statistical and economic analysis, industry and community development, and industrial marketing. Dr. Ullmann has written more than eighty books, monographs, and articles in professional journals. His books include his most recent *The Prospects of American Industrial Recovery,* as well as *Quantitative Methods in Management, Social Costs in Modern Society, The Suburban Economic Network, The Improvement of Productivity* and *Manufacturing Management.* He is a Fellow of the New York Academy of Sciences and member of the Labor Panel of the American Arbitration Association.

1.1.   ENGINEERING DEPARTMENTS

The function of engineering is carried out in two major ways. The first is that of the engineering department of an organization having broader purposes of one sort or another. The other, which is discussed in Section 1.2, is that of the engineering firm or other organization in which engineering is the main activity. Engineering departments carry out staff activities (see Chapter 5), whereas engineering firms or similar organizations have engineering as their line activity.

Engineering departments are by far the most numerous of the two groupings. Most engineering departments are found in manufacturing, but they are also a necessity in mining enterprises; utilities in the fields of energy, transportation and communications; government agencies at all levels, from local highway departments to the armed services; and other sectors with quasimanufacturing operations, including maintenance and repair services.

1.1.1.   Categories of Work

The work done by engineering departments generally falls into five categories. The proportion of work in each category depends on the nature of the product and managerial objectives (see Section 1.1.2). In approximate ascending order of sophistication the five categories of work done by engineering departments are:

1. *Technical support for present products.* This includes responding to customers' problems, making minor design changes, and assessing competitive factors relating to product design. Spare parts or spare-part kits and operating supplies for the product would also be included.

2. *Process development and production systems.* This relates to new methods of making the product. The work may focus on small incremental changes in tooling, operation sequencing, and responses to minor product design changes. This is something of a routine activity in many engineering departments of manufacturing firms. At the other extreme the work may involve design of radically new production facilities, including experimental units, pilot plants, and the like. Some of the more advanced and research-oriented work done by engineering departments almost invariably has a production systems component as well as one related only to the product. For instance, a major change in materials used, almost always involves a change in production methods as well.

3. *New models of present products.* In a sense this category also is something of a routine function in engineering departments of manufacturing firms. The scope of the work depends on how extensive the changes are. In practice they are often mainly cosmetic, such as new sheet metal designs in cars or other superficial changes that may serve as starting points for new advertising campaigns. New models of present products may be developed in response to technical changes made by suppliers or to a need for cutting costs in manufacturing processes or production materials. Such changes probably account for the majority. At the other end of this category are more substantial modifications of design that incorporate some technological innovations. The most advanced stage would be that of essentially fulfilling the function of the original product by means of a drastically different design. At this point the tasks would begin to shade over into the fourth category.

4. *A new product.* This may be an item totally new in function or an item designed to fulfill the function of an existing one by a design or chemical formulation so different as to require major changes in the production process as well. New products may require new production lines, substantial relayouts, or entirely new machinery or processing units.

5. *Basic research.* Basic research is defined by the National Science Foundation (NSF) as an "original investigation for the advancement of scientific knowledge. This investigation does not have a specific commercial objective, although such investigation may be (and usually is) in a field of present or potential interest to the company"[1] or other entity which the engineering department serves.

At this point it is also useful to set forth the NSF's definitions of applied research and development. According to the NSF, applied research consists of "investigations directed to the discovery of new scientific knowledge having specific commercial objectives with respect to products or processes. This definition differs from that of basic research chiefly in terms of the objectives of the . . . company." Development consists of "technical activities of a nonroutine nature concerned with translating research findings or other scientific knowledge into products and processes. Does not include routine technical services to customers or other activities excluded from the above definitions of research and development."[2]

The aforementioned definitions leave enough to individual interpretation to suggest that distinctions between various kinds of engineering work lie to a large extent in the eye of the beholder. Still it is clear, in terms of the NSF categories, that applied research and development is carried out under categories 2, 3, and 4, while basic research is carried out under 5. Category 1 is specifically excluded in the definition of development. In a typical engineering department several kinds of jobs are usually done simultaneously. Basic research may not play much of a role, but even an advanced literature search may have elements of it.

1.1.2.   Engineering and Corporate Objectives

The importance of engineering in the company structure depends first on the objective characteristics of the product. The technological content varies enormously, ranging from high technology products in which it really is the main ingredient to such industries as apparel in which there is little technological change and the problems are not primarily technical. Several areas, including some services, have seen an increase in technical content (for example, computer-related work), leading to a corresponding rise in the importance of engineering in corporate councils.

On the other hand, the growth of multiproduct conglomerates that had been put together for financial rather than operational reasons or product relationships has been associated with an emphasis on short-term results, and, more important from the viewpoint of engineering, which must often work with lengthy time horizons, this growth has been linked to the short-term deployment or redeployment of corporate assets. Companies merged into a conglomerate found themselves being spun off or even liquidated when they no longer fit into the financial structure. In other cases companies used their own resources to fund wider corporate objectives or other acquisitions. This often happened to manufacturing firms or their units, as manufacturing in general experienced a decline in the United States from about 1960. In this context the importance of engineering suffered in many organizations; even the share of chief executives with engineering backgrounds (never large) fell.

It is clear that the role of engineering within top management is not always determined by its own internal performance. Yet its importance is manifest; engineering managers are thus under the necessity of justifying their activities in a broader context than mere budgetary infighting or rivalries with such staff departments as research or styling which may have very similar survival problems.

1.1.3.   Methods of Financing

The general question of budgeting for engineering is addressed in Chapter 16. It is proper to note at this stage, however, that the financing of engineering in a business organization takes place in two ways. The first is as a regular staff activity, the expense of which is part of the regular business obligations of the company. The second is to seek out separately compensated engineering activities by way of research contracts. This method is regularly encountered in military procurement where research and production contracts are often awarded separately, sometimes to different firms. However, based perhaps on that example and on the availability of government funds in such new aspects of product design as safety and environmental problems, firms in ever-wider areas are seeking to have their engineering or research functions paid for separately by outside agencies. The major deterrent is potential government ownership of innovations which were originally paid for by public funds.

1.1.4. Outside Help

The complex and varied nature of modern engineering often requires that companies seek outside assistance with their engineering problems. This is obviously true when patent rights for products and processes must be obtained. But even in the general purchase of industrial equipment, vendors are obligated by contract as well as custom to render substantial technical service to their customers. This is a major part of the technical support discussed in Section 1.1.1. The recipients of such help benefit from not having to expend engineering resources of their own for such services.

It is a grave mistake, however, for companies and their engineering managers to rely too intensively on such services. Whatever the quality of such services may be, they are rendered in order to further the interests of the suppliers; these may overlap substantially with those of the equipment users, but studies of economic factors in product design, reveal that reducing the manufacturing costs is a key determinant. This bears importantly on such aspects as product-life or durability in which producers and users may have sharply divergent concerns.[3]

A potentially worse problem is created when the innovative portions of engineering activities are left predominantly to outsiders. The rationale here is that the company can always buy whatever it needs from actual or would-be suppliers of capital equipment. For example, the steel industry of the United States has bought new processes and methods from equipment suppliers instead of making any significant effort toward creating its own process development. The result has been its relative technical backwardness as compared with its international competitors.[4]

These caveats are very important. Nevertheless, the obvious need to avoid reinventing the wheel or other functional advantage may well require that the question of make or buy be resolved in favor of the latter. If deemed necessary, however, this should not absolve the client from careful evaluation in terms of what is being bought and how the job is progressing.

1.2. ENGINEERING FIRMS

Firms that sell engineering services as their principal or only product fall into two major categories. The first of the two is that of construction-related activities. The second is a much more variegated group that engages in many different other kinds of engineering and related services.

1.2.1. Construction-Related Firms

The firms of this group are generally part of what the Standard Industrial Classification (SIC) of the United States Department of Commerce calls Industry 8911, "Engineering, Architectural and Surveying Services." A much smaller and sharply fluctuating volume of work is done by engineering job shops (SIC 7362) which essentially furnish temporary engineering help and are used for overloads and similar purposes in the same way as office temporaries. In fact, SIC 7362 covers all such temporary employment. The reason for combining engineering with architecture and surveying is that these services are often found in combination, especially the first two, or alternatively, the latter two. The group does, of course, include many firms which specialize in one area, such as heating, ventilating and air conditioning, and electrical work.

Though in important respects consultative, the scope of the work of these firms is often managerial in character. Such firms are in charge of overall designs and layouts and assist in getting bids on the equipment to be ordered. The firms evaluate the bid, propose or authorize purchases, and then carefully examine the drawings and specifications submitted by vendors. Their job is delimited to a degree by the amount of technical and design work done by equipment suppliers. While there is obviously a need for the services offered by these firms, the same warnings on excessive reliance as those discussed in Section 1.1.4 apply here. Engineering firms are often construction managers as well, offering surveillance for the duration of a construction job. Some large organizations offer design and construction together, thus serving as engineers, architects, and general contractors.

1.2.2. Other Engineering Organizations

Other engineering organizations engage in a wide variety of research, consulting, and design services. They include research laboratories (SIC 7391, if for profit, SIC 8922, if nonprofit) and management consultants (SIC 7392; this wide grouping includes "management engineers," a synonym for industrial engineers).

All these industries may engage in work with a substantial engineering content. In some cases, as in testing laboratories, substantial engineering work may be involved in devising the testing apparatus itself. Other engineering firms in this section do contract design work of the specialized kind described in Section 1.1.4, or exist to exploit a particular design or process in which they have proprietary or patent rights. There are also university units and government-owned firms such as Mitre or Rand Corporation which carry on engineering functions of a predominantly military nature for the government. Consultants in computer-related activities (SIC industry group 737) may also have to do significant technical work in their assignments.

The aforementioned is clearly a highly varied group. Many of its members deal with newer areas of technology or technical services. Although still a minority, consultants in computer-related activities are a steadily growing sector within engineering firms.

1.3. SPECIALIZATION

Specialization in engineering has a profound effect on its organization, both in engineering departments and engineering firms. The basic question is what branches of engineering are required for the job to be done. Engineering departments may themselves call for a relatively narrow range of work, enough to service the product being made. A firm making fabricated metal products without moving parts may only need mechanical engineers expert in that area. Machinery, however, almost invariably calls for electrical or electronic engineers as well; in fact, plastic parts may call for further specialists.

As to engineering firms, some may only work in one branch of engineering, but they are likely to be the exception. Most firms involved in construction have staff in some or all of several branches including architecture; civil, mechanical, and electrical engineering; air conditioning and other climate control; interior design; and landscape architecture.

There are two basic modes of organizing a firm: (1) by project or (2) by specialty. In a sense this parallels the division of manufacturing systems into product or process layouts or mass production and job shops. In a project-type organization, teams are assembled from various specialties and work together to complete the job, with the project engineers or managers exercising full line supervision over their staff. In an organization along specialty lines, each branch of engineering has its own internal supervision, with specific jobs for the projects done by individuals within each branch. The project engineer or manager then essentially only exercises a coordinating function. This may appear somewhat convoluted, but project engineers and managers are busy people who have to keep in close contact with clients and their representatives. It is easier for them if the day-to-day supervision of their staffs is someone else's responsibility.

It is not easy to specify which is the better mode of organization. Both have advantages and disadvantages, the import of which may well be mainly determined by some of the personalities involved. A project-type organization has the great advantage of small group flexibility; hence even large and otherwise strongly hierarchical organizations often resort to it when the project is discrete and separable. However, problems may arise whenever a job is finished or nearing completion. Projects may retain people just in order to maintain internal power relationships or people may remain simply by inertia. Meanwhile other projects have hired people and are fully staffed; therefore, when the first job is over, its participants are hard to place elsewhere. This may exacerbate the feast-or-famine nature of much engineering employment.

A specialty-type organization avoids some of this; the head of, say, electrical engineering, may reassign members of the department to various projects as the need arises, and can thus (if the job is done well) improve the overall deployment of departmental resources. However, when different assignments are made, the close familiarity with the job, often found in project-type organizations, may be impaired. There is further scope for internal dissension if project engineers and the chief engineers of specialties get into turf battles.

1.4. THE SOCIAL SETTING OF ENGINEERING

Engineering, technology, and their contiguous branches of other sciences are obvious key determinants of the way of life in human societies. One does not need to subscribe to the many self-congratulatory writings on this subject to recognize this basic truth—whether it be for good or ill. The responsibilities of those who manage the engineering function are thus correspondingly broad as they involve ethical and sociopolitical elements as well as technical ones. Such issues as environmental impact and the military preemption of technical work quickly come to mind. Finally, any discussion of engineering and its management cannot lose sight of the fact of deterioration in many technically oriented industries where markets and research initiatives are lost to competitors and the very plants affected are demolished. In such conditions the quality of management of the engineering function clearly becomes crucial and calls for understanding not only among members of the profession itself, but of the public at large, in the setting of national purpose and objectives.

NOTES

1. National Science Foundation, *Research and Development in Industry 1974,* NSF76–322 (Washington, D.C.: U.S. Government Printing Office, 1976), p. 17.

2. *Ibid.*

3. J.E. Ullmann, *The Prospects of American Industrial Recovery* (Westport, Ct.: Quorum, 1985), pp. 143–158.

4. *Ibid.,* Ch. 7.

2

THE STRUCTURE OF ENGINEERING

Donald G. Weinert

DONALD G. WEINERT became Executive Director of the 75,000-member National Society of Professional Engineers (NSPE) in 1978. Earlier, he had had a highly successful career in the United States Army Corps of Engineers, attaining the rank of Brigadier General. He helped establish and manage a wide array of construction, operation and maintenance activity, relating to water management, pollution abatement, and other areas. He is a veteran of the Korean and Vietnam conflicts, gaining three awards of the Legion of Merit, a Meritorious Service Medal, two Bronze Stars and four Army Commendation Medals. A graduate of the U.S. Military Academy, he received his Masters in Engineering at Purdue University and attended the U.S. Army War College and the Army's Advanced Management Training Program at Northwestern University. His Professional Engineer registration is in the State of Texas. Mr. Weinert is also secretary of NSPE's Educational Foundation, president of its MATHCOUNTS Foundation, and president of the Junior Engineering Technical Society (JETS) in 1986–87. In 1983–1985, he served on a National Research Council committee to study the education and utilization of engineers in the United States.

2.1. DEFINITIONS

Engineering as a human activity—manipulating nature to produce something needed or desired—has taken place since the beginning of human history and has been well-documented in the literature of all ages. However, the existence of engineering as a profession is a relatively recent phenomenon. Attempts to apply formal definitions to engineering and to the engineer parallel the emergence of a definable engineering profession.

The "discovery" of fire and how to use it, the fashioning of the first primitive wheel, early irrigation channels for cultivated crops, the masterly works of ancient Persia, Egypt, Greece, and Rome all involved engineering. The men and women responsible for those "discoveries" and enterprises were in every respect engineers. The physical laws and the properties governing the forces and materials they worked with were at first unknown. Accident, curiosity, experimentation, and trial and error led to the discovery and ultimate codification of engineering knowledge. Modern engineering involves those same ingredients and also benefits from a highly developed and recorded body of knowledge plus a formal educational process which was not enjoyed by our early ancestors.

In coming to terms with definitions we find that what is at issue is not so much what engineering work is, although its diversity and complexity certainly present definitional challenges. Instead, the central point of contention seems to be who is entitled to be called an engineer. The contemporary basis for that contention ranges from concern that unqualified practitioners might harm public health, safety, and welfare to a somewhat elitist rejection of those perceived as not holding the "proper" credentials. These are generally defined by those advocating stringent theoretical definitions as graduation from an accredited engineering program and/or some type of legally recognized licensing or certification process.

2.1.1. Origins of the Word "Engineer"

A review of the historical evolution of the terms "engineer" and "engineering" used by scholars, writers, engineering academicians, and professional engineering organizations is instructive in coming to grips with the definitional challenge.

The word "engineer" stems from the Latin "ingenium," meaning natural talent or capacity; it also means a clever invention. The words ingenuity and engine also stem from ingenium. In addition, the word "engine" now has as obsolete meanings "ingenuity" and "wile." Its current meanings include "something used to effect a purpose; agent, or instrument." People acting as agents or instruments to effect a purpose are quite correctly referred to as "engines" of that purpose. Thus the current meaning of the word "engine" goes beyond the commonly thought of mechanical device or machine used for converting various forms of energy into mechanical force or motion. Further, the word "engineer" is just as close to the word ingenuity as it is to the word engine. Thus the modern words "engineer," "engineering," "ingenuity," and "engine" are all related through common ancestry.[1] "Engineer" does not relate solely to the word engine in its narrower meaning of mechanical device or machine. For an excellent discussion of the origin of the word "engineer" see the paper prepared in 1914 by Hunter McDonald, who was president of the American Society of Civil Engineers (ASCE) at the time.[2]

Unfortunately, early use of the word engineer specified almost exclusively a military engineer. History reports that Tertullian used the word "ingenium" in its meaning of "a clever invention" to describe a battering ram. Through common use it came to apply to any kind of military engine. The words ingeniator, ingeniarius, and ingegerus were all applied to builders of military machines in historical documents. Hence the word "engineer" is today associated with machines and generally used for a variety of functions associated with machinery and mechanical equipment. The significance of this evolution is that the terms "engineer" and "engineering" have never been, nor can they ever be, the sole province of the engineering profession as we define it today.

2.1.2. Formal Definitions

In discussing definitions it is important to note that the words "engineer" and "nature" are etymologically related. Remember, ingenium means "natural capacity or talent," and relates to the inherent nature of people and things. This has special significance when one studies the many definitions of engineering over time. Those definitions have consistently contained three principal elements, either explicitly or implicitly. First, they link engineering with "the forces of nature"; second, they refer to "the use or good of man"; and third, they specify or imply a special knowledge and skill relating to natural or physical phenomena. The latter element is implicit in early definitions and descriptions of engineering activity and explicit as "a knowledge of the mathematical, natural, or physical sciences" in later descriptions as a recognizable body of engineering knowledge began to take form.

Early definitions also frequently describe engineering as a branch of science—that branch in which scientific knowledge is applied to create useful goods and services. The terms "applied science" and "engineering" were (and still are) often used interchangeably. However, as the body of engineering knowledge, its educational process, and its work became more distinct from those of science, engineering took on its own independent identity.[3]

A compilation of engineering and scientific society leaders' views published by the National Society of Professional Engineers (NSPE) in 1963 demonstrates the preoccupation with making the distinction between engineering and science.[4] It also reveals consistency with earlier definitions of engineering wherever "engineering" is separately defined. It is very interesting to note that while the engineering and scientific functions are well-covered in the 1963 compilation, there are very few attempts at defining an engineer or a scientist except in terms of the function. Clearly, as stated by W. L. Everitt, who was at that time Dean of Engineering at the University of Illinois, "it is easier to distinguish between the 'scientific function' and the 'engineering function' than to distinguish between the man who should be called a scientist and who should be termed an engineer. Many men perform both functions, and do it very well. . . ."

One of the first formal written definitions was put forward in 1828 by Thomas Tredgold in the charter of the Institution of Civil Engineering in England. Tredgold defined engineering as "the art of directing the great sources of power in nature for the use and convenience of man." That definition contains the three classic definitional elements: the "great sources of power in nature"; "use and convenience of man"; and the special knowledge and skill relating to physical and natural phenomena implied by the phrase "the art of directing the great sources. . . ." Most early definitions and those in modern dictionaries contain the same three elements.

As alluded to earlier, the real definitional difficulties began when the engineering profession, as represented by its leaders and the engineering societies, began to add to the traditional definitions of engineering. Most notably they added a fourth element, as in the Engineers' Council for Professional Development (ECPD) definition, specifying that the knowledge and skill explicitly or implicitly required for engineering should be acquired by "study, experience or practice." It was then a short step to defining an engineer first in terms of the type of knowledge and skill required; second, by how the knowledge is acquired; and finally, by what type of evidence is necessary that it has been acquired. Evidence of "study" translates to graduation from an accredited engineering program and evidence of "experience" and "practice" are in part reflected in licensing and certification procedures. Some engineering organizations, notably NSPE, even advocate use of licensing as a means to show evidence of study, experience, and practice.

As noted at the outset, adding the academic credential to the definition of an engineer and, when applicable, the practice/experience credential, has complicated the business of describing the engineering profession. Those credentials exclude those with educational backgrounds in science and those without either a four-year accredited engineering degree or a license, who are nonetheless performing what has traditionally been described as an engineering function.

The emergence of the engineering technologist with a four-year Bachelor of Engineering Technology degree exacerbated the definitional dilemma and provided further impetus in some circles for tightening definitions because of the similarities between the educational programs for the engineer and the technologist.

2.1.3. The Engineering Team

In the late 1970s, under the umbrella of what was then ECPD and is now Accreditation Board for Engineering and Technology (ABET), many of the engineering societies participated in a comprehensive review of definitions, which included engineering, the engineer, engineering technologists, and technicians. Their report, titled "The Engineering Team," was approved by the ECPD Board of Directors in 1979. It contained the following definitions and explanatory notes.

ENGINEERING

Engineering is the profession in which a knowledge of the mathematical and natural sciences gained by study, experience, and practice is applied with judgment to develop ways to utilize, economically, the materials and forces of nature for the benefit of mankind.

Engineer—With a strong background in mathematics, the basic physical sciences, and the engineering sciences, the engineer must be able to interrelate engineering principles with economic, social, legal, aesthetic, environmental and ethical issues, extrapolating beyond the technical domain. The engineer must be a conceptualizer, a designer, a developer, a formulator of new techniques, a producer of standards—all to help meet societal needs. The engineer must plan and predict, systematize and evaluate—must be able to judge systems and components with respect to their relation to health, safety and welfare of people and to loss of property. Innovation must be central to the engineer.

The engineer will normally have received the first professional degree from an accredited engineering program, which requires a minimum of one-half year of mathematics, beginning with differential and integral calculus. Education in engineering analysis and synthesis shall prepare the engineer to enter the profession with potential for further development in research, design, development, management, establishment of systems, and translation of concepts into realities. An engineering education is the principal route to professional licensure.

ENGINEERING TECHNOLOGY

Engineering Technology is that part of the technological field which requires the application of scientific and engineering knowledge and methods combined with technical skills in support of engineering activities; it lies in the occupational spectrum between the craftsman and the engineer at the end of the spectrum closest to the engineer.

Engineering Technologist—The engineering technologist must be applications oriented, building upon a background of applied mathematics through the concepts and applications of calculus. Based upon applied science and technology, the technologist must be able to produce practical, workable results quickly; install and operate technical systems; devise hardware from proven concepts; develop and produce products; service machines and systems; manage construction and production processes; and provide sales support for technical products and systems.

Normally, the engineering technologist will hold a four-year degree from an accredited engineering technology program. Because of the key role as an implementer, the engineering technologist must be prepared to make independent judgments that will expedite the work without jeopardizing its effectiveness, safety, or cost. And the technologist must be able to understand the components of systems and be able to operate the systems to achieve conceptual goals established by the engineer.

Engineering Technician—With a minimum of two years of post-secondary education, ideally in engineering technology, with emphasis in technical skills, the engineering technician must be a doer, a builder of components, a sampler and collector of data. The technician must be able to utilize proven techniques and methods with a minimum of direction from an engineer or an engineering technologist. He/she shall not be expected to make judgments which deviate significantly from proven procedures.

The technician should expect to conduct routine tests, present data in a reasonable format, and be able to carry out operational tasks following well-defined procedures, methods, and standards.[5]

Furthermore, the 1979 report contains a matrix depicting the engineering team which appears in Table 2.1.

The 1979 definition of engineering in the ECPD (ABET) report remained unchanged from that first developed by ECPD in 1961 and varies little from that put forth by Tredgold in 1828 except for the addition of the "by study, experience and practice" element. However, the definitions of engineer, technologist, and technician became much more detailed and began to include a great deal of explanatory material. Emphasis was put on the differences in academic credentials among the three. As to function, it is clear that all three fall within the definition of engineering, albeit at different levels. Hence we have the term "The Engineering Team."

TABLE 2.1. The Engineering Team

Concepts	Engineer	Engineering Technologist	Engineering Technician	Skills
	....Predict .. Anticipate .. Implement .. Test .. Operate .. Build .. Maintain			
		Preparation		
	The first professional degree from an accredited engineering program	Four-year degree from an accredited engineering technology program	An associate degree in an accredited engineering technology program	
	One-half year of mathematics beginning with differential and integral calculus	Applied science and mathematics (through concepts and applications of calculus)	Mastery of technical skills	
	Basic physical sciences		Training in specific instruments/equipment	
	Engineering sciences	Technical sciences and specialty areas		
	Interrelate engineering principles with economic, social, political, aesthetic, ethical, legal, environmental, etc., issues	Field orientation	Perform operational tasks, following well-defined procedures, to support engineering activities	
		Apply technological methods and knowledge, with technical skills, to support engineering activities		
		Career Goals		
	Research	Hardware design and development	Drafter	
	Conceptual design	Product analysis and development	Laboratory operations	
	System synthesis and development	System operation	System maintenance	
	Product innovation	Process management	Machine operations	
	Operations management	Technical sales and services	Data collection	

19

TABLE 2.1. (Continued)

.... Predict .. Anticipate .. Implement .. Test .. Operate .. Build .. Maintain ....

Concepts	Engineer	Engineering Technologist	Engineering Technician	Skills
		Descriptors		
	Conceptualizer	Operator of systems	Performer of operational tasks	
	Innovator	Translator of concepts into hardware and systems	User of proven techniques and methods	
	Planner/Predictor	Director of engineering technicians and craftsmen	Builder of components	
	Designer	Implementer	Operator	
	Developer	Applier of established techniques and methods	Tester	
	Systematizer	Maintainer of systems	Collector of data	
	Judge	Producer	Maintainer of components	
	Decision maker (beyond technical domain)	Analyzer	Preparer of technical drawings	
	Producer of standards			
	Formulator of techniques and methods			
	Synthesizer			

Source. Engineers' Council for Professional Development, "The Engineering Team" (1979).

20

2.1.4. National Council of Engineering Examiners' (NCEE) Definition

Another definition of note is that of the National Council of Engineering Examiners Model Law, 1984 revision. Section 2, Chapter I, contains the following definitions:

1. *Engineer*—The term "Engineer," within the intent of this Act, shall mean a person who is qualified to practice engineering by reason of his special knowledge and use of the mathematical, physical and engineering sciences and the principles and methods of engineering analysis and design, acquired by engineering education and engineering experience.

2. *Professional Engineer*—The term "Professional Engineer," as used in this Act, shall mean a person who has been duly registered or licensed as a Professional Engineer by the board.

3. *Engineer-in-Training*—The term "Engineer-in-Training," as used in this Act, shall mean a person who has qualified for, taken and has passed an examination in the fundamental engineering subjects, as provided in this Act.

4. *Practice of Engineering*—The term "Practice of Engineering," within the intent of this Act, shall mean any service or creative work, the adequate performance of which requires engineering education, training and experience in the application of special knowledge of the mathematical, physical and engineering sciences to such services or creative work as consultation, investigation, evaluation, planning and design of engineering works and systems, planning the use of land and water, teaching of advanced engineering subjects, engineering studies, and the review of construction for the purpose of assuring compliance with drawings and specifications; any of which embraces such services or work, either public or private, in connection with any utilities, structures, buildings, machines, equipment, processes, work systems, projects, and industrial or consumer products or equipment of a mechanical, electrical, hydraulic, pneumatic or thermal nature, insofar as they involve safeguarding life, health or property, and including such other professional services as may be necessary to the planning, progress and completion of any engineering services.

   A person shall be construed to practice or offer to practice engineering, within the meaning and intent of this Act, who practices any branch of the profession of engineering; or who, by verbal claim, sign, advertisement, letterhead, card, or in any other way represents himself to be a professional engineer, or through the use of some other title implies that he is a professional engineer or that he is registered under this Act; or who holds himself out as able to perform, or who does perform any engineering ser-

vice or work or any other service designated by the practitioner which is
recognized as engineering.[6]

Quite understandably, the focus of the NCEE definitions is on qualification and on the licensed engineer—who is referred to as "Professional Engineer." The definition of the "Practice of Engineering" is also substantially
more detailed since it attempts to define the many types of engineering work
covered by the Model Law. The introduction of the term "Professional Engineer" to describe only licensed engineers has further confused the definitional picture in that to some it implies that nonlicensed engineers may
somehow be unprofessional.

2.1.5. National Science Foundation Definition

Finally, confronted with the practical challenge of collecting and analyzing
data on the scientific and engineering professions, the National Science
Foundation set up eight criteria for determining who should be counted as a
member of a given field of science or engineering. The NSF criteria place
emphasis on two factors: (1) attainment of an appropriate academic degree
and (2) occupation in a scientific or engineering field. The criteria are listed
in the NSF publication, "U.S. Scientists and Engineers 1980," and are
quoted below:

1. had earned a master's degree or higher in a coincident field of study and
    who regarded oneself, based on one's total education and experience, as
    having a coincident profession;

2. had earned a Ph.D. in any field of social or natural science and was em-
    ployed in a coincident occupation;

3. had earned a bachelor's degree or higher in a coincident field of study,
    and was employed in a coincident occupation;

4. had earned a bachelor's degree or higher in any field of study, was em-
    ployed in a coincident occupation, and regarded oneself as having a coin-
    cident profession;

5. whose highest degree was in a coincident field of study at any degree level
    and who was employed as a college president, college dean, or manager or
    administrator of research or development, production, or operations, or
    had earned a bachelor's degree or higher in a coincident field of study,
    was employed in a related occupation, and regarded oneself as having a
    coincident profession;

6. has earned a bachelor's degree in a coincident field of study since 1969
    and who regarded oneself as having a coincident profession;

7. had earned a bachelor's degree or higher in any field of science and was employed as a college president, college dean, or administrator or manager of research or development, production, or operations and who regarded oneself as having a coincident profession; or

8. whose highest degree was in a related field of study and who was employed in a coincident occupation and who regarded oneself professionally to be a college president or a dean, or an administrator or manager of research or development, production or operations.[7]

To sum up the definitional issue, several points are clear. First, the definition of engineering is extremely broad and can accommodate a wide range of practitioners. Second, that range involves level of function (from simple to complex), area and type of practice, job titles, academic background, and experience. Third, to adequately portray and understand the engineering enterprise in the United States, all of those substantively involved in that enterprise must be accommodated in the definitional frameworks adopted, whatever the level and type of academic, experience, or practice credentials.

2.2. NUMBERS OF ENGINEERS, MAJOR ENGINEERING ACTIVITIES, AND SPECIALTIES

2.2.1. The Engineering Profession in the United States—Size and Trends

The engineering profession is the second largest in the United States. According to the Bureau of Labor Statistics there were 1.6 million persons employed in Engineering in 1982. Only the number of teachers is greater among the "learned professions." The medical and legal professions, with which engineers often make comparisons, number only about 500,000 each. (There are many more lawyers than 500,000, but they are not involved in the practice of law.)

As shown in Figure 2.1, the number of engineers in the United States almost doubled from 1960 to 1982, from 800,000 to more than 1.3 million. The figure also reveals that the rate of increase has accelerated since 1976. Engineers as a percent of the total work force also increased from 1.2% in 1960 to a high of 1.6% in 1970 and to 1.4% in 1980 (see Figure 2.1). These figures signify that the number of employed engineers grew faster than the total employed population during the period 1960 to 1970. The slight decrease during the 1970s, was due to the rapid increase in total employed population during that period.

Accurate employment data on engineering technicians and technologists

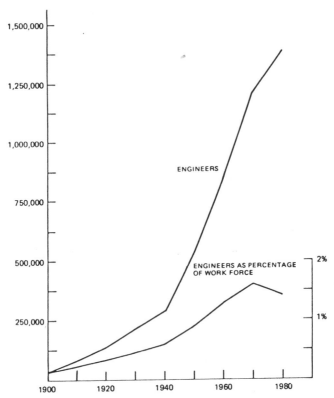

FIGURE 2.1.   Employed scientific/engineering manpower, 1960–1982 —engineers. Source: National Research Council Study committee "The Education and the Utilization of the Engineer."

are difficult to obtain. This is because data collection involves self-reporting by the individuals who frequently classify themselves as engineers even though they may not qualify for such classification. It also involves employer reporting which is inconsistent among employers. Some classify technicians and technologists as engineers if they have what the employer has classified as engineering positions. The best available data on the number of technicians are shown in Figure 2.2.[8]

There are no available employment data on the number of engineering technologists (those with four-year technology degrees). The field is relatively new, having its real beginning during the period of dramatic decline in engineering enrollments following the aerospace dislocations in the late 1960s and early 1970s. As reported in *Engineering Manpower Commission Bulletin 73* in 1984, the number of four-year bachelor technology graduates

NUMBERS, ACTIVITIES, AND SPECIALTIES

25

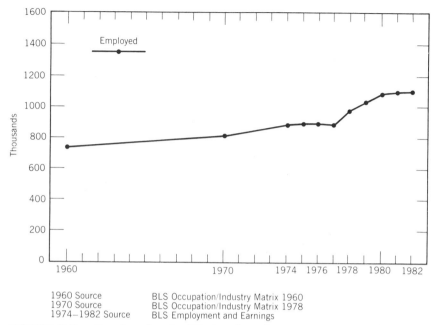

1960 Source — BLS Occupation/Industry Matrix 1960
1970 Source — BLS Occupation/Industry Matrix 1978
1974–1982 Source — BLS Employment and Earnings

FIGURE 2.2. Employed scientific/engineering manpower, 1960–1982 —engineering technicians. Source: National Research Council Study committee, "The Education and the Utilization of the Engineer."

in 1983 was 9200. This is about 13% of the annual output of BS Engineers which was reported by the Engineering Manpower Commission to be 72,500 for 1983.[9] Considering the annual output and the recent origins of the field, it is unlikely that in 1985 there were more than 100,000 engineering technologists. Many of them were probably reported as either engineers or technicians.

2.2.2. Major Specialties and Types of Engineering Activity

The distribution in 1981 of employed engineers in the six largest disciplines is given in Table 2.2. Those fields accounted for just over 1 million of the 1.5 million employed engineers at that time. Figure 2.3 shows the growth in the major disciplines since 1960. Most dramatic growth has occurred in electrical/electronic engineers (many are in computer- and communications-related work) and in industrial and mechanical engineers. Note that beginning in 1978 there was a sharp increase in the number of engineers in other than the six major disciplines. This reflects a dramatic growth in recognized engi-

TABLE 2.2. Distribution of Engineers Employed in Six Largest Disciplines, 1981

Discipline	Engineers Employed	
	Number	Percent[a]
Electrical/electronic	279,200	18.9
Mechanical	249,500	16.9
Civil	200,300	13.5
Industrial[b]	143,000	9.7
Chemical	79,400	5.4
Aero/astro	50,200	3.4

Source. National Research Council Study Committee on "The Education and the Utilization of the Engineer," 1983.

[a] Totals do not add to 100% because of the large number of smaller disciplines.
[b] Based on 1980 data adjusted upward.

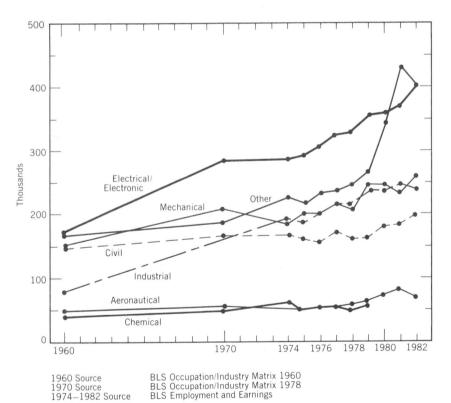

FIGURE 2.3. Employed scientific/engineering manpower, 1960–1982 —engineers. Source: National Research Council Study committee, "The Education and the Utilization of the Engineer."

TABLE 2.3. Primary Activities of Employed Engineers, 1982[a]

Activity	Women Engineers (%)	All Engineers (%)
Research	10.9	4.7
Development[b]	15.2	27.9
R&D management	3.4	8.7
Other management	16.6	19.3
Teaching	7.3	2.1
Production/inspection	13.6	16.6
Other[c]	33.0	20.7

Source. Unpublished tabulations, NSF. Based on 1982 Post-censal Survey of Scientists and Engineers, July 1984.

[a] These data are compiled by NSF's National Science Board from a variety of sources, including employer surveys and engineer (self-reporting) questionnaires. Thus they reflect a considerable degree of subjectivity and inconsistency in the definition of activities.
[b] This category includes "design" activity.
[c] Includes: consulting; reporting, statistical work, computing; other.

neering specialties and is reflected in part by the large variety of engineering organizations, as discussed in the next section.

In terms of areas of employment, available data from the National Science Foundation[10] and the Bureau of Labor Statistics show that in 1980 40% of all employed engineers were in the durable goods (manufacturing) sector. Almost 20% were in the service sector. Government and utilities accounted for just under 10% each and nondurable goods about 8%. The remaining engineers are distributed over all other sectors. Since 1970 the number in the durable goods sector has declined and the number in services has increased. Primary activities of employed engineers are reflected in Table 2.3. Those in consulting engineering practice are included in the "Other" category. A large percentage is classified as management (8.7% in Research and Development and 19.3% in other management).

2.3. ORGANIZATION OF THE PROFESSION, ENGINEERING SOCIETIES, AND ASSOCIATIONS

There are over 60 individual societies, institutes, and associations at the national or international level which represent the interests of and provide support to engineers and engineering.[11] Typical of the support provided are development and dissemination of technical information; publication of often highly prestigious journals; continuing education seminars, symposia, home study; salary surveys and employment guidelines; general news and information about the profession, a specific technology, or area of practice;

college scholarships in engineering; precollege guidance; advocacy and representation of engineering interests in public policy before legislatures and government agencies; public information about engineers and engineering achievements; honors and awards for engineers and engineering; employment referral services; setting technical standards for engineering practice; personal and business services (insurance, car rental, and so on); development and enforcement of standards for engineering education and licensing; services to the public (community action); and many others depending on the interests of members. A partial listing is given in Table 2.4.

The engineering societies and associations fall in four major groupings. First, there are those focused primarily on established or emerging engineering disciplines. The American Society of Civil Engineers (ASCE), American Society of Mechanical Engineers (ASME), Institute of Electrical and Electronics Engineers, Inc. (IEEE), American Institute of Chemical Engineers (AIChE), and American Institute of Mining, Metallurgical, and Petroleum Engineers (AIME), frequently referred to as the "Founder Societies," are the foremost examples of this first group. Such societies traditionally have been most concerned about promoting development and exchange of technical information in the discipline concerned. Concurrently, they have engaged in technical and professional activities of interest to their members, which have included establishing technical standards, setting standards of professional conduct, promoting the public image of engineers and engineering, insuring quality of engineering education pertaining to their discipline and others, depending on the interest of their members at a given time.

The second group of engineering societies and associations is focused on practice in a broad occupational field. Examples in this category are the Society of Automotive Engineers, American Institute of Aeronautics and Astronautics, American Society of Agricultural Engineers, American Society of Naval Engineers, and American Institute of Plant Engineers, among others. This group develops and promulgates technical and nontechnical information about engineering practice within the occupational area concerned, but also engages in other technical and professional activities based on the interests of its members.

The third and fastest growing group are those focused on a specific technology, group of technologies, or one of the specific "materials" or "forces of nature" always referred to in classical definitions of engineering. Examples in this group are the American Society for Metals; American Society of Heating, Refrigerating, and Air-Conditioning Engineers; Society of Plastics Engineers; American Nuclear Society; American Welding Society; American Society of Safety Engineers; Society of Manufacturing Engineers; Asso-

TABLE 2.4.   Partial Listing of National Engineering Organizations

Abrasive Engineering Society
Acoustical Society of America
Alliance for Engineering in Medicine and Biology
American Academy of Environmental Engineers
American Association of Cost Engineers
American Association of Petroleum Geologists
American Consulting Engineers Council
American Engineering Model Society
American Institute of Aeronautics and Astronautics, Inc.
American Institute of Chemical Engineers
American Institute of Mining, Metallurgical, and Petroleum Engineers, Inc.
American Institute of Plant Engineers
American Nuclear Society
American Railway Engineering Association
American Society for Engineering Education
American Society for Engineering Management
American Society for Hospital Engineering
American Society for Metals
American Society for Quality Control
American Society of Agricultural Engineers
American Society of Body Engineers
American Society of Civil Engineers
American Society of Gas Engineers
American Society of Heating, Refrigerating, and Air-Conditioning Engineers, Inc.
American Society of Lubrication Engineers
American Society of Mechanical Engineers
American Society of Naval Engineers, Inc.
American Society of Plumbing Engineers
American Society of Safety Engineers
American Society of Sanitary Engineering

Association for Systems Management
Association of Conservation Engineers
Association of Consulting Chemists and Chemical Engineers
Association of Consulting Management Engineers, Inc.
Association of Energy Engineers
Association of Engineering Geologists
Association of Iron and Steel Engineers
Association of Soil and Foundation Engineers
Audio Engineering Society, Inc.
Illuminating Engineering Society of North America
Institute of Electrical and Electronics Engineers
International Association for Shell and Spatial Structures
Operations Research Society of America
Society for Computer Applications in Engineering, Planning and Architecture, Inc.
Society for Experimental Stress Analysis
Society of Allied Weight Engineers, Inc.
Society of American Military Engineers
Society of American Value Engineers, Inc.
Society of Automotive Engineers, Inc.
Society of Broadcast Engineers
Society of Carbide and Tool Engineers
Society of Explosives Engineers
Society of Fire Protection Engineers
Society of Logistics Engineers
Society of Manufacturing Engineers
Society of Mining Engineers of AIME
Society of Motion Picture and Television Engineers, Inc.
Society of Naval Architects and Marine Engineers
Society of Packaging and Handling Engineers
Society of Petroleum Engineers of AIME
Society of Photo-Optical Instrumentation Engineers
Society of Plastics Engineers, Inc.
Standards Engineering Society

ciation of Energy Engineers; and many others. Societies in this group engage in activities to promote development and sharing of the body of engineering and scientific knowledge necessary to their specific technologies and, as do the others, pursue other technical and nontechnical goals in accordance with the interests of their members.

The final group is composed of those associations and societies formed either by individual engineers or by groups of societies to accomplish a specific purpose. The National Society of Professional Engineers (NSPE) was formed to promote the professional and nontechnical interests of engineers and the profession with emphasis on professional standards (registration and ethics), the image of engineering, the quality of engineering educational practice, and involvement in public policy. The American Society for Engineering Education has an essentially interdisciplinary concern with training and educational practice in engineering itself and its contiguous scientific fields, including the social sciences. In a related area, the Accreditation Board for Engineering and Technology (ABET) was formed to accredit engineering programs and serve as the quality control mechanism for engineering education. The National Council of Engineering Examiners (NCEE) was formed to coordinate the state licensing process. From time to time attempts have also been made to form an overall umbrella organization to represent the entire profession, much as the American Medical Association (AMA) is seen by some as representing the entire medical profession. The most recent of these is the American Association of Engineering Societies (AAES) which was formed in 1980.

The more than 60 national or international engineering societies are estimated to represent about one million of the 1.4 million practicing engineers in the United States. However, this is merely an aggregation of membership totals of individual societies and does not adjust for considerable overlapping. It is not uncommon for some practitioners to belong to five or more while many others have no memberships at all. There has not been any significant attempt to adjust for this.

2.4. PROSPECTS FOR ENGINEERING—FUTURE NEEDS AND SUPPLY

2.4.1. Demand

Employment projections for engineers from 1980 to 1990 have been made by the Bureau of Labor Statistics.[12] Total employment and expected percent changes are given in Table 2.5. Note that the expected percentage increase for engineers ranges from a low of 27.7% to a high of 37.9%. This compares with lows of 17.1% and highs of 25.3% for all occupations and a low of 19.9% and a high of 26.4% for professional, technical, and related workers. Clearly the future demand for engineers is seen as very strong by the Bureau of Labor Statistics.

The National Science Foundation has also projected increasing requirements from 1982 to 1987.[13]

Those projections show requirements for engineers growing between 2.6% and 4.5% per year. If economic recovery and defense spending remain strong, the growth should be near the top of the range. That rate (near 4.5%) would be consistent with the 4.7% annual growth rate recorded over the 1977–1982 period. Comparable increases in demand growth are projected by the NSF for engineering technicians.

The explosion of new technologies and the growing dependence on technology in daily life point to a strong long-term demand for technically educated and trained persons. There is every evidence of a continuing strong demand for engineers into the next century. Further, evidence of long-term demand for engineers can be found in long-term unemployment rates for engineers. Unemployment statistics published by the Bureau of Labor Statistics show that between 1963 and 1982 the unemployment rate for engineers stayed close to 1% for most of the period. It exceeded 2% in only four years and hit a peak of 2.9% in 1971 at the time of the aerospace dislocations. The rate for engineers in 1980 was 1%, well below the national rate of 7.1%.

2.4.2. Meeting the Demand

On balance, studies by the Bureau of Labor Statistics, the National Science Foundation, and the Engineering Manpower Commission show that over the long term, demand for engineers has been satisfied despite short-term fluctuations due to changing economic, political, and technological patterns. The engineering education system has responded well in the past although strains have been evident in faculty shortages and in deficiencies in laboratory equipment and other facilities. When demand is high there have also been shifts back into engineering of qualified persons who hold engineering degrees and have left engineering for other fields. Spot shortages and surpluses in particular specialties or geographic areas have been evident as the supply system adjusts to new demands. For instance, the most pronounced imbalance in 1984–1985 involved electrical/electronic engineers and those in computer-related engineering fields. In these areas, available supply has not been able to keep up with the demand. Conversely, downward trends in the petroleum and chemical industries resulted in temporary excess in the supply of chemical engineers.

TABLE 2.5.   1980 & Projected 1990 Employment & Percent Change, 1980–90 in Occupations with 1980 Employment of 5000 or Higher

Occupation	Employment, All Industries (thousands)				Percent, change		
	1980	1990 (Low)	1990 (High I)	1990 (High II)	1980–90 (Low)	1980–90 (High I)	1980–90 (High II)
Total, all occupations	102,107.3	119,591.1	127,908.4	121,448.5	17.1	25.3	18.9
Professional, technical, and related workers	16,395.2	19,662.3	20,727.6	19,917.4	19.9	26.4	21.5
Engineers	1,177.8	1,504.3	1,624.3	1,531.2	27.7	37.9	30.0
Aeronautical and astronautical engineers	68.0	97.6	103.6	100.1	43.4	52.3	47.2
Chemical engineers	55.9	68.4	73.1	70.0	23.2	31.6	26.0
Civil engineers	165.4	207.9	217.2	210.3	25.7	31.3	27.1
Electrical engineers	326.7	441.2	479.9	449.2	35.1	46.9	37.5
Industrial engineers	115.9	145.7	159.3	148.3	25.8	37.5	28.0
Mechanical engineers	212.9	273.9	300.0	279.1	28.7	40.9	31.1
Metallurgical engineers	15.4	20.4	22.0	20.8	32.4	42.5	34.7
Mining engineers	6.1	8.4	9.2	8.7	37.7	51.2	42.6
Petroleum engineers	17.9	26.0	27.6	25.8	45.7	54.2	44.3
All other engineers	193.9	214.6	232.5	218.9	10.7	19.9	12.9
Life and physical scientists	253.8	300.2	317.3	305.6	18.2	25.0	20.4
Agricultural scientists	19.8	21.6	22.7	22.1	9.1	14.3	11.5
Biological scientists	44.8	51.2	54.1	52.5	14.1	20.7	17.0
Chemists	93.6	112.9	119.5	115.0	20.6	27.6	22.8
Geologists	39.8	51.7	54.9	52.0	30.1	38.2	30.7
Medical scientists	8.1	9.4	9.7	9.4	15.6	19.4	15.8
Physicists	20.5	23.1	24.4	23.6	12.5	19.0	14.8
All other life and physical scientists	27.1	30.3	32.0	31.1	11.6	18.1	14.5

Mathematical specialists	52.0	61.7	65.8	63.3	18.7	26.5	21.6
Actuaries	7.8	10.9	11.6	11.3	39.5	48.3	44.8
Mathematicians	12.7	14.4	15.2	14.7	13.2	20.0	15.8
Statisticians	26.5	30.9	33.2	31.6	16.7	25.2	19.3
All other mathematical specialists	5.0	5.6	5.8	5.7	10.9	16.3	13.2
Engineering and science technicians	1,267.9	1,578.0	1,700.8	1,610.1	24.5	34.1	27.0
Broadcast technicians	16.5	17.8	18.4	17.8	7.5	11.4	7.7
Civil engineering technicians	24.6	30.7	31.6	31.1	24.8	28.4	26.2
Drafters	321.6	411.1	445.3	418.8	27.9	38.5	30.2
Electrical and electronic technicians	359.5	466.3	513.8	480.3	29.7	42.9	33.6
Industrial engineering technicians	32.4	40.2	43.6	40.8	24.0	34.4	25.7
Mechanical engineering technicians	48.5	60.9	67.1	62.2	25.5	38.2	28.1
Surveyors	61.3	72.9	78.0	75.4	18.8	27.2	22.9
All other engineering and science technicians	403.4	478.0	503.0	483.8	18.5	24.7	19.9
Computer specialists	432.8	683.1	733.9	697.2	57.8	69.6	61.1
Computer programmers	228.2	339.9	366.0	346.6	48.9	60.4	51.8
Computer systems analysts	204.6	343.2	367.9	350.6	67.7	79.8	71.4

Source. Bureau of Labor Statistics, "Occupational Projections and Training Data," 1982.

NOTES

1. *Random House Dictionary of the English Language.* Unabridged Edition.

2. H. McDonald, "Origin of the Word 'Engineer,' " *Transactions of the American Society of Civil Engineers* 77: 1737 (1914).

3. D.G. Weinert, "The Definition of Engineering and Engineers in Historical Context," Supplemental report prepared for the National Research Council Study Committee on the Education and Utilization of the Engineer (1984).

4. National Society of Professional Engineers, "The Function of the Engineer and the Scientist" (January 1963): NSPE File No. 2111.

5. Engineers' Council for Professional Development, *The Engineering Team* (New York: Report to the ECPD Board of Directors, 1979).

6. National Council of Engineering Examiners, *Model Law 1984 Revision* (Clemson: NCEE, 1984).

7. National Science Foundation, *U.S. Scientists and Engineers, 1980* (Washington, DC, U.S. Government Printing Office, 1982), NSF 82–314.

8. U.S. Bureau of Labor of Statistics, *National Industry Occupational Matrices* (Washington, DC, U.S. Government Printing Office, 1960, 1978, 1981).

9. Engineering Manpower Commission, "Engineering Manpower Commission Bulletin 73" (1984).

10. National Science Foundation, *Changing Employment Patterns of Scientists, Engineers and Technicians in Manufacturing Industries: 1977-80* (Washington, DC, U.S. Government Printing Office, 1982), NSF 92–331.

11. American Association of Engineering Societies, "Directory of Engineering Societies and Related Organizations" (May 1984).

12. U.S. Bureau of Labor Statistics, *Occupational Projections and Training Data* (Washington, DC, U.S. Government Printing Office, 1982).

13. National Science Foundation, *Projected Response of the Science and Engineering and Technical Labor Market to Defense and Non-Defense Needs: 1982-87* (Washington, DC: U.S. Government Printing Office, 1984), NSF 84–304.

3

INTEGRATING THE ENGINEERING FUNCTION

Ercole Rosa

ERCOLE ROSA is a senior instructor at the IBM Corporation and adjunct professor of Management at the C.W. Post Center of Long Island University, where he develops and teaches courses concerned with the interaction of technology, systems, and management. He joined IBM in 1956, after having served as director of technical assistance for the Economic Development Administration of Puerto Rico. He is vice-president of the North American Management Council, CIOS, and former vice president, Region III, of the American Society of Mechanical Engineers. He has also served as chairman of the Management Division, ASME and chairman of the Henry L. Gantt Medal of Honor Committee, which is jointly sponsored by ASME and American Management Association. A book he coauthored with Samuel J. Kalow, *Office Systems,* was published in 1984 by Prentice-Hall. He earned his B.S. and M.S. degrees in Industrial Engineering at Columbia University.

3.1. INTRODUCTION

There are several important questions that must be resolved when the goal is the effective integration of the engineering function within the organization. Some of these are subject to objective analysis, such as the cost of the engineering operations as a ratio of the gross income produced by the organization. Other equally important issues involve questions of style and intuition such as the nature of the reporting relationships and latitude provided to professionals and managers in initiating projects or allocating resources.

The process of integrating the engineering function is a design process which requires orderly definition of the problem, evaluation of the alternatives, generation of group support, systematic implementation, and continuing review and support.

These issues are most important in those firms where the engineering function is a critical process in the organization and there exists a close relationship between the effectiveness of the engineering function and the growth and profitability of the organization. In those instances where this is not the case, it is appropriate to consider alternative means for acquiring the engineering services.

In planning for the integration of the engineering function, the organization should be recognized as a system in which each function must be managed with full regard for the impact on other functions, corporate goals, and performance.

Effective performance depends upon the organization's ability to bring together the resources of the firm to meet the customer's needs and to respond to changes in the environment. This means that each individual must appreciate the contribution of each function to the challenges encountered by the firm. This also means that time and attention are devoted early enough to identifying opportunities, challenges, and potential crises so that the individuals and the groups can be committed and make the appropriate contribution.

In recent years there has been a tendency to broaden the scope of the engineering function to include the process of technology management. In turn, the definition of technology has been broadened to include a wide range of activities and issues related to the generation of knowledge concerned with the development of products and services.

The management of technology involves analysis at three levels. The first requires a monitoring and development activity in the areas of science that affect the firm's activities. The second requires forecasting, evaluation, and implementation of new technologies. The third requires analysis and design of the elements of organization and management that will affect the firm's performance. The goal is to establish high exposure for the engineering function and enhance the contribution of the engineering group to the corporate goals and profitability.

There are five challenges which serve as the design objectives for the process of integrating the engineering function.

1. To develop a technical organization that works together, renews itself, and demonstrates that it can respond
2. To participate effectively in the corporate planning and decision-making process

3. To identify developments in science and technology that will generate opportunities
4. To keep close to the customer and to the customer's environment to identify requirements
5. To develop relationships with other groups in the firm so there is communication and sharing of goals and experience

These goals provide the basis for planning for the long term, at least for a five-year horizon or longer. These goals also provide the basis for developing a set of principles and rules that will make the engineering function an effective participant in the internal corporate marketplace. The ultimate goal is to have the engineering function rated as the most cooperative and effective function in the organization.

The design of the engineering function requires consideration of the environment, a process of interfacing, examination of opportunities and resources, definition of priorities, relating to corporate goals, supporting the engineering profession, and anticipating the future. Each of these activities will be discussed in the pages that follow.

Consideration of the environment involves examination of the many factors that affect the operations of the firm. This includes the changes in the demands that have been put on the engineering function. The process of interfacing involves the activities that enable the engineering function to coordinate and interact with the other functions. Examination of opportunities and resources involves those activities which enable the managers of the engineering function to define the areas where the greatest contributions can be made to corporate growth and profitability. Definition of priorities requires the use of project management procedures to assure that the resources are well utilized. Relating to corporate goals requires a commitment to being involved in activities that will keep the goals of the engineering function in harmony with the corporate goals. Supporting the engineering profession is necessary to assure the generation of knowledge and training of new professionals. Anticipating the future involves the use of tools and procedures to provide early-warning systems as the basis of corporate and engineering planning. Each of these activities represents an important element of a program to achieve effective integration of the engineering function within the corporate structure.

3.2. EXECUTIVES AND ENTREPRENEURS

What is the proper perception of engineering management that will serve as the basis for planning into the late 1980s and early 1990s? Is it one of the models that has served in the past or will it be necessary to develop a new model that will incorporate responses to conditions and opportunities created by the developments of the recent years?

It is apparent that many of the conditions and relationships that have defined the environment for the engineering function have changed in recent years. The rate of introduction of new technologies and products has accelerated. The scope of markets and range of competition has expanded greatly. The flexibility of choice for both the individual and commercial consumer has changed considerably. The distribution of innovation and production across the mature industrialized nations and newly industrialized countries such as Taiwan, Brazil, and Korea has diminished the significance of the United States as the major source of production capacity and know-how. The growth of the postindustrial society, with its emphasis on the "information" industries and "service" industries, has changed the level of dependence on traditional engineering functions. The rise in emphasis on financial- and capital-oriented activities has tended to diminish the role of engineering and production in the corporate management process. Directing the engineering functions in the future will require taking into account the impact of these continuing changes in society and ways that industry and commerce are structured and processed.

The engineering function has changed radically over the years as well. Many engineers who practiced in the 1800s were great technicians and combined this with great entrepreneurial skills. Whether it be in civil engineering with bridge building or railroad building, or in mechanical engineering with the building of the machine tool industry or the development of power machines, the role of the lead engineer combined that of the developer and businessman. It was not enough to develop the idea; the innovation had to be converted into working designs and into reality. This required organization and financial support. This era came to a close at the end of the nineteenth century when the engineering function became institutionalized within the larger corporate structure.

An important milestone in the evolution of engineering management occurred in 1886 when Henry R. Towne[1] presented the paper, "The Engineer as Economist," in which he suggested that the engineer should be active in developing and organizing the experience gained in managing operations. His words led to the continuing effort to rationalize production operations by such pioneers as the Gilbreths, Taylor, Gantt, and others. This included the analysis and definition of tasks, planning and scheduling of operations, use of incentive programs, use of tools for economic analysis such as the break-even chart, and emphasis on the human factor elements in work organization. Over the years this trend has resulted in the use of formal methods in management analysis including operations research and management science. In some instances this process resulted in the separation of planning from doing, with a heavy emphasis on specialization.

By the late 1950s the number of white-collar workers matched the number of blue-collar workers. This occurred as the result of greater use of capital equipment and resulting increases in productivity; the shift to activities in service industries; and increased emphasis on planning, financial analysis, and control as management functions. In 1967 Peter Drucker[2] focused attention on this shift by declaring that executive decisions could be made by any knowledge worker whose choices would have a significant impact on the performance and results of the whole organization. Drucker defines "knowledge workers" as the entire employed educated middle class not engaged in manual work, that is a wide array of professional and semiprofessional employees. Full acceptance of the implications of this concept was slow in coming until examples of effective Japanese management demonstrated the value of larger participation in planning and improvement activities—in the form of quality circles, for example. More recently increased use of computer-based communications systems has facilitated the sharing of planning and control information throughout the organization with more opportunities for on-line interaction. The linking of computer-aided design and computer-assisted manufacturing is an example of the contribution of the computer to the improvement of engineering–manufacturing communication.

Another trend that was already well under way by the late 1950s was the effort to create situations where the individual, typically an engineer, was encouraged with corporate assistance and support to branch off and start independent businesses to take advantage of a new technological development. The General Electric Company was a leader in this development. In 1964, Peter Drucker[3] provided the blueprint for the evolution of the executive into the entrepreneur. He suggested that the executive focus attention on opportunities as well as on problems, and on the results and conditions outside the business. He also emphasized the need for constant review of current operations to determine their contribution to the profitability of the firm. In 1982, John Naisbitt[4] indicated his expectation that the trend toward entrepreneurialism will continue with considerable momentum. This is particularly evident in the electronics industry, although there are many other examples.

There are examples in the electronics industry where the emphasis on entrepreneurial initiatives have tended to disintegrate the original firm. In 1983, Rosabeth Kanter[5] recognized this as a potential major problem and suggested that corporations work to develop an integrative environment which could balance the need for individual initiative with the benefits of group interaction, support, and continuity. This environment would be marked by the existence of ambiguities, overlaps, decision conflicts, and other violations of traditional organization theory. Groups would be encouraged to compete, challenge, and support one another in innovational activities, and occasionally be reminded that there was still one overall organization. The challenge to the corporation was to provide support and encourage interaction with minimum interference and encumbrance while providing a sense of direction and a system of values.

It is becoming more apparent that programs that support innovation within the firm must be matched by national programs that support investment, development, and the expansion of markets. Engineering executives have the responsibility to be actively involved in the dialogue designed to generate more productive national policies. A wealth of information concerning the opportunities for improving the economic environment for innovation is becoming available. In 1983, Robert Reich[6] outlined the choices that management can make which will result in a more productive organization. He proposed that firms be more flexible in responding to changes in the environment and more willing to contribute to raising the level of understanding of the impact of particular programs on the firm's ability to innovate and compete. In the same year J. Morton Davis[7] presented a set of recommendations for changes in the national economic policies that will be needed to assure that the nation's resources are allocated in ways that support innovation and competitive strength. It is likely that examination and discussion of these issues will require more management attention in the future.

Systems theory provides some guidance for the analysis of these competing demands upon the time and attention of engineering management. The appropriate model is one of overlapping and interacting systems. Initially, the structure and characteristics of each system must be defined and documented. The appropriate variables and range of behavior must be identified. The steady-state conditions and the likelihood of excessive variation should be determined. The response pattern of the systems to external influence and the opportunities for exerting control should be defined. Formal and informal programs for simulating the behavior of these systems should be developed. This may include computer models and preparation of scenarios. Management-development programs and role-playing exercises will be useful in improving the skills needed for dealing with the complex situations that may evolve. There should be continuing observation and involvement to identify tendencies and possibilities and encourage further development of the group interactions. When appropriate, provision should be made for the use of consultants to support the analysis of the operations and the direction of the development programs.

Systems theory further suggests an integrative approach in which management should define a higher-level system that includes each of the direct subsystems as well as others within the firm and takes into account the conditions and relationships in the environment. The next sections will examine these issues in more detail.

3.3.   INTERFACING

Managing the interfaces between the engineering function and other components of the corporation is the critical process involved in integrating the engineering function within the firm. The goal is to develop flexibility and responsiveness in the interactions while providing a sense of continuity and support for the individuals. Matrix organization and taskforces represent two of the procedures used to increase the flexibility in the use of the human resources. Matrix organization enables the individual to participate in projects with individuals from other functional groups while maintaining the fundamental association with his own group. Taskforces bring representatives from different functional areas together for typically short periods of time to create new programs or resolve crises. These organizational techniques bring benefits but can also create problems.

The traditional organization chart (Figure 3.1) clearly demonstrates the difficulties one faces when trying to integrate the engineering (or any other) function in a corporate environment. The organization chart portrays the structure and the functions of the organization. It represents stability and order. It shows the patterns of authority and responsibility. It conveys a

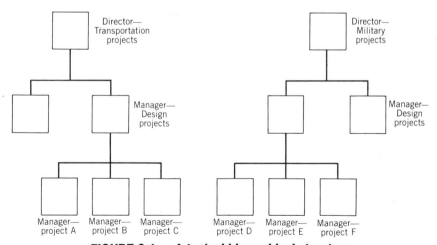

FIGURE 3.1.   A typical hierarchical structure.

sense of the formal organization, but it lacks the ability to show how the informal organization works. The rapid change and complexity, which are characteristics of the current development environment, require a better balance of the formal and informal systems.

An effective firm is a system which supports and directs communication and interaction among the components of the organization. The variety of interactions between just two departments is shown in Figure 3.2. A problem in meeting a sales requirement may require the joint effort of engineering, production, and marketing. The quality problem may require the combined efforts of engineering, production, and personnel. A product promotion problem may require the combined efforts of marketing, advertising, and finance. In each of these cases an executive must recognize the full scope of the problem and identify the potential contribution of the relevant corporate functions. When this happens, the structure of the organization must be modified to bring the appropriate resources to bear on the situation.

In the traditional organization chart the flow of action and communication tends to be vertical; it is important that the senior executive responsible for the function be informed and involved in all interactions involving the particular function. An interaction between the engineering function and

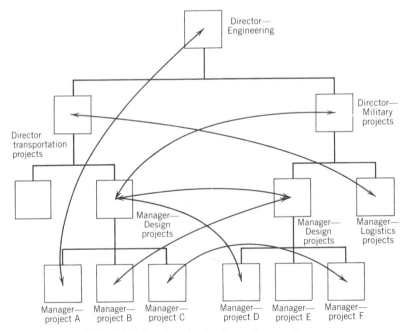

FIGURE 3.2.   Typical interaction patterns.

the marketing function could, under these circumstances, require that the senior executives for engineering and marketing be directly involved. The questions are how much and how often.

A systems approach to this condition is to develop a set of principles and rules for dealing with the range of situations where this type of multifunction interaction would be required and a responsibility center to implement the procedures with a manager of interfaces. Managing the interfaces requires consideration of many factors including the following:

1. Scope and impact of the situation
2. Exposures created by the situation
3. Precedents for the situation
4. Groups involved
5. Relation to other current situations
6. Time involved

The interface manager receives the requests for the allocation of resources, negotiates the specifications for the project or taskforces, establishes the assignments, communicates the pertinent information, and monitors the progress. In some organizations this work is done by a project office. The senior executive for the function may sometimes assume this responsibility.

What information does the interface manager need to operate effectively? First, the interface manager must have a complete inventory and understanding of the resources of the department, their allocation, and capabilities. Second, an awareness of the goals, priorities, and plans for the department is needed. Third, an early indication of the types of situations that may develop that require cross-department interaction and cooperation is necessary. Fourth, the interface manager needs a history of previous situations and knowledge of the level of performance of resources drawn from other groups. It also helps the interface manager to know which individuals and groups work well together. Much of this information could be collected on a systematic basis and incorporated into a computer-based support system. Access to a computer-based information system will provide a wider range of options for dealing with these situations. The information retrieval and electronic mail capabilities of these systems makes it easier to identify resources and communicate information about arrangements and status.

One of the major difficulties of managing the interfacing process is the need to find the proper level of trade-off between formal and informal procedures. There is a wide range of multifunction interactions that can occur. Some are simple requests for information; others require participation in major taskforces. The emphasis should be on informal procedures which are well understood and are reviewed regularly. It is difficult to prepare and maintain detailed documentation of formal procedures required to deal with the range of situations that may be encountered. A simple statement will suffice which supports the need for these interactions and provides the specifications for a pattern of response and interaction which is compatible with the goals and values of the organization.

The traditional means for developing performance plans do not work too well in the flexible environment required to support innovation and responsiveness. The concept of a long interval between performance reviews is not compatible with dynamic conditions. The focus must be on recognizing the need for interaction between the individual and other individuals in all parts of the organization in order to develop a pattern of support and reinforcement while providing for the opportunities of leadership which produce results. Each individual represents a resource which can be committed or shared. The challenge to management is to develop each individual's potential contribution and to provide a context where the contributions can be allocated to appropriate opportunities. The appraisal process must support this allocation process as a joint activity involving the individual and the superior. Success may come from a new process developed by an individual with little involvement from others. Success may come from a project supported by a group from another area. Success may come from an individual accumulating support from several areas to achieve an important objective. Managing interfaces deals with the interactions across organizational boundaries. The preparation of performance plans and appraisal reviews can be an effective tool for supporting the interfacing process.

The corporate environment can support or inhibit the interfacing process. Corporate goals that focus on growth and opportunities for innovation and improvement are positive elements in such an environment. Corporate goals and values that stress respect for the individual and commitment to individual opportunity and support reinforce the individual sense of commitment. Corporate goals that contribute to the improvement of the organization's stature lead to a greater sense of cohesiveness among the members of the organization. Each corporate goal, policy, or practice must be reviewed to determine if it contributes to breaking down barriers for cooperation and enhances the effort to achieve the best balance of individual and group interaction across the whole organization.

The responsibility for achieving an effective interfacing program must be shared by all members of the organization. The Drucker dictum that each knowledge worker is a potential executive and therefore is able to affect the corporate outcome, provides a sound foundation for developing a shared sense of the need for having all of the components of the organization work together. Communication programs which reinforce the need and benefits of effective cross-organization interaction and which demonstrate procedures for accomplishing these goals constitute an important element in this effort. Conferences and other meetings which bring people together can also contribute to this end. Sharing of experiences, participation, and support from corporate leaders can be the major factor in the effort to achieve effective interaction across organizational boundaries.

The success of the program of managing the interfaces is measured by the reality of the performance of the total organization and by the perception of each individual that he or she is making a full contribution to the corporate goals.

3.4. OPPORTUNITIES AND RESOURCES

The process of interfacing establishes the requirements for communication with the other parts of the organization. These relationships define a pattern of interaction that represent a significant aspect of the environment for the engineering function. The management of the engineering function involves the direction, support, and control of the processes, interactions, and resources of the engineering function. The goal is to achieve the highest level of contribution to the organization's growth and profitability with the available resources.

It is difficult to determine what the potential output of 100 or 1000 individuals can be. Studies have shown that most organizations are able to tap 20% or less of this potential. One element of improving this level is the managing of opportunities and resources. The challenge is to find the potential value of each individual and resource and bring these individuals and resources together in ways that will enable the engineering organization to capitalize on the most significant opportunities that can be identified.

Opportunities are represented by situations where the firm can provide a product or service that is valued by the customer and produces a significant return to the firm. Resources are represented by factors or conditions which will support an activity that can fulfill an opportunity. This process of expanding the contribution of the engineering function works best when opportunities are sought on the widest horizon and the set of resources is expanded to include all sorts of factors that can produce a positive effect on the function's ability to produce benefit.

Managing opportunities and resources involves more than the traditional issues of effectiveness and efficiency. Effectiveness is a measure of doing the right thing. Efficiency is a measure of doing the thing right. Opportunities and resources are dynamic and exist in a dynamic environment. The practice of managing opportunities is built on the premise that the process of change for each of these factors can be anticipated and to a considerable degree controlled. The process of managing opportunities involves orderly scanning of the environment for situations that can be converted into gain for the customers and firm. It also involves identifying and building up factors that can be organized to help accomplish gainful situation. It involves creating an organizational unit where these processes can be performed well. The aim is to have an effective and efficient engineering unit as the first requirement for achieving true integration of the engineering function within the firm.

Opportunities and resources exist in symbiotic relationship because an abundance of opportunities will have a positive effect through providing a justification for expanding the value and contribution of the resources. In the basic case where the resource is the individual, a major opportunity will motivate the individual to achieve and create at a high level. Similarly, competent and motivated individuals have a way of selecting and responding to situations which represent major opportunities. A fortunate consequence of this interaction is systematic improvement of the development process. The goal is to make this a continuing process.

The process of organizing the engineering function with the emphasis on opportunities and resources starts with an inventory of products and services provided by the engineering function and an inventory of resources that are available to the engineering function. In a general sense the ratio of output to input is the equivalent of traditional return on investment. But much more is involved here. Both outputs and inputs are variables which can be expanded. From one period to the next, the output from the engineering function can be changed to enhance the value of the contribution to the firm. From one period to the next, the value of the resources can be enhanced to increase the potential for improved output. In the best case, engineering managers are finding new products and services that represent an expanding set of opportunities, while at the same time they are selecting, training, developing, and motivating the staff to achieve higher levels of competence and productivity. While this focuses on the staff as the major resource, there are many other elements that must be included within the domain of resources. These will include the physical facilities, computer-based information systems, data bases, accumulated experiences, customer good will, a network of relationships with suppliers and associated professionals, patents and other design assets, and a set of goals and practices which bring all of these factors together into a working system.

The inventory of outputs starts with a review of projects that have been completed for the last several years. What contribution did each of these projects make to the firm's growth or profitability? What were the opportunity costs for each of these projects? What would have been the results if the resources, time, and attention devoted to these projects had been allocated to other activities or other opportunities? Was there a substantial list of other opportunities to which these resources could have been applied? The mark of a vigorous engineering organization is the abundance of opportunities that have been identified and are being developed.

Finding new opportunities for the established engineering function, or for a new one, is an exercise in creativity, communication, and data collection. For the most part, opportunities exist in the future because profitable situations which already exist and are recognized today, may either disappear before they can be capitalized upon or will be in a competitor's domain. The greatest advantage is produced when a situation that has the potential of becoming an opportunity is anticipated through the use of forecasting procedures with sufficient lead time for appropriate resources to be developed and the program prepared to convert the opportunity into gain. The second source of advantage includes those opportunities which have not yet been recognized by others and can be converted into gain. The third source includes those opportunities which others have recognized but have not been converted into gain.

There are many agencies and procedures that can be utilized to identify opportunities. The most important is direct interaction with the actual or potential customers. The knowledge, imagination, experience, and contacts of members of the staff are a rich source of ideas. Market research groups are able to survey user requirements and probe for potential requirements which have not yet been recognized. Forecasting techniques can be used to identify and extrapolate trends and generate hypothetical situations which, if they should evolve, will create opportunities. Joint studies with users, vendors, universities, and research institutes can help identify opportunities. Thorough examination of the firm's experience and requirements can provide leads for new opportunities. Internal research and development activities can generate opportunities. The journals, newspapers, and other media can be sources of ideas. The goal is to utilize all available means for stimulating and supporting the recognition, generation, and development of situations which can produce gain.

Accumulating an inventory of opportunities is the first step. The next step is the cataloging and evaluation of each gain situation. How do these opportunities relate to each other? How do these opportunities relate to the resources that are currently available? What must be done to develop the present or new resources in order to bring the opportunities within the reach of fulfillment? The process of matching opportunities and resources is a principal management activity. The scoping of projects is the means for matching opportunities and resources. The project can include all or part of an opportunity with the required resources. The project can refer to a new product, service, or market which must be examined in more detail. The project specifications define a set of inputs and outputs. The projects specifications contain estimates of the costs of inputs, value of outputs, as well as time and activities required to complete the project. The engineering function is organized to generate project specifications; its role is to evaluate, gain support for, and implement projects, as well as evaluate the results produced by completed projects.

Establishing the priorities for projects and determining the sequence in which they should be implemented requires considerable management attention and skill.

3.5. PRIORITIES

One aim of planning is to assure that there are more than enough opportunities and almost enough time to capitalize on them. The result is that there will always be the need to prioritize the choices. The key to success is the ability to put all the conflicting choices into perspective and apply to them a reasonable scale which will result in acceptable short-term and long-term gains. A priority becomes an issue when there are too many things that need to be done without sufficient resources to take care of all the options that have value.

The primary focus of managing opportunities is to create the largest possible number of situations that can contribute to the growth and profitability of the firm. Each opportunity situation can then be specified as an element in a potential action program, maintained in an appropriate opportunity inventory system, and ranked in terms of feasibility and value. The larger the pool of opportunity situations, the more flexibility the firm will have in plotting a course for future corporate growth. This involves a continuing examination of changes in technology and needs and conditions of the market. The focus on the resource end is to reduce low-potential factors to assure that the cut-off point for opportunity situations is always high.

A sound priority system depends upon a consistent procedure for establishing projects, dealing with opportunity situations, and relating costs and payoffs for the project. Each project relates an opportunity situation with the resources needed to satisfy the opportunity. The project definition must be specific and well-communicated. Each project candidate should be structured to permit reasonable comparison and evaluation against its peers. Since this is a measurement process, the specifications of the project should also include indications of tolerance toward each of the elements needed for comparison. There should be upper- and lower-limit boundaries for the revenues, time duration of the revenue stream, resources required, and time required to implement the project. This setting of bounds on the variation of the critical factors must be supplemented with appropriate applications of sensitivity analysis to determine the effect of variation of single factors on the total project result. This can be summed up as requiring an engineering approach to the control of projects.

In the long run, the value of each project and its ranking are determined by the payoff over the life of the project and its resulting product or system. Part of the problem with this is due to the fact that the concept of payoff may include elements which are objective and some which are intuitive. There is uncertainty involved in the estimate of the life of the project or product. There is uncertainty related to the point at which the product is replaced by another technology or by a competitor's product. There is uncertainty connected to the estimate of the customers' acceptance of the product and the impact upon existing company products and services. As each of these and other sources of variability in the product estimates are identified, there are means available which can be used to narrow the tolerance to acceptable limits.

Some opportunity situations involve a single product while others may incorporate the possibility of generations of products. It is apparent that the product stream will generate a longer revenue potential, but it will also entail more risk. Each product strategy will require consideration of the underlying technology structure, strength and flexibility of the competition, and composition and duration of the customer demand for the product sequence. The scope of the issues to be examined demonstrates the need for input from the engineering function, marketing function, and groups that monitor the changes in the structure of the market and competition. Each input contributes data that enhance the precision of the estimates required to properly evaluate the viability of the opportunity situation.

It is not likely that two major product-opportunity situations will arise at the same time so that the allocation decision can become a strict comparison of two candidates. In the base case each project or opportunity situation must be evaluated against some allocation standard. This may involve achieving a specified share of the market, specified financial return, or some combination of these parameters. The value of these parameters will rise or fall with changes in the marketplace, strength and strategies of competitors, availability of capital, or availability of resources or facilities.

There is a tendency to put more emphasis on the financial elements in these decisions while downplaying the engineering or product development elements. There can be significant nonfinancial reasons for pursuing a new product or a new technology which should be factored into the allocation decisions. This is one reason why the engineering function should have major input into the preparation of the corporate goals; an issue that will be discussed later. This also demonstrates the need for frequent interaction with the financial functions within the firm. This interaction includes deliberations concerning methods for evaluating and allocating financial resources, for planning future resource requirements, and gaining company-wide support for engineering projects.

There are many algorithms for evaluating project investments, including payback period, average return on investment, and net present value of the revenue stream. Frequently the different procedures will assign unique rankings to the same project, depending upon the estimates of the operating environment for the duration of the project and then its installed life. The initiative for determining which of these methods to be applied, in general and specific instances, typically resides in the financial function. The choice of the evaluation method will often determine which project is supported which, in turn, means that the engineering manager will want to have substantial input in the process of selecting the evaluation algorithm.

Once the priority ranking system has been established, it is imperative that all members of the engineering function be familiar with its requirements and procedures and support its application. With the availability of computer-based decision support systems, it is possible to program the evaluation and allocations procedures with input to appropriate models and simulation systems that facilitate the assignment of priority status to competing projects. These systems could also be programmed to provide the data which can be used to monitor the execution of the project and revenue stream produced by the project result.

The sense of urgency that is necessary to make a priority system work most effectively is created when the engineering function is operated as a profit center which provides a visible connection between the contribution of each individual or team to the results generated by the total engineering function. Typically such an approach provides a more realistic foundation for the allocation of engineering resources, enables the engineering managers to relate more directly to the corporate goal-setting activities, and provides a broader exposure and orientation for the engineering managers which prepares them for progression into corporate management. Given the greater effect of technology-based considerations on commercial operations in the future, there are sound reasons for encouraging individuals with technical backgrounds to become involved in the senior decision-making levels of the corporate structure.

The effectiveness of a priority system depends upon the review of the results produced by executed projects and the forecasting of conditions that will exist in the future environment for the projects that will be implemented. This requires a corporate attitude that considers the collection of historical data as an essential phase of good planning and management. It requires close cooperation between the engineering and financial functions.

It also requires a management structure which relates project results with points of accountability, or, at least, to a specific decision process. While this procedure may be needed to assess responsibility, it is more useful as a factor contributing to the learning process of the firm which enables it to improve its procedures for allocating resources. Rapid turnover of managers tends to reduce the effectiveness of this feedback process which is one reason why such procedures as career planning and management development can affect the ultimate effectiveness of the priority system. The set of issues concerning the priority system pertains to many parts of the firm and is properly a major element in the corporate activity for setting goals.

3.6.   SETTING CORPORATE GOALS

The corporate goals provide the marching orders for the organization. They tell where the firm should be going and how it will get there. The statement of corporate goals can be short and general or it can be long and detailed. The important thing is that it should provide a sense of direction, system of values, and be understood and accepted by the members of the organization. The corporate goals should provide a cohesive planning envelope for the organization.

The statement of goals for the engineering function will be subordinate to the corporate goals and produce the same sense of direction for the engineering function. This statement is a logical extension of the corporate goals and expands to include the issues significant to the operation of the engineering function.

There are three critical stages in the life of the statement of corporate or engineering goals. These include the time when the statement is being prepared, period after it has been adopted, and period when it is to be modified. Most organizations start with a statement of goals prepared by the founders who have the opportunity to develop this statement. Those who join the organization become familiar with the goals and make them an integral part of their operating philosophy. The open issue is just how much influence the corporate goals will have on the day-to-day operations of the firm and how their impact will be reinforced and measured. In a well-managed firm the corporate goals have considerable influence on the operations and are regularly reviewed and reinforced. They represent a strong bonding force for the members of the organization. This is true for both the corporate goals and the engineering goals.

To the extent that the firm is dependent upon effective operations in such areas as product development, productivity, quality, and service, the contribution of the engineering function represents a major factor in the success of the firm. The reverse of this is also true; because the goals, direction, and environment determined in the statement of corporate goals have such an influence on the opportunity for the engineering function to operate at a high level of effectiveness, the engineering manager has a high stake in seeing that the statement of corporate goals supports and reinforces the engineering goals.

At this level the engineering manager becomes more than the supervisor of the engineering function. His task becomes one of providing effective input to the corporate planning- and goal-setting processes with the aim of generating understanding and awareness at the corporate level that will allow effective engineering activity. The need for this type of involvement and interaction changes over time, reflecting among other things, the changes in the background and interests of senior corporate managers and requirements of the business in relation to the marketplace and other elements in its operating environment. In recent years the participation of individuals with technical backgrounds in corporate management has diminished as the primary focus at that level has been on legal and financial issues rather than on issues of productivity, product development, and marketing. It is likely that this orientation will shift as more individuals with technical backgrounds take an active interest in the corporate management process and as the business environment puts more emphasis on the production and marketing aspects of the firm's operations.

One element of the corporate goals is the expectation for growth. This is an area where the engineering function can have a valuable contribution. Growth depends upon the development of new products, identification and development of new applications and markets, and implementation of productivity and quality programs that increase the value of products to customers. The ability of the engineering function to meet its obligations in support of reasonable growth objectives can provide an upper limit to the level of projected growth. If there is a gap, the firm can look to the engineering function to expand its capabilities or to grow through acquisitions and other sources of the necessary capabilities. In both the short and the long run, it is better for the engineering function to assume responsibility for organizing and conducting its operations in support of the growth goals of the firm.

Another element of the corporate goals affects the internal environment of the firm. This includes those parts that refer to the rules and values which structure the corporate culture. These can enhance the contribution of the engineering function or introduce frictions and concerns that increase transaction costs within the business. The aim of corporate goals should be to reduce these transaction costs through emphasis on joint use of resources, cooperation and communication, respect and recognition for the individual, delegation and decentralization, achieving a reasonably stable corporate environment, and on programs that foster group cohesiveness and individual identification with the corporate goals. This set of considerations may affect the members of the engineering function in particular ways, which suggests that the engineering managers have the obligation to provide effective input for the preparation of the components of the corporate goal that are critical to the achievement of the engineering goals.

The statement of engineering goals addresses additional items that will more precisely determine the structure, direction, scope, and intensity of the engineering activity. Among the items that are relevant are those that deal with the relationships with other functions such as research, production, finance and marketing; those that define the areas of development to be encouraged; those that specify the level of "state of the art" to be supported; and those that support and enhance the spirit of professionalism in the engineering function. Just as in the case of corporate goals, the involvement of the individuals in the preparation and support of the engineering goals is an important objective. Communication and demonstration of the significance and utility of the goals needs to be a continuing endeavor. A well-formed statement of engineering goals will serve internal needs and can be a useful vehicle for communicating with external groups, including current and prospective customers and clients, and professionals who are being recruited.

The engineering manager, because of background and activities, will be particularly alert to many aspects of corporate activity upon which the firm will need guidance. These inputs to the corporate planning and goal-setting activities could include those respecting the impact of the company's operations on the environment, the impact of changes in technology, patent, regulation, and structure of the markets served by the company. The relations with professional societies and schools is another area where the engineering manager will want to provide guidance for those preparing the corporate goals. The significance of these areas to the success of the firm and of the engineering function suggest that the engineering manager will have a real interest in developing and enhancing the means of communication with representatives of other functions and individuals involved in the preparation of the corporate goals and strategies.

These activities related to the preparation of goals represent a set of continuing tasks that supplement the process of leading the engineering function and require that the engineering manager maintain information systems that scan the environment and support the preparation of forecasts for the future. These decision support systems will rely on the use of computers and communications systems, but will also require constant human observation, monitoring, and review. These activities can also be used to develop the skills of the individuals in the engineering function as a phase of professional and management development.

3.7. ANTICIPATING THE FUTURE

The issues that have been discussed are fundamental to effective integration of the engineering function. Any action concerning them will affect operations in the future and in turn will be affected by conditions in the future. The challenge then is to develop adequate procedures for identifying changes and conditions that are likely to exist in the future and relate these to programs and operations in the present.

The increasing concern with anticipating the future has resulted in the development of many new procedures, formal and informal, for making estimates and judgments about events and relationships in the future. Some of these are procedures for pooling group estimates, as in the Delphi method. Model building and simulation using computers requires serious analysis of alternative relationships in the future and provides the means for testing different sets of values within these alternative models. Scenario building encourages the analyst to get involved with possible events of the future and develop alternative sequences of events as a means for developing the consequences of particular courses of action.

The concern here is not so much with the specific tools of technology forecasting and assessment, but with the means for organizing the selection and implementation of the appropriate tools and incorporating them within the planning and project activity. There is a tendency to consider the technology forecasting and assessment tools as ends and not means. These procedures are fascinating and can be quite seductive in stimulating project activity which focuses on the nature of the future environment rather than on the relationship of possible futures to the significant current or planned activity.

The question is how to organize and allocate resources so that possible conditions in the future are regularly reviewed and interpreted to the end that the firm is alerted to both positive and negative possibilities. There is a wide variety of factors that must be examined. These factors define the environment within which the firm will operate and influence the relationships that the firm must be able to control. These factors include:

1. Level of business activity
2. National policies and priorities
3. International conditions

4. Changes in technology
5. Changes in tax laws and financial practices
6. Population changes and movement
7. Levels and trends in consumer spending
8. Levels and trends in business investment
9. Changes in levels of activity from industry to industry and from type of industry

There is a wide range of approaches that can be implemented within a firm for organizing to deal with the future. In some instances the strategic planning function will be centralized at a high corporate level and the conclusions distributed to the individual components of the firm. This represents a "top-down" planning process. Alternatively, the functional areas could be asked to prepare estimates of future conditions for a range of situations with some process for integrating the resulting estimates. This would represent a "bottoms-up" planning process. After the estimates have been combined, the results would be distributed and provide a basis for the planning at the functional level.

The "bottoms-up" planning process is the more rational approach because it is a more open process and improves the communication among the functional areas. It also engages the functional areas at earlier stages in the planning process. In addition, the "bottoms-up" planning process puts a greater obligation on the functional areas to collect and analyze data on future trends in areas that will affect their and the firm's opportunities. Each input from a functional area concerning a trend or a change in relationships can generate a question or challenge for another functional area. This enhances the variety of issues explored and makes the planning process more productive. Fewer possibilities are overlooked.

The time distribution of activity in each functional area will be different. The financial cycle is different from the engineering cycle; the marketing cycle is different from the production cycle. These cycles are affected by the rate of change of factors in the environment that influence the need for response and by the sequence of activities that the functional area must execute to implement a program. This includes the time and effort required to implement changes within a given set of conditions or system and the time and effort required to change from one system to another. It also includes the time and effort required to determine which kind of shift is appropriate.

The tools of technology forecasting provide the basis for dealing with the future in one of two ways. They are (1) exploratory forecasting which takes the present as a given and extrapolates likely conditions in the future and (2) normative forecasting which reaches out to some period in the future, determines a reasonable set of circumstances for that time period, and then asks,

"What changes taking place in the present will produce the expected circumstances?" Normative forecasting builds on other estimates which help define the nature of the environment in the future time period. These could be prepared by the organization or by others. Exploratory forecasting guides the organization into the future by extending the present; normative forecasting drives events to create the future.

It is in the nature of engineers and of engineering to be constantly scanning the environment, looking for alternative interpretations of what is happening and is available. This is the fundamental requirement of forecasting. This suggests that the engineering function can exercise considerable influence on the practice of forecasting within the firm. The issues relating to technology and productivity are very near to the center of the factors affecting future development of the company's business activity. These must be supplemented with data on social changes that are likely to occur and estimates on the level of economic activity. This indicates that the engineering function can work closely with the marketing function in developing statements that describe the possible, and then the most likely, patterns of development.

Scenario building is one activity in technology forecasting that provides the opportunity for representatives from several functional areas to work together to develop descriptions of likely patterns of change. A scenario is much like the script of a play where the events are projected to take place in the future. The activities included in the scenario must appear real and be consistent. If there are significant leaps in the process of development, there must be a reasonable basis for accepting that the projected changes are possible. The scenario can be as detailed or general as desired, but experience shows that the more details included in the preparation of the scenario, the more potential problems and opportunities are identified.

The amount of time and effort devoted to forecasting future events is related to the needs and the expectations of the firm. There has been a consistent trend to devote more resources to long range planning and interpretation and application of planning data. This is partially due to the fact that many projects and development programs have long cycles. There is a built-in expectation that the world at the time that the project is completed may well be significantly different both from the present and from what is currently expected. To increase the likelihood that long-range planning data is incorporated within the short-term planning process, many firms will conduct two planning cycles with some overlap both in time and in participation by individuals from the different functional areas, such as engineering and marketing. The long-range planning cycle may be concerned with events and developments relevant to a time horizon extending 8 to 10 years out into the future. The short-term planning cycle encompasses two years, or the time covered by the detailed budgeting process.

While it is inevitable that there will be heavy emphasis on use of formal methods of analysis, there is still much room for the intuitive judgments of individuals. In many cases intuitive estimates generated by knowledgeable individuals provide the basis for a framework for more formal, detailed analysis to determine if the intuitive judgment is viable. Trend analysis includes that use of statistical procedures for estimating significant parameters and limits to growth and change. There is, however, a low emphasis on the analysis of causes and relationships which can provide the basis for more intelligent forecasting of the process of growth and change. The success of all the forecasting activity will be largely determined by the firm's ability and inclination to devote resources to anticipating and capitalizing on the process of change, and on its realization that occasionally a forecast will be wrong. Good planning anticipates this consequence and makes provisions for minimizing the down-side risks associated with long-range plans and estimates. Part of the investment in forecasting and planning includes training for the people involved and the use of resource people who are able to bring different perceptions to the planning process.

There will always be conditions that cannot be foreseen and whose consequences cannot be controlled. The firm that devotes resources to forecasting and planning will be able to identify the conditions which can be positive and those which carry the elements of difficulty. The activity of planning can provide early warning of possible exposures that could adversely affect the operations of the firm. This generates lead time that can be used to develop procedures for dealing with the anticipated situation. Each year the firm will evaluate its investment in forecasting and planning to determine how the procedures should be changed and improved and to identify the factors that were not detected in the preparation of earlier projections. This then establishes a continuing learning process.

The problems of technological and related business forecasting have given rise to an extensive literature which deals both with substance and methodology. A selected set of references is given at the end of the chapter. See notes 8–16.

3.8. DESIGNING THE ORGANIZATION

The major activities that are unique to the engineering manager have been defined as including interfacing, managing opportunities and resources, developing and applying priority systems, setting goals, and anticipating the future. Some of these activities may be significant in other functional areas. Where this is the case, the engineering manager is concerned with the execution of these practices in areas of company operations affected by developments in technology and productivity. The next challenge to the engineering manager is to design the organizational structure as well as the team that will execute these major activities and meet the needs of the firm in relation to effective communication, control, and support.

The conceptual elements of the process of designing and developing an organization tend to be apparent to managers with an engineering background. This process involves analysis, design, and development. Loss of efficiency comes with determining the right balance of discipline and control as opposed to flexibility and decentralization. In order to manage engineers as professionals, one must accept the professional individual's requirement for the room to be creative and exercise judgment and initiative within his or her realm of competence. Traditional organizational structures include constraints and specifications for order left over from earlier days of bureaucracies. Current and future demands for engineering effectiveness call for innovative approaches to the definition of engineering tasks and the grouping of these activities into units that can be managed with elegance. These patterns will need to balance the requirements for direction and professional latitude.

Recent experience has produced many new organizational forms for meeting these requirements. The matrix form of organization, where the individuals are assigned to groups of professional specialties and then temporarily assigned to projects as the need arises, is one such development. The extensive use of ad hoc task forces is another organizational approach for dealing with the need to bring varied resources together to satisfy a short-term project requirement. The extensive investment in communication systems, computer-based planning, and control systems is another significant development that supports greater flexibility in meeting the need for adaptive organizational relationships. The acceptance of dual lines of development, managerial or professional, has also provided the means for letting professionals rise in organizational stature and responsibility without accepting the full range of obligations set forth in the typical manager's job description. A highly competent professional in this environment can assume leadership responsibilities for a project without being completely divorced from deep involvement in the technical elements of the project activity. There will be other patterns of organizational structuring and access to professional talent developed to meet the needs for engineering productivity in the future.

There are two phases in the development of an organizational structure. The first is a process of task analysis and allocation that provides the "theoretical" structure. This takes the range of activities to be executed, divides the activities into tasks, groups the tasks into bundles that can be defined as individual positions which, in turn, can then be grouped into units which can be managed. In this phase all individuals in each professional or managerial category are considered to be homogeneous and equivalent; not a very realistic assumption. In the next stage, the distribution of tasks and individuals is refined to produce the "working" organization. Here the characteristics and interests of the individual managers and professionals are taken into account to provide an organizational structure that capitalizes on and reinforces the unique potential of each individual.

Experience shows that the completion of this two-phase process may not yet result in a fully functioning organization. So far the organization works well on paper; there are interpersonal adjustments that must be accepted to achieve effective working relationships and some managers and professionals may not accept the offered pattern as being responsive to their own perception of the unique potential. Aspirations and expectations are factors to be considered. Senior managers have been known to establish unusual reporting patterns for individuals who possess unusual capabilities or interests when the need or the potential gain can be justified. In the engineering environment better results are achieved when the organization structure is accepted rather than imposed.

Complexity is a common characteristic of the modern organization. It was explained earlier that the traditional hierarchical organization chart does not adequately represent the pattern of interactions that are necessary in an effective firm. The organization chart is an attempt to represent multiple levels of relationships. In a bureaucratic organization these relationships all flow vertically. In the typical modern, complex organization, the relationships involving reward and punishment; direction and guidance; concurrence; communication and exchange; and support and interaction may flow in different directions, have different origins, and reflect different initiatives. Provision must be made for the effect of corporate politics, ability of one unit to withhold support or approval vital to another unit, dependence of one unit on the productivity of another unit, nature of the data collection systems which support the reporting systems, length of the development and production cycles, and range of management styles. The "rational" organization structure is superseded by the "informal" organization structure. While it may be difficult to represent the "informal" organization graphically, it is not difficult to perceive and operate effectively within this organization, for this is the "real" organization.

Those engineers who will move into the manager line of career development are the individuals who not only can operate within the context of the "real" organization but are willing to accept the frustrations and machinations necessary to achieve significant effect on the company's operations. The need for and limits on these elements of the manager's activities are defined by the organizational culture and value systems. These in turn are influenced by the statement of corporate goals and the management styles of the senior corporate and engineering managers. The situation is not unlike the environment of a free market in the economic sense. The corporate culture establishes the rules for the conduct of the business of the firm, especially the interactions within the firm. These rules define the limits of internal contention that are acceptable and the point at which cooperation becomes essential. They provide support for challenge of established relationships and occasional unsuccess. Here the traditions and verbal history of the firm may exert considerable influence upon the behavior of the members of the organization.

The most successful organizations are those that devote considerable time and attention to matters affecting the careers and opportunities of the members of the organization. The aim is to delegate and decentralize the decisions and actions affecting the throughput functions of the organization such as engineering, production, and marketing, and to concentrate the focus of senior management on matters that affect the well-being and future of the organization as a whole. These activities have been defined as executive functions which, although they are most apparent at the higher levels of the organization, will also be represented by decisions and actions taken at all levels of the organization. Managers in the engineering function will have many opportunities to contribute to the formulation of corporate policies and to the definition of the corporate culture. It will be through this personal involvement at the corporate level and in corporate-wide activities that the engineering function will most effectively be integrated within the corporate environment.

NOTES

1. H.R. Towne, "The Engineer as Economist," *Transactions of the American Society of Mechanical Engineers* (1886): pp 428–432.

2. P.F. Drucker, *The Effective Executive*, (New York: Harper & Row, 1967).

3. P.F. Drucker, *Managing for Results* (New York: Harper & Row, 1964).

4. J. Naisbitt, *Megatrends* (New York: Warner Books, 1982.)

5. R.M. Kanter, *The Change Masters* (New York: Simon & Schuster, 1983).

6. R.B. Reich, *The Next American Frontier* (New York: Times Books, 1983).

7. J.M. Davis, *Making America Work Again* (New York: Crown, 1983).

8. W. Ascher, *Forecasting* (Baltimore: Johns Hopkins University Press, 1978).

9. E. Jantsch, *Technological Forecasting in Perspective* (Paris: Organization for Economic Co-Operation and Development, 1967).

10. H. Kahn and A.J. Wiener, *The Year 2000* (New York: Macmillan, 1967).

11. N.M. Kay, *The Innovating Firm* (New York: St. Martin's Press, 1979).

12. S.B. Lundstedt and E.W. Colglazier, Jr. (eds.), *Managing Innovation* (New York: Pergamon, 1982).

13. E. Mansfield et al, *The Production and Application of New Industrial Technology* (New York: W.W. Norton, 1977).

14. J.E.S. Parker, *The Economics of Innovation* (New York: Longmans, 1978).

15. F.A. Rossini and A.L. Porter (eds.), *Integrated Impact Assessment* (Boulder: Westview Press, 1983).

16. D. Sahal, *Patterns of Technological Innovations* (Reading, MA: Addison-Wesley, 1981).

4

PROFESSIONAL PRACTICE AND ORGANIZATION OF ENGINEERING FIRMS: THE REGULATORY FRAMEWORK

Milton F. Lunch

MILTON F. LUNCH, General Counsel, National Society of Professional Engineers, is a native of Washington, D.C. He has been associated with NSPE since September 1946, with primary duties extending to legal matters, legislation, registration, interprofessional relations, ethics, and government relations. He served in the Corps of Engineers with an engineer combat battalion during World War II. During the Korean War he was recalled to active duty in the army and served from January 1951 to May 1952 as chief of Industrial Relations Branch, office of the Judge Advocate General, and labor advisor to the Under Secretary of the Army. He was admitted to the Bar of the District of Columbia in 1947 and is admitted to practice before the Supreme Court of the United States. Mr. Lunch is a graduate of National University Law School with an LL.B. (1948) and an LL.M. (1949).

4.1. TYPES OF FIRMS

The traditional form of practice of firms offering engineering services to the public was either as an individual (solo) self-employed professional or as a partner under a partnership agreement with others legally qualified to engage in the same field of professional practice. This concept was in accord with the state licensing laws governing the engineering profession which provided only for the licensing of individuals, and remains true today.

However, because of economic and tax-related aspects of engineering practice, and as engineering firms flourished in both numbers and scope of practice, there developed a movement to expand the forms of practice to meet the more complex legal and economic requirements of firms. From this early recognition on the part of many that some change was needed there emerged one of the major debates within the engineering profession, centering on the question of whether the state registration laws should be changed to allow for corporate practice of engineering. The battle between the "traditionalists" and the "modernists" reached its peak in the years following World War II and the resultant expansion of engineering firms into more complex project assignments, as well as the emergence of a large number of new firms.

As far back as 1929 the basic position in favor of allowing for corporate practice was stated by T. Keith Legare who was then the executive director of the National Council of Engineering Examiners and was one of the pioneers in promoting state engineering registration laws:

> With the development of large engineering companies and corporations, and the growth of large state, municipal, public utility, railroad and other engineering organizations, the practice of engineering is becoming less and less a field for the individual practitioner. Whatever may be thought of this form of organization, it seems that it will be the agency through which practically all of the engineering work of the future will be done. Unlike such professional service, for example, as that of the physician, which is and will probably always be a service of an individual to an individual, engineering lends itself to group service.
>
> (quoted in "Corporate Practice of Engineering," by William Britton Stitt, *The Business Lawyer*, July 1959)

The issue was, technically, not whether corporations should be allowed to engage in the practice of engineering, but whether engineers legally qualified to render such services should be allowed to do so through the corporate form of organization. During the heat of the battle the opponents of allowing corporate practice brought a test case to make the point that a corporation could not be licensed as an engineer. The court upheld that argument, as the sponsors expected, stating under the definition of "professional engineer" in the state law "... manifestly only a natural person can come within this definition. It cannot under any circumstances apply to an artificial person," and further noting that an applicant for registration "... must be a person of good character and repute ... obviously only a natural person can comply with this requirement." Further, the decision made the obvious points that a corporation could not take the required examination or have the necessary experience, and education required for licensure. "It is clear to the Court that the statute must be construed by its very terms as inapplicable to corporations and applicable to natural persons only" (*Potomac Engineers v. Walser, et al.,* 127 F. Supp. 41 (D.D.C. 1954), aff'd 223 F.2d 96 (1955).

In addition to the concerns of the opponents because of a departure from tradition, their more pragmatic arguments were that corporate practice would tend toward larger firms at the expense of solo practitioners or smaller groups of engineers acting in a partnership status. They cited the fact that the other professions, such as law and medicine, did not allow corporate practice (which is no longer quite as true), and that corporate practice would allow nonengineers to control engineering judgments and policies, making the engineering function that of a "hired hand."

However, these arguments were gradually overcome by the impact of the larger context of engineering practice, as predicted by Mr. Legare in 1929, and even more so by the unfavorable consequences of the tax laws faced by individual practitioners compared to those whose ownership interest was in the form of stock ownership in corporations. This was particularly connected with certain fringe benefits and pension plans available on more favorable terms to engineer-stockholders of corporations. Later the pension plan disadvantage to individual practitioners was overcome to a large extent through enactment of the self-employment retirement act (Keogh Act) and its liberalization over the years to allow self-employed professionals and others to set aside on a tax-deferred basis earned income to be placed in a retirement fund. As of January 1, 1984, the limit for a set-aside has been increased to 20% of earned income, with a limit of $30,000.

The outcome of the "great debate" over corporate practice has now been largely resolved through a carefully tailored limited form of provision in the state registration laws, with some exceptions, principally in New York, which still bars corporate practice except for "grandfather" corporations organized and existing on April 15, 1935. The main points of the current approach as expressed in the "Model Law" of the National Council of Engineering Examiners are:

1. Registered engineers may practice "through" a corporation.
2. One or more of the corporate officers designated as being responsible

for the engineering activities of the corporation is registered under one of the state registration laws.

3. All personnel of the corporation who act in its behalf are registered in the state.
4. The corporation has been issued a certificate of authority by the state registration board.
5. All final drawings, specifications, plans, reports or other engineering documents involving the practice of engineering must be signed and sealed by those who prepared them or under whose immediate and responsible direction they were prepared.
6. The corporation must file with the registration board each year the names of all officers and board members, and those in responsible charge of engineering activities.
7. Neither the corporation nor its individual engineer officers or employees are relieved of responsibility for the conduct or acts of its officers or employees by reason of holding a certificate of authorization.
8. The registration board may revoke or suspend the certificate of authorization if any officer or director is found guilty of any conduct which would authorize revocation or suspension of their certificate of registration for violation of the board's rules of professional conduct.

Within these broad parameters, the state registration laws provide the specific requirements for corporate practice, with some of the major differences being requirements that certain percentages of officers or directors be registered engineers, or that a certain percentage of shareholders be registered engineers. In each instance a firm intending to practice in the corporate form should check the precise wording of the state law. A summary of the state law provisions appears in Appendix A.

Closely connected with the historical development of corporate practice of all the professions was the emergence of a somewhat different, and more limited form of corporate practice known as "professional corporations," or in some cases, "professional associations." This development came principally from the medical and legal professions which faced similar concerns over the tax consequences and pension limitations of individual or partnership practice prior to enactment of the Keogh Law and its later extension for larger benefits.

The "P.C." or "P.A." form of practice is available to engineering firms, but has been used relatively little because of the more favorable latitude found in the engineering corporate practice laws. However where regular corporate practice is not permitted (such as New York) or where the engineering law is found to be too burdensome for other reasons, the "P.C." or "P.A." form is used by some engineering organizations.

Major differences between the regular engineering corporation laws and the "P.C." or "P.A." approach, in most instances, are that the latter in all states require that all shareholders be licensed in the particular profession (this can be a major problem where the firm consists of a combination of engineers, architects, land surveyors, and other disciplines), and at retirement or death the interest of the shareholder must be transferred to another person licensed in the particular profession. The "P.C." and "P.A." laws specify that the professional shareholders and officers bear personal liability for the acts of the corporation, but this is also true of engineering corporations as applied to those who sign and seal the engineering documents, or who are in responsible charge of the engineering activity.

4.2. REGISTRATION QUALIFICATIONS

Every state has an engineering registration law which, while varying in a number of details, basically requires that a person offering or performing engineering services to the public or clients must be licensed by the state registration board.

In general, the requirements for registration involve completion of a four-year approved engineering curriculum, four years of acceptable engineering experience, successful completion of an eight-hour examination in the fundamentals of engineering, and passage of an eight-hour examination in the principles and practices of engineering. The written examinations are prepared by the National Council of Engineering Examiners in designated fields of practice and are administered by the state boards.

Prior to completion of the experience requirements, an engineering graduate may take the first eight-hour examination in fundamentals and thereafter be certified as an "engineer-in-training" (engineer intern in some states).

In addition to the standard registration procedures indicated above, most of the state laws have provisions which allow for registration on the basis of "long-established" practice. Usually those provisions permit a person with an engineering degree (or its equivalent), 12 years of acceptable experience in engineering, and a character deemed by the board to indicate competence, to be registered without examination. However, the trend has been to use this provision sparingly, and more often than not the state board will require the applicant to pass the practice portion of the written examination.

A few states still provide for registration without examination on the basis of "eminence," a term which is broadly defined to apply to those who have acquired outstanding reputations of national or international standing and who have made exceptional contributions to the engineering profession. The "eminence" provision is hardly ever used—the last notable use was by Iowa to register Herbert Hoover.

A provision of more practical importance is the "reciprocity" or "comity" clause under which a person registered in one state may be registered without further examination in another state. The usual caveat in such provisions is that the original registration must have been in accord with the same qualifications of the state in which reciprocity is sought. Thus, if State A (the state of original registration) did not require a degree, and the law in State B (in which reciprocity is sought) does require a degree, it will likely be held that the nondegree applicant is not entitled to reciprocal registration because the qualifications are not the same. Or if State A registered the applicant without examination and State B requires passage of a written examination, reciprocity will likely be denied.

No two state laws are exactly alike. In each case the person desiring registration should examine the appropriate state law to determine the precise requirements. A summary of state requirements is given in Appendix B: the names and addresses of state and other registration boards are shown in Appendix C.

4.3. EXEMPTIONS

The state registration laws include a variety of exemptions which range from federal employees (who are exempt on constitutional grounds whether in the state law or not) to certain types of buildings (for example, farm structures), or buildings below a stated number of square feet or not exceeding a stated height. The practice of other related professions, such as architecture, or other professions, is exempt. The architectural registration laws likewise exempt the practice of engineering. Over the years there have been a number of court cases in which the architectural registration board has challenged the right of an engineer to design buildings. The latest such case followed the pattern of most earlier cases, holding that there is an overlap between the two design professions, and that either may be the prime professional for design of a building, provided that the engineer does not identify himself as an architect, or that the architect may not designate himself as an engineer. (See *Alabama v. Jones,* 267 So.2d 247 (1972).

The major exemption, over which much controversy swirled for a number of years, is the so-called industrial exemption. The specific wording of this kind of exemption varies considerably, but in the main it provides that engineering services provided in the research, development, or manufacture of products does not require registration. A typical industrial exemption clause, taken from the Michigan law, reads:

> A designer of a manufactured product, if the manufacturer of the product assumes responsibility for the quality of the product.

Furthermore, many of the exemption clauses refer to activities in interstate commerce, which in today's legal world applies to virtually all commercial activities.

For a time there was an extensive effort made by various engineering societies at both the national and state level to eliminate the industrial exemption clauses. The argument was that the design of products, such as power mowers, automobiles, airplanes, and even stationary products, such as ladders, may affect the public health and safety just as well as the design of a bridge or a building.

There is considerable evidence to support the logic of the argument, as one may read almost daily, in connection with the failure of products attributable to faulty design.

Industry mounted strong objections to removal of the industry exemption clauses, arguing that a registration requirement was impractical, particularly in connection with complex products of many hundreds or thousands of parts, which are designed, fabricated, and assembled in a number of states. The objections also included problems related to products imported from other countries which would not be subject to U.S. state laws.

The heart of the industry's objections, however, was that state control of employee's qualifications would be an unnecessary and burdensome imposition because the industrial company is responsible (and liable) if it puts out a poorly designed product which leads to injury of users.

Whatever the merits of the contending views over a period of years, efforts to remove the industrial exemptions have largely ceased. To a remarkable degree a number of leading industrial companies have subscribed to a voluntary program of encouraging the registration of its engineering staff, particularly those at higher levels. In many cases industrial employers actively promote registration through payment of examination fees, time off to take refresher courses and the examination itself, and recognition in company publications of those who achieve registration status.

4.4.   ENFORCEMENT OF REGISTRATION LAWS

Although not widely recognized as such, the registration laws of the states are criminal laws. That is, a person found guilty of practicing or offering to practice engineering without a license may be fined (or even imprisoned— but that penalty is rarely, if ever, used).

The more practical enforcement tool is through an injunctive procedure whereby the state, through the attorney general, brings a suit to have the court order the offender to cease offering or performing engineering services. A violation of the court order, of course, may result in a contempt of court finding, followed by further fine or imprisonment. In some states, notably in Texas and Wisconsin in recent years, there have been a substantial number of enforcement cases ordering unqualified firms to delete "engineering" from the firms' names in order to erase the implication that the firm is qualified to provide engineering services.

For those who do provide engineering services to clients, but who are not qualified under the state law, the more meaningful penalty may be a loss of ability to collect fees due the individual or firm. It is a well-established legal doctrine in all states that the courts will not lend themselves to those who seek to enforce a contract made in violation of the law. Such persons are deemed to have no "standing" to use the courts for the economic benefit.

How far this may go is illustrated by a Texas case in which an engineer who had obtained his P.E. license in 1938 and had failed to pay his annual renewal fee (required in all states), brought suit to collect a fee for services performed while his license was not current. The state supreme court, while recognizing that failure to pay the renewal fee did not touch on the technical qualifications of the engineer, nevertheless held that he could not collect his fee through court action because,

> The object of the renewal obligation is both to supply an obviously important part of the revenues on which the whole operation of the law depends and to keep an up-to-date record of practicing engineers. Unless there is some substantial penalty to compel compliance, the burden of supporting the Board falls entirely on those who comply while the others get the benefits of the law but evade its obligations.
>
> (*M.M.M. Inc. v. Mitchell,* 265 S.W.2d 584 (1954), Texas Supreme Court)

By contrast, the Hawaii Supreme Court in dealing with essentially similar facts held that an architect could recover his contract fee on the basis that the renewal requirement was designed purely to raise revenues; therefore the performance during the lapse period did not make the architect incompetent to provide services affecting the public health and safety (*Wilson v. Kealakekua Ranch, Ltd.,* 551 P.2d 525 (1976)).

In a 1976 decision by a federal court it was held that an incorporated Maryland firm could not enforce its contract because it had not qualified under New Jersey law, even though the plans and specifications for the building had been sealed and certified by an employee who was licensed in New Jersey (*Dalton, Dalton, Little v. Mirandi,* 412 F. Supp. 1001 (1976), U.S.D.C., N.J.).

Similar cases are noted regularly over the years, including those in which construction companies are denied monies due them because they failed to comply with the state contractor licensing law of the state in which the construction took place.

Not all the courts however have been quite that strict with the "letter of the law." There are a series of cases in which the courts held that while they will not enforce the contract as such, they will permit the nonqualified person or firm to collect on a "quantum meruit" basis; that is the fair value of the work provided the client, but less than the contract price.

In such a case the court said that an out-of-state firm was entitled to a "quantum meruit" award "where a contract is not evil in itself," and where in such a case total denial of the fee would amount to an additional penalty above the penalty prescribed in the law for those practicing without a license. (*Food Research & Engineering, Inc. v. Alaska,* 338 F. Supp. 342 (1975), U.S.D.C., Alaska).

Those individuals or firms faced with this kind of potential problem when operating outside of their home state, may seek relief by one of several procedures. First, the individual or firm may undertake to become licensed, or obtain a corporate certificate of qualification where that option exists, in the other state in which the project is located. If the individual does not qualify under a reciprocity clause, and is unwilling to undergo the examination process, he or she may elect to "associate" with another person qualified in the state and have that other qualified person sign and seal the plans and specifications. That would obviously entail a fee arrangement between the two. The same kind of arrangement may be made between corporate firms in the form of a joint venture or by having the legally qualified firm sign the contracts and have its registered personnel sign and seal the plans and specifications, with a side-agreement between the two firms over the division of work.

Second, the state registration laws provide for a temporary permit for those licensed in other states. An application is made to the state registration board of the state where the project is located and a nominal fee is required for the temporary permit. Some state laws limit the temporary permit to a stated period of time (not over one year), while others issue the temporary permit for a specified single project regardless of time. If a registered person moves to another state to open an office, he or she may protect his or her position by securing a temporary permit based on the initial registration, which is valid while the application for registration in the other state is being considered.

4.5. THE VIRTUES OF REGISTRATION

Over and above the legal or economic reasons for a qualified engineer to seek state registration, it is a curious fact that a large number of engineers do not obtain such status at the earliest possible time in their professional careers.

Even for those who do not contemplate the need for registration for legal or economic reasons early in their careers, when passing the examination is most opportune, there are personal advantages in having a P.E. status. Large numbers of engineers who did not seek registration because they were industry-exempt or government-exempt and expected to remain in that type of employment throughout their career, can testify to their "mistake" in later years when their situation changed and they entered into other employment requiring registration, such as employment with a consulting firm.

Even those in state or local government, may find that their advancement is curtailed at a certain point in the engineering heirarchy if they are not registered. In 1984 the attorney general of Oklahoma ruled that state employees responsible for engineering activities must be registered, and even that subordinates with engineering duties may not use "engineering" in their job titles unless registered (Opinion 83-266, January 6, 1984).

Apart from employment status, engineers in industry and particularly those in some forms of sales work may find less than a friendly reception when dealing with registered engineers in other firms.

Almost all engineers, without visible exception, share a common feeling that they are entitled to full professional status in the eyes of the public and among their contemporaries in other professions. To achieve that recognition it is necessary that there be a common denominator—a standard by which others can determine that professional status. This is equally true of the other learned professions, law, medicine, dentistry, accounting, and so on, all of which have established the single standard of state licensure. The extent to which engineering acquires similar recognition turns in large part on the willingness of its members to repair to a common standard. Today, roughly half of all engineers are registered under state laws. Although those not registered may be as truly "professional" as others, both technically and ethically, by being outside the circle of public acceptance through the laws created to protect the public health and safety, they delay the day engineers will no longer feel compelled to proclaim their professional status. When all qualified engineers are registered, the professional status of engineering will be a fact, not a claim.

APPENDIX A: SUMMARY OF STATE LAWS GOVERNING CORPORATE ENGINEERING PRACTICE

Alabama (Ala. Code §34-11-9)

A corporation may engage in the practice of engineering provided that persons connected with the firm acting in a professional capacity and in responsible charge of the practice of engineering are registered in the state. The practice of professional engineering to or in connection with production manufacturing may be carried on by any firm or corporation if such engineering services are practiced by or under the direction of a professional engineer in conformity with state law.

Alaska (Alaska Stat. §08.48.241)

A corporation may engage in the practice of engineering provided the firm files an application for a certificate of authorization which contains a board of directors' resolution designating an engineer registered in the state as being in responsible charge of the practice of engineering by the corporation with full authority to make final engineering decisions on behalf of the corporation with respect to work performed by the corporation. The corporation must also designate those persons who are registered and who are in responsible charge of each major branch of engineering activities in which the corporation specializes. Final engineering decisions must be made by the specified engineer in responsible charge or another registered engineer under his direction and supervision. The corporation must also file a statement of the types of engineering to be practiced in the state and evidence of its professional competence and that of its employees and personnel. The corporation's bylaws must contain a provision that all engineering decisions pertaining to activities in the state will be made by the engineer designated by the corporation's board of directors as being in responsible charge or a registered engineer under his direction and supervision.

Arizona (Ariz. Rev. Stat. Ann. §32-141)

A corporation may engage in the practice of engineering provided the work is under the full authority and responsible charge of a registered engineer who is also an officer of the corporation. Corporations must identify the responsible registered engineer. The corporation must designate those responsible officers and describe the services the firm is offering to the public on a form provided by the board of registration. The corporation must notify the board on a form provided by the board, of any changes occuring in the list of corporate officers within 30 days of the change.

Arkansas (Ark. Stat. Ann. §71-1019)

A corporation may engage in the practice of engineering provided it does so through its officers, agents or employees. One of the owners or incorporators must be a registered engineer. The practice of engineering must be done under the supervision and direction of a registered engineer.

California (Cal. [Bus. & Prof.] Code §6738)

Civil, electrical and mechanical engineers may offer to practice engineering through the medium of a corporation provided a registered civil, mechanical or electrical engineer is a directing officer in charge of engineering practice of the corporation. All plans, specifications and reports must be prepared under the direct supervision of a registered engineer in the appropriate branch of professional engineering. The corporate name must not contain the name of any person who is not registered by the board in a branch of professional engineering, or registered as an architect or geologist. Any holding-out by the corporation of any individuals to the public as members of the corporation other than by the use of the names of such individuals in the corporate name must clearly and specifically designate the license status of the individuals. No person may practice, or offer to practice, civil, electrical, or mechanical engineering either as an officer or employee of a corporation who is not registered in the appropriate branch of professional engineering. (The above provisions do not apply to or prevent the use of the name of any corporation engaged in rendering civil engineering services which were lawfully in existence on or after September 30, 1947, provided that all civil engineering plans, specifications and reports are prepared by or under the direct supervision of a registered civil engineer who is a permanent employee of the corporation. Nor do the above provisions apply to or prevent the use of the names of any corporation in rendering electrical or mechanical engineering services which were lawfully in existence on or after December 31, 1967, provided all electrical or mechanical engineering plans, specifications and reports are prepared by or under the direct supervision of a registered electrical or mechanical engineer who is a permanent employee of the corporation. The above provisions do not prevent the use of the name of a corporation engaged in rendering professional engineering services, which was in existence on or after September 30, 1947, by any lawful successor in interest or survivor if the registration board, upon written application informing it of the proposed method of carrying on the business and of the changes, if any, in the personnel in charge of the engineering activities of the successor or survivor, determines after investigation that the actual operation of the corporation is substantially carried into and becomes an operating part of the successor or survivor and that the public health, safety and welfare will not be impaired. Other branches of engineering are not permitted to practice in the corporate form.

Colorado (Colo. Rev. Stat. §12-25-103)

The corporate practice of engineering is permitted provided that person in responsible charge of engineering activities of the corporation is a professional engineer.

Connecticut (Conn. Gen. Stat. Ann. §20-306a [West])

The corporate practice of engineering is permitted provided personnel acting on the corporation's behalf as engineers are registered, and provided the corporation is granted a certificate of authorization by the board of registration for professional engineers. Each corporation must file with the board of registration a designation of an individual or individuals registered to practice engineering in the state who will be in charge of engineering by the corporation in the state.

Delaware (Del. Code Ann. Title 12 §2821)

The practice or offer to practice engineering for the public by a corporation is permitted provided that one of the officers, or one of the employees of the corporation is designated as being in responsible charge of the engineering activities and the engineering decisions of the corporation, and is a registered engineer. All personnel of the corporation who practice engineering must be registered. A certificate of authorization must be issued to the corporation by the state registration board.

District of Columbia

Statutes do not mention corporate practice; a past court decision has construed the engineering registration law to forbid a corporation from holding a license to practice professional engineering. Observers have construed this decision to prohibit the corporate practice of engineering.

Florida (Fla. Stat. Ann. §471.023 [West])

The practice or offer to practice engineering by registrants through a corporation offering engineering services to the public, or by a corporation offering services to the public through registrants under general licensing examination procedure, as agents, employees or officers is permitted provided one or more of the principal officers of the corporation and all personnel of the corporation who act in its behalf as engineers are registered as professional engineers in the state. The corporation must obtain a certificate of authorization by applying to the state registration board.

Georgia (Ga. Code Ann. §43-15 23)

The corporate practice of engineering is permitted provided one or more of the officers of the corporation, and all personnel of the corporation who act in its behalf as professional engineers, are registered. The corporation must obtain a certificate of authorization from the state registration board. The corporation must file with the board the names and addresses of all officers and board members of the corporation including the principal officer duly registered in the state, and individuals who are in responsible charge of the practice of engineering by the corporation.

Guam (Guam Gov't. Code §47021)

The practice or offer to practice engineering by individual engineers registered under the territorial registration law through a corporation as officers, employees or agents is permitted provided that one or more of the corporate officers of the corporation is designated as being in responsible charge for the engineering activities and decisions of the corporation and is a registered professional engineer, either under the laws of Guam, a state, the District of Columbia or another territory. All personnel of the corporation who act in its behalf as professional engineers must be registered professional engineers. Corporations must obtain a certificate of authorization from the registration board.

Hawaii (Hawaii Rev. Stat. §464-12)

A corporation may engage in the practice of professional engineering provided the person or persons connected with the corporation directly in charge of the professional work is duly registered, and provided the name or names have been filed with the board of registration.

Idaho (Idaho Code §4-1235)

The practice or offer to practice professional engineering through a corporation is permitted provided that all personnel of the corporation who act in its behalf as professional engineers are registered. The corporation must obtain a certificate of authorization by applying to the state registration board.

Illinois (Ill. Ann. Stat. Ch. 111 §102 et seq. [Smith-Hurd])

The corporate practice of engineering is permitted provided an officer or managing agent in charge of the engineering activities of the corporation is a registered professional engineer. A certified copy of a resolution of the board of directors of the corporation designating such officer or managing agent and investing him with full authority to make all final decisions involving engineering work must be filed with the registration board. (In the case of a corporation organized after August 30, 1963, a majority of the board of directors of the corporation must be registered professional engineers in Illinois in addition to the other requirements.)

Indiana (Ind. Code Ann. §25-31-18 [Burns])

The corporate practice of engineering is permitted provided the practice is carried out under the responsible direction and supervision of a registered professional engineer who is an officer of the corporation. The name of that designated person must appear whenever the firm name is used in the professional practice of the corporation.

Iowa (Iowa Code Ann. §114.26)

Corporations engaged in designing buildings or works for public or private interests must have all principal designing or constructing engineers registered in the state. This does not apply to corporations engaged solely in constructing buildings and works.

Kansas (Kan. Stat. Ann. §74-7000 et seq.)

The corporate practice of engineering is permitted provided one or more of the corporate officers is designated as responsible for the activities and decisions relating to the practice of engineering and is licensed to practice as a professional engineer. All personnel of the corporation who act in its behalf in the practice of engineering must be licensed professional engineers, or must be exempt from the Kansas Registration Act. The corporation must designate the branch or branches of the professional which it is authorized to practice, i.e., engineering, architecture, land surveying. The corporation must also obtain a certificate of authorization from the state registration board.

Kentucky

The statute prohibits licensing of corporations, but an opinion of the attorney general rendered May 2, 1957, to the counsel for the state board of registration holds that corporations may enter into enforceable contracts to furnish professional engineering services, provided an officer who is a licensed professional engineer has responsible charge of all phases of engineering work.

Louisiana (La. Rev. Stat. Ann. §37:689 and Rules for Professional Engineers and Professional Land Surveyors, LAC 19-3:4 et seq.)

The corporate practice of engineering is permitted provided all professional services are executed by or under the direct supervision of a registered engineer designated by the firm as being a supervising professional. The registered professional must be a full-time, active employee whose primary occupation or employment is with the corporation. The firm must also designate those registered engineers who will be providing professional services on behalf of the corporation. No registered professional can be designated as a supervising professional of more than one firm. Letterheads, business cards, advertisements and other identifying items issued by the firm must reflect that the firm has full-time employees who are registered in the state. The corporation must also obtain a certificate of authorization from the state registration board which must be renewed annually.

Maine (Me. Rev. Stat. Ann. Title 32 §1253)

The corporate practice of engineering is permitted provided that the practice is carried on by professional engineers who are registered.

Maryland (Md. Ann. Code Art. 75-½ §19A)

The corporate practice of engineering is permitted provided that the practice is carried on by professional engineers who are registered. The law permits the work of an employee subordinate or an employee practicing lawfully if such work is done under the direct responsibility and supervision of a person holding a certificate of registration or a person practicing lawfully.

Massachusetts (Mass. Ann. Laws Ch. 112 §81R)

The corporate practice of engineering is permitted provided a nonregistered employee of the corporation does not perform final design work or make decisions for the corporation. All professional work must be performed under the direct responsibility, checking and supervision of a person holding a certificate of registration. In addition, the person or persons in charge of the corporation must be a licensed professional engineer.

Michigan (Mich. Stat. Ann. §18.425 [2010])

A firm may engage in the corporate practice of engineering provided that at least ⅔ of the officers and directors of the firm are registered. All nonlicensed officers and directors must apply to and receive approval from the state. The corporation must employ a person in responsible charge in the field of professional service offered at each place of business where a service is offered by the firm except at a field office which provides only a review of construction.

Minnesota (Minn. Stat. Ann. §326.14)

A corporation may engage in work of an engineering character provided the person or persons connected with the corporation in responsible charge of the work is or are licensed as a professional engineer.

Mississippi (Miss. Code. Ann. §73-13-43)

A corporation may engage in the practice of professional engineering provided the person or persons connected with the corporation in charge of the designing or supervising which constitutes such practice is or are registered as professional engineers. A corporation performing engineering services to the public must include in each agreement for such services the name and registration number of the engineer who will be primarily responsible for the work.

Missouri (Mo. Ann. Stat. §327.011 et seq.)

The corporate practice of engineering is permitted provided the corporation obtains a certificate of authorization from the state registration board. The application for the certificate of authorization must state the practice of engineering as the purpose, the names and addresses of directors and the fact that the directors have assigned responsibility for engineering activity to registered engineers. and the names of engineers in charge of engineering activities. If the engineer in charge of engineering activity is not a full-time employee, the corporation must submit a copy of the written contract which defines the responsibility to the state registration board.

Montana (Mont. Rev. Codes Ann. §37-67-103)

The practice of professional engineering by a corporation or by its officers or employees on its behalf is permitted provided that each person personally supervising and in direct charge of all activities of the corporation which constitutes the practice of professional engineering holds a certificate of registration.

Nebraska (Neb. Rev. Stat. §81-854)

A corporation may engage in the practice of professional engineering provided such practice is carried on by persons registered as professional engineers.

Nevada (Nev. Rev. Stat. §625.240)

A corporation may engage in the practice of professional engineering if those immediately responsible for engineering work performed in the state are registered as professional engineers. Every office or place of business of any corporation engaged in the practice of professional engineering must have a registered professional engineer in residence and in direct responsible supervision of the engineering work conducted in the office or place of business. (This provision does not apply to corporations engaged in the practice of professional engineering at offices established for limited or temporary purposes, such as offices established for construction inspection.)

New Hampshire (N.H. Rev. Stat. Ann. 310-A:20)

The practice of or offer to practice professional engineering for others by individual engineers licensed in the state through a corporation as officers, employees or agents is permitted provided that one or more of the corporate officers of a corporation is designated as being responsible for the engineering activities and engineering decisions of the corporation and is a licensed engineer. All personnel of the corporation who act in its behalf as professional engineers must be licensed. An engineer who renders occasional, part-time or consulting engineering services to or for a corporation may not be designated as being responsible for the engineering activities of the corporation. The corporation must obtain a certificate of authorization from the state registration board.

New Jersey (N.J. Stat. Ann. §45:8-27)

The corporate practice of engineering is permitted provided the persons in responsible charge of engineering activity are licensed under state law.

Those persons carrying on the actual practice of professional engineering on behalf of the corporation or those designated by the corporation as "engineers" must be licensed under state law.

New Mexico (N.M. Stat. Ann. §61-23-22)

The corporate practice of engineering is permitted provided the person in responsible charge of the activities of the corporation is a professional engineer who has the authority to bind the corporation by contract and all contracts are signed by the engineer with such authority. A corporation may not use or assume a name involving the term "engineer" or "engineering" or any modification or derivative of such terms unless the corporation is qualified to practice engineering within the requirements of the law.

New York (N.Y. Educ. Law, Article 145 §7209(6) [McKinney])

It shall be lawful for a corporation organized and existing under the laws of the state which on April 15, 1935 and continuously thereafter, was lawfully practicing engineering or land surveying in the state, to continue such practice provided that the chief executive officer and persons carrying on the actual practice of engineering shall be professional engineers licensed under this Article. No corporation shall change its name or sell its franchise or transfer its corporate rights, directly or indirectly to any person, firm or corporation without the consent of the State Education Department.

North Carolina (N.C. Gen. Stat. §89C-24)

A corporation may engage in the practice of engineering provided the persons connected with the corporation in charge of the designing or supervision which constitutes such practice are registered professional engineers. At least two-thirds of the shareholders must be professional engineers registered in the state and not more than one-third of the shareholders may be non-registered employees of the corporation. The corporation must obtain a certificate of authorization from the state registration board.

North Dakota (N.D. Cent. Code §43-19.1-27)

The corporate practice of engineering is permitted provided all of the officers and shareholders are duly registered as professional engineers or are exempt from the registration provisions. In addition, all employees, officers, and agents of the corporation who will perform the practice of engineering must be registered as professional engineers. Each person in responsible charge of the activities of the corporation, which activities constitute the practice of professional engineering, must be a licensed professional engineer. The corporation must obtain a certificate of authorization from the state registration board.

Ohio (Ohio Rev. Code Ann. §4733.16(A))

The corporate practice of engineering is permitted provided the services are performed only through natural persons registered to offer those services and over 50% of the directors or shareholders of the corporation own more than 50% of the corporation's shares and more than 50% of the interests in the corporation, and must be professional engineers, surveyors, architects or landscape architects, or a combination thereof, who are registered. The corporation must obtain a certificate of authorization from each state registration board.

Oklahoma (Okla. Stat. Ann. Title 59 §475.21(1))

The corporate practice of engineering is permitted provided one or more officers designated as being responsible for the engineering activities and decisions is a professional engineer. All personnel of the corporation who act in its behalf as professional engineers must be registered. The corporation must obtain a certificate of authorization from the state registration board.

Oregon (Or. Rev. Stat. §672.030)

The corporate practice of engineering is permitted provided the individuals responsible for the engineering activities and engineering decisions of the corporation are licensed professional engineers. The names of these individuals must be reported annually to the state registration board.

Pennsylvania (Pa. Stat. Ann. Title 63 §153 [Purdon])

The corporate practice of engineering is permitted provided the directing heads and employees of the corporation in responsible charge of its activities in the practice of engineering are licensed and registered in conformity with the requirements of the law. The corporation must obtain a certificate of authorization from the corporation bureau and the state registration board. The corporate name must be approved if it contains some form of the word "engineer".

Puerto Rico

Past court decisions have construed its statutes as prohibiting corporate practice.

Rhode Island (R.I. Gen. Laws §5-8-21)

The corporate practice of engineering is permitted provided all personnel of the corporation who act in its behalf as engineers are registered engineers, or are exempt as temporary permit-holders, employee/subordinates or nonresidents.

South Carolina (S.C. Code §40-21-410)

The corporate practice of engineering is permitted provided the persons acting in a professional capacity or in responsible charge of the corporate practice of engineering are licensed in the state. Work may be performed for the corporation by one who is not a licensed professional engineer provided such work does not include final design or decisions and is done under the direct responsibility, checking and supervision of a person registered as a professional engineer in the state.

South Dakota (S.D. Codified Laws Ann. §36-18-31.1 et seq.)

The corporate practice of engineering is permitted provided all officers, employees and agents within the state in responsible charge of the practice of engineering and all those who will perform the practice of professional engineering within the state for the corporation are registered professional engineers. The corporation must obtain a certificate of authorization from the state registration board.

Tennessee (Tenn. Code Ann. §62-2-601)

The corporate practice of engineering is permitted provided that at least one of the officers of the corporation is in responsible charge of the practice of engineering and is a registered engineer. A resident registered engineer must be in responsible charge at each place of business. The corporation must file a list of all registered and nonregistered officers with the state board of registration. The list must indicate which officer is in responsible charge of the practice of engineering by the corporation.

Texas (Tex. Stat. Ann. §3271(a) et seq.)

The corporate practice is permitted provided the practice is carried on by or under only professional engineers registered in the state. No corporation may hold itself out to the public as being engaged in the practice of engineering under any assumed name unless the corporation is actually and actively engaged in the practice of engineering or offering engineering services to the public and all services are either personally performed by a registered engineer or under the responsible supervision of a registered engineer.

Utah (Utah Code Ann. §58-22-21)

The corporate practice of engineering is permitted provided the practice is carried on by professional engineers authorized to practice.

Vermont (Vt. Stat. Ann. Title 26 §1052)

The corporate practice of engineering is permitted provided such practice is carried on by professional engineers registered in the state.

Virgin Islands (V.I. Code Ann. Title 4 §291)

The corporate practice of engineering is permitted provided all personnel of the corporation who act in its behalf as engineers are licensed.

Virginia (Va. State Board of Architects, Professional Engineers and Land Surveyors §III-3.1 et seq.)

The corporate practice of engineering is permitted provided the corporation files an application for and receives a certificate of authority from the state board of registration. The corporation's bylaws must state that cumulative voting by shareholders is prohibited, that nonlicensed individuals will not have a voice or standing in matters affecting the practice of the corporation which requires professional expertise or professional practice. The state board of registration shall issue two kinds of certificates of authority: general and limited. A general certificate of authority entitles the corporation to practice the professions of engineering, land surveying or architecture. A limited certificate of authorization permits the corporation to practice only the profession or professions shown on its certificate of authority. Capital stock ownership is limited to Virginia-licensed architects, professional engineers, land surveyors and nonlicensed employees of the corporation engaged in the practice of architecture, professional engineering and land surveying or any combination thereof. Nonlicensed employees are not permitted to own more than one-third of the capital stock of the corporation. Any type of joint ownership of the stock of the corporation is prohibited. Ownership of stock by nonlicensed employees does not entitle these employees to vote in any matter affecting the practice of engineering by the corporation. The corporation's board of directors must contain at least one director appropriately licensed in each discipline offered or practiced by the corporation who devotes substantially full-time to the business of the corporation to provide effective supervision and control of the final professional product. Nonlicensed directors are not entitled to exercise control or vote in any matter affecting the practice of engineering by the corporation. All work performed by an individual who is not a licensee must be performed under the direction and supervision of a licensee who is either the employer of the unlicensed individual or an employee of the same firm as the unlicensed individual. A corporation maintaining a place of business in the state for the purpose of providing or offering to provide professional services must have in responsible charge at all places of business a licensed professional engineer, architect or land surveyor.

Washington (Wash. Rev. Code Ann. §18-43.130)

A corporation may engage in the practice of engineering provided the firm files an application for a certificate of authorization with the state registration board which contains a board of directors' resolution designating an engineer registered in the state as being in responsible charge of the practice of engineering by the corporation with full authority to make final engineering decisions on behalf of the corporation with respect to work performed by the corporation. The corporation must designate those persons who are registered who are in responsible charge of each major branch of engineering activities in which the corporation specializes and those persons who are in responsible charge of each project. Final engineering decisions must be made by the specified engineer in responsible charge or another registered engineer under his direction and supervision. The corporation must file a statement of the types of engineering to be practiced in the state, a current certified financial statement and evidence of its professional competence and that of its employees and personnel. The corporation's bylaws must contain a provision that all engineering decisions pertaining to activities in the state shall be made by the engineer designated by the corporation's board of directors as being in responsible charge of a registered engineer under his direction and supervision.

West Virginia (W. Va. Code §30-13-9)

The corporate practice of engineering is permitted provided such practice is carried on by professional engineers registered in accordance with state law.

Wisconsin (Wis. Stat. Ann. §443.08)

The corporate practice of engineering is permitted provided all personnel who practice or offer to practice in its behalf as professional engineers are registered in accordance with state law. The corporation must obtain a certificate of authorization from the state registration board. The application for the certificate must list the names and addresses of all officers and directors and all individuals in the corporation's employment registered to practice professional engineering who will be in responsible charge of professional engineering being practiced in the state through the corporation.

<h2 style="text-align:center">Wyoming (Wyo. Stat. §33-29-108)</h2>

The corporate practice of engineering is permitted provided the person or persons connected with the corporation in active charge of the investigation, design or construction of any work which may constitute such practice is a registered professional engineer under state law.

APPENDIX B: SUMMARY OF REGISTRATION PROVISIONS BY STATES AND OTHER JURISDICTIONS

The following bases for registration are discussed below:

EIT (or EI in some cases)—Engineer-in-training or engineering intern (indicating passage of the fundamentals portion of the two-part examination process, but not authorizing the practice of engineering)

PE—Professional engineer (entitled to offer and perform engineering services to the public)

LEP—Long-established practice (providing same status as PE)

E—Eminence (providing same status as PE)

ALABAMA

EIT: —4-year approved engineering degree (or 4 years of experience) and passage of a written fundamentals of engineering examination: or
—4-year engineering degree and 2 years of experience and passage of the fundamentals of engineering examination.

PE: —4-year approved engineering degree, 4 years of experience, and passage of a written fundamentals and a principles and practice of engineering examination; or
—8 years of experience and passage of a written fundamentals and a principles and practice of engineering examination.

ALASKA

EIT: —4-year approved engineering degree (or 4 years of experience) and passage of a fundamentals of engineering examination.

PE: —4-year approved engineering degree, 4 years of experience (of which 2 years have been in responsible charge), and passage of a fundamentals and professional engineering examination; or

—8 years of experience (of which 2 years have been in responsible charge) and passage of a fundamentals and a professional engineering examination.

LEP: —20 years of experience (of which 2 years have been in responsible charge) and passage of a professional engineering examination.

ARIZONA

EIT: —4-year approved engineering degree (or 4 years of experience) and passage of a basic engineering examination.

PE: —4-year approved engineering degree and 4 years of experience (or 8 years of experience), and passage of the fundamentals and principles and practice of engineering examination. Should the applicant fail to satisfy the Board that he or she is qualified, the applicant will be required to submit additional data or may be required to submit to an oral or written examination.

LEP: —4-year approved degree, 12 years of experience, and passage of the principles and practice of engineering examination.

ARKANSAS

EIT: —4-year approved engineering degree (or 4 years of experience) and passage of an examination.

PE: —4-year approved engineering degree, 4 years of experience, and passage of an examination; *or*
—8 years of experience and passage of an examination.

CALIFORNIA

EIT: —4-year approved engineering degree (or 4 years of experience) and passage of the fundamentals of engineering examination.

PE: —4-year approved engineering degree, 2 years of experience, and passage of a fundamentals and a principles and practice of engineering examination; *or*
—6 years of experience and passage of a fundamentals and a principles and practice of engineering examination.

COLORADO

EIT: —4-year approved engineering degree (or 6 years of experience) and passage of a fundamentals of engineering examination; *or*
—4-year approved engineering technology degree, 6 years of experience and passage of the fundamentals of engineering examination; *or*
—4-year engineering degree (or related science degree), 10 years of experience and passage of the fundamentals of engineering examination.

PE: —4-year approved engineering degree, 4 years of experience, and passage of a fundamentals and a principles and practice of engineering examination; *or*
—4-year engineering degree (or related science degree), 8 years of experience, and passage of the fundamentals and the principles and practice of engineering examination; *or*
—8 years of experience and passage of a fundamentals and a principles and practice of engineering examination; *or*
—4-year engineering degree, 16 years of experience (of which 10 years

have been in responsible charge), and passage of a principles and practice of engineering examination.

LEP: —10 years of experience indicating professional competence and passage of the fundamentals and the principles and practice of engineering examination.

CONNECTICUT

EIT: —4-year approved engineering degree (or 6 years of experience) and passage of part 1 of the examination.

PE: —4-year approved engineering degree, 4 years of experience, and passage of parts 1 and 2 of the examination; *or*
—10 years of experience (part of which indicates knowledge equivalent to a 4-year-approved engineering degree) and passage of parts 1 and 2 of the examination.

LEP: —20 years of experience and passage of parts 1 and 2 of the examination (part 1 may be waived by the Board); *or*
—4-year approved engineering degree, 8 years of experience, and passage of parts 1 and 2 of the examination (part 1 may be waived by the Board).

GUAM

EIT: —4-year approved engineering degree and passage of an examination; *or*
—6 years of experience and passage of an examination.

PE: —4-year approved engineering degree, 4 years of experience, and passage of a written examination; *or*
—10 years of experience and passage of a written examination.

LEP: —4-year science degree, 20 years of experience (of which 10 have been in responsible charge), and passage of an oral, written, or oral and written examination.

HAWAII

EIT: —No provision for EIT.

PE: —4-year approved engineering degree, 3 years of experience, and passage of a written, oral, or written and oral examination; *or*
—12 years of experience and passage of a written, oral, or written and oral examination.

IDAHO

EIT: —4-year approved engineering degree and passage of an examination; *or*
—Evidence of knowledge approximating that attained through graduation from a 4-year approved engineering curriculum, 4 years of experience to indicate applicant is competent to enroll as an EIT, and passage of an examination.

PE: —4-year approved engineering degree, 4 years of experience, and passage of an examination; *or*
—Evidence of knowledge approximating that attained through graduation from a 4-year approved engineering curriculum, 8 years of experience, and passage of an examination.

ILLINOIS

EIT: —4-year approved engineering degree and passage of a written fundamentals of engineering examination; *or*

—4-year engineering degree (or 4-year science degree), 4 years of experience, and passage of a written fundamentals of engineering examination.

PE: —4-year approved engineering degree, 4 years of experience, and passage of a fundamentals and a principles and practice of engineering examination; *or*

—4-year engineering degree (or 4-year science degree), 8 years of experience, and passage of a fundamentals and a principles and practice of engineering examination.

INDIANA

EIT: —4-year approved engineering degree and passage of part 1 of the examination; *or*

—4 years of engineering education and experience indicating knowledge approximating that acquired through graduation from a 4-year approved engineering program, and passage of part 1 of the examination.

PE: —4-year approved engineering degree, 4 years of experience, and passage of parts 1 and 2 of the examination; *or*

—8 years of combined education and experience and passage of parts 1 and 2 of the examination.

IOWA

EIT: —4-year approved engineering degree (or 8 years of experience), and passage of a written, oral, or written and oral examination in the fundamentals of engineering.

PE: —4-year approved engineering degree, 4 years of experience, and passage of a written, oral, or written and oral examination in the fundamentals and the professional practice of engineering; *or*

—12 years of experience and passage of a written, oral, or written and oral examination in the fundamentals and the professional practice of engineering.

LEP: —4-year engineering degree, 25 years of experience, and passage of the examination in the professional practice of engineering.

KANSAS

EIT: —4-year approved engineering degree and passage of a fundamentals of engineering examination.

PE: —4-year approved engineering degree, 4 years of experience, passage of a written fundamentals of engineering examination, and a professional examination; *or*

—8 years of experience, passage of a written fundamentals of engineering examination, and a professional examination.

LEP: —4-year approved engineering degree, 12 years of experience, passage of an oral or written examination, and passage of a professional examination. Any applicant qualifying under this section and demonstrating 25 years of experience will be given the option of an oral examination in lieu of a written examination.

KENTUCKY

EIT: —4-year approved engineering degree and passage of a written examination in the fundamentals of engineering.

PE: ——4-year approved engineering degree, 4 years of experience, and passage of a written examination in the fundamentals and the principles and practice of engineering; *or*
——8 years of experience and passage of a written or written and oral examination designed to show knowledge approximating that attained through graduation from a 4-year approved engineering curriculum. Applications will be accepted under this provision until July 1, 1983. After that date a 4-year approved engineering degree will be required for PE registration.

LOUISIANA

EIT: ——4-year approved engineering degree (or 4-year engineering degree and approved masters degree, or 4-year engineering degree or related science degree and 4 years of experience), and passage of the fundamentals of engineering examination.

PE: ——4-year approved engineering degree, 4 years of experience, passage of a written fundamentals and principles and practice of engineering examination; *or*
——4-year engineering degree (or 4-year related science degree, or 4-year engineering technology degree), 8 years of experience, and passage of the fundamentals and the principles and practice of engineering examination.

*LEP: ——20 years of experience (of which 12 years have been in responsible charge) and passage of the principles and practice of engineering examination.

MAINE

EIT: ——4-year approved engineering degree and passage of a fundamentals of engineering examination; *or*
——8 years of experience and passage of a fundamentals of engineering examination.

PE: ——4-year approved engineering degree, 4 years of experience, and passage of a fundamentals and a principles and practice of engineering examination; *or*
——12 years of experience and passage of a fundamentals and a principles and practice of engineering examination; *or*
——15 years of experience (of which 10 have been in responsible charge) and passage of an oral or written principles and practice of engineering examination.

MARYLAND

EIT: ——4-year approved engineering degree and passage of a fundamentals of engineering examination; *or*
——4-year engineering degree, 4 years of experience, and passage of a fundamentals of engineering examination.

PE: ——4-year approved engineering degree, 4-years of experience, and passage of a fundamentals and principles and practice of engineering examination; *or*
——4-year engineering degree, 8 years of experience, and passage of a fundamentals and a principles and practice of engineering examination; *or*

* Louisiana LEP provision expired 12/31/84

—12 years of experience (of which 5 years have been in responsible charge) and passage of a principles and practice of engineering examination.

MASSACHUSETTS

EIT: —4-year approved engineering degree and passage of a written fundamentals of engineering examination; *or*

—4-year engineering or science degree, 4 years of experience, and passage of a written fundamentals of engineering examination.

PE: —4-year approved engineering degree, 4 years of experience, and passage of a written fundamentals and principles and practice of engineering examination; *or*

—4-year engineering or science degree, 8 years of experience, and passage of a written and oral fundamentals and principles and practice of engineering examination; *or*

—12 years of experience (of which 5 years have been in responsible charge) and passage of a written and oral fundamentals and principles and practice of engineering examination; *or*

—20 years of experience (of which 10 have been in responsible charge) and passage of an oral or written principles and practice of engineering examination.

MICHIGAN

EIT: —4-year engineering degree and passage of part 1 of the examination.

PE: —4-year degree in engineering, 4 years of experience, and passage of an engineering theory and an engineering practices examination; *or*

—4-year approved engineering degree, 4 years of experience (of which 2 years may constitute engineering education), and passage of an engineering theory and an engineering practices examination.

MINNESOTA

EIT: —4-year approved engineering degree and passage of a fundamentals of engineering examination.

PE: —4-year approved engineering degree and 4 years of experience and passage of a fundamentals and the principles and practice of engineering examination.

MISSISSIPPI

EIT: —4-year approved engineering degree (or 4 years of experience) and passage of a written examination in basic engineering.

PE: —4-year approved engineering degree, 4 years of experience, and passage of a written or written and oral examination; *or*

—8 years of experience and passage of a written or written and oral examination.

MISSOURI

EIT: —4-year approved engineering degree and passage of part 1 of the engineering examination.

PE: —4-year approved engineering degree* and 4 years of experience which in-

* Missouri allows only one year of experience credit for all post-graduate work.

dicates that the applicant is qualified to be placed in responsible charge of engineering work and passage of the fundamentals and the principles and practice of engineering examination.

LEP: ——4-year approved engineering or science degree, 50 years of age, 20 years of experience, and passage of an oral examination; or
——50 years of age, 20 years of experience, and passage of a written examination.

MONTANA

EIT: ——4-year approved engineering degree and passage of a written examination in the fundamentals of engineering; or
——4-year engineering degree or science degree, 4 years of experience, and passage of fundamentals of engineering examination.

PE: ——4-year approved engineering degree, 4 years of experience, and passage of a fundamentals and principles and practice of engineering examination; or
——4-year engineering or science degree, 8 years of experience, and passage of a fundamentals and a principles and practice of engineering examination; or
——4-year engineering degree (or 4-year science degree), 20 years of experience of which 10 have been in responsible charge, and passage of a principles and practice of engineering examination.

NEBRASKA

EIT: ——4-year engineering degree (or 4 years of experience) and passage of a fundamentals of engineering examination.

PE: ——4-year approved engineering degree, 4 years of experience, and passage of a written examination in the fundamentals and the professional practice of engineering; or
——5-year approved engineering degree, 3 years of experience, and passage of a written examination in the fundamentals and the professional practice of engineering; or
——8 years of experience and passage of a written examination in the fundamentals and the professional practice of engineering.

NEVADA

EIT: ——4-year approved engineering degree and passage of an oral or written examination; or
——4 years of experience and passage of part 1 of the examination.

PE: ——4-year approved engineering degree, 4 years of experience, and passage of parts 1 and 2 of the examination; or
——4-year approved engineering degree, 8 years of experience, and passage of part 2 of the examination; or
——8 years of experience and passage of parts 1 and 2 of the examination; or
——4-year related science degree, 6 years of experience, and passage of part 1 and 2 of the examination.

NEW HAMPSHIRE

EIT: ——4-year approved engineering degree and passage of a fundamentals of engineering examination.

PE: ——4-year approved engineering degree, 4 years of experience, and passage of a written or written and oral examination in the fundamentals of engineering and a written professional practice examination; *or*

——8 years of experience and passage of a written examination designed to show knowledge approximating that attained through graduation from a 4-year approved engineering program and passage of a written professional practice examination.

LEP: ——40 years of age, 15 years of experience (of which 5 have been in responsible charge), and satisfaction of the Board that applicant is qualified. Should the applicant fail to satisfy the Board that he or she is qualified, the applicant would be required to successfully complete an oral or written or oral and written examination.

E: ——33 years of age, 10 years of experience, and a record of outstanding qualifications.

NEW JERSEY

EIT: ——4-year approved engineering degree and passage of a written fundamentals of engineering examination; *or*

——4-year science degree, 2 years of experience, and passage of a written fundamentals of engineering examination.

PE: ——4-year approved engineering degree, 4 years of experience, and passage of a written or written and oral examination; *or*

——4-year approved science degree, 6 years of experience, and passage of all parts of a written or written and oral examination; *or*

——4-year approved engineering degree (or 4-year approved science degree), 15 years of experience, and passage of a specialized portion of a written or written and oral examination.

NEW MEXICO

EIT: ——4-year approved engineering degree and passage of a written fundamentals of engineering examination; *or*

——4-year engineering degree (or 4-year science degree), 4 years of experience, and passage of a written fundamentals of engineering examination.

PE: ——4-year approved engineering degree, 4 years of experience, and passage of a written examination in the fundamentals and the principles and practice of engineering; *or*

——4-year engineering degree (or 4-year science degree), 8 years of experience, and passage of a written examination in the fundamentals and the principles and practice of engineering; *or*

——4-year engineering degree (or 4-year science degree), 20 years of experience (of which 10 years have been in responsible charge), and passage of a written examination in the principles and practice of engineering.

EIT: ——4-year approved engineering degree (or 12 years of experience) and passage of an EIT examination.

PE: ——4-year approved engineering degree and 4 years of experience, (or 12 years of experience), and passage of an examination.

E: —15 years of experience and possession of established and recognized standing in the engineering profession.*

NORTH CAROLINA

EIT: —4-year approved engineering degree (or 4-year approved science degree) and passage of a written fundamentals of engineering examination; or
—4-year engineering degree (or 4-year science degree or equivalent education and experience), 4 additional years of experience, and passage of a written fundamentals of engineering examination.

PE: —4-year approved engineering degree, 4 years of experience, and passage of a written fundamentals and a principles and practice of engineering examination; or
—EIT certificate, 4 years of experience, and passage of a principles and practice of engineering examination; or
—4-year engineering degree (or 4-year science degree or equivalent education and experience), 8 additional years of experience, and passage of a written fundamentals and principles and practice of engineering examination.

LEP: —20 years of experience and passage of a written principles and practice of engineering examination.

NORTH DAKOTA

EIT: —4-year approved engineering degree and passage of a written fundamentals of engineering examination; or
—4-year engineering degree, 4 years of experience, and passage of a written fundamentals of engineering examination.

PE: —4-year approved engineering degree, 4 years of experience, an EIT certificate, and passage of a principles and practice of engineering examination; or
—4-year engineering degree (or 4-year science degree), 8 years of experience, and passage of a written fundamentals and a principles and practice of engineering examination; or
—20 years of experience (of which 10 years have been in responsible charge) and passage of a written principles and practice of engineering examination; or
—4 years of teaching in an approved engineering program, 2 years of experience, and passage of a written principles and practice of engineering examination.

OHIO

EIT: —4-year approved engineering degree and passage of part 1 of the examination.

PE: —4-year approved engineering degree, 4 years of experience, and passage of a written or written and oral examination; or
—4-year engineering degree (or 4-year science degree), 8 years of experience, and passage of a written or written and oral examination.

* While the NY eminence still exists in law, the statute establishes that it may be administered "on recommendation of the board"; however, since January 1979 the Board has followed a policy requiring satisfaction of the examination requirement by all licensure applicants. Qualification under the "eminence clause" is therefore no longer available for licensure in NY.

LEP: —Recognized standing, 4-year approved engineering or physical science degree, 45 years of age, 20 years of experience (of which 10 years have been in responsible charge), and passage of a special written or written and oral examination.

OKLAHOMA

EI: —4-year approved engineering degree and passage of a written fundamentals of engineering examination; *or*
—4-year engineering degree (or 4-year related science degree), 1 year of experience, and passage of a written fundamentals of engineering examination.

PE: —4-year approved engineering degree, 4 years of experience, and passage of a written fundamentals and principles and practice of engineering examination; *or*
—4-year engineering degree (or 4-year related science degree), 6 years of experience, and passage of a written fundamentals and a principles and practice of engineering examination.

LEP: —12 years of experience of which 8 have been in responsible charge and passage of a principles and practice of engineering examination.

OREGON

EIT: —4-year approved engineering degree (or 4 years of experience) and passage of a written or written and oral examination devoted to basic engineering subjects.*

PE: —4-year approved engineering degree, 4 years of experience, and passage of a written or written and oral examination (one part devoted to basic engineering, the second part devoted to practical engineering problems); *or*
—8 years of experience and passage of a written or written and oral examination (one part devoted to basic engineering subjects and the second part devoted to practical engineering problems); *or*
—Certification as an EIT, 4 years of experience, and passage of a written or written and oral examination devoted to practical engineering problems.

PENNSYLVANIA

EIT: —4-year approved engineering degree (or 4 years of experience evidencing knowledge approximating that attained through graduation from a 4-year approved engineering degree program) and passage of a fundamentals of engineering examination.

PE: —4-year approved engineering degree, 4 years of experience, and passage of a fundamentals and principles and practice of engineering examination; *or*
—12 years of experience of which 8 have been under supervision of a professional engineer and passage of a fundamentals and principles and practice of engineering examination.

---

* Board regulations provide that under certain circumstances EIT certificates may be gained by passing an examination devoted to basic engineering subjects without meeting educational and experience requirements.

PUERTO RICO

EIT: —4-year approved engineering degree and passage of a written fundamentals of engineering examination.

PE: —4-year approved engineering degree, 4 years of experience, passage of a fundamentals of engineering examination, and satisfactory proof to the Board that the applicant is qualified to practice as a professional engineer. Should the proof fail to satisfy the Board, the applicant may be required to submit additional evidence.

E: —International renown for achievement in engineering.

RHODE ISLAND

EIT: —4-year approved engineering degree (or 8 years of experience) and passage of a written fundamentals of engineering examination.

PE: —4-year approved engineering degree, 4 years of experience, and passage of a written fundamentals and principles and practice of engineering examination; or
—4-year approved engineering degree, 12 years of experience, and passage of a written principles and practice of engineering examination; or
—12 years of experience and passage of a written fundamentals and principles and practice of engineering examination.

LEP: —20 years of experience (of which 10 have been in responsible charge) and passage of a written principles and practice of engineering examination.

SOUTH CAROLINA

EIT: —4-year approved engineering degree and passage of a written examination; or
—4-year approved engineering degree (or 4-year science degree), 4 years of experience, passage of written examinations designated to show knowledge approximating that attained through graduation in an approved engineering program, and passage of an additional written examination.

PE: —4-year approved engineering degree, 4 years of experience, and passage of an oral or written examination; or
—4-year approved engineering degree (or 4-year science degree), 8 years of experience, and passage of a written or written and oral examination.

SOUTH DAKOTA

EIT: —4-year accredited engineering degree and passage of a written fundamentals of engineering examination.

PE: —4-year accredited engineering degree, 4 years of experience, and passage of a written or written and oral fundamentals and principles and practice of engineering examination; or
—5-year accredited engineering degree, 3 years of experience, and passage of a written or written and oral fundamentals and principles and practice of engineering examination; or
—12 years of experience and passage of the written or written and oral fundamentals and principles and practice of engineering examination; or
—M.S. or Ph.D. from an accredited engineering program, 3 years of experience, and passage of a written or written and oral fundamentals and principles and practice of engineering examination; or
—4-year accredited engineering technology degree, 5 years of experience,

and passage of a written or written and oral findamentals and principles and practice of engineering examination; *or*

—4-year engineering degree (or 4-year engineering technology degree), 6 years of experience, and passage of a written or written and oral fundamentals and principles and practice of engineering examination; *or*

—2-year vocational, trade, or correspondence program certificate in engineering, 9 years of experience, and passage of a written or written and oral fundamentals and principles and practice of engineering examination.

LEP: —1-year accredited engineering degree, 20 years of experience, and passage of a written or written and oral principles and practice of engineering examination.

TENNESSEE

EIT: —4-year approved engineering degree and passage of a written fundamentals of engineering examination; *or*

—4-year approved science degree, 4 years of experience, and passage of a written fundamentals of engineering examination.

PE: —1-year approved engineering degree, 4 years of experience, and passage of a written fundamentals and principles and practice of engineering examination; *or*

—1-year approved science degree, 8 years of experience, and passage of a written fundamentals and principles of engineering examination.

LEP: —4-year approved engineering degree (or 4-year science degree), 12 years of experience, and passage of a principles and practice of engineering examination.

TEXAS

EIT: —4-year approved engineering degree and passage of a written fundamentals of engineering examination; *or*

—8 years of experience and passage of a written fundamentals of engineering examination; *or*

—4-year engineering technology degree, 6 years of experience, and passage of a written fundamentals of engineering examination.

PE: —4-year approved engineering degree and 4 years of experience; *or*

—8 years of experience and passage of a written or written and oral examination designed to show knowledge approximating that attained through graduation from a 4-year approved engineering program.

LEP: —35 years of age and 12 years experience of which 5 have been in responsible charge.

UTAH

EIT: —4-year approved engineering degree and passage of a written examination on the basics of engineering; *or*

—4 years of experience and passage of a written or written and oral examination on the basics of engineering.

PE: —4-year approved engineering degree, 4 years of experience, and passage of a written or written and oral examination; *or*

—8 years of experience and passage of a written or written and oral examination.

E: —Established and recognized standing in the profession, engaged in practice for 12 years (of which 5 have been in responsible charge), and 35 years of age.

VERMONT

EIT: —4-year approved engineering degree and passage of a written fundamentals of engineering examination; *or*
—8 years of experience and passage of a written fundamentals of engineering examination.

PE: —4-year approved engineering degree, 4 years of experience, and passage of a written examination; *or*
—12 years of experience and passage of a written or written and oral examination.

LEP: —20 years of experience (of which 10 have been in responsible charge) and passage of a written or written and oral examination.

VIRGIN ISLANDS

EIT: —4-year approved engineering degree; *or*
—4 years of experience and passage of a fundamentals of engineering examination.

PE: —4-year accredited engineering degree, 2 years of experience, and passage of a fundamentals and a principles and practice of engineering examination; if the applicant was a bona fide resident of the Virgin Islands prior to undertaking a course of study toward the 4-year accredited engineering degree, no examination will be required; *or*
—8 years of experience and passage of a fundamentals and a principles and practice of engineering examination.

LEP: —12 years of experience and passage of a principles and practice of engineering examination.

VIRGINIA

EIT: —4-year approved engineering degree and passage of a fundamentals of engineering examination; *or*
—4-year engineering degree (or 4-year science degree or 4-year approved engineering technology degree), 2 years of experience, and passage of a fundamentals of engineering examination; *or*
—6 years of experience (part of which is equivalent to a 4-year engineering curriculum) and passage of a fundamentals of engineering examination.

PE: —4-year approved engineering degree, 4 years of experience, and passage of a fundamentals and a principles and practice of engineering examination; *or*
—4-year engineering degree (or 4-year science degree), 6 years of experience, and passage of a fundamentals and principles and practice of engineering examination; *or*
—10 years of experience (part of which is equivalent to a 4-year engineering curriculum) and passage of a fundamentals and a principles of engineering examination.

LEP: —4-year engineering degree (or 4-year science degree), 20 years of experience (of which 10 years have been in responsible charge), and passage of a principles and practice of engineering examination; *or*
—4-year engineering degree (or 4-year science degree), 30 years of experi-

ence (of which 20 years have been in responsible charge), eminent qualification, and passage of a special oral examination.

WASHINGTON

EIT: —4-year approved engineering degree (or 4 years of experience) and passage of a fundamentals of engineering examination.

PE: —4-year approved engineering degree, 4 years of experience, and passage of a written, oral, or written and oral examination; or

—4-year approved degree, 6 years of experience, and passage of a written, oral, or written and oral examination; or

—8 years of experience and passage of a written, oral, or written and oral examination.

WEST VIRGINIA

EI: —No statutory provision for an EI*

PE: —4-year approved engineering degree, 4 years of experience, and passage of an examination; or

—4-year approved science degree, 6 years of experience, and passage of an examination; or

—10 years of combined education and experience in engineering (part of which is equivalent to a 4-year approved engineering or science curriculum) and passage of an examination.

WISCONSIN

EIT: —4-year approved engineering degree (or 4 years of experience) and passage of a written or written and oral examination; or

—4-year approved degree, 2 years of experience, and passage of a written or written and oral examination.

PE: —4-year approved engineering degree, 4 years of experience, and passage of a written or written and oral examination; or

—4-year approved degree, 6 years of experience, and passage of a written or written and oral examination; or

—12 years of experience and passage of a written or written and oral examination.

LEP: —12 years of experience and 35 years of age.

WYOMING

EIT: —4-year accredited engineering degree (or 4 years of experience) and passage of a written basic engineering examination.

PE: —4-year approved engineering degree, 4 years of experience, evidence of competence, and passage of a written or written and oral examination. Where the evidence does not appear conclusive, applicant may be required to submit further evidence; or

---

* WV; Board regulations provide for registration of "Engineer Interns" (EIs). Registration as an EI in WV requires a 4-year approved engineering degree and passage of a fundamentals of engineering examination. Nonengineering graduates must meet the same experience and education requirements as are mandated for PE licensing before they may sit for the fundamentals examination and become registered as an EI. Further, all applicants for licensing as PEs must first meet all requirements for registration as EIs.

—8 years of experience and passage of a written or written and oral examination designed to show knowledge approximating that attained through graduation from a 4-year approved engineering curriculum.

APPENDIX C: STATE ENGINEERING REGISTRATION BOARDS

Alabama State Board of Registration for Professional Engineers and Land Surveyors
750 Washington Avenue, Suite 212
Montgomery 36130
(205) 832-6100

Alaska State Board of Registration for Architects, Engineers and Land Surveyors
Pouch D
State Office Building, 9th Floor
Juneau 99811
(907) 465-2540

Arizona State Board of Technical Registration
1645 W. Jefferson Street, Suite 140
Phoenix 85007
(602) 255-4053

Arkansas State Board of Registration for Professional Engineers and Land Surveyors
P.O. Box 2541, 1818 W. Capitol
Little Rock 72203
(501) 371-2517

California Board of Registration for Professional Engineers and Land Surveyors
1006 Fourth Street, 6th Floor
Sacramento 95814
(916) 445-5544

Colorado State Board of Registration for Professional Engineers and Professional Land Surveyors
600-B State Services Building
1525 Sherman Street
Denver 80203
(303) 866-2396

Connecticut State Board of Examiners for Professional Engineers and Land Surveyors
State Office Building
165 Capitol Ave., Room G-3A
Hartford 06106
(203) 566-3386

Delaware Association of Professional Engineers
2005 Concord Pike
Wilmington 19803
(302) 656-7311

District of Columbia Board of Registration for Professional Engineers
614 H Street, N.W., Room 910
Washington 20001
(202) 727-7454

Florida State Board of Professional Engineers
130 North Monroe Street
Tallahassee 32301
(904) 488-9912

Georgia State Board of Registration for Professional Engineers and Land Surveyors
166 Pryor Street, S.W.
Atlanta 30303
(404) 656-3926

Guam Territorial Board of Registration for Professional Engineers, Architects and Land Surveyors
Department of Public Works
Government of Guam, P.O. Box 2950
Agana 96910
646-8643

Hawaii State Board of Registration for Professional Engineers, Architects, Land Surveyors and Landscape Architects
P.O. Box 3469 (1010 Richards Street)
Honolulu 96801
(808) 548-7697

Idaho Board of Professional Engineers and Land Surveyors
842 La Cassia Drive
Boise 83705
(208) 334-3860

Illinois Department of Registration and Education
Professional Engineers' Examining Committee
320 W. Washington Street, 3rd Floor
Springfield 62786
(217) 785-0872

Indiana State Board of Registration for Professional Engineers and Land Surveyors
1021 State Office Building
100 N. Senate Avenue
Indianapolis 46204
(317) 232-1840

Iowa State Board of Engineering Examiners
State Capitol Complex—1209 East Court Ave.
West Des Moines 50319
(515) 281-6566

Kansas State Board of Technical Professions
214 West 6th Street, Second Floor
Topeka 66603
(913) 296-3053

Kentucky State Board of Registration for Professional Engineers and Land Surveyors
Rt. 3-96, 5 Millville Road
Frankfort 40601
(502) 564-2680 & 564-2681

Louisiana State Board of Registration for Professional Engineers and Land Surveyors
1055 St. Charles Avenue, Suite 415
New Orleans 70130
(504) 568-8450

Maine State Board of Registration for Professional Engineers
State House
Augusta 04333
(207) 289-3236

Maryland State Board of Registration for Professional Engineers
501 St. Paul Place, Room 902
Baltimore 21202
(301) 659-6322

Massachusetts State Board of Registration of Professional Engineers and Land Surveyors
Room 1512 Leverett Saltonstall Building
100 Cambridge Street
Boston 02202
(617) 727-3088

Michigan Board of Professional Engineers
P.O. Box 30018 (611 West Ottawa)
Lansing 48909
(517) 373-3880

Minnesota State Board of Registration for Architects, Engineers, Land Surveyors and Landscape Architects
5th Floor, Metro Square
St. Paul 55101
(612) 296-2388

Mississippi State Board of Registration for Professional Engineers and Land Surveyors
P.O. Box 3 (200 S. President Street, Suite 516)
Jackson 39205
(601) 354-7241

Missouri Board of Architects, Professional Engineers and Land Surveyors
P.O. Box 184 (3523 North Ten Mile Drive)
Jefferson City 65102
(314) 751-2334

Montana State Board of Professional Engineers and Land Surveyors
Department of Commerce, 1424-9th Avenue
Helena 59620-6521
(406) 449-3737, Ext. 30

Nebraska State Board of Examiners for Professional Engineers and Architects
P.O. Box 94751 (301 Centennial Mall, South)
Lincoln 68509
(402) 471-2021, or 471-2407

Nevada State Board of Registered Professional Engineers and Land Surveyors
1755 East Plumb Lane Suite 102
Reno 89502
(702) 329-1955

New Hampshire State Board of Professional Engineers
77 N. Main Street, Room 214
Concord 03301
(603) 271-2219

New Jersey State Board of Professional Engineers and Land Surveyors
1100 Raymond Boulevard
Newark 07102
(201) 648-2660

New Mexico State Board of Registration for Professional Engineers and Land Surveyors
P.O. Box 4847
Santa Fe 87502
(505) 827-9940

New York State Board for Engineering and Land Surveying
The State Education Department
Cultural Education Center, Madison Avenue
Albany 12230
(518) 474-3846

North Carolina Board of Registration for Professional Engineers and Land Surveyors
3620 Six Forks Road
Raleigh 27609
(919) 781-9499

North Dakota State Board of Registration for Professional Engineers and Land Surveyors
P.O. Box 1264 (609 N. Broadway)
Minot 58701
(701) 852-1220

Ohio State Board of Registration for Professional Engineers and Surveyors
65 South Front Street, Room 302
Columbus 43215
(614) 466-8948

Oklahoma State Board of Registration for Professional Engineers and Land Surveyors
Oklahoma Engineering Center, Rm 120
201 N.E. 27th Street
Oklahoma City 73105
(405) 521-2874

Oregon State Board of Engineering Examiners
Department of Commerce, Room 403
Labor and Industries Building
Salem 97310
(503) 378-4180

Pennsylvania State Registration Board for Professional Engineers
P.O. Box 2649
(Transportation & Safety Bldg., 6th Floor,
Commonwealth Avenue & Forester Street)
Harrisburg 17105-2649
(717) 783-7049

Puerto Rico Board of Examiners of Engineers, Architects, and Surveyors
Box 3271, (Tanca Street, 261, Comer Tetuan)
San Juan 00904
(809) 725-7060

Rhode Island State Board of Registration for Professional Engineers and Land Surveyors
308 State Office Building
Providence 02903
(401) 277-2565

South Carolina State Board of Engineering Examiners
2221 Devine Street, Suite 404
P.O. Drawer 50408
Columbia 29205
(803) 758-2855

South Dakota State Commission of Engineering and Architectural Examiners
2040 West Main Street, Suite 212
Rapid City 57701
(605) 394-2510

Tennessee State Board of Architectural and Engineering Examiners
546 Doctors' Building
706 Church Street
Nashville 37219
(615) 741-3221

Texas State Board of Registration for Professional Engineers
1917 IH 35 South
P.O. Drawer 18329
Austin 78760
(512) 475-3141

Utah Representative Committee for Professional Engineers and Land Surveyors
Division of Registration
P.O. Box 5802
160 East 300 South
Salt Lake City 84110
(801) 533-6628

Vermont State Board of Registration for Professional Engineers
Div. of Licensing and Regulations
Pavilion Building
Montpelier 05602
(802) 828-2363

Virginia State Board of Architects, Professional Engineers, Land Surveyors and Certified Landscape Architects
Department of Commerce
Seaboard Bldg., 5th Floor
3600 West Broad Street
Richmond, Virginia 23230-4917
(804) 257-8512

Virgin Islands Board for Architects, Engineers and Land Surveyors
Submarine Base, P.O. Box 476
St. Thomas 00801
(809) 774-1301

Washington State Board of Registration for Professional Engineers and Land Surveyors
P.O. Box 9649
(3rd Floor, 9th & Columbia Building)
Olympia 98504
(206) 753-6966

West Virginia State Board of Registration for Professional Engineers
608 Union Building
Charleston 25301
(304) 348-3554

Wisconsin State Examining Board of Architects, Professional Engineers, Designers, and Land Surveyors
P.O. Box 8936
(1400 East Washington Avenue)
Madison 53708
(608) 266-1397

Wyoming State Board of Examining Engineers
Barrett Building
Cheyenne 82002
(307) 777-6156

PART

ORGANIZING TO MANAGE EFFECTIVELY

5

THE MANAGER'S JOB

Ralph DiPietro

RALPH DIPIETRO is professor of Marketing and Management and chairman of the Marketing Department at Montclair State College's School of Business Administration. A graduate of City College of New York, he returned there for his M.B.A. degree and received his Ph.D., with distinction, at New York University. He is an active marketing and management consultant to several public and private organizations such as Bristol Meyers; Bustop Shelters Inc.; The City of New York, Department of Personnel; The Institute of Retail and Management, Management Training Director; Fortunoff's; Sharp Electronics; Manufacturers Hanover Trust Company; Bally of Switzerland; and The Battus Corporation (Saks Fifth Avenue Stores). He is the author of a monograph, "Managerial Effectiveness: A Review and an Empirical Testing of Model," published by the American Psychology Association. He has also written several articles for several professional journals. Among them are: *The Operations Research Society of America, The Institute of Management Sciences, The Quarterly Journal of Management Development,* and *The Australian Society of Operations Research.* He is also a contributor to the *Handbook of Executive Communication,* published by R. D. Irwin. Dr. DiPietro is a member of The International Association of Applied Psychology, American Academy of Management, American Marketing Association, and Omicron Delta Epsilon, Honor Society of Economics.

5.1   INTRODUCTION TO THE MANAGEMENT PROCESS

Organizations may be defined as groups of two or more people with a common purpose or direction and a social structure with role relationships among members. Anthropologists argue that civilization began when man moved from nomadic food-gathering societies to sedentary food-producing societies and began to record knowledge in permanent form. Hence organizations predate civilization itself, for even cave dwellers lived in groups characterized by division of labor, social structure, and common purpose. Modern man is essentially organizational man, since organizations are omnipresent and affect almost every aspect of human life. While organizations can be marvelously efficient engines for the satisfaction of human needs, they have only been studied systematically for just over 100 years.

The study of organizations and the ways of managing them more effectively is increasingly becoming a serious concern to many of us in a world of relentless, ever increasing change. Consequently researchers in the administrative and behavioral sciences are turning their efforts toward the study of organizations and the individuals who manage them.

5.2   ENTREPRENEURIAL MANAGEMENT

In the last century business organizations were primarily managed by those who owned them. These entrepreneurial visionaries (Matthew Josephson calls them robber barons)[1] painted on giant canvasses in broad strokes creating large organizations which they managed in "hands-on" fashion. They seemed to have certain common insights. Recognizing the burgeoning mass markets that were forming, they designed the mass production and distribution systems necessary to serve these markets. Carnegie, Singer, Rockefeller, Armour, Swift, and later the classic production genius, Henry Ford, all exemplify these developments. Logically this period is referred to as the production era. The route to success was engineering and production efficiency; the ability to reduce unit cost to a price that the mass market could afford. This product concept, a good product at a fair price, dramatically changed life in society. Material goods and services were made affordable for a larger segment of society than ever before in history.

This focus on what has now come to be known as industrial engineering was the beginning of modern management theory. Since most of the first managers were primarily engineers, early management training essentially emphasized production engineering techniques. Today, the Japanese have reactivated this focus on production efficiency in managing their companies with enviable success.

5.3   PROFESSIONAL MANAGEMENT: THE BERLE-MEANS THESIS[2]

As the early corporate giants grew larger to accommodate and benefit from the mass markets, their need for capital escalated rapidly. The early entrepreneurs who focused on production, had little sympathy for the financial controls instituted by the Morgans and Mellons in the second stage of managerial focus, that is, the ascendancy of Wall Street and finance capitalism in general. As capital requirements grew, owner-managers found it all but impossible to maintain complete control of their organizations. Equity in the corporation became increasingly diffused among numerous stockholders who delegated responsibility for the management of their assets to a professional managerial elite. By the 1920s the separation of ownership from control in large companies was manifest.

Managers and management became a distinct class apart from the labor they supervised and the owners (through their representatives, boards of directors) to whom they reported. According to Peter F. Drucker, "management may be the pivotal event of our time. Rarely, if ever, has a new basic institution, a new leading group, a new central function, emerged as rapidly as management."[3] In their work on the criteria of corporate success, Peters and Waterman report that in many successfully managed companies, "everyone is considered part of management," and the separation of "planning from doing," a postulate of classical management, is deemphasized. Indeed, workers are encouraged to purchase equity in the firm which serves to reinforce their motivation and commitment.[4]

5.4.   THE MANAGER'S JOB

In studying the manager's job, a multidimensional perspective is required to understand the complex nature of this activity.

5.4.1.   The Nature of Managerial Work

The physical sciences have historically viewed the behavioral and administrative sciences as art rather than science. This view results from the fact that while it is remarkably difficult to make a rocket "conjunct with Mars," the study of human behavior is still more complex and unpredictable, rendering it difficult to apply experimental methodology.

When people are asked to distinguish art from science, they consistently view biology, physics, chemistry, and engineering as "science." Medicine receives a wavering "science," psychology, a skeptical "science" or "not sure," while management is universally perceived as "art."

Science is not an area of study but rather a methodology; a systematic and rigorous way of gathering data and "manufacturing" (or translating) it into information. An area of study is more aptly termed a discipline and all disciplines are varying combinations of art and science. During the Renaissance when engineers, who had absorbed a body of scientific knowledge, designed new architectural forms, it was art, that is, the creative application of a body of knowledge. Similarly, the sketches of DaVinci and the paintings of Raphael reflect advances in the physical sciences.

In a sense, the controversy over "art versus science" may be viewed at best as academic and at worst as counterproductive and divisive to an interdisciplinary body of knowledge. Management is an area of study; a discipline concerned with how we organize scarce resources and human effort to achieve specified goals and objectives. A manager may indeed be viewed as a chemist whose task is to blend scarce resources in the most effective manner so as to achieve the stated aims of the owners of those resources.

5.4.2. Management Skills

Leadership (management or stewardship) has historically been considered an innate ability. The expression "a born leader" was the accepted norm. This was reflected in the great importance attached to the concepts of "bloodline" and "birth right" and even extended to entire social classes with the advent of Social Darwinism.[5]

By the end of the nineteenth century, however, a new perspective began to take hold. It was believed that leadership and management involved certain skills that could be developed and were transferable. This view signaled the beginning of management as a discipline involving universal functions, activities, and skills that could be taught; witness the recent proliferation of business schools and the highly revered MBA degree.

The first attempts at implementing this new view involved studying successful managers and developing a profile of their specific skills and traits which would then be taught to aspiring managers. Unfortunately, no consistent profile of specific traits and skills was found to be predictive of good managers in all management positions. Recently, researchers have identified more general skills areas which are believed to be basic to all management positions. One such study discovered that all leadership positions require the following three basic skills (1) administrative skills which include planning, decision making, organizing, and controlling,[6] (2) human skills which involve social intelligence, communicative ability, and sensitivity to others (i.e., the ability to develop rapport and positive, productive relationships with others) and (3) technical skills which entail knowledge of the specific activities the manager is responsible for supervising, for example, in the case of the manager of a civil engineering firm or department, the extent to which the manager has mastered the knowledge of engineering.

The study also found that the proportion of these three skills changes as one moves up through the hierarchy of the organization, from first-line supervisor or foreman, through the vice-presidential level, and ultimately to the president or chief executive officer. To illustrate, in first-line supervision the research showed that while all three skills are necessary, technical skills are the most crucial, closely followed by human skills, and then by administrative skills. This is explained by the requirement that a foreman must often instruct an operator on the proper handling of a particular aspect of a task. The foreman must also be on hand to perform the task personally in peak operating periods or in the absence of the employee. Human skills are at least as important at this level, especially in today's environment where employees demand more personal and social satisfaction from work than was expected by prior generations of workers. The first-line supervisor, a strategic link between management and labor, must translate management's aims to workers and insure their understanding and commitment. In so doing he or she is understandably perceived by the workers as "management."

As one moves up the chain of command, administrative skills rise to prominence, closely followed by human, particularly communicative, skills. In a small firm technical and human skills may be more useful than administrative skills. When the firm grows to a substantial size, both administrative and human skills become crucial.

Robert L. Katz has developed a similar taxonomy which consists of three basic skills. Again, technical and human skills are recognized as well as what he refers to as conceptual skills, that is, the ability to discern relationships, solve problems, and make decisions; to analyze data and exercise balanced judgment.

Consistent with earlier findings by others, Katz discovered that the proportion of these three skills fluctuates as one moves up in the management hierarchy. While human skills are found consistently important, technical skills diminish relative to conceptual in higher-level positions.[7]

5.4.3.   The Roles of a Manager

A third perspective required to understand the managerial job is to become aware of the multitude of roles demanded of the position. One of the more widely read empirical researchers in the field of management, Henry Mintzberg, has identified a number of roles required of the manager.[8] These roles are clustered in the following three areas:

1. Interpersonal roles, which involve titular or symbolic activity of a formal, legal, or ceremonial nature. Interpersonal roles also include the leadership responsibility for guiding and motivating, as well as the liaison activity necessary in providing linkages to other units of the organization.

2. Informational roles in which a manager must gather and analyze relevant on-line information and communicate it to subordinates as well as appropriate others outside his area of responsibility.

3. Decisional roles, considered by many to be the raison d'être of a manager, is the last category of roles in Mintzberg's classification. These roles include entrepreneurial activities such as initiating change, coping with exigencies, and other nonroutine issues. Included also are decisions involving the efficient allocation of resources and those involving negotiating in both internal and external environments.

5.4.4. The Functions of Management

Perhaps the most widely accepted recognition of the manager's job is that certain universal functions exist in which all managers must engage, whether they are managing in a hospital, an oil refinery, an engineering department, or university. While there may be some semantic differences in the labeling of these functions, a common ground has generally been agreed upon.

It has been suggested that all organizations must, through managers, engage in the process of planning, organizing, staffing, directing, and controlling their activities. While some scholars treat coordinating as a separate function, it is inherent at each stage and throughout the management process. Others include motivating, which in the text is subsumed under the directing function. Finally, decision making is frequently incorporated in discussions of the planning process. The remainder of this chapter devotes itself to outlining these universal functions and serves to introduce the chapters that follow, where these topics will be covered in greater detail.

Planning.   This activity involves making decisions today with consequences for the future. Planning can only take place with a clear understanding of where the organization is at present. In other words, planning requires a situational audit. Armed with information, the organization can plot a course of action toward its predetermined goal. Planning involves making forecasts of what the future will be and develops plans of action that are compatible with that future. The capacity for sensing the past, assessing the present, contemplating the future, and making appropriate preparations may be unique to humans. This futuristic process is a bridge-gapping activity. In fact, one recently developed planning tool is referred to as gap analysis.

When we speak of management as a process, it is implicit that the various functions are interdependent. This fact may be lost to the reader of a textbook in which these functions are treated in sequential chapters, giving the illusion that the process is linear rather than curvilinear; Figure 5-1 illustrates this point.

That planning and controlling are Siamese twins is clearly exemplified by the budgeting activity, which most often is treated as a control mechanism. Yet it may reasonably be viewed as a plan for the allocation of some scarce resources over a future period of time. Alternatively, when one is developing a plan consideration must be given as to how organizing, staffing, directing, and controlling will be executed under that plan. In order for planning to be comprehensive, specific questions must be answered, such as what, where, when, how, and by whom the plan is to be implemented. Too often one or more of these areas are left to chance or ignored entirely, resulting in poor implementation.

The different types of planning may be categorized in chronological order or according to where in the hierarchy of the organization the activities most often take place. From the chronological perspective the most nebulous type of planning is referred to as an organizational philosophy or climate and more currently as a corporate culture or mission. This mood reflects the values and attitudes, indeed the "personality" of the organization, and profoundly affects the entire planning process in much the same way as an individual's self-concept and personality affect behavior. It is of utmost importance that a manager be intimately aware of this culture in order to have the appropriate role perception of the job and "acculturate" properly into the structure.

While organizations have always had climates, a major shift has recently occurred. Today this mood, culture, or mission is more clearly defined in a

FIGURE 5.1.

formal manner, optimized by its top echelon executives and communicated to the various publics the organization addresses (specifically its other channel members, customers, employees, competitors, various governmental groups, the community, and society as a whole). Most successful firms have a clear understanding of their mission and culture and communicate them effectively, as evidenced by such companies as Hewlett-Packard, IBM, General Electric, McDonald's, and so on. With a clear sense of corporate or organizational identity, the manager is in a better position to search with an educated eye for the appropriate goals and objectives.

The development of long-term goals, as well as the short-term and intermediate objectives instrumental in achieving them, has been an area of major focus for scholars and practitioners alike. Traditionally, in financial or accounting parlance, "short term" refers to the fiscal year; "intermediate term," to 2 to 3 years; and "long term" to 5 years or more From a strategic planning perspective, how far into the future we plan is predicated on such factors as availability of information and the nature of the industry. To illustrate, an electric utility may be required to estimate population demographics, economic productivity, and consequent energy needs 50 years into the future, while a fashion house need only look two to three years ahead. What is intermediate-term planning for the utility is long-term planning for the fashion house.

Many practitioners and scholars believe that with the advent of computer technology and operations-research techniques, there has been an overemphasis on quantitative standards of job performance to the detriment of the qualitative aspects of job activity and goal development. Ideally goals and objectives should provide tangible purpose and direction for a manager. Quantification is best seen and utilized as the natural route to such rational goal setting.

Unfortunately the less measurable aspects of job performance tend to be ignored. Productivity is too often viewed in terms of how many units produced, sales calls registered, and new products designed, rather than in terms of the quality of those units produced, calls made, and products designed. Conceivably 900 Toyotas per day may register more productivity at 0.5 defects per vehicle than 1000 at 5.5 defects per vehicle—a harsh reality that is facing American automobile manufacturers today. In current writings, the term "quality" is sometimes being used as a synonym for "productivity." Qualitative goals need not be considered "soft" if they are expressed with enough specificity to give the necessary focus to a manager's energies and efforts.

For instance, a chief engineer may aim to improve efficiency in his department by breaking this nebulous goal into a series of specific objectives, each with its own timetable. For example, to improve efficiency of the department, we may agree to set a specific objective of developing an ongoing follow-up program for all new employees, utilizing a "buddy system." This system would substitute for the usual "sink or swim" method of orienting new trainees. A second objective might be to set up a company newsletter with a section devoted to each department. Such objectives may become company-wide policy or may be purely departmental depending on their suitability.

The manner in which goals are set is of utmost importance. The concept of Management by Objectives has been in use for over 20 years. Management by Objectives (MBO) involves both superordinate and subordinate working out and agreement upon a set of challenging and meaningful objectives for a stipulated period of time. This practice enables the superior to play the role of a coach, counsellor, and supportive agent. Clearly, Management by Objectives is both a philosophy of management and a leadership style in the manner in which it views people. (See Chapter 12.) An integral part of the approach is the appraisal interview in which objectives and performance are compared. (See Chapter 29.)

Having developed a sense of direction through goal setting, a strategic roadmap, blueprint, or game plan must be devised to accomplish these goals. The term "strategy" implies being core or central to the survival of the system. A strategic weapon in a defense system is a nuclear warhead; a strategic error in a chess game inevitably leads to checkmate; a strategic weakness would be one which can destroy the system. The term strategy itself has a particular meaning in contemporary business jargon. In government and the political sciences the term "policy" was traditionally considered a synonym for what we refer to today as strategy. A comprehensive strategy pulls together and coordinates all of the various resources, activities, and the interrelationships necessary to achieve the objectives previously developed.

Program Evaluation Review Technique (PERT), may be viewed as a graphic, quantitative representation of a stratagem. This planning tool will be discussed in detail in Chapter 17.

In the process of translating the strategy from the drawing board to the marketplace, unforeseen problems inevitably arise. No strategy can be developed which is capable of anticipating any and all exigencies. If a particular problem is perceived to be one which will continuously reoccur, policy formulation is appropriate. Too often in bureaucratized organizations policies are set without regard to whether a problem is recurrent in nature. This leads to structures that are cumbersome, overly complicated, and rigid in operational design. Thus policies are set to avoid repeated analysis so that we may practice Management by Exception.[9] Management by Exception is based on the notion that a manager's time is spent most efficaciously by concentrating on major problems that are in some way emergencies or otherwise unusual, exceptional, or unexpected, realizing that much of the remaining work is in the form of accustomed, expected, and routine events for which policies and procedures may be established. Since these create ways of dealing with many contingencies, the task of doing that part of the manager's job is more readily left to subordinates; in a sense the decisions "make themselves."

Policies, however, are not designed to be rigid (or at least they should not be in a well-run organization). They are guidelines or parameters for acceptable behavior and give the decision maker a degree of flexibility or discretion. For example, in the civil engineering field, subcontractors usually report promptly with a full complement of resources in the early stages of a project, but may tend to shunt these resources to new clients once they are on the job. This is a common problem in the field, and therefore, a civil engineer may set a policy of sending a certified letter of default to the subcontractor. If a substitute for the subcontractor cannot be found, that policy, at the decision maker's discretion, probably will not be enforced.

Much debate has centered around which form of planning must be developed first; policy or strategy. Applying the scenario traced above, policies are formulated in the process of implementing strategy. However, it can be readily seen that policy formulation will, in turn, affect new strategy. Thus the relationship between them is interactive. A strategy is indeed a living, organic document which is modified and adapted as a result of feedback from the implementation process.

Procedures further narrow the decision maker's discretion. A procedure is a stepwise, sequential manner of carrying out a routinized activity. For example, approval of an engineering change order might require adherence to a procedure like the following: A need for a change may come about because an original design, once thought feasible, proves physically impossible to construct. To rectify this problem the designer is given the opportunity to correct the design. However he must first state what was wrong with the original and give his recommendations for a feasible solution. Such a rule may seem rather obvious, but in practice, many an engineering project has experienced great, if not catastrophic cost overruns precisely because engineering change orders were not tightly controlled. Note that a full procedure would also have to establish who in the company is responsible for approval and other controls at various stages of a change procedure.

Lastly, rules and regulations afford no discretion to the decision maker and are usually accompanied by a penalty for nonconformance. As an example, on a construction site various penalties may exist for not wearing a hard hat in the "hard hat area."

We may view the planning process from a hierarchical perspective. (see Figure 5.2). Tactical or functional planning is the manner in which the vari-

FIGURE 5.2.   Hierarchial perspective of planning.

ous functional departments will carry out their part of the total strategem. Operational planning concerns itself with the day-to-day planning decisions and first-line supervisory activities.

If the organization is to have a cohesive blend of the three levels of management planning; strategic, tactical, and operational, everyone must be familiarized with the total game plan and have some degree of input into the total strategy. This calls for an approach sometimes called "consensus management" in which, successively, all layers of the organization are enlisted in formulating company plans and strategies. This is seldom done well in organizations; the concept is actually one that derives in part from Japanese practice, although American firms, such as JC Penney, Hewlett-Packard, and Manufacturers Hanover Trust, are also successful practitioners. It is not yet widely accepted in engineering firms as such. Considering that they are often small and that their fortunes may fluctuate sharply, it often surprising how little communication between management and employees on such issues actually exists.

Planning in its most implicit sense is decision making, often considered to be the distinguishing characteristic of a manager. There has been a continuous debate over two basic approaches to decision making which has led to the development of two models. One model referred to as lineal, systematic, or analytic, and in more polemic terms, as rational, logical, or scientific, has gained increasing favor during the past 20 years. The second model has often been condescendingly referred to as primitive, illogical, or nonrational, and perhaps more objectively as intuitive or holistic. It should be noted that any model is only as valid as the assumptions upon which it is built and how accurately it describes the reality it purports to explain. It appears that the controversy swirling about these two models has strong elements of divisiveness and counterproductivity with respect to understanding the basic process of decision making. A review of the literature shows that most scientists and engineers are at a loss to explain in "rational" terms how they make successful decisions in problem areas which have significant degrees of uncertainty. To illustrate, Rand Corporation scientists have concluded that "no decision mechanism can be devised that will escape the basic uncertainties and complexities that plague larger problems."[10]

Many complex problems require both intuitive and what we refer to as scientific approaches to decision making. Indeed the human brain is bicameral in nature. The left hemisphere of the brain deals with what is commonly referred to as the conscious mind—encompassing language, rational knowing, a sense of time, mathematical and linear functions. The right hemisphere appears to favor nonverbal thinking, spatial orientation, intuition, and the direction of many artistic activities.[11]

These considerations lead to the recognition that intuitive and rational approaches to complex decision-making are both appropriate and productive depending on the nature of the problem we are facing and the amount and kind of information that we have. The rational model is based on two assumptions: First, the decision maker is viewed as a streamlined logico-deductive computer, processing information with minimal perceptual distortions. Secondly, the decision maker is viewed as a maximizer, for example, he or she would choose the alternative that would lead to the optimal solution to a problem. When we attempt to dissect the term, "rational," problems immediately arise. As Herbert Simon has noted, rationality implies acting upon available information rather than complete information. Therefore it is more appropriate to speak of bounded rationality, that is, rationality delimited by available information.[12]

This may be better understood by viewing the term "Operations Research," the discipline concerned with various mathematical analyses of the deployment of technical and economic resources in several areas. Operations Research has lent itself most readily to operational planning, that is, day-to-day routinized planning where there is considerable information relating to the problem area and a sufficient degree of consistency. It has failed conspicuously when the information was incomplete, poorly interpreted, or when the task of measuring important variables and defining their relationship was poorly done. This has been experienced especially with respect to the social impact of operational changes in public services, construction, and environmental policy.

The rational model is thus based on the two assumptions of rationality and maximization, and involves the following six steps:

1. Statement of objective
2. Statement of the problem

3. Tracing out the alternatives
4. Evaluation
5. Choice
6. Feedback and follow-up

While the logic of the rational model is manifest, implementing the process in organizational life is obstructed by a number of realities. For example, power plays may create a situation where the alternative chosen reflects the power structure of the organization rather than maximizes the objective function. A second example, uncertainty avoidance, involves choosing that alternative whose outcome we are most sure of, rather than the one which seems to be most effective on paper. A third example, conflict resolution, involves choosing the alternative that minimizes conflict among those parties in the decision process.

Before the planning process can evolve intelligently, an internal-external or "WOTS-UP" analysis of the company must be performed. An acronym for the weaknesses, opportunities, threats, and strengths underlying and affecting planning activities, the WOTS-UP analysis attempts to identify the potential opportunities as well as the threats that exist in the environment relative to the company's strengths and weaknesses.

Thus the manager's goal in performing this analysis is to discern the market opportunity among many, that is, that particular opportunity whose requisites for success the firm has in abundance such that it would possess an advantage or leverage over a potential competitor. A continuous, systematic evaluation of the company's management and marketing skills, corporate image, technical and financial resources, and competitive market position, as well as the social, economic, political, competitive, and legal forces in the organization's environment is vital to achieving this end.

The manner in which this analysis is approached is critical to the process. If an internal analysis of the company's strengths and weaknesses is performed in too rigorous a fashion and occurs prior to the assessment of the environment, a narrow, rigid profile of potential opportunities may result. This may cause the firm to lose sight of profitable opportunities that could, with minor modifications, be successfully exploited. Conversely, looking to the marketplace without a clear sense of corporate capabilities and vulnerabilities involves an inefficient utilization of time, energy, and resources in evaluating opportunities irrelevant to the organization.

Alternately for the analysis to be a valuable phase in the planning process, a palatable synthesis of the two approaches should be utilized. Having executed a cursory analysis of the organization's strengths and weaknesses, a firm is able to search the environment with a discerning eye (considering the organization's profile as well as the threats associated with available opportunities) for those opportunities that would prove profitable if pursued.

Organizational Design.   Having developed purpose or direction for the system, an *organizational design* or structure must be developed to realize the goals and objectives of the company. Ideally the structural design developed should be reflective and supportive of the attainment of those goals. In his classic work, Alfred Chandler used the phrase, "form follows function."[13]

The organizing function may be viewed in terms of two basic subsystems: departmentation and decentralization. Departmentation involves breaking up the activities and work to be performed into manageable units, an issue of increasing importance as organizations grow in size and complexity. Decentralization involves the breaking up of managerial work into effective units of authority throughout the chain of command. Organizing might best be understood by viewing the individual job position. Job analysis deals with organizing at an atomistic level, breaking down each job position in terms of both the job description and the job specification. A job description states the title of the job and describes all of the activities necessary to perform that position properly. Job specification entails enumerating those qualifications the incumbent must possess in order to perform the job description. These qualifications include education, experience, physical qualifications, and so on.

Staffing.   The *staffing* function encompasses those activities which influence the effectiveness of an organization in achieving the objectives of both the firm and its employees. Staffing is the core or central activity of the personnel or human resource function. As a result of changes in economic conditions, inflation, societal and legal forces, and the expectations of workers today, the staffing function has become a strategic problem and a major issue in the 1980s. The effective management of human resources in an organization today is proactive rather than reactive, focusing on the individual. In their study of the most successful American companies, Peters and Waterman have found that these companies view people as their primary source of quality, productivity gain, and organizational well being.[14]

Employment planning involves accurately determining and forecasting labor needs; recruiting and attracting an appropriate reservoir of candidates; selecting, placing and orienting; training; and performance appraisal. Accurate determination of manpower needs is a critical first step in the staffing process. A widely utilized tool for analyzing and projecting manpower needs is the manning table, whose elemental unit is the "three position plan." Each job position is viewed not in isolation but in terms of the job which leads into it and the job that logically follows from it (see Figure 5.3). The concept of lead time is a particularly important feature of manning tables. Lead time is the most reasonable forecast of the amount of time it would take to find a

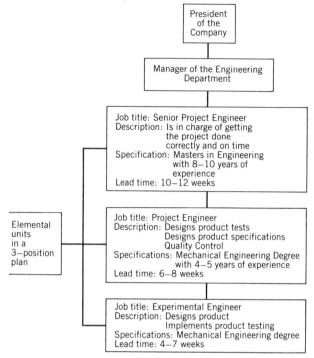

FIGURE 5.3.   A Section of a manning table.

replacement for a position should the current holder be lost for any reason. Effective staffing involves selection and recruiting, training, and performance appraisal. These are important aspects of management and are discussed in detail in Chapters 18–36.

Directing or Leadership.   The *directing* or *leadership* function is closely associated with the motivational process. In its most basic sense, leadership involves creating an environment where individuals are motivated and committed to pursuing the organization's goals. Peter Drucker has succinctly defined management as "getting things done through people."

Good leadership has always been a scarce commodity. Historically, management theorists have viewed the leadership function from two perspectives, the universal versus the contingency approach.

The universal approach argues that there is one best style of leadership, focusing on the leader's concern for people or production. Within the universal framework controversy over the optimal managerial style exists. Supporters of the classical approach argue that an autocratic manager, one who is task oriented and practices Theory X, is most effective in all situations and environments. Advocates of the participative approach, however, purport that a democratic manager, one who is people oriented and practices Theory Y, will be the most successful leader. (See Section 6.3.1 for explanations of Theory X and Theory Y.)

These two universal theories were debated until the Ohio State Leadership studies broke the controversy in the late 1940s. In this comprehensive study it was found that no single style of leadership is best in all situations. At this point the contingency or situational approach to leadership began to take hold. Advocates of this approach argue that the most effective leadership style is contingent upon the particular situation that the manager is facing. Fred E. Fiedler, who first defined contingency management, further argued that attempting to remake a leader's style is a herculean if not impossible task. Instead it is advisable to place the manager in that situation where his style of leadership will work best.[15] Other scholars contend that style of flexibility in leadership are vital and all managers should have the ability to be flexible in the leadership position depending on the situation. Robert J. House, in his Path Goal Theory, makes the point that the leader should use that particular style appropriate to facilitating the attainment of his subordinate's career goals. House emphasizes this supportive counselling role by stating that, "The motivational function of the leader consists of increasing personal payoffs to subordinates for work-goal attainment, and making the path to these payoffs easier to travel by clarifying it, reducing roadblocks and pitfalls, and increasing the opportunities for personal satisfaction en route."[16]

The concept of style flexibility has been further developed in the Hersey and Blanchard Model.[17] In this model leadership style is based on the amount of guidance or directiveness and the amount of social distance, in effect, social–emotional support given by the leader. This is determined after an analysis of the subordinate's job maturity. Job maturity is defined as the degree of ability or skill an individual possesses with respect to the job to be performed as well as psychological maturity of the individual, which is defined by willingness to pursue the goals of the organization. As seen in Figure 5.4, when an individual has reached a psychologically high maturity level and high job maturity, then low direction and support are appropriate. If job and psychological maturity are low, then that subordinate requires high directive supervision while the social distance factor between the individual and the mentor must remain high. For the individual who is not at either of the extreme levels, it is necessary to combine different aspects of each of the above leadership styles and formulate a congruent style for the specific needs of the individual. For example, the "plateaued" engineer, someone who knows his job well but has become apathetic toward his work,

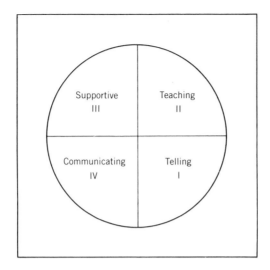

Maturity level:

I Low maturity	II Low to moderate maturity	III Moderate to high maturity	IV High maturity
unable and unwilling or insecure	unable but willing or confident	able but unwilling or insecure	able/competent and willing/confident

Appropriate style:

I Telling	II Teaching	III Supportive	IV Communicating
high task and low relationship behavior	high task and high relationship behavior	high relationship and low task behavior	low relationship and low task behavior

FIGURE 5.4.   Leadership styles. Adapted from Paul Hersey and Ken Blanchard.

needs to be given warmth, empathetic listening, and understanding, but little job direction (see Figure 5.4). In summary the key to successfully practicing situational leadership lies in the assessment of the leader, the subordinate, and the situation.

Controlling.   Controlling encompasses those activities which measure and correct the performance of subordinates to make certain that the company's goals and objectives and the strategy devised to achieve them are being accomplished. While planning deals with setting objectives and goals, controlling is concerned with assuring that these goals and objectives are being achieved. The controlling process is comprised of the following three steps: establishing standards, measuring actual performance against these standards (a monitoring device), and identifying and correcting deviations from these standards.

The control function is often viewed negatively by most managers. Visions of constraints upon a manager's creativity and performance are too often the norm. When a company is small controls tend to be more informal, if not lax, rather than rigid or stifling, providing sufficient flexibility for creative management and measured risk taking. However, as a company grows, control can become the instrumentality for exorcizing the entrepreneurial spirit from organizations and driving managers toward risk aversion rather than measured risk taking. By setting parameters for effectuating goals, the control function tends to act as the reins on the system. Almost universally, managers argue that although goals and objectives are set, sufficient resources are not allocated for effective implementation.

Budgets are a classic example of the control function in action. The weak link in the budget process is the lack of sufficient flexibility to accommodate the uncertainties inherent in that future world for which the control mechanism is developed. With the kinds of uncertainties and changes which occur in that future world, contingency planning and budgeting should be the norm. The organization must develop forecasts on a contingency basis, that is, more than one plan is devised so that the firm will be prepared for several possible eventualities. For future activities, three basic forecasts should be developed: a most optimistic, most likely probabilistic, and pessimistic (see Figure 5.5). Budgeting plans should be developed for each eventuality so that no matter what next year's scenario, the firm is prepared. There is a danger of the deliberate wasting of resources in order to consume an allotment in a given year and so assure a continued resource allocation to a particular department or function. Zero-base budgeting, so popular during the Carter administration, is an attempt at justifying allocation of resources for a future period of time based on demonstrable activities and objectives that we are attempting to accomplish.

Having looked at the negative stigma controls possess, a self-regulating mechanism should be designed, wherever possible, to alleviate this problem. Cybernetics, first defined by Norbert Wiener, rests on the proposition that the same principle of control seen in organic systems can and should be applied to mechanistic systems. Just as a thermostat in a house regulates the heating system so that the temperature is maintained as desired, so must a firm regulate itself to maintain its level of productivity and quality.[18]

This concept of self-regulation is needed in today's rapidly changing world. An organization no longer has the luxury of doing a periodic audit to

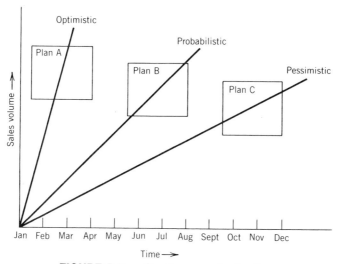

FIGURE 5.5.   Contingency budgeting.

detect problems; the firm must now be a continual Management Information System (MIS). Management Information Systems constantly monitor
environmental factors and adapt the organization to meet these changes.
The realization of such systems has proved more difficult than once expected. Aside from system design problems, the capital and operating costs
of the required computer installations have not always been justifiable in
terms of verifiable results. A more modest goal of maintaining an adequate
and consistent set of records and performance measures, called Decision
Support Systems (DSS) or Data Base Management, has been advocated, but
the choice between the two is as yet unresolved. (See Chapter 11.)

A newly emerging process for assisting organizations in addressing this
issue of continual monitoring is Organizational Development. The organization should ideally be in a constant state of flux and modification in order
to adapt to predicted changes in the environment. Two subsystems of Organizational Development are "structural intervention" and "process intervention." Structural intervention introduces change through new formal
guidelines, procedures, and policies; the emphasis resting upon planning,
organizing, and controlling. Having made these structural changes, the firm
begins process intervention, which deals with the human element: training,
communication, nurturing commitment and motivation, and selection and
staffing. The wave of corporate mergers, acquisitions, divestitures, and take-
overs in the 1980s required much organizational development. Where this

was done badly, dire consequences to the people and organizations affected helped fuel opposition to the whole trend.

5.5. SUMMARY

Although organizations predate civilization itself, the twentieth century has seen organizational power reach a zenith. Recognition of the significance of the form of twentieth century power and the individuals who wield it has energized the movement toward a more intensive analysis of management as a discipline. The rise of a professional managerial elite or class, has been a concurrent phenomenon of equal importance.

In this chapter we have described the general behavior of those who practice management. This description has been undertaken from a series of perspectives. The nature of management is viewed, as with other disciplines, as a synthesis of art and science. In addition, managers are seen as playing a number of roles: as conduits of communication, decision makers, leaders, coordinators, and linkages to the larger environments of which their organizations are a part. It has been argued that successful managers possess general skills areas such as human, technical, and administrative, or conceptual, and that the proportion or mix of these skills differ, depending on the situation in which the management takes place.

The remainder of the chapter concerns itself with tracing out the universal functions that all managers must engage in, whether they have responsibility over a small partnership, department within a division, or control of a major corporation. Planning and decision making, organizing and staffing, directing, and controlling are the essential elements of the art of management; they were described in broad outline.

In reviewing these functions a common theme surfaces, whether it be in planning, organizing, or any other functional activity: People are a major element in the management of organizations. While the human variable may be less predictable than others, often irritatingly so, this factor of unpredictability is the source of imagination, creativity, and tremendous energy which, when focused productively, usually makes the difference between threshold organizations and dynamic leaders of most industries.

In Chapter 6 this theme will be pursued as organizational theories, concepts, and tools of analysis are presented to help determine how to best structure organizational units through the development of systems and procedures. It will become apparent that the imposition of constraints on human behavior, through the design of rules, policies, and procedures, reduces itself to a concept of balance, that is, the appropriate degree of structure necessary to insure adequate consistency without denuding the human factor of its potential for imaginative and reasoned judgment.

NOTES

1. M. Josephson, *The Robber Barons* (New York: Harcourt, Brace, and World, 1934).

2. A. Berle and G.W. Means developed the theory of separation of ownership from control and the ascendancy of a professional managerial elite in *The Modern Corporation and Private Property* (New York: Macmillan, 1933).

3. P.F. Drucker, *Management: Tasks, Responsibilities, Practices* (New York: Harper & Row, 1974).

4. T. J. Peters and R. H. Waterman, Jr., *In Search of Excellence* (New York: Harper & Row, 1983), Ch. 7.

5. R. L. Carneiro, *The Evolution of Society: Selections from Herbert Spencer's Principles of Sociology* (Chicago: University of Chicago Press, 1967).

6. B.S. Georgopoulous and F.C. Mann, *The Community General Hospital* (New York: Macmillan, 1962).

7. R.L. Katz, "Skills of an Effective Administrator," *Harvard Business Review*, 52 (September–October 1974): 90–102.

8. H. Mintzberg, *The Nature of Managerial Work* (New York: Harper & Row, 1973).

9. W.J. Reddin, *Effective Management by Objectives: The 3-D Method of MBO* (New York: McGraw-Hill, 1971).

10. J.D. McDonald, "How Businessmen Make Decisions," *Fortune* (August 1955): 84–87.

11. R.E. Ornstein, *The Psychology of Consciousness* (San Francisco: W. H. Freeman, 1972).

12. J.G. March and H.A. Simon, *Organizations* (New York: Wiley, 1958); pp. 136–171.

13. A. Chandler, *Strategy and Structure* (Cambridge: MIT Press, 1962).

14. Peters and Waterman, *op. cit.*, Ch. 3.

15. F.E. Fiedler, *A Theory of Leadership Effectiveness* (New York: McGraw-Hill, 1967), pp. 22–32.

16. R.J. House, "A Path-Goal Theory of Leader Effectiveness," *Administrative Science Quarterly* 16 (September 1971): 321–38.

17. P. Hersey and K. Blanchard, *Management of Organizational Behavior: Utilizing Human Resources* (Englewood Cliffs, NJ: Prentice-Hall, 1982).

18. N. Wiener, *Cybernetics* (Cambridge: MIT Press, 1948).

6

SYSTEMS AND PROCEDURES

Ralph DiPietro

6.1. BACKGROUND: THE ROLE OF THEORY IN MANAGEMENT AND ORGANIZATION

Engineers are particularly appreciative of the value of theory construction and model building in understanding and reacting to professional issues. In this chapter three models of organization are described which trace the evolution of management theory and establish a foundation for understanding the role of systems and procedures in organizations.

As with many disciplines the constructs of management theory fall into a macro–micro–macro typology. The earliest model of management, referred to as the classical or scientific school, describes a functional theory of organization. This was followed by the behavioral or human relations model which focused additional attention on micro, that is, internal aspects of organizational theory by identifying and making explicit the assumptions of the classical school. The third, a macro model, is the modern systems or contingency theory of organization. This framework focuses on still other characteristics of organizations. It is developed from a different set of assumptions which are presented later in this chapter.

All three models have made significant contributions to our understanding of organizational management. The tools and techniques derived from them are useful in improving the efficiency and effectiveness of management operations.

6.2.   THE CLASSICAL SCHOOL: A UNIVERSAL APPROACH

Classical theory development was perhaps the first systematic attempt at building a set of universal principles of management. This general theory "grew up" in the Newtonian world of the nineteenth century. Newton's mechanistic view of the universe influenced the thinking of early management scholars in their attempts to replace rule-of-thumb management with scientific observation. Their goal was to design a smooth-running piece of precise scientific machinery, preprogrammed, predictable, and depersonalized.

Whether they focused on shop management (time and motion study), as did Frederick W. Taylor and Frank and Lillian Gilbreth, or clerical work, as did Max Weber or the development of a "handbook" on administrative management composed of universal functions and a set of guiding principles, as did Henri Fayol, James Mooney, and Alan Reiley, the basic goal was the same: To improve efficiency of operations through scientific observation aimed at determining the best way of handling each activity in the system. A schematic illustration of their areas of concern is presented in Figure 6.1.

6.2.1.   Scientific Management

The emphasis on organizational efficiency, through industrial engineering, especially in production was particularly congruent with the needs of the

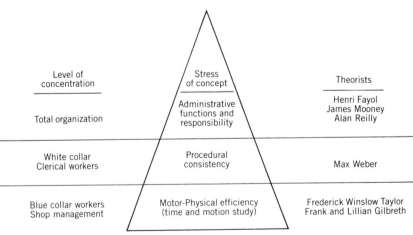

FIGURE 6.1.   Hierarchic view of classical theorists' contribution to organization.

marketplace of that period. The ability to drive down per unit costs and reflect this efficiency in a price the mass market could afford was thought to be the route to success. Taylor experimented in time and motion study in his classical examination of pig iron loading, with its stress on psycho-motor skills and manipulative task design. However, his work was soon succeeded by Fayol's articulation of the universal functions of management as planning, organizing, staffing, directing, and controlling. Some examples of these principles illustrate the essence of this approach.[1] (See Table 6.1.)

In the span of control principle we begin to sense the essence of the classical model. In the earnest desire of management theorists to systematize, clarify, and formalize the new discipline, rigidities inexorably arose despite the caveat that these principles were guidelines rather than immutable laws. V.A. Graicunas went so far as to develop a mathematical formula to "prove" that the optimum span of control, that is, number of people to be supervised, for any manager was not less than four or more than seven employees.[2] His formula was $R = n(2^{n-1} + n - 1)$, where R is the number of bilateral relationships and n the number of employees to be supervised. This gives the following selected results:

$n =$	3	4	5	6	7	8
$R =$	18	44	100	222	490	1080

Graicunas considered the number of bilateral relationships as the essential limiting factor; an R value of 18 was too low and 1080 too high. Today it is

TABLE 6.1. Classical Principles of Management

Principle	Explanation
Division of labor	Work should be broken down into basic elemental units in the interests of simplicity and efficiency.
Coextensiveness of authority and responsibility	When a task or an objective is assigned the employee should be given sufficient authority to carry out his responsibility.
Unity of command	Each subordinate should be held accountable to one superior to prevent the confusion that may arise from conflicting orders issued by different superiors. This is a common problem in small and midsize businesses, especially partnerships.
Scalar principle	A clear line of authority of "chain of command" should exist from the chief executive officer down to the first line supervisor or foreman level.
Span of control	There exists an optimum number of subordinates that each manager can most effectively supervise.

accepted that the nature of the task being supervised as well as the characteristics of the subordinates and their superior all affect the span of control appropriate in any situation. Specifically, organizational relationships are unique and cannot always be assigned equal importance. The formula also represents only potential relationships which are influenced by many situational factors including worker's training, initiative, interdependence, and type of task being performed.

6.2.3. Bureaucracy

To most individuals the term bureaucracy engenders visions of red tape, inefficiency, and lack of responsiveness. From an organizational viewpoint nothing is accomplished without bureaucracy; no system can function without some degree of consistency or predictability. The crucial issue becomes one of determining the appropriate degree of systematic and procedural structure for a given situation.

The concept of bureaucracy introduced into organizational theory a degree of systematic and procedural arrangement relatively new for the nineteenth century. Max Weber, a German sociologist, observed with disdain the "spoils system"[3] characterized by nepotism, social class discrimination, and the frailties of unpredictable human emotion which held sway in military, government, and church organizations of that period. It was against this backdrop that he developed an "ideal form" of organization which was aimed at holding the human variation in check by an elaborate set of rules, regulations, and procedures. This would insure that a smooth-running, preprogrammed piece of precise machinery, depersonalized and objective, could accomplish its predetermined mission with minimal dysfunction. Weber developed a set of guiding principles which in many respects are similar to those of Fayol, Mooney, and other classicists of the time (division of labor, specialization, hierarchy of authority, and so on).

In other aspects of his model Weber broke new ground and is considered the father of the modern civil service system. The need to insure consistency in the organization was reflected in most features of his approach but particularly with respect to power. He scorned the traditional power found in tribal societies and monarchies of his period. Charismatic power based on the strength of the leader's personality was also considered inappropriate for organizations, since it was grounded in the emotions and feelings of subordinates. Weber believed that the job, position, or office, as objectively designed, should hold that degree of authority necessary to accomplish the task, rather than relying upon the particular incumbent's personality characteristics. This rational-legal power, as he referred to it, was based on proven or demonstrated expertise and the legal nature of a "constitutional" form of organization.

The essence of the rational-legal form of power is that it is based on information and circumscribed by law and that no individual is above that law. Weber's formulation is an attempt to use science and objectivity to determine managerial behavior in general, whether it be at shop, clerical, or administrative levels of the organization.

Objective examinations were to be designed based on job descriptions, and those who scored highest in open competition would be assigned the position with full job security. This would avoid the "spoils-system" mentality, which in earlier periods inhibited the growth of a professional managerial class. (It has been argued that a concept of balance with respect to job security may well be transcended in many civil service positions today!)

One never ceases to be amazed when studying the early "sedentary merchants" (John Jacob Astor, and so on,) who managed vast enterprises by jotting down the barest of notes on the backs of envelopes and scraps of paper. It is also quite remarkable that as late as 1916 most major companies were still controlling their activities by single-entry bookkeeping. Along with helping develop the modern civil service system, Weber also helped introduce the modern office filing system. Predictability, consistency, and continuity were to be assured through an elaborate system of written documents which were preserved and updated, giving the organization continuity as individuals entered and left the company.

Weber moved away from the wage method of renumeration when he proposed a regular pecuniary compensation—a salary and guaranteed security in old age by providing a pension. This salary was not measured like a wage in terms of hours logged or piece work done, but was measured according to the job title or status, that is, according to the type of function (the rank). This salary also took into account the length of loyal service. Weber's approach was aimed at assuring that two people holding the same job title would not receive different rewards based on favoritism or discrimination.

These guiding principles appear to be rather reasonable, objective, and democratic approaches to addressing the problems of managing. Why then does the term bureaucracy conjure up such negative imagery? If the answer were to be reduced to its most elemental property, it is that a concept of balance has been lost. It appears that Weber was attempting to impose a machine-machine environment on a man-machine system. In its attempts to circumscribe and control the most unpredictable of management resources, the human element, the bureaucracy inevitably becomes overly administered, undersupervised, and depersonalized, rendering it threatening and unpleasant to human beings. When an individual is put in this sterile situation, rebellion or apathy often ensues. Victor Thompson has coined the term "bureaupathology"[4] to describe the "bureautic" behaviors which arise when human beings are thrust into an environment which is so totally antiseptic and controlled. The dysfunctions include:

1. Overdependence on rules and regulations—As a result of the manner in which power is obtained in a bureaucracy, the bureaucrat is driven to compulsive rule. The individual's personality and interactional skills are reduced to a minimum in this system which allocates power solely to job positions. As a result the manager's self-confidence is undermined to the point where he or she uses the force of law (title and position power) exclusively to effectuate influence.

When so much overdependence on rules and regulations is structured into the system, the individual is actually punished for thinking creatively, using his or her imagination, or deviating from the established rule, even where it makes no sense, is downright harmful, or causes economic loss.

2. Exaggerated aloofness and insistence on the rights of office—One never has difficulty in distinguishing a colonel from a sergeant in the army or police captain from patrolman in law enforcement agencies. Certain kinds of organization create an exaggerated degree of status difference and social distance between superordinate and subordinate with the superordinate clutching the mantle and trappings of office as a means of insuring compliance with his or her dictates. The key to the executive washroom, chauffeured car, larger office, and "fruit salad" on the general's chest are clear indications of social distance that distinguish one level from the other. Again because this system attempts to depersonalize the role of the superordinate and emphasize the job position, individuals become overly dependent on their positions to gain respect and compliance.

On most army bases there is an enlisted man's club, N.C.O. club, and officer's club; social distance is the norm. This leads to an overly administered, but undersupervised power structure. Superordinates are fearful of getting too close to subordinates. While there is a glut of rules, regulations, and administrative procedures, there is little in the way of coaching, counseling, and supportive behavior on the part of the bureaucrat.

There have recently been attempts to diminish social distance in many modern corporations. Andrew S. Grove, president of Intel Corporation, has developed a "one-on-one" concept to guide managers in their dealings with subordinates.[5] He has also denied himself many executive privileges, such as a reserved parking space, in order to stress equality in the firm. Fewer companies have executive dining rooms today.

3. Suboptimization—The bureaucracy clearly defines departments and

groups and then clusters then into air-tight compartments, each reporting directly to the next higher level. This arrangement often leads to separate power blocs; small baronies that contribute to isolation and lack of coordination. The existence of separate power blocs is sometimes referred to as subunit chauvinism because each department is consumed by its own self-interest and looks to further its own ends rather than working in coordinated fashion to accomplish the organization's goals and objectives.

4. Means become ends—In a variation on the theme of the ends justifying the means, in bureaucracies means become ends unto themselves. Bureaucrats are most often characterized by their concern and attention to the means by which they obtain their objectives rather than by the accomplishment of the objectives.

5. Committeeization and Buckpassing—Bureaucracies are vulnerable to the use of forestalling tactics, particularly in periods of crisis. When an emergency arises the temptation is to form a committee or a commission to study the problem as a way of postponing or perhaps avoiding action.

6. Inbreeding and Resistance to Change—Inbreeding refers to the practice of effectively screening out individuals whose attitudes, values, and behavior patterns are not congruent with those of the established hierarchy within the system. This is most often seen in military and paternalistic organizations.

Traditionally one had to have an Annapolis or West Point education to be considered seriously for an upper echelon position in the military. The highly structured atmosphere in these institutions attracted certain personality types and certain of their traits were further reinforced during training. It is such a combination of prior disposition and subsequent reinforcement that encourages and facilitates inbreeding. The personality tests used in industry in some cases formally weed out personality types considered unsuitable; the well-known difficulties of some major firms and industries, such as automobile manufacture, have been attributed to such practices. In small organizations, including many engineering firms, the tendency of owners to pick subordinates in their own image may have a similar effect. For discussion of resistance to change see Section 6.6.

6.3. THE HUMAN RELATIONS (BEHAVIORAL) SCHOOL

The classical school's contribution to management theory is seen most vividly in its focus on formal structure and procedural arrangement. Particular emphasis was given to the technical aspects of efficiency which were to be achieved through scientific analysis. Employees, although not ignored, were simplistically viewed as being motivated purely by financial considerations.

Workers were seen as "appendages of the machine" and were to be integrated nondifferentially with the other elements in the production functions.

With this "economic-man" perspective it was logical to break jobs into highly specialized, simple, repetitive tasks which allegedly insured a more productive employee. A portion of the increased profits could then be passed onto workers in the form of higher wages, thus eliminating any potential conflict between management and labor.

This practical outcome of the classical approach has often been used by contemporary management theorists in evaluating the classical production-engineering model. It is argued that while its authors made a significant contribution to systematizing the management of organizations, they ignored the human, informal systems that help determine organizational effectiveness.

Perhaps it might more objectively be asserted that many of the early industrialists ignored the caveat that these principles were guidelines rather than rigid rules. Many industrialists did pick up the techniques of scientific management, such as time and motion study, and pursued them with inadequate attention to human factors, partly because these techniques showed a quick return in reducing costs and increasing profits.[6]

One must evaluate a theory or concept within the context and times in which it was developed. The human expectations of the nineteenth century were quite different in respect to job satisfaction than they are today (or for that matter 50 years ago). Many employees viewed work as envisioned in the Bible, that is, as "toil," rather than as a vehicle for satisfying self-esteem or even self-actualization needs.

During the 1920s and 1930s however, a series of economic, legislative, and cultural changes began to occur which significantly affected the environment in which work took place. Factories increasingly became mechanized, leading to even greater degrees of job specialization and interdependence of activities. The westward movement was no longer an effective "safety valve"[7] for workers; it had been acting as a countervailing force when capital pressed against labor or political restraints impeded their freedom.

On the other hand and with many a backward step along the way, the role of government in labor-management relations shifted from probusiness interest to active support of labor's right to unionize. Instead of a laissez-faire attitude, government became increasingly involved in economic matters. Antitrust legislation was aimed not against union organizing but at the huge industrial monopolies for which it had originally been enacted. Social Darwinism began to be questioned as minimum wage laws, women's right to vote, and electing senators by direct popular vote became realities.

6.3.1.   The Hawthorne Studies

A significant early contribution to the human relations or behavioral approach to management resulted from the Hawthorne studies conducted at Western Electric Company's Hawthorne Works in Chicago from 1927 to 1932. Elton Mayo, a Harvard psychologist, began the studies and was joined by Fritz Roethlisberger and William Dickson. The objective of this research was to demonstrate the uniqueness of the human element in the production process.

The first series of experiments took place in the Relay Assembly Test Room, isolated from the influence of the assembly line. Two groups of workers were involved. The experimental group worked under varying degrees of illumination while the control group was furnished with a constant light intensity. Both groups surprisingly exhibited marked increases in productivity. Furthermore, as intensity was decreased, production continued to rise. In fact productivity did not show any significant declines until the experimental group was kept in almost total darkness. This phenomenon became known in the social sciences as the "Hawthorne Effect." That is, when human beings perceive that they are being observed, their behavior is altered. This phenomenon presents behaviorists with an obvious problem when trying to apply rigorous, scientifically controlled experimentation aimed at establishing causal relationships.

In a second series of studies, the Bank Wiring Room Experiment, a researcher was "camouflaged" as a worker to avoid the Hawthorne effect. It was hoped that he could determine the cause of the department's nonresponse to piece-rate pay, a system which assumed that a worker's production was directly dependent upon increased incentive pay.

Based upon the researcher's experience, as a deviant producing more units than fellow workers, it was discovered that individual behavior was strongly influenced by an informally established group norm. The majority of workers, fearful of administrative and technological change, including speedups and cuts in piece rates, attempted to maintain an atmosphere of stability in workplace procedures. Workers who did not conform to the group's standard output and those who either overproduced or underproduced became subject to harsh peer pressure tactics. Included in these tactics were social ostracism and punching offending workers so as to numb their hand and arm movements temporarily. In this way it was concluded that acceptance by fellow workers could prove to be more of a motivating force than monetary incentive.

This "discovery" of an informal structure composed of informal groups, leaders, and communication systems, (grapevines) is perhaps the most significant result of the study (and indeed of the early human relations movement). It was concluded that this social (or informal) system profoundly affects organizational performance and should be identified and engineered into formal organizational planning and design rather than being ignored or eliminated as was the goal of the bureaucratic model.

The Hawthorne Studies resulted in a shift in managerial focus toward understanding individuals and groups with less emphasis placed on rules, procedures, and the structuring of the organization. Improved productivity was to result from sensitivity to people and the creation of a work environment that would engender high job satisfaction. New and aspiring managers had to learn such human relations skills as motivation, communication, and democratic-participative leadership techniques. A happy worker, it was assumed, would be a productive worker.

While these are unexceptionable objectives, the Hawthorne studies themselves have been subject to an increasing volume of criticism over the years. Elton Mayo and his associates conducted some 21,000 interviews, inquiring into all manner of intimate attitudes, hopes, and fears of the employees, under conditions which, most certainly today, but even in earlier years, raised very serious questions of propriety and employee privacy; the issue is now part of a general controversy regarding the use of human subjects in scientific experiments. Some of the participants in the Hawthorne studies came to share these misgivings.[8]

The contrast between the classical and human relations approach to the interpretation of human behavior in organizations is basically one of differences in assumptions. Douglas McGregor has formalized the contrast in his Theory X (classical or traditional) and Theory Y (contemporary) models. Table 6.2 summarizes his formulation of the two theories; McGregor's view was that a systematic application of Theory Y was the proper approach.

Jobs should be designed with appropriate consideration for human needs and allow for a sense of purpose as well as a means of overcoming boredom. Each job has two dimensions: range and depth. Range concerns itself with the number of different activities necessary to perform a particular job. Depth refers to the amount of autonomy or control which the incumbent has over a particular job. Modern management theorists argue that wherever possible these two dimensions should be expanded to meet the human needs of the individual performing them. A job too narrowly defined provides little opportunity for higher-order, human-need satisfaction; that is, self-actualization. Ideally, any job position in the organization should allow the incumbent some degree of influence in planning, organizing, and controlling his work. Today's labor market, with its high job turnover, places insufficient stress on quality and productivity. A focus on range and depth may

TABLE 6.2.   Two Sets of Assumptions about People

Contemporary Theory (Y)	Traditional Theory (X)
People are naturally active; they set goals and enjoy striving.	People are naturally lazy; they prefer to do nothing.
People seek many satisfactions in work: pride in achievement; enjoyment of process; sense of contribution; pleasure in association; stimulation of new challenges, and so on.	People work mostly for money and status rewards.
The main force keeping people productive in their work is desire to achieve their personal and social goals	The main force keeping people productive in their work is fear of being demoted or fired.
People normally mature beyond childhood; they aspire to independence, self-fulfillment, responsibility.	People remain children grown larger; they are naturally dependent on leaders.
People close to the situation see and feel what is needed and are capable of self-direction.	People expect and depend on direction from above; they do not want to think for themselves.
People who understand and care about what they are doing can devise and improve their own methods of doing work.	People need to be told, shown and trained in proper methods of work.
People need a sense that they are respected as capable of assuming responsibility and self-correction.	People need supervisors who will watch them closely enough to be able to praise good work and reprimand errors.
People seek to give meaning to their lives by identifying with nations, communities, churches, unions, companies, causes.	People have little concern beyond their immediate material interests.
People need ever-increasing understanding; they need to grasp the meaning of the activities in which they are engaged; they have cognitive hunger as extensive as the universe.	People need specific instructions on what to do and how to do it; larger policy issues are none of their business.
People crave genuine respect from their fellow man.	People appreciate being treated with courtesy.
People are naturally integrated; when work and play are too sharply separated both deteriorate; "The only reason a wise man can give for preferring leisure to work is the better quality of the work he can do during leisure."	People are naturally compartmentalized; work demands are entirely different from leisure activities.
People naturally tire of monotonous routine and enjoy new experiences; in some degrees everyone is creative.	People naturally resist change; they prefer to stay in the old ruts.

141

TABLE 6.2.   *(Continued)*

Contemporary Theory (Y)	Traditional Theory (X)
People are primary and seek self-realization; jobs must be designed, modified and fitted to people.	Jobs are primary and must be done; people are selected, trained, and fitted to pre-defined jobs.
People constantly grow; it is never too late to learn; they enjoy learning and increasing their understanding and capability.	People are formed by heredity, childhood and youth; as adults they remain static; old dogs don't learn new tricks
People need to be released and encouraged and assisted.	People need to be "inspired" (pep talk) or pushed or driven.

Influence of the Organizational Setting

If people are treated as inferior, lazy, materialistic, dependent, irresponsible, they tend to become so. People in authority-obedient systems become X-minded.

If treated as responsible, independent, understanding, goal-achieving, growing, creative people, they tend to become so. People in this kind of organizational setting become Y-minded.

Source: Douglas McGregor, *The Human Side of Enterprise* (New York: McGraw-Hill, 1961).

provide for an environment conducive to reaching both organizational and individual goals.

An engineering job might thus be defined by the following range elements:

1. Responsibility for the preliminary design and proposal of the product
2. Coordinating and planning of project tasks
3. Overseeing the project design
4. Establishing test requirements
5. Writing test specifications
6. Conducting the actual tests
7. Overseeing the product manufacturing
8. Customer relations

Within each of these there is a set of tasks and relationships which in turn define depth. A specific job essentially consists of various depths in each of the above elements. Depending on its level, a job may be limited or involve highly intensive or extensive control.

Wherever possible the job environment is modified to be more congruent with the employee's lifestyle, which increases the potential for job satisfaction, commitment, and loyalty with, one hopes, no additional costs to the firm. The range of each job should be increased so that a sense of purpose is inherent in every position. Job rotation is a technique where individuals perform a family of job positions in order to increase variation in work. Job enlargement expands the number of duties assigned to each individual. Application of these techniques helps to minimize boredom and fatigue by giving a greater sense of involvement and perspective to employees so that they more clearly see how their job fits into the larger scheme of the organization. Job enrichment, associated with the depth dimension, endeavors to provide the worker with more autonomy in respect to planning, organizing, and controlling the job.

Quality Circles, which have been implemented in Japanese firms, have recently been applied in several American companies with considerable success. Quality Circles involve the setting aside of specified blocks of time for employees to meet and discuss micro aspects of their work. Suggested procedures analyzed by workers which increase productivity, product quality, or reduce costs are adopted by the company and commensurate reimbursement is given. This is in contrast to the traditional separation of planning from doing, which is most noticeable in the time-honored method of utilizing an efficiency expert to achieve the aforementioned results. (Everyone is management in healthy organizations.)

Much of the early human relations focus made its main impact in symbolic, highly visible displays, particularly with respect to the physical proxemics of the work environment. Muzak, pastel-colored walls, the company bowling team, and the annual picnic when upper-echelon executives brush elbows with the workers, replaced the austere autocratic "sweatshop" environment. All too often, however, the contents of the suggestion box in the cafeteria were remanded to the circular file. When employees sense that they are being manipulated or given only the illusion that they are meaningful participants whose contributions are valued, they react more negatively than if they are not asked to participate at all.

Despite the charge that much of the early human relations concepts were at best simplistic (a happy worker is not always a productive worker) and at worst manipulative and illusory, the behaviorist approach did draw attention to the human potential for imagination and creativity when people are intelligently managed. The realization that human activity could be overspecialized to the point of mental fatigue or that in an attempt to design a smooth-running, predictable piece of scientific machinery, the organization could be rendered arthritic through overuse and misapplication of rules, regulations, and procedures, was a significant accomplishment in the progressive development of management theory.

6.4. SYSTEMS THEORY AND ITS APPLICATIONS TO MANAGEMENT AND ORGANIZATION

Today systems theory affects much of our thinking as we refer to systems analysts, management, and marketing information systems and view organizations systemically, that is, as living, dynamic, information-processing social systems. In actuality, systems theory is not a theory at all, a theory being a purported cause and effect explanation of a phenomenon, but rather a metatheory, framework, vehicle, or tool for analyzing phenomena. The biologist, Ludwig von Bertalanffy, is generally credited with developing this approach to studying organisms and then making the applications to social structures.[9]

The relevance of systems theory in studying management and organizations is not that this framework will solve a problem on an executive's desk, but rather that it provides an appropriate focus to the analyst's attention in understanding, and ideally improving upon the phenomenon in question. (Many academicians argue that encouraging students to use a systemic approach to problem solving may be the best way to achieve that most ambitious goal often expected of educators—to teach students how to think.)

The basic theme of the systems approach is that every system is made up of subsystems and is itself a subsystem of a greater system. Such a focus draws the analyst's attention to the boundaries and action points in a problem area. For example, General Motors may be viewed in terms of its largest subsystems, its motor divisions (Pontiac, Chevrolet, Cadillac, Oldsmobile, and Buick). These divisions may be further segmented into marketing, finance, engineering, and production departments and ultimately to that smallest subsystem—the individual job position.

Building outward, as large as General Motors is, it must be viewed in the context of the automobile industry; this in turn is a part of the transportation industry, and furthermore the economic, legislative, cultural, and technological subsystems, among others, which create the sociotechnological system of the United States. The relevance of this seemingly cumbersome framework is quickly appreciated when we move from a relatively static environment to one with increasing degrees of change.

When the environment is stable and certain there is less necessity for constant monitoring. A stable environment may be comprehensively studied, resulting in a set of systems, procedures, and rules which when set in motion are capable of processing vast amounts of homogeneous (interrelated) activity. (The United States Postal Service exemplifies this.) The assumptions underlying this scenario however, are becoming less realistic in today's world of ever increasing change, more interdependent among environmental subsystems and larger organizational structures. A systems focus stresses the need for developing appropriate linkages between the parts and further concentrates on how that system interacts with the environment in which it functions. When each department understands its relationship to the whole and effectively communicates with the other departments, the phenomenon of "synergy" may be appreciated—the whole is greater than the sum of its parts. Too often the parts of a business system do not adequately communicate with each other and may even intentionally withhold or provide misinformation because of the highly competitive attitudes present in many companies.

When emphasis is placed on the larger environment of which the organization is a part, the manager's attention is drawn toward the changes taking place in other relevant subsystems (technological, cultural, and so on). Rather than being passively exposed to that environment, the organization is better positioned to predict change and ideally to initiate change—to view change as an opportunity rather than a threat.

In summary, a healthy organization, from the systems perspective, is one whose parts have proper communication linkages, a clear sense of purpose or mission, and are appropriately focused on the relevant environmental changes which affect the organization. We saw this stress reflected in the current attention given to defining an organization's culture, formalizing, and communicating it to all employees in Chapter 5. Furthermore, a well-functioning system must be able to respond constructively to change.

6.5. ORGANIZATIONAL RESISTANCE TO CHANGE

In studying resistance to change the question has been posed: Do human beings resist change? Actually every individual *requires* a degree of change. Total absence of change leads to monotony, boredom, and stagnation in human beings or organizations. Indeed individuals do seek out change and heterogeneity in their environment for survival and growth.

A system which views change as the natural order of things and is able to predict and ideally initiate change, rather that being a passive recipient of its environment, is a healthy and productive system. Twentieth century technology has been the driving force in producing a greater degree of change than ever before. If individuals do not resist change per se, but rather seek out change, why is the phenomenon of resistance to change so prevalent in organizations?

Each individual has a unique threshold for the amount of change which he or she finds palatable. This threshold is a function of one's values, attitudes, needs, and expectations—personality. A particular degree of change in one person's environment may bring about uneasiness, boredom, or apathy in another person. A second major force affecting resistance to change depends on how it is perceived and implemented. A number of factors have been identified which exacerbate the problem of resistance to change:

1. **Economic factors.** In the work environment, threats to economic security are a primary cause of resistance to change. The introduction of new machinery and equipment which is perceived as a possible threat to economic security engenders fears which, however antiprogressive they may be considered by outsiders, frequently appears fully justified to those affected.

2. **Inconvenience.** As mundane as it seems, employees will often reject changes simply because of the inconvenience involved in learning new procedures or techniques. This causes them to give up habitual behaviors with which they have become comfortable.

3. **Uncertainty.** Lack of understanding often breeds fear—which in turn engenders resistance to the unknown. We see this in new product introductions as evidenced by the marketing of the personal computer. A major factor inhibiting the diffusion of this innovation is the fear related to its inherent complexities.

4. **Symbolic change.** Individuals are particularly sensitive to symbolic change. Altering a symbol quite often is more intimidating than the actual change itself.

5. **Violation of expectations.** Resistance to an unanticipated change often occurs when an individual's psychologic set or expectation is disturbed.

Methods of reducing resistance to change include:

1. **Economic incentive.** When introducing a change assure the employees that it will have no negative effects on economic security and that the change will be beneficial to them. For example, introducing new machinery will make them more productive as employees and thus allow a bonus or wage increase without impairing job security.

2. **Two-way communication.** Thoroughly briefing the recipients of a change before its implementation is an excellent tactic for overcoming fears of the unknown. Discussion of the change and the answering of all questions are preferable to a one-way written memo or communique.

3. **Participation.** One of the more effective ways of reducing resistance to change is to involve those individuals who will be affected by the change in the decision process. It is important for individuals to understand the necessity for change so that it is not perceived as simply

a nuisance. Individuals who have played an active role in instituting a change inevitably find it very difficult to resist that which they have been a party in developing.

4. **Holding the symbol constant.** Wherever possible symbolic change should be kept to a minimum so that the actual changes will not be perceived as inordinate. For example, keeping the job title the same and making minor revisions in the job function is less threatening than changing the job title.

5. **Making change tentative.** When we clearly explain the need for change and have arrived at a possible solution to a problem, assuring employees that the change will be instituted on a tentative basis until it is proven effective acts to reduce resistance to trial.

6. **Building on the past.** Showing the employee that most new changes are based upon past procedures tends to reduce cognitive dissonance. For example, when introducing office personnel to new word processing equipment it is important to make them aware of the similarities between electronic keyboards and typewriters. In addition we might emphasize the extra editing features that will ease the workload.

7. **Ceremony.** The introduction of change may be more easily presented in a positive light through the use of pomp and circumstance. Ribbon-cutting ceremonies and festivities are common methods of setting the mood for adaptation.

8. **Slow versus immediate introduction of change.** Much debate has centered around whether a change should be introduced immediately or slowly in stages. If it is assured that the change will have a positive effect on the subject, implementing it in stages is usually the best tactic. Conversely, if negative effects are expected, a quick implementation may preempt the mobilization of organized resistance.

9. **Truth.** The possible strategies previously mentioned are potentially manipulative. It should, but unfortunately does not, go without saying that those involved, at whatever level of the organization, should be told the truth and, one should add, the originators of change should beware of self-deception as well. Trust in an organization is very difficult to recreate, once it has been significantly impaired.

6.6.   SYSTEMS, PROCEDURES AND ORGANIZATIONAL DESIGN

Having developed purpose and direction for the organization, management must determine a suitable structure that will reinforce or support implementation of the planning stage. Managers often do not have the luxury of developing a new structure but must live within the constraints of or modify an existing organizational design (which manifestly puts limitations on the types of environments to which the firm may effectively respond). However, when the opportunity to develop an organizational design exists, a structure should be conceived which is sufficiently flexible to accommodate the imagination and creativity of the human element which must function within it. This concept must be viewed as a basic theme in the development of systems and procedures for any organizational unit.

Ideally, strategy dictates structure and form follows function; but as indicated above, an existing structure may preordain the manner in which decisions are made, goals set, controls established, and supervisory or leadership styles elected. Formal organizational design is an extremely powerful, highly visible device and deserves considerable attention, given its affect on all other aspects of the management process.

As the organization grows, systems and procedures must be designed to assure adequate levels of coordination and predictability. The critical issue becomes determining the extent to which formal structures should be imposed on the various groups in that system. This is of particular interest in engineering where small, informal project teams are a typical and frequent form of organization. This presents problems, especially when a formerly small, informal firm is swallowed up in a larger and much more formally run organization. However, even when a small firm is successful by itself, it grows larger. The elusive goal now becomes how to benefit from size without losing the advantages of small, entrepreneurial, close knit systems. How do we make the transition from entrepreneurial to managerial systems without losing a healthy entrepreneurial spirit? Part of the answer lies in excessive care in the development of systems and procedures.

The firm must exercise tight yet loose control—a seemingly paradoxical statement. Peters and Waterman[10] refer to top management's establishment of a clear corporate culture and mission which still provides sufficient autonomy to the subsystems to implement their view of that mission. General Electric has clustered activities around what it referred to as strategic resource units in an attempt at achieving sufficient flexibility without sacrificing overall common purpose and direction. In its most recent successful product introductions, IBM, otherwise a highly structured and centrally controlled company, followed a decentralized procedure that tried to create a relatively autonomous project team structure.

Finally, military work is subject to a highly centralized control structure which is established by law and government regulation. Managers, in short, are not free agents in designing their organizations, nor indeed in many other managerial precincts. The problems of secrecy, detailed controls, and red tape are burdensome enough. Yet these safeguards have not prevented the well-known cost overruns and product failures in that branch of the technical industries.[11]

6.7. THE CONTINGENCY MODEL APPLIED

A convenient way of summarizing the three models of management and reasonably applying them would be to look at organizations mechanistically (classically) and organically (behaviorally) and then determine where each model's assumptions are found to operate most effectively, that is, specify the systems-contingency.

As shown in Table 6.3, the mechanistic organization tends to view the management process from a perspective built on the classical or scientific management model. In this model there is a close adherence to formal organizational design with particular stress on the chain of command and formal written top-down communication through that chain. A functional division of work, through which the tasks and problems facing the concern as a whole are broken down into specialized activities, is the norm. Jobs are narrowly defined in terms of range and depth with a detailed description providing precise definition of authority and methods of performing tasks. Interaction between employees is primarily of a vertical nature, that is, superordinate–subordinate rather than horizontal. Span of control is narrow,

TABLE 6.3.   Contingency Approach to Organizing

Characteristics	Type of Organization	
	Mechanistic	Organic
Type of environment	Stable, predictable	Innovative, uncertain
Comparable to	Classical organization	Behavior organization
Adherence to chain of command	Close	Flexible—chain of command often bypassed
Type of departmentalization	Functional	Divisional
How specialized are jobs	Specialized	Unspecialized—jobs change daily, with situation
Degree of decentralization	Decision making centralized	Decision making decentralized
Span of control	Narrow	Wide
Type of coordination	Hierarchy and rules	Committees, liaisons, and special integrators

leadership style autocratic, and decision making highly centralized with a clear separation of planning and doing.

In the biologic-organic organization the perspective grows out of the assumption of the human relation or behavioral model, and as a result, the managerial process is viewed very differently. The formal organization design and chain of command are not really adhered to; for example, communication is more personal, face-to-face, and often occurs outside formal channels. The content of communication is more informational and advisory rather than instructional and directive with more lateral or horizontal communication occurring than purely vertical.

Functional departmentation is softened with less use of high specialization; job design escalates in terms of range and depth. Jobs are not clearly defined in advance, but rather are adjusted and redefined as the situation demands. Employees do not respond to requests for help saying "it's not my job . . . go to window seven." Decision making is more decentralized and leadership style tends to be more democratic-participative. Self-control rather than externally imposed mechanisms are the norm; this results from the belief that organization tasks and goals motivate people.

The characteristics of the environment determine which model of management is appropriate. When the environment is stable, predictable, and unchanging with respect to such factors as technology, government policies, and competition, the mechanistic model's assumptions may result in quite efficient operations. However as the environment becomes increasingly uncertain, the bureaucracy with its elaborate systems and procedures may effectively inhibit the organization's adaptive capability.[12]

An analogy might be made between the muscular weight lifter who has built a massive physique for the purpose of lifting a vast amount of dead weight and the flexible, fluid body of the tennis player whose environment requires quick responses to a more changing and unpredictable environment. The systems model attempts to apply such biological analogies to organizational design.

NOTES

1. H. Fayol, *General and Industrial Administration* (London: Pitman, 1949), is the most comprehensive statement of Fayol's work in English.
2. V.A. Graicunas, "Relationship in Organization" in L. Gulick and L. Urwick (eds.), *Papers in the Science of Organization* (New York: Institute of Public Administration, 1937), pp. 181–187.
3. M. Weber, *Economy and Society: An Outline of Interpretive Sociology* (New York: Bedminster Press, 1968).
4. V. A. Thompson, *Modern Organization* (New York: Knopf, 1961), p. 23.

5. A. S. Grove, *High Output Management* (New York: Random House, 1983).
6. H. Koontz, C. O'Donnell, and H. Weihrich, *Management,* 8th ed. (New York: McGraw-Hill, 1984).
7. F. J. Turner, *The Frontier in America Today* (New York: Holt, Rinehart & Winston, 1920), p. 259.
8. For the full account of the Hawthorne experiments, see E. Mayo, *The Human Problems of an Industrial Civilization* (New York: Macmillan, 1933) and F.J. Roethlisberger and W.J. Dickson, *Management and the Worker* (Cambridge: Harvard University Press, 1939). For the reservations by some of the participants, see J. L. Wilensky and H. L. Wilensky, "Personnel Counseling: The Hawthorne Cases," *American Journal of Sociology* 57 (November 1971): 265–280.
9. L. von Bertalanffy, *General Systems Theory* (New York: Braziller, 1969).
10. T.J. Peters and R.H. Waterman, Jr., *In Search of Excellence* (New York: Harper & Row, 1983), p. 318.
11. J.E. Ullmann, "The Pentagon and the Firm," in J. Tirman, ed., *The Militarization of High Technology* (Cambridge, MA: Ballinger, 1984) Ch. 6.
12. T. Burns and G.M. Stalker, *The Management of Innovation* (London: Tavistock, 1961).

7

MARKETING FOR ENGINEERING FIRMS

Gerre L. Jones

Actively engaged in consulting, writing, and professional education for more than 25 years, GERRE L. JONES is president of Gerre Jones Associates, Inc., Washington, DC, and editor-publisher of the newsletter *Professional Marketing Report.* An author of six books about marketing professional design services, Mr. Jones also has designed, produced, and led several hundred marketing workshops for engineers and architects in recent years. Trained in engineering by the Army in World War II at Virginia Military Institute, he went on to postwar undergraduate and graduate studies in journalism and law at the University of Missouri. He is a frequent lecturer on marketing subjects in many U.S. colleges and universities.

7.1. INTRODUCTION

For professional services, as in most businesses, the most important role of management is customer and client development. Somewhat similar to an experienced realtor's advice about land investments ("Location! Location! Location!"), is the veteran practitioner's rule for success in engineering: "Get the job! Get the job! Get the Job!"

Without client commissions to create cash flow, occupy staff, and create profits for owners, all else in the practice of engineering is vanity. Whether your firm has a staff of 5 or 5000; whether you have been in business 50 weeks or 50 years; and whether you are headquartered in Gnawbone, Indiana, or Los Angeles—someone, or several people, must regularly sell your professional services to prospective and past clients.

Marketing in engineering firms today is important to the growth of firms and maintenance of the desired practice mix, that is, the proportion of different kinds of work. An active, continuing, and professional marketing program enables firms to select their markets, instead of the other way around.

Earlier in this century most design firms practiced various forms of "reactive selling"—principals were order takers from clients who came to the design office. A few firms still practice this type of passive selling—and have little or no control over their professional and business destinies.

7.1.1. No Secrets in Marketing

There are no secret avenues to success in marketing and selling; marketing is at best a soft science. Some of those who have been in marketing for a while question whether it is *any* kind of science. These are some of the not-so-secret weapons of successful marketers of engineering services:

An encyclopedic knowledge of the firm—its principals, staff capabilities, and project experience

A genuine interest in dealing with all types of people (clients are people)

A good general knowledge of the basics of behavioral psychology and an understanding of what motivates prospects to select one design firm over another equally qualified firm

A capability for sustained hard work—long hours automatically go with marketing

Creativity, enthusiasm, flexibility, and above all, the ability to think and react rapidly, positively, and professionally, often under considerable stress (as in formal presentations)

In general, successful marketing requires organization; planning; a knowledgeable, disciplined, and enthusiastic selling approach, and a lot of hard work.

Unfortunately, the formal education of most engineers does not include practical courses in office management, marketing, bookkeeping, financial management, and the like. Whether or not such courses *should* be part of an engineer's formal education is a subject of growing debate, but as of now, most engineers' knowledge of marketing and sales techniques must be gained primarily from observation, discussions with others, specialized workshops, and self-study such as this book.

Here are two more important points to keep in mind:

1. No marketing program can be expected to go anywhere without a serious and meaningful commitment of money and time from the firm's owners and top management.

2. No marketing program will ever realize its full potential without a detailed, written marketing plan. The marketing plan should be continually reviewed and revised, with a major update made at least annually. (You can easily relate the importance of a written marketing plan to the need for detailed plans and specifications for a design project.) Without a written plan your firm has no way of knowing (or documenting) where it has been, is now, or should be headed in marketing.

7.2. THE MARKETING PLAN

Practical marketing plans do not depend on massive documents—or endless documentation. Some plans, particularly those for large, multidiscipline firms, *do* run to several hundred pages, but most marketing plans can be adequately covered in 10 to 25 pages.

A veteran marketer suggests a three-step program to develop a marketing plan:

1. Determine what kind of work you want to do.
2. Decide who your potential clients are
3. Work out tactics and strategies for selling services determined in Step 1 to prospects identified in Step 2.

The answer to the first point depends largely on present capabilities and experience of your staff. The third step can be a fairly complex and technical one, often requiring specialized marketing knowledge.

The success of the second step—identifying potential clients (defining markets)—rests on the thoroughness, accuracy, and depth of your marketing research.

7.2.1. A Typical Plan

A typical marketing-plan outline might look like this:

I. Introduction
II. Executive summary
III. Goals and objectives
    A. External and internal assessments
    B. Target markets
IV. The formal plan—this is the action plan; essentially who does what to whom and when. Specific tasking of staff members who will be involved in marketing is an important part of this section
V. Promotion—basically this section covers public relations and advertising in support of the overall plan
VI. Budget
VII. Monitoring and evaluation procedures
VIII. Conclusions

7.3. MARKETING BUDGETS

How much should you budget for marketing? There are no pat answers to this question. If you normally market to other professionals—if your work is essentially interprofessional in nature—your outlay for such items as four-color brochures and public-relations activities may be smaller than that of firms appealing to a broad range of public and private clients. Or if your discipline offerings are in relatively rare fields and you have little competition, your marketing costs will probably be less than those for a firm having many disciplines under one roof.

Budgets for marketing are usually based on some percentage of past or expected gross fee income. This method is fairly simple, has a ring of logic, and enables you to make certain comparisons with marketing expenditures of firms similar to yours.

The task method of budget setting, in which careful considerations are given to the probable costs of entering new markets, holding or increasing traditional client types, maintaining the desired annual growth rate, and meeting other goals and objectives as set out in the marketing plan, is the only logical approach to budgeting for the marketing operation. As a practical matter the marketing budget will have at least some relation to total projected annual billings.

7.3.1.   What Should You Spend on Marketing?

Various surveys taken among engineering and architectural firms over the years consistently yield an average budget figure of about 6.5% of a firm's annual gross billings. The range is generally 4–11% of gross billings. This is reportage and not a recommendation for any one firm. Your firm may be justified in adopting a marketing budget of 2% or 15%.

Don't do as one principal did in looking for a quick, easy answer: He surveyed other engineering firms in his market area and found their marketing budgets averaged 5.5%. He then set his at 6.5%, figuring the extra 1% would give him an automatic edge on all the competition. Such simplistic approaches often give misleading answers.

A new firm, in its first year to 18 months of operation, should plan on spending up to 18% of its projected billings for start-up marketing costs. From an accounting standpoint many of these costs can be amortized over several years, but such items as brochures, project photographs, slides, and audiovisual projection equipment must be paid for on delivery. After this initial heavy investment period, marketing costs should fall more in the ranges set out above.

7.4.   MARKET RESEARCH

Market research may be viewed as a tool to aid in and supplement executive judgment. We collect and analyze pertinent information so as to reduce uncertainty about the consequences of our planning decisions.

Some of the potentially most helpful sources of market research are:

Departments and agencies of the federal, state, and local governments

Banks, including the Federal Reserve system and its ten member banks

Trade, professional, and technical associations

Trade journals for the industries you serve

Commercial research companies (Frost & Sullivan, Predicasts, and the like)

Libraries (public and private)

Newspapers (*Wall Street Journal, Barron's, New York Times, Washington Post, Los Angeles Times,* plus local newspapers)

Magazines—especially the more business-oriented ones such as *Forbes, Fortune,* and *Business Week*

To be effective, market information collection should be a continuing process. Once you achieve a marketing mind-set, that is, think of information as a constant source of potential opportunity, you will find good market research data literally everywhere.

7.5. LEAD FINDING

According to legend, engineers used to wait patiently in their offices for clients to walk in and shower design commissions on them. If that ever really was once the case, it is no longer so. Such a passive outlook on marketing a firm's services is a certain road to low (or no) profits, an uncontrolled (usually uninspiring) practice mix, and even bankruptcy.

Prospect leads come from either live or printed sources—or some combination of both. Among traditional live lead sources are:

Lost, past, and present clients

Your own office staff

Friends and relatives of office staff and principals

Bankers, lawyers, accountants, and other professionals

Developers

Contractors

Other design consultants

Suppliers

Newspaper business and real estate editors and reporters

Chambers of Commerce

State and local industrial development commissions

Printed lead sources include:

Newspapers and magazines

Clipping services

Commercial lead tip sources

Corporate annual reports

Press releases from governmental and quasigovernmental agencies and departments

Justification data for construction authorization submitted annually to Congressional appropriation committees by all federal agencies

Most of these lead sources should be self-explanatory, but a couple of them may be somewhat obscure. "Lost clients" refers to those prospect pursuits in which a firm makes the short list and is interviewed by client representatives—but not selected for the job. "Commercial lead-tip sources" refers to the growing number of organizations that make it their business to find and qualify prospects for engineering firms (usually by telephone) and to sell the information gathered to a client or subscriber list. Some lead-search services issue the information in newsletter form, while others use special, periodic reports.

7.6. HIT RATES IN MARKETING

What sort of hit or kill rate should an engineering firm expect from its sales efforts? How many raw leads or suspects are required to net one job?

For most firms the answers to these and related questions are meaningful only when considered as broad guidelines. If yours is a highly specialized consulting service, with only two real competitors in the entire country, for example, the odds are you'll win 33% of the time—even if you don't try very hard.

7.6.1. The 50:5:1 Ratio

Since most of us can identify many more than two competing firms, the 50:5:1 ratio may help in planning. The three ratio elements mean:

> 50 marketing initiatives =
> 5 semifinalist positions =
> 1 job won by your firm.

A "marketing initiative" is any kind of marketing or sales contact in pursuit of a project. Such initiatives include telephone cold calls to find or qualify leads, personal visits with past clients, letters to prospects previously made aware of your interest in working for them, and attendance at a trade show or professional meeting to make contacts on behalf of your firm.

Fifty such initiatives should result in five invitations to present your qualifications in detail. The five semifinal spots may be in the form of requests for proposals or formal interviews, or both, depending on the way the selection process is organized.

From the five presentations you should net one job; a 20% win record at that stage is a relatively conservative goal for most firms. Given a stable or growing economy and good presentation skills, most firms should beat the 50:5:1 hit ratio. But to know how you compare with others in this respect, certain internal records are an obvious necessity. If your hit rate falls below 2% (an average of 2 projects from every 100 marketing initiatives) your marketing program needs attention before your work load and cash flow become twin disasters.

7.7.   MARKET COMMUNICATIONS

Since most people tend to be persuaded by the printed word, the majority of marketing tools used by design professionals are in printed form. The primary printed marketing tools for selling engineering services are:

General capability brochures
Qualifications brochures
Proposals
Newsletters
Reprints of articles and speeches by members of the firm—or about the firm
News releases to the media
Standard Forms 254 and 255 (for government projects)

General capability brochures take many forms, ranging from individual project or fact sheets in a simple folder to hard-cover books in full color. Ideally brochures should be completely customized for every prospect. As a practical matter we aim for as much flexibility as possible within our marketing budget—allowing for some degree of customization in most cases.

7.7.1.   Brochure Systems

Some firms develop brochure systems that are a graphic and content-related series of around 2 to 5 small brochures (4 to 12 pages each). In the system approach one brochure is typically devoted to each in-house discipline. As the system develops additional brochures are produced for primary project types such as airports, wastewater treatment plants, highway and bridge projects, and the like.
The goal of any brochure system is to be able to use off-the-shelf components to give as customized an appearance as possible to the final product. Brochures do not in themselves win jobs. They are tangible evidence of a firm's existence and may be considered a success if they motivate a prospect to open a file on your firm.

7.7.2.   Qualification Brochures

Qualification brochures, as the name implies, are to present certain experience and abilities—usually for a specific project or to another design consultant. They are much more focused in content than are general capability brochures. Letters of qualification, with or without brochure backup, are often used for the same purpose.

7.7.3. Proposals

Proposals are normally solicited from design professionals through formal requests for a proposal (RFPs) by a prospective client. Most of a proposal's content and organization are governed by what is asked for in the RFP. Selectivity and attention to detail are particularly important in proposal writing.

This list of 22 possible proposals elements may be of help:

Cover

Letter of transmittal

Title page (usually optional)

Table of contents (also optional except in very lengthy proposals)

List of illustrations and tables (optional)

Executive summary

Terms of reference, scope of services, and statement of work to be accomplished

Proposed execution (methodology)

Schedules

Exceptions, if any, to the RFP

Company history

Facilities

Related past experience

Project management

Staff résumés for the nominated project team

Associates, joint venturers, subcontractors, and consultants to be used

Visual aids—charts, graphs, maps, photos, diagrams—within reason, the more, the better the proposal. One proposal consultant recommends a ratio of text to visual pages of 2:1

Cost

Terms and conditions

Explanation of the cost estimating techniques used

Certifications and representations

Draft contracts—when requested in the RFP

Summaries should be used freely throughout the proposal. This list of possible elements is not necessarily in order; certainly not all of the elements shown are required in every proposal.

7.7.4. The Most Important Selling Elements

Because of their visibility and chance of being read by decision makers (rather than only by proposal evaluators), the cover, letter of transmittal, and executive summary are the three most important selling elements of most proposals.

This is not, of course, to say that the remainder of the proposal is unimportant. Technical content, visuals, and writing styles should all be of the highest possible quality throughout, but once a proposal has passed the scrutiny of the evaluators, few others will take time to read the entire document.

Always try to customize the proposal cover. The prospect's name and project title should get prominent display; in type at least as large as that used for the name of your firm or joint venture. Some sort of graphic recognition of the prospect or the project, or both, is desirable—a logo, a seal (for government prospects), a map, or a pertinent photograph or drawing will help add to the cover's general graphic interest.

Pay attention to the cover's graphics and layout. An eye-pleasing design, readable type, and a good grade of paper stock signal your overall attention to detail—a signal not lost on most prospects.

In spite of the fact that transmittal letters are the opening round in a written sales salvo, many still sound as if they were written by a nineteenth century lawyer; full of "whereases," "herewiths," and "yours of the 19ths." It seems unlikely that so many transmittal letters could be this dull, uninteresting, uninformative, and unserving by accident.

What belongs in a letter to transmit a proposal? Other than a few standard reference items such as the name and number of the RFP; conditions, if any, contained in your response; and the type of contract proposed, the purpose of a transmittal letter is to set out as much information, in the clearest possible writing style, about your firm's and the designated team's direct and related project experience with the subject of the RFP as you can squeeze into two pages. Nothing else belongs in the letter.

Most rules have exceptions of course. Occasionally, for good and valid reasons, a proposal transmittal letter *should* exceed two pages, but such exceptions are rare.

Executive summaries in a proposal are sometimes headed *introduction, foreword,* or *summary.* A true executive summary is an abstract or recap of the total proposal that follows; not an introduction or a foreword. Since most secondary readers of proposals, the majority of whom are decision makers, no doubt think of themselves as executives, it seems only logical to title this section in a manner calculated to draw attention to it.

The executive summary, which is one of the last proposal sections written, should be used to highlight, emphasize, and reinforce the most significant selling points about your firm, nominated project team, and your direct and related project experience. Use it also to summarize and restate the three, four, or five primary reasons your firm should be retained to handle the project.

Since the executive summary has the potential to be one of the most important sales elements of the proposal, it deserves extra effort to make it brief, pertinent, interesting, and graphically attractive to potential readers. Generous use of a variety of graphic elements—bullets, underlines, subheads, white space, indented sections, and the like—will help pull readers into the summary text. Some proposal managers use different texture, color of paper, or both to call attention to the executive summary.

7.8. FORMAL PRESENTATIONS

Most clients begin their consultant selection process with a list of from 4 to as many as 30 (or more) design firms. Over a period of weeks to months the initial list is cut in stages to the short list. The short list usually consists of between three and five firms.

In most cases these top-ranked firms are then invited to make formal presentations to the client selection committee. Obviously, all the early marketing and sales efforts are geared to the important subgoal of being asked to make a presentation. The primary goal, naturally, is to be selected for the job.

No firm should ever go to an interview with a client group without holding at least one dry run of what they intend to discuss and show at the interview. The more experienced the marketing staff, the more thorough and realistic the rehearsals will be. As the cost of closed-circuit television (CCTV) systems has gone down, their regular use in practice sessions has grown in popularity among design firms.

Remember that the formal presentation is usually the last opportunity you will have to do a meaningful selling job. Nothing belongs in the presentation that is not directly related to interests of the prospective client and to the project itself.

Kurt Meyer has some good advice for presenters, based on his experience in interviewing many design firms as a member of the Los Angeles Community Redevelopment Agency:

Once you are in a situation where you will be interviewed for a project, you are one of the finalists! You can forget about the platitudes of the teamwork and the efficiency of the organization, the number of awards you have won, and the showing of all too many beautiful slides of all the projects you have designed. Any time you are in the finals we are already convinced you can do all those things. Now we want to get to know you as a person.[1]

7.9. THE "SOMETHING ELSE" THEORY OF MARKETING SUCCESS

In *The Successful Professional Practice,* management consultant Robert Levoy points out that professional success is not based solely on technical competence:

> Skill, technique and method are as important as ever to successful practice. They are in fact its basic dimension. But, contrary to what we learned in professional school, competence per se is not the only dimension for a successful practice. The practitioner who attains twice your practice growth, twice your success, does not have twice your technical competence. Nor does he have twice your intelligence, or twice your convictions, or twice your dedication. Call it an extra-technical dimension, if you will, but he does have "something else."[2]

At least part of what Levoy refers to as "something else" and an "extra-technical dimension" is a basic factor of progressive, professional marketing. It is, among other things, a people-oriented outlook, coupled with a good understanding of basic behavioral psychology.

A pretty good short guide to successful marketing can be found in Professor Harry Overstreet's *Influencing Human Behavior:* "First, arouse in the other person an eager want. He who can do this has the whole world with him. He who cannot walks a lonely way."[3]

7.10. INTERPROFESSIONAL SELLING

Interprofessional relationships in the design profession usually cover engineers' contacts with architects and with other engineers, that is, those you get work from or, as prime contractor, provide with work.

Until a few years ago many prime contractors were willing to give their consultants an essentially free ride in marketing. Most interprofessional firms spent little, if any, on marketing their services. As marketing became more intense, professional, and costly, many design firms began to expect their consultants to share in the costs of getting work.

Today, before beginning serious pursuit of a project with one or more consultants, the prime firm clearly spells out a reasonable split of the antici-pated sales costs, to cover both time and materials.

Many interprofessionals are convinced they need to do more than share the costs of getting a job the prime turned up. Consultants qualify and pass along leads as they find them to prime contractors with whom they work regularly or want to work.

A reputation for bringing in leads and helping to sell the project is helpful to any interprofessional's practice.

7.11. MARKETING/PRODUCTION ACTIVITY MATRIX

Marketing is the collective or umbrella term for the many activities required to move services (or goods) from producer to client or customer. Marketing is *not* the same as sales; marketing defines, outlines, and sets the stage for sales. Marketing creates, expands, and enhances opportunities for selling.

In outline form the process looks like this:

I. Marketing
    A. Intelligence gathering
    B. Marketing support
    C. Sales
        1. Personal selling
        2. Paper work
        3. Courting the prospect
        4. Proposal submission
        5. Interview
        6. Follow-up
        7. Client relations

The overall marketing process, and its interrelation to both marketing and production staffs is further illustrated in Table 7.1, the Marketing/Pro-duction Activity Matrix.

TABLE 7.1. Marketing/Production Activity Matrix

Activity	Responsibility		Explanation
	Marketer	Project Manager	
		Nonselling Phase—1	
Intelligence gathering and qualification	Primary responsibility.	Secondary responsibility.	*Open search*—cold calls, general inquiries, following general and specialized publications, and lead tip sheets.
	How: Use all printed and live lead sources; heavy on personal and telephone contact (cold calls); some correspondence.	Assist with marketing when indicated. Review all prospects with marketing for appropriate project manager input and possible contacts.	*Clients*—Lost, past, and present.
			Referral—Suspects and prospects from others—family, associates, contractors, friends, interprofessionals, suppliers, and the like.
		How: Project managers (and all others in production) supplement marketing's efforts in intelligence gathering. Pass along all printed and word-of-mouth leads to marketing for follow-up. Assist in marketing research as requested; pass on comments (good and bad) from past and present clients to marketing.	Use all internal/external reference sources: marketing files, files on past and present clients, and the like. You'll need all possible information at this point to decide *go* or *no go*.
			1. Is it a job?
			2. Is it a job for our firm?
			Necessary information at this point includes the status of funding; site selection stage; start, completion, and occupancy schedules; budgets; names of decision makers; your probable competition; and details about the prospect's selection process.

Selling Phase—2

2-A Preparation and early paper work — Primary responsibility

Secondary responsibility; usually limited.

Should make input and assist marketing in all possible ways at this stage.

Marketing files on the the prospect are opened. (These files, or large segments of them, will form the nucleus of the project manager's job file when the project is won.)

Success of this activity depends largely on the thoroughness of the intelligence gathering activity. Based on information collected in Phase 1, a total strategy for pursuing the work is developed. Brochures (general capability, special, and qualification) and proposals are organized.

The outline for structuring proposals (when required) and the formal presentation is written. (NOTE: You cannot have too much information about prospects and projects at this point.)

2-B Advanced paper work — Primary responsibility.

Secondary responsibility; which is always significant at this stage.

Here marketing and production staffs cooperatively produce proposals in response to RFPs. The proposal manager may be the project manager for the job—and *may* be involved in the contract negotiations after the project is awarded.

167

TABLE 7.1. (Continued)

Activity	Responsibility		Explanation
	Marketer	Project Manager	
2-C Personal selling	Primary responsibility How: Telephone, correspondence, personal contact.	Secondary responsibility; usually significant. How: Supplement marketing efforts as appropriate and requested.	Continue intelligence gathering on prospect and project. First selling meeting is scheduled. Technical input and support may be requested of production staff in the initial one-on-one session; such support usually is required in subsequent meetings with the prospect. The project manager can assist in followup activities for each contact, under marketing's guidance.
2-D Courting	Primary responsibility How: Through followup meetings, sending special brochures and selected reprints; all types of printed materials are brought into use now, including speeches, special papers and technical articles, newsletters, and the like. Some third party endorsements and other references from past and present clients may be furnished to the prospect at this point. Visits to sites of similar projects are scheduled.	Secondary responsibility. How: Suggestions to marketing for pertinent printed items. Accompany prospect representatives on project visits when arranged by marketing.	This an especially critical stage in selling. The prospect's attention has been gained through the activities outlined in 2-A and 2-C; now rapport, trust, and confidence in your firm must be generated and nourished in a low key, productive, and professional manner. Many jobs are won or lost at this point. The major goal now is to make the prospect's short list and be invited to make a formal presentation. All efforts must be directed to that goal.

2-E Presentation	Shared responsibility. How: Assists in preparation of the presentation; supervises rehearsals; may participate in the presentation.	Shared responsibility. How: Assists in preparation; participates. May lead the presentation. Depending on the firm's organization and marketing strategy, the project manager may act as a closer-doer in the presentation.	Marketing and production share in the selection of media to be used (slides, flip charts, models, photo boards, overhead projector, movies, videotape, blackboard, etc.). Use the outline developed in 2-A for structuring the presentation. The entire presentation *must* demonstrate your understanding of, and concern for, prospect's interests and project requirements. Unrelated data, graphics, and discussion have no place in the presentation.
2-F Follow-up	Primary responsibility.	Secondary responsibility (usually). If the project manager acts in a closer-doer role he or she may assume primary responsibility at this point. Participates in the debriefing.	This is the "awkward period" in selling professional design services. The prospect's interest and attention must be maintained—and belief in your firm's experience and nominated design team reinforced—until the job is awarded. What is learned in the debriefing—especially when the job is awarded to another firm—is important to future marketing approaches and strategy.

TABLE 7.1. (Continued)

Activity	Responsibility		Explanation
	Marketer	Project Manager	
		Contract Phase—3	
Negotiation	Limited.	Primary responsibility.	Most contracts require at least some negotiation of fee, schedules, and personnel to be assigned. Some of the points to be negotiated will have been raised in the proposal and in the interview; now they must all be resolved to everyone's satisfaction.
			One of several project promotion possibilities occurs in Phase 3—the contract signing. (See also promotion possibilities in Phases 4 and 5.)
		Production Phase—4	
Production	Limited or nonexistent	Complete responsibility.	Potential promotion possibilities in Phase 4 are the bid opening, award of the construction contract, and groundbreaking.
	How: The marketer may occasionally sit in as an observer in client conferences during concept and working drawing stages. These are "production presentations" rather than "marketing presentations"—and are every bit as important to getting future business and referrals from the client.	How: The project manager keeps marketing informed about the project status, problems—major or minor—that may arise, and the like.	

Follow-up Phase—5

Follow-up (client relations)	Primary responsibility.	Secondary responsibility; significant.
	How: Identifying and following up on promotional possibilities of benefit to both consultant and client: Topping out Dedication Occupancy or first use Open houses	Marketing concerns are now: 1. Repeat business from the client. 2. Referrals. 3. Letters of reference and other overt, usable evidence of client satisfaction as the project closes out. In Phase 5 marketer and project manager again become essentially equal partners in the marketing process. Newly acquired clients must be continually mined for information about planned projects of their own and of others. They must become full-fledged partners in the intelligence gathering and qualification phase (1). A schedule for continuing contacts should be set up for marketers and project managers, with all keeping others informed of pertinent developments.

Source. Marketing/Production Activity Matrix. Copyright 1983 by *Professional Marketing Report.* Used by permission.

NOTES

1. K. Meyer, "How to Structure and Give Persuasive Presentations," *Professional Marketing Report* (August 1977): 4.
2. R.R. Levoy, *The Successful Professional Practice* (Englewood Cliffs; NJ: Prentice-Hall, 1970), pp. 19–20.
3. H.O. Overstreet, *Influencing Human Behavior* (New York: People's Institute, 1925), p. 49.

8

ETHICAL ENGINEERING

Samuel E. Gluck

SAMUEL E. GLUCK is a consultant in New York City and an associate professor of Business and Economics, College of Mount Saint Vincent. He is both an engineer and a professional philosopher. His forty years of industrial experience include many years as Vice President for Research and Public Affairs of the Bonded Scale & Machine Company, a manufacturer of mineral and chemical processing machinery and material handling systems. He has authored more than 100 publications in engineering and humanities.

8.1. INTRODUCTION

The primary purpose of this chapter is to encourage the reader to develop two personal qualities: intellectual self-confidence in dealing with moral issues, and through that self-confidence, to achieve moral autonomy. If this sounds "philosophical," indeed it is, but it also has hardheaded practical and professional aspects. This, then, is a discussion about asking the right questions. Although we are especially concerned with the ethics of engineering, our professional moral lives cannot be lived differently from our lives as whole persons in the world at large. The same, of course, must be said for "medical ethics," "legal ethics," "business ethics," and any other occupation that must have specialized (not necessarily "higher") ethical guidelines for the doing of deeds, making of things, and governing of men including ourselves. What follows is a brief discussion of a few selected issues in ethical engineering, *not* a survey of all the issues or a review of all points of view on the questions that are discussed. It is not a list of do's and don'ts, nor a synthesis of the codes of ethics of the various engineering societies. Instead, we shall attempt to shed light on (but not give final answers to) questions such as these:

1. How can we recognize ethical (moral) problems?

2. Are there fruitful ways of thinking through a moral dilemma?

3. How can I relate general principles to specific cases? How, for example, can an engineer be involved in designing a cut-price, less durable version of a product, "good for what it is"? Are there *moral* lines to be drawn even if there is no dramatic issue of product safety?

4. How do I recognize a conflict of interest? Am I required to *anticipate* conflicts of interest, and if so, where does this projection into the future end?

5. What constitutes legitimate self-interest? What should I do when self-interest conflicts with other moral obligations?

6. Does a person have moral (as distinguished from legal) obligations to society? Does an engineer, because of his expertise, have additional and special obligations? If so, what are they?

7. Several engineering society ethical codes insist that my obligation to society, whatever that may be on any particular occasion, is my "paramount" obligation? Can I really accept this? For example, is it morally imperative that I blow the whistle knowing that I will lose my job, probably be blackballed in the industry, and be unable to support my family?

8. Are there any general principles to guide me in ranking conflicting moral obligations?

9. Are certain demands made by codes of professional ethics simply wrong-headed? Are some of them unfair? Are all aspects of professional conduct really moral aspects?

The scope of ethical engineering is far broader than the decisions and acts of professional engineers. It embraces the decisions and acts of everyone involved in an engineering project or product and deals with the relations between those persons. It includes technicians, research scientists, inspectors, and factory workers. It embraces the owners and managers who made the economic decisions about product design. It includes the political decision makers and the lobbyists who influence them in such matters as safety regulations and the promulgation of standards. The very concept of establishing a *minimum* standard of safety or performance, whether a legal standard or an industry consensus standard, is an act that carries great moral responsibility. When standard-making bodies such as ANSI committees are organized, there are rules for assuring that all "interests" are represented; principally manufacturers, those who use the product by incorporating it into other systems, and representatives of the industries in which the product or process is used. These interests are almost always economic ones—so much so that any mention of safety is sometimes excluded from the statement of purpose of the committee. There is also fear on the part of the members individually, and the governing body of the standard-sponsoring organization, that addressing safety issues will expose them to product liability and other litigation should the standard be adopted as a "recommendation" and accidents still occur.

Today, because of the widespread consequences of major technological innovations, engineers must also confront the practical and moral problems of "informed consent" on the part of those who will be most immediately affected by an engineering decision. Should there be wider requirements for environmental impact statements? Should there be more economic impact statements required for what few OSHA, FDA, and other regulations still remain? Are there responsibilities to future generations? If the answers to these questions are in the affirmative, how are they to be generated in the most efficient manner so as to eliminate harmful delays? How are negative answers in particular cases morally justified?

8.2. DEFINING SOME PHILOSOPHICAL TERMS AND IDEAS

In everyday language we usually use the word "ethics" in connection with professional and business conduct and "morality" in connection with the other aspects of our lives, especially, for example, sexual conduct. In ordinary talk we also distinguish between ordinary run-of-the-mill morality and the immorality that violates special expectations or trust in important relationships in which others are dependent upon the expert knowledge or highly regarded reputation of an individual. A man who seduces women may be regarded by some as immoral. That same person, should he be a physician who seduces his patients or a superior in an organization who sexually harasses a subordinate, is said to have violated professional ethics and is so regarded even by persons who would not, under "ordinary" circumstances, call his general behavior immoral. To both the professional and the man in the street, there is a special dimension to professional ethics; it applies to conduct peculiar to a domain of *trusted expertise,* and carries with it responsibilities precisely because of that trust and expertise. In a common sense way this distinction serves us well, but when we begin to think critically and analytically about moral matters, especially when we are faced with moral dilemmas—painful decisions in the moral realm, we need to clarify our thinking.

8.2.1. Some Basic Philosophical Terminology

Morality is concerned with *conduct* and *motives,* right and wrong, and good and bad *character. Ethics* is the philosophical study of morality; it is *moral philosophy.* When we do moral philosophy, that is, when we practice the philosopher's craft, we are subjecting the questions *of* morality to other critical and analytical questions *about* morality.

"Morals" is a set of rules of conduct and standards of evaluation that a culture uses to guide its individual and collective behavior and direct its judgments. Codes of professional moral conduct are specialized subsets of these rules and standards. The standards of a culture are frequently examined by historians and social scientists who observe how the rules function in preserving the society's way of life and the way individuals and groups adapt to them, attempt to change themselves or the rules, or fail to adapt. For example, engineers may answer questions, often anonymously, about whether they believe certain acts are right or wrong and also whether they have ever done or, under certain circumstance, would do those acts. More elaborate surveys of attitudes and beliefs usually include thumbnail sketches of cases. The respondents are asked how they would make the moral choice. Such surveys are familiar to readers of professional journals and technical trade magazines. Similarly the public is often sampled as to its beliefs about the morals of the professions: For example, are engineers honest? Are they more or less honest than physicians, accountants, politicians, clergymen, or insurance salesmen? Although these surveys may stimulate us toward self-scrutiny, they do not tell us how we *ought* to act unless we blindly follow convention, authority, or mass opinion.

By contrast philosophers study morality with different goals and purposes. Philosophers have a *normative* goal and an *analytical* goal. The aim of normative inquiry is to show us how to construct a universal, consistent system of norms—a system of rules of conduct and standards of evaluation that can apply to everyone. The other aspect of philosophical ethics is *analytic* ethics or *meta-ethics,* and on this level the philosopher's task is two-fold. It is first to analyze the meanings of terms used in moral discourse: how do such terms as "good," "right," "duty," and "ought" function in moral language? The second task is the analysis of the logic of moral reasoning. The philosopher Paul W. Taylor lists pairs of questions which illuminate this two-fold inquiry.[1]

Is there a valid method by which the truth or falsity of moral beliefs can be established?

If so, what is this method and on what grounds does its validity rest?

Are moral statements verifiable?

If so, what is their method of verification?

Is there such a thing as knowledge of good and evil, right and wrong?

If so, how can such knowledge be obtained?

Is there a way of reasoning by which moral judgements can be justified?

If so, what is the logic of such reasoning?

Can we claim that the reasons we give in support of our moral judgments are good (sound, valid, acceptable, warranted) reasons?

If so, on what grounds can we make this claim? What are the criteria for the goodness (soundness, validity, etc.) of a reason?

In other words, analytic ethics examines the presuppositions of normative ethics. Philosophers undertake this examination in many ways and there are many disagreements amongst them as to some of the details of the methods and content of moral norms themselves. Our aim in the present discussion is not the manifestly impossible one of "every engineer a philosopher in his own right," but as Taylor puts it:

> The ultimate purpose of normative and analytic ethics is to enable us to arrive at a critical, reflective morality of our own.[2]
>
> It is this sort of person—one who can think for himself and make decisions on the basis of his own thinking—who is the true individualist. Even if he ends up ... in general agreement with ... his society, ... [his norms] are of his own choosing. ... [I]t is his critical reflection ... that makes him an individualist.[3]

A historical example familiar to every college freshman will illuminate this issue. In ancient Athens, Socrates was accused of "corrupting the youth" because he taught them to question authority. He would not recant, actually baited his accusers, and was condemned to death. His friends planned his escape but he refused, and he is usually quoted as saying that "The unexamined life is not worth living." A more accurate translation from the Greek yields a crucially different statement: "The life untested by criticism is not worth living."

8.3. ETHICAL THEORIES

An ethical theory is a systematic attempt to provide a sound system of normative ethics and a universally applicable touchstone for determining right and wrong. Broadly speaking there are three categories of ethical theories: goal-based theories, duty-based theories, and rights-based theories. Each one has its insights, intellectual and emotional appeal, and shortcomings. Just as many philosophers have tried to synthesize them into a system that will take "some good from each," most of us manage with a patchwork that works pretty well until we face decisions vital to our lives.

8.3.1. Goal-Based Theories

Sometimes called consequential theories, these assert that the rightness or wrongness of an act is determined by the consequences that result from it. The traditional philosophical term for such theories is *teleological,* from the Greek word "telos" meaning "purpose" or "end."

Two of the most familiar goal-based theories are *utilitarianism* and *ethical egoism.* Of the two, utilitarianism has been the most appealing and influential, despite its difficulties. There are two versions of utilitarianism. Rule-utilitarianism says that we should obey those *rules* which if followed would produce the greatest good for the greatest number of people. Although some individuals may suffer in particular instances, our experience with the rule confirms that it is the best course of action. Act-utilitarianism asks that we judge each situation, to see which *act* will produce the greatest good for most people. Both types of utilitarianian theory have as their goal the goodness that is derived.[4] The difficulty, of course, is that we must define what we mean by good, and in the mean-ends chain we must distinguish between instrumental goods and intrinsic goods. Even if we could know, quantitatively, what the greatest number is within the relevant population, how would we measure amounts of good and compare them? Are all instrumental goods equal? Are all intrinsic goods of the same *kind?* How far into the future do we project? What happens to minority rights? What if we just happen to be wrong: Can we correct ourselves? In the United States and other democracies we legislate many utilitarian rules, but we also modify them by adopting policies and passing laws to protect those outside what we perceive to be the majority. However, modern conceptual tools for quantitative prediction, combined with computer technology, tempts us more and more into utilitarianism, especially rule-utilitarianism. The potential danger is increasing because we have become more and more convinced of the "truth" of such projections. The larger the scale on which we operate, both in contemporaneous scope and future time spans, the more likely it is that we will generate self-fulfilling prophecies with incorrectable momentum.

8.3.2.   Ethical Egoism

This is a goal-based theory of "rational" self-interest, which states that we ought to do what is in our own interest all of the time. It argues that although "little injustices" may appear to be committed from time to time, the ultimate good will be achieved because "rational" self-interest does, from time to time, entail cooperation, charity, and even friendship. A certain amount of "rational" concern for others is part of looking out for ourselves. Today, ethical egoism is expounded in its narrowest sense by the economist Milton Friedman and the novelist Ayn Rand.[5]

8.3.3.   Duty-Based Theories

These reject consequence as the criterion for judging an act. Rather, they assert that there are moral imperatives that we must obey, the consequences notwithstanding. They are what the German philosopher Immanuel Kant (1724–1804) called *categorical imperatives.* They have no exceptions: For example, we must tell the truth though the heavens fall. How do we recognize a categorical imperative? A categorical imperative is universal in that it can be applied to everyone. It is a moral principle, binding only if we are willing to have everyone, including ourselves, act in accordance with it. For Kant, the Categorical Imperative was to always treat mankind, whether in the person of another or oneself, as an end in itself, never as a means. Each of us has intrinsic worth; we are all, in Kant's words, citizens of the "kingdom of ends." To steal, lie, or break promises, is to treat humanity as mere means, not as an end.[6]

Most of us, however, would find it difficult to accept an ethical theory in which we could not recognize that there are exceptions; that some duties take precedence over other duties. The philosopher W.D. Ross (1877–1971) sought an orderly solution to this problem by proposing that there are in fact *prima facie duties* which we intuitively know are higher than others when we are faced with such choices.[7] But the problem remains that even prima facie duties may be in conflict, and there are other problems with claims that there is intuitive moral knowledge. Kant had asserted that even in a situation where we see a traveller on the highway being "pursued by ruffians," it is our duty to tell them which way he went. Many people would wonder which was more important: to save "mankind" from falsehood or to save the life of a person from mortal danger.

The contemporary philosopher John Rawls argues that autonomy and universality are possible because there are two basic principles about which all rational persons would agree. Put roughly these are (1) each of us is entitled to the maximum freedom compatible with equal freedom for others and (2) differences in social and economic benefits are only justified when they are likely to benefit everyone.[8]

8.3.4. Rights-Based Ethical Theories

Proponents of these theories claim that there are certain fundamental human rights and that moral obligations arise in the context of those rights. Are they "natural" rights or do they arise out of human association (not some particular association but societal association in general)? This is one of the many questions about which philosophers disagree. In the United States we have a strong tradition that leans toward popular belief that there are both natural rights and derivative rights that stem from them, although what specific rights there are, especially derivative ones, are matters for dispute. The Declaration of Independence speaks of "self-evident" truths: "that all men are created equal, and that they are endowed by their Creator with certain unalienable rights," those of "life, liberty, and the pursuit of happiness." This states that certain rights are natural; not rights to be earned but rights to be protected. The English philosopher John Locke (1632–1704), an intellectual ancestor of our Founding Fathers, stated that Man's rights are life, liberty, and property (the fruits of our labor); this last "right" being much narrower than that of the pursuit of happiness. Other rights are derivative rights. The Bill of Rights is derivative from those set forth in the Declaration. One may believe in natural rights and still believe that they can be forfeited. They start out as givens—we each begin life having them. Other rights come about by circumstances in different cultural settings. A contract or a promise carries with it both the duty and the right of fulfillment. The derivative right of freedom of speech carries with it the duty to speak responsibly and truthfully, but also carries with it the speaker's right to protection when he does speak and the duty on the part of others who can partake of that right themselves to support that protection.

8.4. SOCIAL RESPONSIBILITIES

The two areas of responsibility that most readily come to mind are the engineer's responsibility to his or her employer and society at large. Most professional codes of ethics acknowledge these and state that the latter takes moral precedence over the former. What does each include? Are there other responsibilities, both professional and personal, that may complicate the ethical decisions of engineers?

8.4.1. Responsibilities to One's Employer

In an important and wide-ranging paper,[9] John E. Ullmann (the editor of this volume) discusses both of these types of responsibility. His major point is that employer interests and public interests are not necessarily in opposition to each other but can overlap when all parties realize they have interest in common. With regard to the engineer's responsibility to his employer, Ullmann calls for the engineer to act, when necessary, in what might be called "loyal opposition." He gives examples of major engineering failures that had serious social consequences either because of the failure of engineers to be critical, their employers' refusal to take heed; or in some cases, where engineers themselves, enamored of their own projects would listen to no one else. He condemns those engineers who say that they are only the implementors of the goals of others and therefore need not, or should not, attempt to influence policy formulations and decisions. They are convinced that they can live in a state of moral neutrality, with no responsibilities, and can thus with impunity design weapons systems, boondoggling public works, and the gas chambers and crematoria of concentration camps.

Ullmann also points out that there are others who do wish to speak up when they believe a mistake, moral or technical, is being made, but they are understandably influenced by whether they are "young, single, and mobile; married and in debt; or old and well provided for."[10] They know that an employer can always "find underlings pliable enough to do his bidding."[11] He suggests that one way to protect engineers and other professionals from retaliation could be a system akin to academic tenure. Unfortunately, with the increasing pressure to abolish academic tenure, and the long-standing reluctance of the courts to enter this "private" institutional domain, he holds little hope for such a system.

Although our attention is most often drawn to decisions affecting the present and future well-being of the community or of mankind at large, the *principle* of loyal opposition underlies many narrower and less dramatic cases. It is an engineer's duty to assert his reasoned and verifiable objections to an awkward or inefficient product design (not necessarily a dangerous one) or argue for or against expenditures. Some would argue that he has both a duty and a right to do so, especially in matters of public welfare and safety. The question is not always what is *morally* "best" because there are many decisions that do not involve moral issues. But even in the design of a new household gadget, the engineer has a moral duty to meet the good faith expectations of his employer by doing his job well, and that includes giving his best intellectual and professional support.

8.4.2.   Safety: Risk, Benefit, and Cost

When we call a process or a product "safe" we mean, again in a common-sense sort of way, that its users will not suffer harm because of it. As Martin and Schinzinger point out, those who use it, especially if they are of the general public who do not understand its intricacies, are entitled to rely on the experts who designed, built, and recommended the use to which it is put.[12] All persons meaningfully involved with a product or process share in the responsibility for its safety. Martin and Schinzinger point out that one major component of the process of engineering is *social experimentation.* Beyond the various controlled experiments involving materials, processes, design, and testing, "each engineering project taken as a totality may itself be viewed as an experiment."[13] First, any new development has uncertainties and is carried out in partial ignorance. Second, the final outcomes are generally uncertain not only on the immediate level of whether it will work, but in terms of future consequences. This is especially true of large projects such as power plants, nuclear reactors, reservoirs, highway systems, electronic devices, and the like. Some technological advances initially produced for social good in mind may at later times be employed for the opposite effect. New materials of fabrication or construction and innovative designs are ultimately tested only by public use. Therefore, an ongoing system of monitoring, reporting, analysis, and correction must be planned in advance.[14] An engineer is acting responsibly toward society when he is sensitive to the experimental nature of his work and accepts "a primary obligation to protect the safety of and right of consent of human subjects."[15]

We judge something to be safe relative to the extent to which we understand the risks involved and to the degree that our values place the benefits of use higher than the risk. One example will illustrate these points and also demonstrate the weakness of such an approach. It is now some years since several DC-10 aircraft crashed because of the defective design of their cargo doors. At present DC-10's are used in the United States and elsewhere, and passengers who know of the old trouble apparently assume that it has been corrected. A more sophisticated passenger, who knows about United States government enforcement procedures (the Federal Aviation Agency), would be more confident of the accuracy of his risk assessment on a United States carrier, but less confident when faced with a DC-10 flight on a foreign carrier, especially one belonging to a "developing" country. He would have no *direct* evidence upon which to base his anxiety, but would base his risk assessment on what he knows about expertise, worker supervision, and attitudes in certain cultures. Prudence would dictate not to fly, but he might anyway if the alternative was a costly detour. Would that be an irrational choice? Does the passenger have a "wrong" system of values, a confused system of values, or has he simply suspended his judgment of the risks? It would appear from their public statements that many employees of chemical plants who are fully aware of the dangers of explosion, toxicity, and genetic damage, "don't think about it" because they do not have or believe that they do not have an alternative. One must not confuse this with having accepted the risk. To "accept" a risk, in the sense that we speak of levels of safety and probability of disaster, implies that there are *nondesperate* alternatives. Despairing acceptance is not acceptance in a way or type that can be legitimately used in risk analysis.

In the March 4, 1985 issue of *Chemical Engineering*, there appeared a reader survey questionnaire about hazardous wastes. One of the questions was: How much money would be "reasonable compensation" to pay each family within a two-mile radius of a facility for allowing it to be located in their immediate vicinity? The facilities listed included various kinds of hazardous waste dumps, a nuclear power plant, a refinery, an automobile factory, and a shopping center. The amounts proposed were: none; $1000; $10,000; $100,000; $1 million, and "no amount of money would be sufficient." The questionnaire was addressed to chemical engineering professionals and included other questions to determine the basis of expertise of each respondent. The answers to these questions, eventually to be published, will not reveal the objective level of risk, but will reveal how experts assess the risks relative to each other. They have been asked to place a dollar value on discomfort, disease, deformity, and death. The questionnaire, no doubt seeking to avoid leading or biased questions, does not ask, "How much money is your family worth?" No probability estimate will predict the next event in a series. Would you stake your life that, in an indefinitely prolonged series of independent tosses of a coin, the probability of heads (or tails) converges toward one-half? Would you stake your life on the next toss?

8.4.3. Cost-Benefit versus Risk-Benefit

It is not possible to discuss all aspects of safety analysis, but we can point to our two different conceptual approaches. *Cost-benefit* analysis is essentially an economic concept that attempts to reduce all costs of a project, even what we often call human costs, to economic terms. The cost of providing safety devices for a machine is balanced against the beneficial results of lower insurance premiums and fewer lawsuits. One benefit of the 55 mile per hour speed limit, fewer accidents, is alleged to be offset by the increased cost of transporting goods at a lower speed; and besides, it is argued, so few lives have been saved compared to the number lost to drunken driving. It has been alleged that the cost-benefit analysis reveals that the investment of money in curbing drunken driving would be a better alternative to enforcing the lower speed limit. Being enamored of cost-benefit analysis, we have ignored the risk-benefit approach and the element of social experimentation; this has created irreversible hazards related to the disposal of nuclear waste. Cost-benefit analysis does not deny the value of human life, but reduces it to monetary value based on the social-class factors of earning potential in relation to occupation and age.

Risk-benefit analysis is also concerned with reducing monetary costs and finding ways to measure and reduce costs and risks, but its criteria are broader. It is, first of all, an attitude which requires a more sensitive moral outlook that in principle finds no level of risk to be "acceptable." Yet it deals with the reality that the world will never be risk-free and that "acceptability" is a criterion forced upon us. If we define acceptability as the level at which informed individuals acknowledge that they are at risk but conduct their lives essentially as if they were unconcerned, then risk-benefit analysts have a starting point. We sign consent forms for medical treatment, in which we acknowledge that we are "fully informed" as to the risks of the procedure and the alternative of having no treatment. The risk-benefit approach seeks to reduce the risk itself, not just its costs. Cost-benefit decisions "satisfice" because they are based on the assumption, made some years ago by Herbert Simon, the originator of the satisficing concept, that we "have not the wits to maximize."[16] The ethics of the risk-benefit approach acknowledge that we do satisfice but do not always and merely satisfice. Random arrivals at gasoline pumps and hospital emergency rooms are both queueing problems, but the latter is not "just another" queueing problem. Risk-benefit analysis is never satisfied and is not reductionist or simplistic. As the philosopher John Stuart Mill wrote, it is "better to be a human being dissatisfied than a pig satisfied; better to be Socrates dissatisfied than a fool satisfied."[17] No amount of quantitative reductionism must intimidate us to believe otherwise.

8.5. LOYALTY

Loyalty is a moral concept in most cultures. We have a broad vocabulary relating to it: We speak of blind loyalty, misplaced loyalty, party loyalty, loyalty to one's country, to principles and beliefs, to one's own principles—to oneself, being in loyal opposition; and, loyalty to the local athletic team. In all except the last, there is an assertion of duty, and in most usages there is implied a mutuality and reciprocity of protection. Even loyalty to one's principles or oneself implies that a person thereby derives inner strength to be a morally autonomous individual, provided he has truly examined that to which he has been committed. In summary, there are two different concepts of loyalty. When defined as the commitment to one's obligations to persons or institutions (whatever those obligations are believed to be), loyalty itself is a moral good and one has a duty to be loyal. When defined simply as commitment *qua* commitment, it does not carry a moral imperative; each instance must be examined on its merits for the object and content of loyalty. In a case where loyalty *prevents* a person from meeting moral obligations, one of those obligations is to be *disloyal.*

8.6. WHISTLE BLOWING

The concept of loyalty figures prominently in all discussions of whistle blowing. Detractors level accusations of disloyalty, while supporters speak of loyalty to a higher cause. Employers usually believe in loyalty to loyalty. College textbooks on "business and society" take widely differing views on the subject. Some (especially those published in the 1970's) consider public disclosure a supravital good in itself. Others take the position that even where a whistle blower's allegations are true, the courts should not protect him from employer retaliation if there is evidence that he acted out of revenge, frustration, or malice; or that unless the whistle is blown out of pure moral indignation, the charges should not even be considered. How the judicial system is able to read one's mind or heart is not made clear.

In whistle blowing are focused many of the issues of modern society: technology, responsibility (corporate, governmental, and individual), political and economic power, and the characteristics of an organizational–managerial society. The general public is more sympathetic to whistle blowing than business and some professions. In practice today there is little adherence to the genteel tradition of not publicly criticizing another member of one's profession, and the public is very much aware of any attempt to suppress freedom of expression, or coverups in general.

One philosopher makes a distinction between justifiable whistle blowing and obligatory whistle blowing, and also attempts to offer some guidelines for resolving moral dilemmas in this area. According to Richard T. De-George,[18] whistle blowing is morally justifiable when there is impending danger and a concerned employee has "made his moral concern known" to his immediate superior who has subsequently failed to act. When that happens, advises DeGeorge, a concerned employee should take his or her complaint upward through company channels, if necessary, to top management. After all internal efforts have failed, public disclosure is justifiable. For whistle blowing to be obligatory as well as justifiable, two more conditions must be met—first, that the employee have documentation or other hard evidence (else his chances of success are slim); and second, that he "must have good reason to believe that by going public he will be able to bring about the necessary changes."[19]

Many engineers, perhaps most people, would agree *in principle* with this approach. In practice however, there are the realities of organizational power, external political and economic interests, time consuming bureaucracies both in the firm and in government, and a matter of "self-interest"— called survival. As soon as an engineer complains to his or her immediate superior, the potential that the complaint will be carried to a higher level in the organization is revealed. Up how many corporate channels does the engineer swim against the current and how long will it be before he or she realizes the futility of the situation. If rebuffed at the next level, this engineer has perhaps already put his or her job at risk.

How far does one proceed in laying a trap for oneself? As for documentation or other evidence, there may be none at the initial stages of the engineer's awareness that something is amiss. Sometimes the "evidence" is knowledge that, although a defective part (e.g., in an airplane) is replaced, it is simply replaced by another just like the first, and several such replacements have been made with no investigation as to cause or plans to redesign the part. If an engineer waits for the "evidence" in the form of a fatal crash, has his or her behavior been moral or immoral? Sometimes the evidence is a document that could not be copied or a conversation accidentally overheard. These are leads, "clues" from which inferences can be made and inquiries initiated. Finally, it is impossible to "know" what the reaction of outsiders will be, although within the organization the engineer may well "know" what will happen.

The strongest point in DeGeorge's position is that he acknowledges that we are entitled to consider our own interest before we put our jobs and professional lives in jeopardy. Martyrs are dead by definition. Related to this is an additional moral issue: As W. David Ross puts it, are there other prima facie duties that override not only obligations to society, but even one's "duty to oneself?"[20] How does one morally justify propelling family or friends into an otherwise autonomous and unilateral decision? How much loyalty can be demanded of others who are truly innocent bystanders and whose lives are intimately bound up with one's own life?

8.7. "CONFLICT OF INTEREST": WHOSE CONFLICTS, AND WHOSE INTERESTS?

In professional life, business, and politics, "conflict of interest" is a familiar accusation. It comes to our attention when there is a scandal and the accused person has profited from the situation or has shielded himself, a friend, or family member from criticism or loss. In almost all cases one of the interests in conflict with that of an employer, the government, a client, or the public, has been *self-interest*. That is, however, a different kind of self-interest from the survival problems associated with whistle blowing.

8.7.1. Gifts

The most commonly discussed alleged conflicts are those arising from a gift relationship. Many organizations prohibit employees from entering into gift relationships with anyone with whom they deal *or may deal in the future* on behalf of the organization. These include employees who can make decisions about suppliers or who have responsibility for compliance with the law. The underlying assumption is that at the very best every man has his weaknesses and at the worst every man has his price. At the very least, so the argument goes, it may be impossible to be unyieldingly professional (or businesslike) toward someone who has given you season tickets to every home game. If the current shipment of components does not pass inspection, you may still pound the table and demand replacement, but—it is feared—you will not have the heart to invoke the penalty clause in the purchase contract, at least "just this one time."

Some companies set a dollar limit on acceptable gifts, such that major league tickets are excluded. But how is the dollar limit arrived at? Apart from employer-imposed constraints, what should a person do from a moral point of view? Not all gifts are bribes or potential bribes. In fact, most people would be shocked and deeply hurt if a genuine gesture of thanks or appreciation were interpreted as dishonest. These are, as the French writer Antoine de St. Exupéry said in another context, among "the things that make life kind to men."

Each of us tries to establish his or her own set of working guidelines, realizing that the boundaries are not rigid. The present writer tries to adhere to the following ones. First, neither give nor receive anything that is nonconsumable. If you do, make it something that has personal value for the recipient—not a market value. One may accept a bottle of spirits, smoked goose at Christmas, box of golf balls, or can or two of tennis balls, but not a silver champagne bucket, golf clubs, or tennis racquet. One may give or receive a "prestige" pen, but only if it has the giver's corporate emblem on it. The emblem is a kind of public announcement that more than one person is eligible to receive it, and even though you may know that the giver could choose the recipients, it lacks the *private* implications that the same pen would carry if it were engraved with the recipient's initials and had no emblem. For many years I have received a Christmas gift from an insurance broker who profits considerably from our dealings. The gifts have included a little "weather alarm" radio, a travel alarm clock, and sometimes an adult game. What was the retail value? It was very small. The cost, in quantity was even smaller. The gift was not the object, but the obvious care with which the gift was chosen to be a gesture of appreciation, useful to the recipient, and without any implication other than a "thank you." Gestures of hospitality are more difficult to handle. It may be best to find a polite excuse not to bring your family for a weekend to someone's house on the lake.

8.7.2.   Moonlighting and "Spinning Off"

There are, however, matters which may be more serious. Is an engineer who moonlights at *any* job (not necessarily an engineering job) in conflict with his employer's interest because he diverts thought and energy to other gainful employment? We would probably say not, unless it resulted in lower performance on his primary job. If he or she moonlighted for a competitor, it would indeed be a conflict of interest. Suppose, however, that the extra work was engineering work on a freelance basis? If the work did not *compete* with the employer, would it still be a conflict of interest?

Let us shift the conditions a bit. Suppose you work for a consulting firm that puts a high markup for overhead and profit in all of its quotations. A prospective client of the firm finds that it cannot afford to go ahead with the project at those fees and you offer, *after* they have decided to abandon it, to do the work privately. You would have been the one who would have worked on it for your firm. You explain that the difference in fee (let us say 20%) lies in the profit structure of the consulting firm. You agree to work nights and weekends, but it will take four to six weeks rather than ten days. The client agrees, Are you in conflict of interest? Disloyal? After all, you did not steal the client. Suppose that the next job comes to you directly, unsolicited; and moreover, you have been recommended to another firm as well. Should you refer this second firm to your employer first?

Over the course of a year you have still other personal clients, and soon you must decide whether to stay with your employer or strike out on your own in association with two colleagues from your own firm who are interested in establishing a consulting group. What is the moral status of your situation? When you decide to make the move, you prepare a modest and straightforward brochure which you send to those clients of your present employer with whom you have personally worked while an employee. You also send this brochure to a large group of potential clients with whom your present firm has had no contact.

Are you acting unethically, or only in acceptable self-interest? Are you stealing clients? Should you only hang out your shingle and wait? Is it any different from the physician's notice to his patients: "Rusti Bohnsau, M.D., and I. O'Newt, M.D., formerly of the Park Bench Medical Group, announce the opening of their office. . . ." Is it any different from the barber or hairdresser announcing that he or she is moving to another shop? If so, why is it different? A veteran salesman is hired because he has "a following."

8.7.3. Acting on "Insider" Information

There is the old story about the apartment shortage in New York City: A person saw someone drowning in the river and asked him for his address. Leaving the victim to drown he raced away only to find that someone had already rented the apartment. "How is this possible?," he wanted to know. "I'm the one who threw him in," was the reply. Is it a conflict of interest if a person acts for his own benefit based upon what he knows? There are many kinds of insider information and ways to act upon it. What, if anything, do they have in common? Insider activity is usually condemned on the basis of it being "unfair"; that is, that not everyone has had an equal chance. What is the ethical import of the concept of "fairness"? It is a question which is hardly discussed except by a few rule-utilitarians, although there is a vast corpus of law, regulation, and custom concerning fairness in employment and use of financial information.

Who is "everyone"? When is information public property? For example, in 1957, Edwin Land, the inventor of Polaroid, was the recipient of an award from the Franklin Institute in Philadelphia. That evening he gave a slide-illustrated lecture and demonstration of the Polaroid camera and process. Polaroid was of course "public knowledge," but the several hundred people at the banquet, listening and watching under the kindly gaze of Benjamin Franklin's giant statue were in a sense privileged. We were being given new and sophisticated information on the basis of which some of us bought Polaroid stock and others did not. To what extent, if any, did our membership in the Institute and our presence on Medal Day in 1957 make us "insiders"? A few years earlier an agronomist colleague invested heavily and profitably in a producer of agricultural chemicals on the basis of hearing a paper at a scientific meeting given by a chemist who worked for that firm.

With these two examples in mind, we may wish to think about the following situations. Are moral questions involved? If so, what are they and how are they to be answered?

Suppose I work for a giant research institute engaged in a project for a client to develop an alternative to metal cans for certain food products. We are successful and prior to the introduction of the product I buy stock in the client's company. To whom am I being unfair? To my employer? To the client? To the millions of investors who *might* want to buy the stock when "everyone" finds out?

Suppose I work for the client company as chief liaison engineer to the research institute. I see the implications for expansion. Toward whom am I being unfair if I buy my own company's stock? In either case does it make a difference how many shares I buy out of the millions on the market?

Suppose I work for an auto maker that has not acted on reports of faulty design in its new "hot" line of compact cars. After twenty years with the company I own several thousand shares, some purchased through the employees' participation plan, the balance on my own. I have decided to quit the firm because I cannot morally sustain a connection with a company that hides fatal flaws in its products. I sell all my shares and resign. Am I dishonest, unfair, or in "conflict" with my employer? Does the fact that half the shares came to me at an employee's discount make a difference? Why or why not?

Quite by accident I learn that my employer, a chemical company, has been dumping toxic wastes in an area where there is a noticeable rise in both the incidence of cancer and the frequency of birth defects. I also learn that next week this will be made public because a group of concerned citizens and an environmental protection group have been quietly preparing an exposé and suit against the company. Monday morning I sell my stock amounting to 1500 shares, and an hour later, after my broker tells me the transaction has been made, I insist on seeing the vice-president, whom I tell what I have learned. On Wednesday the suit is filed.

A philosopher colleague with whom I discussed the stock-sale questions said that such sales would be unethical because selling the stock under circumstances of insider information would be the same as selling my automobile without revealing its defects.

8.7.4. Employment

There is also the question of fairness in employment. I know that there will be a job opening in my firm for a sophisticated specialty. A friend of mine would be perfect for the job. In informal discussions I hear management express regret that he (known by reputation) is not available. I happen to have two coordinated pieces of "insider" information: I know of the opening and I happen to know my friend. I speak up—"I happen to know that he would be available"; I say this knowing that according to law the position must be advertised and on an equal-opportunity basis for all applicants, and knowing also that my friend in Denver would not likely see East Coast newspapers. I call him and he submits his vitae. I am certainly not in conflict of interest with my employer—quite the contrary. Am I in conflict with the "public" or with members of minority groups who may apply because I have introduced into the pool of applicants someone who will almost certainly get the job on merit alone?

8.8. CONCLUDING NOTE

The issues raised in this chapter are as complex as they are varied. Several of them have increased greatly in importance—especially those relating to safety, the environment, or other potential generators of unwanted social costs. As has been noted, engineering is in the middle of such controversies by the scientific nature of its calling as well as by its commercial ramifications and relationships to both the government and public. In a real sense the esteem and public support accorded to the engineering profession and its practitioners depend on how the problems described in this Chapter are resolved.

NOTES

1. Paul W. Taylor, *Problems of Moral Philosophy*, 2nd ed. (Encino, CA: Dickenson Publishing, 1972) p. 9.
2. *Ibid.,* p. 10.
3. *Ibid.,* p. 11.
4. John Rawls, "Two Concepts of Rules," *Philosophical Review* 64 (1) (1955.): 3–32.
5. Miss Rand is known for *The Virtue of Selfishness,* her ideological novels *Atlas Shrugged* and *The Fountainhead,* and her lecture circuit disciple, Nathaniel Brandon. See also William F. O'Neill, *With Charity Toward None,* a critical analysis of her philosophy (New York: Philosophical Library, 1971).
6. Immanuel Kant, *Fundamental Principles of the Metaphysic of Morals* (1785), translated by T. K. Abbot.
7. W. D. Ross, *The Right and the Good* (Oxford: Clarendon, 1930).
8. John Rawls, *A Theory of Justice* (Cambridge, MA: Harvard University Press, 1971).
9. John E. Ullmann, "The Responsibility of Engineers to Their Employers," In J.H. Schaub and Karl Pavlovic, *Engineering Professionalism and Ethics* (New York: Wiley, 1983).
10. *Ibid.,* p. 142.
11. *Ibid.,* p. 141.
12. Mike W. Martin and Roland Schinzinger, *Ethics in Engineering* (New York: McGraw-Hill, 1983), pp. 96–97.
13. *Ibid.,* p. 56.
14. *Ibid.,* pp. 56ff.
15. *Ibid.,* p. 63.
16. Herbert Simon, *Administrative Behavior* (New York: Macmillan, 1947), p. xxiv.

17. J.S. Mill, "Utilitarianism" (1863), in E.A. Burtt (ed.), *English Philosophers from Bacon to Mill* (New York: Modern Library, 1939), p. 902.
18. Richard T. DeGeorge, *Business Ethics* (New York: Macmillan, 1982).
19. *Ibid.,* p. 162.
20. W.D. Ross, *op. cit.*

9

RESEARCH AND DEVELOPMENT AND THE ENGINEERING FUNCTION

Peter V. Norden

PETER V. NORDEN is Program Manager of Technology Transfer for IBM's Engineering/Scientific marketing directorate (NAD). Previously he was manager of the E/S Support Center in White Plains, N.Y. Dr. Norden is also Adjunct Professor of Industrial Engineering and Operations Research at Columbia University's Graduate School of Engineering and Applied Science. Before coming to IBM, Dr. Norden worked as mechanical engineer, and held jobs progressing from draftsman/detailer to Chief Project Engineer with J.A. Maurer, Inc. His education includes B.S. (Math), M.S. (IE), and Ph.D (Operations Research) degrees, all from Columbia University. He is a Registered Professional Engineer (CA). He has published a number of papers and handbook chapters dealing with scheduling, research and development and project management systems, mathematical programming, and computing and information systems.

9.1. THE RESEARCH AND DEVELOPMENT FUNCTION IN THE ORGANIZATION

While the history of mankind is also to a large degree the history of science and innovation, the last four decades are significant because of a fundamental change in our approach to innovation. As Peter Drucker points out, it is planned innovation, that is, purposeful, directed, organized change that characterizes our age.[1] Organized attacks to advance the frontiers of science and technology are the order of the day in the military and civilian sectors of our economy. The result is that today we commonly hear corporation presidents make statements such as, "More than 50% of our revenue comes from products that were not even invented 5 or 6 years ago." The funds expended to muster this massive Research and Development (R & D) campaign have been steadily rising into the billions of dollars as have the number of attendant R & D scientists and engineers. The trend appears to show no sign of abating.

Since World War II we have become accustomed to seeing the words "Research and Development" familiarly paired like "ham and eggs." For some purposes, such as budgeting and finance, this may be fine. In practice, they *are* vastly different. In Section 1.1.1 the accepted definitions of basic research, applied research, and development, as established by the National Science Foundation, were given. Let us restate and elaborate on them slightly to clarify the respective roles of R & D and the rest of Engineering in the company: Applied research consists of "investigations directed to the discovery of new scientific knowledge having specific commercial objectives with respect to products or processes. This definition differs from that of basic research chiefly in terms of the objectives of the . . . company." Development consists of "technical activities of a nonroutine nature concerned with translating research findings or other scientific knowledge into products or processes. Does not include routine technical services to customers or other activities excluded from the above definitions of research and development." Other definitions of engineering activities included technical support for present products; developing new models of present products; developing brand new products; and developing new or improved methods of making the product.

This chapter focuses on the relationships among R & D departments and more conventional engineering functions in manufacturing enterprises. That is, we are confining ourselves to firms that ultimately "make something." These are different from firms whose principal product or service is conducting R & D work, whether or not such work includes building a prototype. R & D firms may perform the same functions as in-house R & D departments, but their relationship to the "customer" and his or her in-house engineering staff is that of a vendor providing service for a price, on a contractual basis. Even under these circumstances the R & D functions of primary concern to the other engineering departments will generally be more "D" than "R". In turn there is commonly far more interaction between Development and Engineering than between Research and Engineering. It should be noted that the development functions—the steps on the way to new and improved products and processes—do not depend on the size of the enterprise. They must be carried out whether the manufacturing firm is large or small: We are not addressing ourselves solely to giant corporations.

9.1.1. The Research, Development, and Engineering Spectrum

To understand the interaction between the engineering and the research and development functions, it is useful to visualize them as segments of a spectrum, or links of a chain, which begins with fundamental research and ends with the manufacturing-engineering function on the factory floor. The sequence is: Fundamental (basic) Research—Applied Research—Product (process) Development—Pilot Production & Manufacturing Engineering—Quantity Production & Tool and Quality Engineering.

One might like to think that the birth of a new product flows smoothly from left to right—from conceptualization and invention in research to reduction-to-practice in the development shop and from there into the factory and the product line. But this is a simplistic view; one rarely finds so direct a transition. Actually, there are many stops along the way; there are delays for economic as well as technical reasons. There are also backtracking and attrition (cases of "back to the old drawing board") and inventions that will never make it into the plant. However, every improved or brand new product must at some time in its genesis pass directly or indirectly through all the segments. What is interesting is that the segments differ from one another primarily in the purposes for which the work done at any stage is undertaken.

Thus the purpose of basic research is generally a quest for deeper understanding of the structure and behavior of entities and phenomena of interest, without a commercial or other comparable objective specified at the time the research activity began. This caveat is key, since often promising insights, "eureka effects", or strokes of serendipity occur in midcourse of a fundamental research project, causing redirection of the study. The project is then frequently transferred to another group, with a commercial or technical specification clearly spelled out, and the project is now Applied Research. Its purpose is to achieve the increment of improvement or technical advance implied by the new specification; and this purpose is again articulated at the start of the applied research program.

As we move from (and including) applied research, through all other segments of the spectrum towards production, the purposes for doing the work are describable a priori. They consist of increasingly detailed and elaborate project specifications or work orders and are usually accompanied by schedules and budgets. Fundamental research is thus uniquely distinguished from all other R, D & E activity in that it is the only segment whose output is totally open-ended at the outset.

After applied research, Development has the purpose of inventing and designing new products or processes. It differs from applied research primarily in the hierarchical scale of the items addressed. That is, to give an example from the electronics industry, applied research might address proving the feasibility of yet faster and tinier circuit elements and their manufacturability, while development concerns itself with packaging circuit boards or even larger devices and subassemblies, combining best state-of-the-art circuit elements into novel, useful, reliable, and marketable end-products. The purposes of prototype and pilot process engineering, and tooling and production/manufacturing engineering are, respectively, to prove out the innovations made in development, and to render the new products or processes manufacturable and economically viable.

9.1.2. Organizational Considerations

The last section set forth the approximate sequence whereby research turns into development and into final engineering. This can create potential organizational problems which deserve special emphasis. They relate to personal and departmental status and not only to the obvious turf battles. It is useful to discuss these problems in two settings: (1) those in which Engineering and R & D are in separate departments and (2) those in which research is carried out by individual members or groups within a single engineering department.

A common element in both is that in many cases there is ample reason why the R & D engineers and scientists are considered internally as an elite group. Pride in one's group has its positive aspects, but it should not become the commercial equivalent of interservice rivalry where the engineers, who, after all, perform the absolutely vital task of providing the technical means of implementing the design, are regarded as a species of lowly mechanics and in turn regard the research group as a bunch of snobs who are excessively remote from reality. At the managerial level such conflict may affect the time and technical stage when the product leaves R & D and becomes the responsibility of the engineering department, with the weaker group seeing its role diminishing. Equally unfortunate would be a situation in which the necessary contact between the departments is impaired, thus reducing the essential feedback which is again an absolute necessity.

If R & D and Engineering are part of the same department and are carried out on a smaller scale, the kind of rivalry noted above occurs between individuals rather than between organizations. Of course it may not need to do so if the whole group is well-managed and if a sense of common interest and shared goals is infused into the group. In this case people are assigned to R & D tasks on the basis of their capabilities and not by actual or perceived favoritism.

9.1.3. Structural Considerations

The most common types of corporate organization structures are the functional and the divisional. The R & D function has a useful place in both. It is generally good to have a key research executive, such as the corporate director of research, report to a key vice-president (or even the CEO) directly. This will assure that the interests of the R & D functions are properly represented at policy levels and, conversely, that coordination and content of R & D plans and programs are properly managed and consistent with current company policies. Divisionalized and geographically dispersed R & D activities are then positioned hierarchically downstream from the key corporate contact.

In many companies the production facilities of the operating divisions are widely distributed, and the products vary extensively within and among locations. There are then three basic alternatives for locating the research and development laboratories:

1. Distribute the R & D activities so that only divisional laboratories or departments exist. In this case the obvious advantage is that one avoids the investment involved in creating a monolithic research center. In addition, the divisional centers could do some corporate research assuming adequate coordination. Last but not least, it allows the R & D people to be close to production facilities, production problems, production people, and market demand; this minimizes the NIH ("Not Invented Here") factor which inhibits technology transfer. Proximity also often permits the use of production facilities for test runs or model making, with the obvious savings in facilities and equipment and the considerable advantage of realism in production processes.

2. Construct a central research center for corporate research. The advantages of this approach are generally taken to be economies of

scale, administrative cohesiveness, and avoidance of the need to duplicate scarce or specialized technical talent in many locations. The
major disadvantages are the reverse of the pluses cited for the dispersed approach.

3. Do both centralized and decentralized R & D. This approach may be
"the best of both worlds," if the firm can afford the expense. It allows
sharing of scarce talent and specialized equipment over a wide range
of problems, processes, and geography; improves communication
among scientists and engineers; makes intramural job transfers easier; but may create a gap between the R & D and production people.

9.2.   ENGINEERING'S INPUT INTO THE R & D PROCESS

While the purposes for which each segment or stage on the R, D & E spectrum is undertaken differ from stage to stage, the same work activities, technical disciplines, or functions tend to occur in all of them. Activities such as
design layout, drafting, some theoretical or physical model building, engineering analysis, testing or experimentation, and so on, are found throughout most R, D & E projects. The stages or cycles differ primarily in the
mixture with which the activities occur in each cycle. Thus the "upstream"
segments such as applied research and the conceptual stage of development
are apt to be richer in theoretical work, experimentation, and analysis. Similarly, the design and release stages will contain far more layout, drafting,
and checking work (whether this is on an old-fashioned drawing board or in
a modern system of computer—assisted design and manufacturing). This
does not imply that drafting is not done in research or that analysis is not
done in the design stage, and so on.

All this is not to belabor the point, but to drive home the fact that it is
futile to try to establish razor-sharp functional boundaries among the research, development, and engineering functions. There is considerable
overlap, and organizational boundaries are at best arbitrary and at their
worst meaningless. Input to R & D from Engineering must be viewed from
the following perspective: Research, Development, and Engineering are
"family." They are the technical heavyweights of the organization, and with
all their individual differences, they are in the sense of industrial sociology,
culturally far closer to each other than any of them is for example, to Marketing. Inputs to R & D from Engineering are thus intramural, and they
generally fall into two categories of "wish lists": (1) quasi-work-order and
(2) feedback on R & D output previously received. In the best of all possible
worlds the interchanges among R & D and Engineering should be productive technical dialogues in which each party provides constructive infor-

mation that stems from their respective fields of expertise (be these theoretical or practical). In reality there are often conflicts—for example, Development may swear that Manufacturing Engineering has completely degraded the wondrous performance of their better mousetrap: In their zeal to shave production costs, those tool and process engineers substituted cheap sheet-metal stampings for Development's die-cast parts. In turn, the engineers have accused Development of designing parts that can't be machined or assembled, such as holes blind at both ends, or screws impossible to reach during assembly.

From this discussion it becomes apparent that we can consider research and development the principal invention, innovation, and discovery arm of the company's technical family, while the engineering functions' mission is the profitable reduction-to-practice of the innovations passed on to them. The work-order-like input to R & D consists of requests for new products or processes which evolving technology now makes feasible. Distinguish this from inputs to R & D from Marketing, which is mostly driven by perceived market needs, requirements, and opportunities. Feedback-like input to R & D from Engineering refers more to the back-to-the-old-drawing-board reactions alluded to earlier. These are requests for product or process improvements, perhaps with some suggestions on how to accomplish them, which the engineers perceive to be needed for enhanced manufacturability and function.

9.3. ESTABLISHING GOALS AND PRIORITIES

9.3.1. What Business Are We Really In?

The definition of the range of interest or sphere of influence of a company necessarily becomes more complex as it increases in size. A further quantum jump in complexity is the result of conglomerates that consist of a group of disparate companies acquired primarily for financial reasons. In the latter case, considerations of redeployment of assets may be the controlling criterion for allocating resources, and parts of the conglomerate not well regarded at headquarters may be kept short of resources and eventually cast aside from a definition of the company's business. Experience over the years indicates that conglomerates have often acquired companies with a strong engineering basis, only to lose interest in them later, and divest themselves of them.

Of course, the key question that may want to be asked is: What business do we want to be in? Very often the growth in corporate scale and diversity, alluded to previously, comes about by accretion and, at least as far as the resulting complexity goes, is largely unintentional. Planning is usually driven by the goal of monotonic growth of the bottom line. The incubi grown in the process periodically surprise management, particularly since habit, tradition, and union pressures militate toward continuing operations long after their contribution to the business has eroded. When he was vice-president for Research and Engineering at IBM a few years ago, Dr. E.R. Piore was asked at a meeting of industrial research directors what he considered to be the chief characteristic distinguishing a Nobel Prize-level scientist from the competent but otherwise pedestrian R & D staff member. Unhesitatingly he answered: "Good taste in his choice of problems!" One can readily draw an analogy to the choice of business a board of directors makes for their company. The criteria applied may be based on offensive or defensive considerations. They may be driven by protective and conservative strategies or to take advantage of emerging opportunities.

A problem in corporate goal setting of this sort is that just as is commonly recommended for corporate charters, it is useful for a company to define its scope widely enough to be able to take advantage of promising developments in a wide area. This is sound policy, but it is also necessary to guard against overexuberance. A company that defines its purpose as being "in communications" or "in the leisure industry" is likely to find out quite soon that these are such enormous and diverse activities that one cannot do decisively well in all of them. This point increases in importance with the degree of technical sophistication of the products. A corporate management geared primarily to financial emphasis will find it difficult to understand the technical points of what is going on and, given the short-range perspective common to such emphasis, will find it difficult to muster the necessary technical understanding and especially the required patience. Yet there are always counterexamples: John F. (Jack) Welch, the CEO of the General Electric Corporation was asked, in the course of an executive visit with engineering students and faculty at Columbia University, what factors at the corporate level made up the "connective tissue" that held together the highly diversified GE Company. He said that in addition to the obvious financial capital, the two key factors were central research and centralized, uniform management training. On this foundation, a great deal of autonomy could be given to a wide variety of largely independent businesses.

9.3.2. The Structure of Innovation

R & D and engineering determine the way in which enterprises respond to innovation; it is therefore useful to examine its nature, as it relates to technical content. Ullmann[2] has proposed four categories; they progress from primary emphasis on basic and closely related applied research to primary emphasis on the routine application of known designs and production technology.

The first category is the true primary or seminal discovery that gives rise to a whole new industry. Such changes are relatively rare in industrial history. The invention of solid-state devices and their derivatives, television, and antibiotics are examples in our time. Plastics played a similar part earlier in this century.

A second group of innovations essentially exists for the purpose of serving the primary discoveries. Computers, for example, would not have become useful without software, nor tape or video recorders without cassettes; the latter wrought changes in the whole culture-service industry.

A third category consists of the small incremental efforts that go on in all technically oriented enterprises. Engineering departments are expected to come up with such design improvements or changes in manufacturing almost as a matter of course, and their efforts are evaluated in this way whenever budget time comes.

Finally, there is a large group of products that are able to meet any foreseeable need of their users. In such cases innovation is constrained because there is no point in making sustained research efforts when the standard products now in use have only limited scope for change and when end runs around present technology are unlikely to yield significant competitive advantage to their creators or are unlikely altogether.

These categories call for changing relationships between technological, scientific, economic, and marketing considerations as well. The problems start out primarily as scientific ones. As the product evolves, the parameters become more and more economic because these determine much of the manufacturing system and, most certainly, the marketability of the product. At the stage of product introduction, marketing considerations become very important, if not dominant. Unless there are a lot of customer complaints requiring a movement back to the old drawing board, marketing considerations tend to retain their importance for much of the remaining product cycle until saturation or stagnation also prompt a return to the laboratory or drawing board for what, it is hoped, will be a competitively decisive breakthrough.

9.3.3. Offensive versus Defensive Research

There is an important distinction between research undertaken to explore wholly new areas with intent to score decisive advantage in existing markets, or enter a new business with such advantage; and that research conducted for countering such efforts or accomplishments by the competition. The former is called offensive research, while the latter is called defensive research.

The difference lies first of all in the degree of discretion at one's disposal. Offensive research is to a large degree elective, that is, it can be engaged in or not, as management wishes. Internally the burden of proof for its sponsors is correspondingly heavier. Because of the increased risks, more opposition is likely from those who believe the resources could be better allocated (sometimes, though certainly not always, to themselves).

In defensive research the threat may be so clear that the need for something to be done is clear to all except those who see the best strategy as getting out of the business altogether. The real intraorganizational conflict tends to come over deciding what to do about it.

Defensive research has also acquired a second meaning which is very important at the present time. It is its role in responding to governmental regulations, notably in safety of products and manufacturing processes, and related environmental effects. Some business leaders use the term pejoratively for this reason. Yet defensive research in response to outside conditions is an old story. In the late 19th century, for instance, the copper industry found it necessary to develop ways of using sulfite ores to replace the oxide ores that were then nearing exhaustion. In a technical sense this was not much different from redesigning a process to meet environmental goals. The change from sulfuric to hydrochloric acid in steel pickling is another and much more recent example; hydrochloric acid may be recycled and thus lessens river pollution. It is also cheaper. The manner in which the term "defensive research" is at times used these days may thus owe more to economic-political value judgments than to technical reality.

9.3.4. Project Selection Criteria

The selection of R & D projects for support and execution and the subsequent assessment of what is needed to sustain them is related to technology assessment which is discussed in part in Chapter 3, which also offers extensive references to the wide literature on the subject. Much of the substance of the problem is beyond our scope, but principal questions would have to include the following:

How much technology is on hand for solving the problem; what major breakthroughs are needed and what are the prospects of getting them?

Accordingly, what are the probabilities of technical success in meeting applicable targets or specifications by a certain time and within a given budget?

What are the economic and other nontechnical parameters of the subject, and what would be required to respond to any constraints they offer?

What are the anticipated growth rates of demand for the product and the anticipated rates of progress in meeting that demand?

What are the anticipated competitive challenges—both domestic and (increasingly important) foreign?

What is the nature of cross-impact, that is, of competitive technical and scientific developments, or of public interventions which might change the prospects of the enterprise, both favorably and unfavorably?

A systematic approach to marshalling the key factors for project selection and way of treating them in a consistent fashion, useful for ranking project against project at any given time and to permit longitudinal comparisons (both historical and projected into the future), was developed by Olin Industries some years ago. While there is no unique formulation used throughout industry, there is a generic formula which incorporates the essence of the approach. This involves establishing an index of attractiveness (I) which one can assign to each project:

$$I(t) = \frac{p(S,t) \times R|S,t}{E(C,t)}$$

where I = index of attractiveness of the project

$p(S,t)$ = the probability of technical success of the project by time t

$R|S,t$ = the dollar return from the project's results, *if* it is technically successful by time t

$E(C,t)$ = the estimated project cost in dollars to time t

A moment's reflection shows that the dimensionality of this index is a pure number, which allows one to rank projects numerically and cut off approval to proceed at any point where the forecast cost exceeds the available budgetary funds. The original Olin index did not have the time factor explicit, but it appears sensible to include it. Just consider what will happen to the expected return if a key competitor succeeds in getting to market well before your own new product is off the drawing board!

A last consideration is that the management style of an organization's R & D policy can be made explicit by means of this index: For a given expected project cost, the success probability factor and the expected dollar return can be conjugated. Thus a minor product improvement, which has a high probability of technical success (since you're not pushing the state of the art but yielding a modest dollar return) and a real technical breakthrough, like a new nylon, which has a low probability of success but huge dollar return could have almost equivalent indices of attractiveness. It takes a watchful review policy to assure that an organization balances its risk-portfolio and does not only go for low-yield sure things, which are generally highly vulnerable to technological obsolescence.

9.4. SELLING RESEARCH IDEAS

Research ideas must often be "sold" internally as well as externally. Internally any major new project must overcome the resistance of vested interests concerned over a potentially adverse change in budgetary preferment or executive interest. The problems discussed in Sec. 9.1.2 are prominent sources of such conflict.

There are also two important external aspects to a company's research activities:

1. Research has important public relations elements. Revealing that a company is on the verge of major breakthroughs can have important beneficial effects, ranging from better internal morale, higher standing in the business community, and greater eagerness of distributors to do their best for the company's products, all the way to a substantial boost for the company's stock price. On the other hand research troubles or delays can undo all these potentially beneficial effects.

2. A research success (as well as a failure) changes the interactions between a company and its industry, affecting trade relationships, defensive measures by competitors, cooperation as in cross-licensing or supplier-customer arrangements, and in other ways.

"Selling" research ideas may also be taken literally. It becomes an obvious necessity when a company tries to secure research contracts as such. When this involves selling to the military, for example, further elements must be considered, notably a highly politicized market environment, a need for secrecy in most cases, and the possibility that some research scientists and engineers may have moral scruples on the subject.

From a direct managerial standpoint, research contracts in general may involve more stringent procedures for accounting and, above all, for report preparation. That may not only be a contractual necessity, but also part of impressing current and potential clients; the former should, after all, be viewed as likely sources of further work. By contrast, internal research reporting may largely consist of periodic meetings with, or demonstrations for top executives, without burdening scientists with the reporting provisions that are frequently a bane of research contracts.

As firms become primarily engaged in research, their managerial styles change; they run by the precepts of research management, rather than engineering management. However, the prescriptions in this volume, involving as they do the deployment of technical professionals in well-designed organizations, would still serve their purpose.

NOTES

1. P.F. Drucker, *Management* (New York: Harper & Row, 1974), pp. 31–32, 782–804
2. J.E. Ullmann, *The Prospects of American Industrial Recovery* (Westport, CT: Quorum, 1985), pp. 98–99.

10

SOURCES OF INFORMATION FOR ENGINEERING MANAGEMENT

Samuel E. Gluck

10.1. INTRODUCTION

"Sources of information" first brings to mind the written word: Books, technical journals, government publications, and other published material. Not only are there many categories of publications that are often overlooked, but there is a broad array of other sources that engineers should use.

The discussion in this chapter is not undertaken from the perspective of the giant corporation. How many firms actually have a "company library"? Even those that do, have budgetary constraints and need to develop criteria for the acquisition of material. How many firms of any kind, whether manufacturing, service, or consulting, have the time, qualified personnel, and money for leisurely and exhaustive scholarship? Most engineers enter upon professional life with their textbooks, some handbooks, a journal subscription or two, and a dictionary. The approach of this chapter is a practical one: What kinds of information do engineering departments require, and how do they go about getting it?

The kinds of information engineering departments require are comprised of the following:

1. Review sources of engineering principles, their applications and scientific foundations

2. Materials and processes of manufacturing, and information about new materials and processes

3. Changes in the characteristics of raw materials on the market

4. Knowledge of new developments in clients' and customers' industries that indicate their emerging needs and herald product-line changes

5. Knowledge of competitors' products and services—what they are, how they are used, their performance claims, the processes by which they are manufactured, competitors' manufacturing costs, their service difficulties and shortcomings, and their points of superiority

6. Internal cost and vendor information, including sources of supply

7. Customer responses, both favorable and unfavorable, regarding the company's products and services—this includes knowledge of actual field experience in the maintenance and repair of one's own products and comparison with competitors' products as to frequency, ease of repair, and maintenance costs

8. Worker safety and occupational health information—both in the industries which the company serves and in the company's own workplace—especially with regard to the potential of product liability in customers' workplaces

9. Applicable industry standards (so-called voluntary consensus standards)

10. Government standards—federal, state and local, and military specifications and standards if relevant

11. New products and techniques available to engineering departments themselves—for example, new computer software for engineering, new testing and measurement devices and methods, and even new materials and instruments for better drafting room productivity

10.2. SOURCES OF SOURCES

The first question facing the engineer engaged on a project is often that of *knowing where to look.* Libraries contain *sources-of-sources.* Among these are regularly updated printed indexes to periodical literature, some of which are also stored by computer bibliographic services that can be accessed either by a university or public library, or through the company's own computer and a modem for telephone linking. Among the indexes available are the following.

10.2.1.   Indexes (and Computer-Access Name, if On-Line)

Applied Science and Technology Index
Science Abstracts (INSPEC): Series A, Physics; Series B, Electrical and Electronic; Series C, Computers
Engineering Index (COMPENDEX)
Chemical Abstracts (CA SEARCH)
Science Citation Index (SCISEARCH)
British Technology Index
Proceedings in Print (conference proceedings)
Index to Scientific and Technical Proceedings
Environment Abstracts (ENVIRONLINE)
Pollution Abstracts (POLLUTION ABSTRACTS)
Energy Index (ENERGYLINE)
Fuel and Energy Abstracts
Energy Information Abstracts (ENERGYLINE)

10.2.2.   Literature (Subject) Guides

Guide to the Literature of Engineering, Mathematics and the Physical Sciences (Baltimore: Johns Hopkins University Press)
Guide to Basic Information Sources in Engineering (New York: Wiley)
Scientific and Technical Information Sources (Cambridge: MIT Press)
Science and Engineering Literature: a Guide to Reference Sources (Littleton, CO: Libraries Unlimited)

10.2.3.   Journal Lists

1. *The Standard Periodical Directory* (New York: Oxbridge Communications)
2. *Ulrich's International Periodicals Directory* (New York: R.R. Bowker)
3. *Business Publications Rates and Data* (popularly called the "Standard Rate and Data Book"). Published monthly by Standard Rate and Data Service, Skokie, Illinois. Any relatively recent issue will list trade journals by field.

10.2.4.   Publishers' Catalogues

Businesses in general and those offering professional services in particular, often receive publishers' catalogues that describe potentially useful books and other sources of information. Many of these come unsolicited because publishers use mailing lists that include professional-society membership rosters and the subscription lists of trade journals. In addition, any publisher will send you a catalogue if requested on your company letterhead. Catalogues will state the publication date of the books, but special sale or bargain offerings often omit that important fact. Sale books are often older editions, and the value of an older book depends upon how fast the field is developing. Unless stated to the contrary, most publishers will pay postage and handling charges on prepaid orders, and many offer a refund if the book is returned within a stipulated number of days. Books should be returned in their original shipping cartons, which are designed to protect their contents. The return privileges may be nullified if the book is examined while eating lunch.

10.2.5.   "Professional Book Clubs"

These should be approached with caution. In the majority of instances, the books, especially those offered as free inducements, are old editions that are on the verge of being supplanted. This does not necessarily lessen their usefulness, but it is always an individual matter. Most "clubs" are subsidiaries of publishers who use them as outlets for surplus stock. Subsequent member discounts tend to be tiny, and the postage and handling charges cancel most of the savings.

10.2.6.   Professional Society Publications

Many societies have large publication programs; for example, the American Society for Metals, American Society For Quality Control, and the various engineering societies. However it is not uncommon for the books of other publishers to be more expensive when ordered through a technical- or professional-society book service.

10.2.7.   Book Reviews

Most society journals and a few trade publications review books, and even those that do not may carry short lists of publications. These reviews and short lists should be filed for future reference as they often are an efficient preliminary means of searching for sources of information. For example, *Chemical Engineering*, published every two weeks, has an excellent book review section, as does *Science,* the weekly journal of the American Association For The Advancement of Science.

10.2.8. Company Research Publications

Many companies with large research programs publish books and monographs on subjects within their field which go far beyond discussions of their own products. Producers of chemicals and chemical raw materials, metallurgical producers, and makers of scientific instruments and measuring devices often publish these types of books. Some firms publish lists of the papers and books published by their employees, with complete bibliographical citations.

10.2.9. Government Publications Lists

1. *Monthly Catalog of United States Government Publications.* Order from the Superintendent of Documents, Government Printing Office, Washington, D.C. (Computer access code: "GPO Monthly Catalog.")

2. *Government Reports Announcements and Index.* Order from National Technical Information services, Washington, D.C., which is a private firm. (Computer access code: "NTIS.")

3. Many federal government agencies and bureaus also have monthly lists, and others maintain mailing lists and will send announcements of publications as they appear. Among these are the Patent Office, Office of Technology Assessment, Department of Agriculture, Bureau of Mines, Geological Survey, Occupational Safety and Health Administration, Mine Safety and Health Administration (located in Alexandria, Virginia), and the National Institute for Occupational Safety and Health (Cincinnati, Ohio).

4. Many foreign governments will send their publications catalogues, and a surprising amount of material is published in the English language—especially in the Scandinavian countries and the Netherlands. To receive British publications lists or to order the publications themselves, write to Her Majesty's Stationery Office, London, U.K.

10.3. HANDBOOKS, SPECIALIZED ENCYCLOPEDIAS, AND OTHER READY REFERENCES

What follows is a short list of these and other references classified into some very broad categories. No such list presumes to be complete, and the absence of a title which the reader deems valuable should not be construed as a negative judgment. It is important to keep several things in mind. First, in daily use many of these references are known by their *titles:* "Mark's Handbook," "Handbook of Chemistry and Physics," "Perry," "the CEMA book," and so forth. Second, books published by industrial companies are rarely listed in *Books in Print.* And finally, many valuable references are out of print, but that does not mean that they are obsolete.

10.3.1. Mathematics and Basic Science

Beyer, W. H., *CRC Standard Mathematical Tables.* Boca Raton, FL: CRC Press, 1976.

Dean, J. A. (ed.), *Lange's Handbook of Chemistry,* 15th ed. New York: McGraw-Hill, 1978.

Gray, D. E. (ed.), *American Institute of Physics Handbook.* New York: McGraw-Hill, 1982.

Heading, J., *Mathematical Methods in Science and Engineering.* New York: Elsevier, 1970.

Johnson, D. E., and Johnson, J. R., *Mathematical Methods in Engineering and Physics.* Englewood Cliffs, NJ: Prentice-Hall, 1982.

Korn, G. A., and Korn, T. M., *Mathematical Handbook for Scientists and Engineers.* New York: McGraw-Hill, 1968.

Markov, A. S. (ed.), *Dictionary of Scientific and Technical Terminology.* The Hague, Netherlands: Martinus Nijhoff, 1984. Available from Kluwer Academic Books, Boston, MA.

McGraw-Hill Editors, *Dictionary of Science and Engineering.* New York: McGraw-Hill, 1984.

Parker, S. (ed.), *McGraw-Hill Dictionary of Scientific and Technical Terms,* 3rd ed. New York: McGraw-Hill, 1984.

Salvadori, M. G., *The Mathematical Solution of Engineering Problems.* New York: Columbia University Press, 1953.

Snedden, I. N. (ed.). *Encyclopaedic Dictionary of Mathematics for Engineers and Applied Scientists.* London and New York: Pergamon Press, 1976.

Weast, R. (ed.). *Handbook of Chemistry and Physics,* 65th ed. Boca Raton, FL: CRC Press, 1985. (revised annually)

10.3.2. Practical Aids for the Working Engineer

Besterfield, D. H. and O'Hagen, R. E., *Technical Sketching for Engineers, Technologists, and Technicians.* Reston, VA: Reston, 1983.

Carroll, P., *How to Chart Data.* New York: McGraw-Hill, 1975.

Steidel, R. F. and Henderson, J. M. *The Graphic Language of Engineering.* New York: Wiley, 1983.

10.3.3. General Handbooks

Heisler, S. I. (ed.), *The Wiley Engineer's Desk Reference.* New York: Wiley, 1984.

Instrument Society of America. *Standards and Practices for Instrumentation,* 7th rev. ed. Pittsburgh: Instrument Society of America, 1983.

Kverneland, K. O., *World Metric Standards for Engineering.* New York: Industrial Press, 1978.

10.3.4. Computer Use and Technology

Artwick, B., *Applied Concepts in Microcomputer Graphics.* Englewood Cliffs, NJ: Prentice-Hall, 1983.

Atre, S., *Data Base Structural Techniques.* New York: Wiley, 1980.

Helms, H. L. (ed.), *The McGraw-Hill Computer Handbook.* New York: McGraw-Hill, 1983.

Ralston, A. (ed.), *Encyclopedia of Computer Science and Engineering.* New York: Van Nostrand Reinhold, 1982.

Synott, W. R. and Gruber, W. H., *Information Resource Management.* New York: Wiley, 1981.

10.3.5. Agricultural Engineering and Technology

American Society of Agricultural Engineers. *Agricultural Engineers Yearbook of Standards.* St. Joseph, MI: American Society of Agricultural Engineers, 1983.

Farm Chemicals Handbook. Willoughby, OH: Meister, 1985. (revised annually)

Krutz, G. W., Thompson, L., and Clear, P., *Design of Agricultural Machinery.* New York: Wiley, 1984.

Richey, C. B., *Agricultural Engineers Handbook.* New York: McGraw-Hill, 1971.

10.3.6. Automotive Engineering

The Technical Committees of the Society of Automotive Engineers, *SAE Handbook,* 4 vols. Warrendale, PA: Society of Automotive Engineers, 1984. (revised annually)

10.3.7.   Chemical Engineering

Blackwell, W. W., *Chemical Process Design on a Programmable Calculator.* New York: McGraw-Hill, 1983.

Chilton, C. H. (ed.), *Perry's Chemical Engineering Handbook,* 6th ed. New York: McGraw-Hill, 1984.

Chopey, N. P., and Hicks, T. G., *Handbook of Chemical Engineering Calculations.* New York: McGraw-Hill, 1983.

Davis, M. E., *Numerical Methods and Modeling for Chemical Engineers.* New York: Wiley, 1984.

Faith, W., *Faith, Keyes and Clark's Industrial Chemicals,* 4th ed. New York: Wiley, 1975.

Kirk-Othmer Encyclopedia of Chemical Technology. New York: Wiley, v.d. (26 volumes under continual revision)

10.3.8.   Civil and Architectural Engineering

Anderson, J. and Mickhail, E., *Introduction to Surveying.* New York: McGraw-Hill, 1984.

Barrie, D. S. and Paulson, B. C., *Professional Construction Management.* New York: McGraw-Hill, 1983.

Bowles, J. E., *Physical and Geotechnical Properties of Soils.* New York: McGraw-Hill, 1979.

Dudley, L., *Architectural Illustration.* Englewood Cliffs, NJ: Prentice-Hall, 1977.

Harvey, J. C., *Geology for Geotechnical Engineers.* New York: Cambridge University Press, 1983.

Koerner, R. M., *Construction and Geotechnical Methods in Foundation Engineering.* New York: McGraw-Hill, 1984.

Kuckein, H. E., *Architectural Illustration and Presentation.* Reston, VA: Reston, 1983.

Merritt, F. S. (ed.), *Standard Handbook for Civil Engineers.* New York: McGraw-Hill, 1983.

Oles, P. S., *Architectural Illustration: The Value Delineation Process.* New York: Van Nostrand Reinhold, 1979.

10.3.9.   Electrical and Electronic Engineering

Fink, D. G., and Beaty, H. W. (eds.), *Standard Handbook for Electrical Engineers.* New York: McGraw-Hill, 1978.

Turner, L. W., *Electronic Engineers Reference Book*, 5th ed. London: Butterworth, 1983.

10.3.10.   Energy and Environmental Engineering

Kaplan, S. A., *Energy Economics: Quantitative Methods for Energy and Environmental Decisions.* New York: McGraw-Hill, 1983.

Rosenberg, S. (ed.), *ASHRAE Handbook*, 4 vols. Atlanta: American Society of Heating, Refrigeration and Air Conditioning Engineers, v.d.

Turner, W. C., *Energy Management Handbook.* New York: Wiley, 1982.

10.3.11.   Industrial Engineering

Dallas, D. B. (ed.), *Tool and Manufacturing Engineers Handbook,* 3rd ed. New York: McGraw-Hill, 1976.

DeGarmo, E. P., *et al., Engineering Economy,* 7th ed. New York: Macmillan, 1984.

DeGarmo, E. P., *et al., Materials and Processes in Manufacturing,* 6th ed. New York: Macmillan, 1984.

Dorf, R. C., *Robotics and Automated Manufacturing.* Reston, VA: Reston, 1983.

Ekambaram, S. K., *The Statistical Basis of Quality Control Charts,* 2nd ed. New York: Asia Publishing House, 1966.

Grant, E., and Leavenworth, R. S., *Statistical Quality Control,* 5th ed. New York: McGraw-Hill, 1980.

Salvendy, G. (ed.), *Handbook of Industrial Engineering.* New York: Wiley, 1983.

10.3.12.   Material Handling

Apple, J. M., *Plant Layout and Material Handling,* 3rd ed. New York: Wiley, 1977.

Engineering Conference of the Conveyor Equipment Manufacturers Association, *Belt Conveyors for Bulk Handling,* 2nd ed. Boston: CBI Publishing Co., 1979. (Or contact CEMA, 152 Rollins Avenue, Rockville, MD 20852, for full CEMA publications list.)

Woodley, E. (ed.), *Encyclopedia of Material Handling,* 2 vols. Oxford, England: Pergamon Press, 1964. (Out of print, but available on order, in photocopy form, from Pergamon Press, Elmsford, NY.)

10.3.13.   Materials Science and Metallurgy

The American Society for Metals (Menlo Park, OH) publishes hand-books, encyclopedias, and reference books, and the society's extensive catalogue is available on request. Publications include multivolume works with the volumes published at intervals of a year or more, often with the earlier volumes being revised before the later ones are published. Among them are:

ASM Metals Reference Book
Engineering Properties of Steels
Metals Handbook (*12 volumes to date*)
Worldwide Guide to Equivalent Irons and Steels
Non-Destructive Testing Handbook

Brady, G. S., and Clauser, H. (eds.), *Materials Handbook*. New York: McGraw-Hill, 1977.

Brown, W. F., Jr. (ed.), *Aerospace Structural Metals Handbook,* 5 vol. Columbus: Metals and Ceramics Information Center, Battelle Memorial Institute, 1963 *et seq.* (An annual updating service was begun in 1974.)

Lynch, C. T. (ed.), *CRC Handbook of Materials Science,* 4 vols. Boca Raton, FL: CRC Publishing Co., 1974–1980. Volume 4 is edited by R. Summitt and A. Slicker.

Polar, J. P., *A Guide to Corrosion Resistance*. Greenwich, CT: Climax Molybdenum Co., 1981.

Schwettzer, P. A. (ed.), *Corrosion Resistance Tables: Metals, Plastics, Nonmetallics and Rubber*. New York: Marcel Dekker, 1976.

Walton, C. F., and Opar, T. J., *Iron Castings Handbook*. Des Plaines, IL: Iron Castings Society, 1981. (Revised at 10-year intervals. Iron Castings Society, 455 State Street, Des Plaines, IL 60016.)

Wieser, P. F., *Steel Castings Handbook,* 5th ed. Des Plaines, IL: Steel Founders Society of America, 1980. (Revised at 10-year intervals. Steel Founders Society of America, 455 State Street, Des Plaines, IL 60016.)

10.3.14   Mechanical Engineering

Baumeister, T. (ed.), *Mark's Standard Handbook for Mechanical Engineers,* 8th ed. New York: McGraw-Hill, 1978.

Bednar, H. H., *Pressure Vessel Design Handbook*. New York: Van Nostrand Reinhold, 1981.

Dally, J. W., and Riley, W. F., *Instrumentation for Engineering Measurements*. New York: Wiley, 1983.

Loomis, A. W. (ed.), *Compressed Air and Gas Data,* 3rd ed., 1982. (Published by Ingersoll-Rand Company and available from Compressed Air Technical Library, 253 E. Washington Avenue, Washington, NJ 07882.)

Machining Data Handbook, 2 vols., 3rd ed., 1980. (Compiled by the Machinability Data Center, Metcut Research Associates, Inc., 3980 Rosslyn Drive, Cincinnati, OH 45209.)

Magyesy, E., *Pressure Vessel Handbook,* 6th ed. Tulsa, OK: Pressure Vessel Handbook Publishing Co., 1983.

O'Connor, J. J., and Boyd, J., *Standard Handbook of Lubrication Engineering.* New York: McGraw-Hill, 1968. (Sponsored by the American Society of Mechanical Engineers.)

10.3.15   Safety and Product Liability

Cralley, L. V., and Cralley, L. J., *Industrial Hygiene Aspects of Plant Operations,* 3 vols. New York: Macmillan, 1982.

General Electric Company, *Material Safety Data.* Schenectady, NY: General Electric Company, v.d. (A looseleaf subscription service.)

Het Veilgheidsinstitutt, *Handling Chemicals Safely.* Amsterdam: Het Veilgheidsinstitutt (Netherlands Safety Institute), 1981. (in English)

Marshall, G. A., *Safety Engineering.* Monterey, CA: Brooks-Cole, 1982.

Sax, N. I., *Dangerous Properties of Industrial Materials,* 6th ed. New York: Van Nostrand Reinhold, 1984.

Tye, E., *The Management Guide to Product Liability* London: New Commercial Publishing Co., 1981.

Weinstein, A. S., and Donoher, W. A., *Products Liability and the Reasonably Safe Product.* New York: Wiley, 1978.

10.3.16   Structural Engineering

American Institute of Steel Construction, *Steel Construction Manual,* 8th ed. Chicago: American Institute of Steel Construction, 1980.

American Institute of Timber Construction, *Timber Construction Manual,* 3rd ed. New York: Wiley, 1985.

Gaylord, E. H., and Gaylord, G. N. (eds.), *Structural Engineering Handbook,* 2nd ed. New York: McGraw-Hill, 1979.

10.3.17   Explosives

DuPont Corporation, *Blasters Handbook,* 16th ed. Wilmington: DuPont Corporation, 1980.

Meyer, R., *Explosives.* Daytona Beach, FL: Verlag Chemie, 1982.

10.4   PRODUCT AND MANUFACTURER DIRECTORIES

The two major categories of product directories are the "buyers' guides" published as specials of, or supplements to, trade journals and independently published product directories. Those published in conjunction with trade journals generally limit listings to the products used in the industries to which they are addressed, and they determine the categories in which products are listed.

Listings are free of charge and the form for listing one's own product usually is sent to the firm's marketing department. If the buyers' guide is a separate publication, it may go only to the marketing department and its existence can remain unknown to the engineering staff. If it is part of one of the regular issues of the magazine, it will come to engineering only if someone in the engineering department is a subscriber.

The general directories of products and manufacturers are used by purchasing departments and should be made accessible to the engineering staff.

Some representative examples of both these types of sources are:

10.4.1.   "Buyers' Guides"

Chemical Engineering Equipment Buyers Guide ("CEEBG"). Annually, by *Chemical Engineering* magazine. New York: McGraw-Hill.

Chemical Week Buyers Guide. Annually, by *Chemical Week* magazine. New York: McGraw-Hill.

Modern Plastics Encyclopedia. Annually, by *Modern Plastics* magazine. New York: McGraw-Hill.

Plant Engineering Directory and Specification Catalog.

Plastics Technology Annual Manufacturing Handbook and Buyers' Guide. New York: Bill Publishing Co.

10.4.2.   General Directories of Products and Manufacturers

Dun and Bradstreet's Reference Book of Manufacturers. Annually.

MacRae's Bluebook, 5 vols. Annually.

Thomas' Register of American Manufacturers, 8 vols. Annually.
U.S. Industrial Directory. 4 vols. Annually.

10.5   TRADE JOURNALS

There are several thousand trade magazines published in the United States. One large group is directed toward readers in particular industries, others address special functions and interests common to many industries, and still others cut across many industries by aiming at audiences of various professional occupations.

Among those serving particular industries are, for example, the mineral industry magazines such as *Engineering and Mining Journal, Coal Age, Coal Mining World, Coal Rock Products, Pit and Quarry, Mining Engineering,* and others. The chemical industry journals, include *Chemical Engineering, Chemical Processing, Chemical Engineering Process, Agricultural Chemicals, Hydrocarbon Processing,* and others. There are journals published for the pulp-, paper-, and forest-product industries, grain industry, agribusiness and food processing, meat packing, glass making, and a host of other industries.

Journals related to particular design elements or categories of equipment include *Power Transmission Design, Hydraulics and Pneumatics, Plant Engineering and Maintenance, Material Handling Engineering, Modern Material Handling, Quality,* and *Occupational Hazards.*

Major journals dealing with broader aspects of engineering products and design include: *Machine Design, Assembly Engineering, Product Design and Development, Tooling and Production, Control Engineering, Production, Design Engineering,* and *Materials Engineering.* Obviously, the previously discussed three categories overlap. There is another type of trade magazine that consists almost entirely of product advertising. Among these are *New Equipment Digest* and *Chemical Product News.*

The most complete listing of trade journals is *Business Publications Rates and Data,* popularly called the "Standard Rate and Data Book." It is published monthly as a guide for placing advertising, industry by industry, but for the engineering department's needs any recent issue will do. If the company's marketing department does not have this directory, it can be found in many public and university libraries and most advertising agencies.

Every staff engineer should know what is happening in client industries and reading their trade journals is a prime source of that information. Competitors' advertising should be read carefully, as should announcements of new products and product literature. Trade journals reveal what is used in the industry and the engineering and marketing departments should be alert to new product possibilities as well as other products that are, or could become, complementary to the company's own products. A few trade magazines have regular book-review sections, and most of them have a calendar of events such as trade exhibitions, technical conferences, and short courses. Not all of them publish sophisticated technical articles, but many do publish items of practical interest to practitioners in the industry and these often yield suggestions that are useful to suppliers' engineering departments. Always preserve the annual or other periodic indexes and the "buyers' guide" issue. Most trade journals are sent free to "qualified" persons in industry and most engineers meet the qualifications of job title and function. Every issue carries a subscription form which requires information about the subscriber and the company.

10.6. COMPETITORS' SALES AND ENGINEERING PUBLICATIONS

This is a source of both engineering and marketing information, and it is important that the engineering and marketing departments bring their points of view together. The most obvious facts to be gleaned from competitors' publications are the specifications, applications, capacities, and other performance characteristics, as well as the construction of their products. One should also attempt to analyze product performance guarantees. Tolerance, accuracy, and even life expectancy of products should be examined, as well as the limitations of competing products, by noting the caveats which concern application and use. Other important information may deal with safety. Parts, service, and maintenance manuals are invaluable.

The marketing information in these publications includes, for example, the areas of application in which one's own products and those of competing products overlap, where the gaps are, where each company's line is incomplete, and especially where the gaps are in the market that no one is filling.

Begin by making a chart which lists all the essential physical and performance characteristic of the product. This should include dimensions, materials of manufacture, and the types and specifications of original equipment manufacturer components (for example, bearings, control mechanisms, and power transmission elements). List these in a vertical column and then, across the page, make a column for each competing item. Make sure that all models and variations are listed, including those of one's own company. Match these as closely as possible. Leave space for additional lines for new aspects that are discovered to be relevant to the comparison.

It is important to know the date of publication of the material you are analyzing and comparing, as well as whether the material is still current. Most companies code their publications with a kind of numerical acronym: for example, "118110M" would mean "November 1981, 10,000 copies printed." If a company is still issuing this bulletin in 1985, it does not prove that there have been no product changes, but it probably indicates that no significant changes have been made in the interim. Stating the source of the document is also important: for example, whether obtained at a trade show, or a reply card request such as the "bingo card" in a trade journal.

Do not routinely discard old information about competing products simply because it is old. If a dating system is faithfully adhered to, one will not fall victim to obsolete information, and keeping old material can supply a history of product changes to compare with the evolution of one's own products. When a new catalogue, brochure, or service manual is acquired, note such information as:

1. Is there more, or less detailed information about particular products? If there is less information than in a previous catalogue, what has been omitted? Is there any hint as to why?
2. What has been added or dropped from the competing product line itself? Can one detect any technical reasons for doing so?
3. Have statements about product applications, performance, or safety been revised? If so, have they become more conservative or have they been expanded?

Many older catalogues, especially the larger ones, are highly prized for their general engineering information sections. This is especially true of the catalogues of manufacturers of process equipment, instruments, and control mechanisms for the basic raw-materials industries; the makers of power-transmission products; and machine-tool manufacturers. The same applies to catalogues of exotic materials such as chemicals, special metals and alloys, protective coatings, and the like. These catalogues often contain formulas, nomographs, charts, and calculation methods, as well as the physical and chemical characteristics of hundreds of natural and manufactured materials. If certain materials are still in broad use, although their production has been discontinued, the older information may be difficult to come by elsewhere. Many of these earlier catalogues include discussions of the basic principles of design and application and the criteria for the choice of products. One reason for eliminating them is, of course, the increased costs of printing and distribution, but there are other very important reasons. Many companies have responded to product liability litigation by requiring closer contact with their customers from the earliest stages of an inquiry, that is, they tell the prospective user *less*, to curtail amateur or incomplete application decisions that may rebound, however unjustified, upon the manufacturer. The fact remains however, that many of the older and more elaborate catalogues are virtual handbooks for the experienced engineer.

Some examples of valuable older catalogues are the *Stephens-Adamson No. 66*, the *Jeffrey Manufacturing Company* (now part of Dresser Industries), *Catalogue 418, Chain Belt* (now Rexnord) *700*, and the highly coveted *Link-Belt* "twins," Catalogues 1000 and 1050. By contrast, one should be extremely cautious about using old information about the chemistry of natural or manufactured raw materials, because it has been discovered that many are carcinogenic, toxic, or otherwise harmful.

10.7. TRADE SHOWS

There are many things one can learn at a trade show, provided one is well organized in advance and has clearly defined objectives. One of the most valuable opportunities supplied by a trade show is the chance to examine displayed items in detail. This is most importantly a chance to observe competing products and scrutinize products that one's own company may wish to use, such as accessory items, control mechanisms, or power transmission components. Trade shows are sources of new product ideas. The decision as to whether to attend should be a joint one. If the show is for a client industry, then engineering and marketing should each send someone. If one is seeking information about plant equipment or materials and components, someone from the manufacturing department should team up with engineering. In the case of a client-industry show, an important factor in the decision to attend is whether competing firms will be exhibiting. If there are few or none, and especially if one's major competitors are absent, it may not be worthwhile to attend. Do not base the decision to attend on the previous year's list of exhibitors.

Mark the advance list of exhibitors for special attention, but do not select certain scattered booths to go to upon your arrival at the exhibition. That is exhausting, time consuming, and may result in missing other important exhibits. Go to the top floor, start at one side of the room, and proceed systematically across and downward. Schedule plenty of time. Often what one sees at one exhibit will suggest a return to other exhibits.

Visiting a competitor's exhibit can be delicate, but it need not be embarrassing, even though one's identification badge may put a damper on conversation. Do not make the serious professional mistake of approaching a competitor under false pretenses. Pick up sales or engineering literature before starting a conversation.

Carry a 35mm camera, and use fast film so that a flash unit is unnecessary. A microcassette tape recorder, slightly larger than a pack of cigarettes, is an efficient substitute for notetaking.

When viewing a competitor's product, note the following points, especially in comparison with the products of one's own firm.

1. Are there significant differences in configuration, power requirements, mounting of suspension, or controls?
2. What manufacturing processes were used?
3. What is the quality of workmanship? Is this a standard model on display, or one that has obviously been manicured to give a good impression?
4. What are the safety devices? How are moving parts guarded? What kinds of warning signs and instructions are attached to the product?
5. Are there features that make service and routine maintenance easy?
6. If the product is a machine and the display is operating, take note of sound level, vibration, and behavior during the starting and stopping ranges.
7. What model or series is on display?
8. Note the brand and model number of OEM component parts.

10.8. SEMINARS AND SHORT COURSES

One is repeatedly cautioned that technology is moving so fast that unless the engineer and scientist continually update their knowledge, they will be left at the post. To facilitate this update there is a vast array of educational programs ranging from conventional graduate degree curricula with evening and weekend schedules for the working engineer, to refresher courses, seminars, intensive short courses, and similar programs. This discussion is devoted to nondegree courses of study and will focus on some of the factors to be considered in making a decision to attend.

From the viewpoint of management, there are, or should be, two distinguishable sets of objectives in deciding whether or not to attend a seminar or similar function. The first purpose which may be served is that of professional training for the staff. This is discussed in detail in Chapters 23, 24, and 25. The second is that central to this chapter, that is, the intelligence gathering which a technically oriented business must maintain. The two objectives may overlap; obviously anyone who attends will get some benefits in both areas. Still, an awareness of these two distinct purposes should inform both the initial decision to attend and the final evaluation of the seminars or courses.

Many courses and seminars are opportunities to *review* knowledge—to see it in the light of new relationships and modes of analysis. Others offer new technical knowledge, especially as related to new processes or developments in instrumentation and control. No matter what one's goal is, the most important element in the decision is the investment in time. If only for that reason, every course announcement should be carefully evaluated. As for the monetary cost, one must include not only the fee, living expenses, and cost of travel, but the salaries of those persons whom the company sends to the course. Courses and seminars, whether sponsored by universities, research institutions, or those private companies that are in the seminar business, are all designed for the purpose of making money. Many are worthwhile, while others are not only fly-by-night but also hit-and-run. Among the points to bear in mind are the following:

1. What does the course or seminar really cover? Is the announcement simple, direct, and to the point? Or is there a staggering list of subheadings that have been proliferated in an attempt to appear comprehensive?

2. Who is sponsoring it? A university or institute can rent space to a private seminar business and have no control over quality.

3. Who are the speakers and what are their qualifications? Does each speaker, or at least the faculty as a group, seem to have a good balance between theoretical background and practical experience?

4. Are the course materials listed in detail? What types of instructional materials will be used? Is it a one-man show with the instructor using only his own book? (Of course that person may be *the* expert in the field, and the author of a definitive treatise.)

5. Exactly how long does each session last? For example, a five-day intensive course that begins at 10:00 A.M. on Monday and ends at noon on Friday is already only a four-day course from which one subtracts coffee breaks, lunch periods, and question periods. Are there evening sessions? If the working days are short and there is heavy emphasis in the brochure on exploring the restaurants and entertainment of Deadman's Gulch or Sin City, one may not be getting full value, even granting the need for relaxation.

6. A well-planned seminar will include a statement of the *prerequisite* knowledge which an attendee should have. Is it too advanced or too elementary for one's own level of knowledge and interests? It is not uncommon to read that the course is meant for engineers, technicians, manufacturing managers, first-line supervisors, managers of

engineering departments, and a wide and varied list of all those whose job descriptions cover the spectrum of a typical company. Be careful: No course can be all things to all people.

7. If hands-on workshops are important to the area of knowledge, they should be included. Discussion periods are a mixed blessing unless an experienced instructor can guide the questions away from trivia.

8. Put little faith in the printed endorsements from past attendees.

9. How many Continuing Education Units (CEUs) are given? If none, the course may not meet the requirements of one's professional society. This does not mean that it is not a good course, nor does the awarding of CEUs guarantee high quality. However, the absence of CEUs is a caution flag.

10. Odd as this statement may appear, beware of political rhetoric substituting for information. For example, a conference billed as one on occupational safety and health, or environmental problems, that is actually an attack on regulatory agencies.

11. Be cautious, also, of seminar programs that have only tentative lists of speakers and topics. For example, if the announcement of a conference only sixty days hence has several topics to be announced or speakers to be announced, it indicates that the sponsors have had difficulty getting qualified experts to commit themselves and may have to settle for very ordinary substitutes.

A conference session should be attended in exactly the same way one attended lectures during school. It is a good idea to arrive early enough to occupy a strategic location in relation to the speakers and audio-visual aids. A small inconspicuous tape recorder is invaluable, together with extra batteries and plenty of tape. A recharger and a spare battery pack are the best combination. Use tapes of sufficient capacity to permit changing them during scheduled recesses, bearing in mind that the longer the tape the thinner it will be and the more care will be required in handling it when transcribing. Number the tapes and their sides and use them sequentially so that there will not be inadvertent double recording. With the recorder listening verbatim, one is free to copy material from the blackboard or screen. During discussion periods try to identify the questioner. If questions or comments come from someone who works for a competitor, what is said may have additional significance for one's own company. Keep a copy of the list of participants with the transcribed notes. Finally, with regard to one's own questions and comments, be aware that others are listening with the same purpose.

10.9.   INTERNAL COMMUNICATION

Effective, well-organized internal communication is one of the engineering department's most valuable sources of information. It would hardly seem necessary to discuss such an obvious point except for the fact that many organizations are beset with interdepartmental rivalries, with weapons that include withholding information. This is the first important source of trouble. A second one is that even where these rivalries do not exist, information is often exchanged only when there is trouble with a product. An example of the first situation is the frequent rivalry between purchasing and engineering departments; and an example of the second is the lack of communication because of inertia or poor coordination between the engineering and marketing departments. Therefore, a brief comment on both types of situations is in order.

10.9.1.   Communications Between Purchasing and Engineering

Because purchasing deals conspicuously with money matters and makes decisions measurable in dollars, it is often perceived by top management as the most vital department. For that reason many firms are in fact not "customer-driven" but purchase-driven; decisions that appear to show how much money is being saved by the purchasing department often work to the long-run detriment of the company, and it may take years to become aware of this problem.

It is likely that the purchasing department will have a better grasp of what materials and components are available because it deals in the marketplace every day. This is especially true with regard to items which are standard and most readily available. On the other hand purchasing decisions, especially when changes are contemplated, should not be made without consulting both the engineers and the factory. One example of a source of conflict is substituting one brand of OEM component for another which has identical *performance* specifications, on the basis of price, without checking to learn if an accompanying change in configuration will create manufacturing or assembly difficulties. If a customer has standardized on a particular item or a particular brand of OEM component, it is the obligation of the sales department to notify both engineering and purchasing and to explain why that particular specification is requested. If cost, delivery, or assembly problems will arise, all interested parties should understand, and the customer notified promptly.

Engineering departments should confer with purchasing as early as possible whenever design changes are contemplated. In turn, when large orders for raw materials or OEM components are to be placed, it is purchasing's obligation to ask the engineering department if any design changes are in the offing. When the purchasing department learns that an item is being discontinued, the engineering department should be notified at once; it does not matter how many months into the future the actual change will take effect. It could take that long to find a substitute that will satisfy design criteria and not require expensive changes in manufacturing procedures. Purchasing's devotion to the economic order quantity must never be so great that it prevents design and manufacturing flexibility to meet customer needs or increased product safety.

Finally, if purchasing expects engineering to be cost-conscious without sacrificing quality, they should make cost information easily and quickly available to the engineering staff.

10.9.2. Engineering and Marketing

Most of us are familiar with the cartoon strip that depicts how various departments conceive the company's product; each department's conception, according to the cartoon, is ridiculous in its own stereotyped way. The last panel depicts "what the customer really wanted": The product is a child's swing, and what the customer wanted was two ropes suspended from a tree limb and attached to a horizontal board. A half-century ago the slogan of one American auto maker (Packard) was, "Ask the man who owns one," and in ancient Athens, the philosopher Socrates cautioned that if one wishes to choose a fine musical instrument, one should seek the advice of the musician and not the instrument maker.

There is a chain of information that begins with the customer, passes to the salesman, and then to the service department. It should, if its links are properly forged, go on to the engineering department and the factory. Salesmen and service technicians not only hear the compliments and complaints, but they also hear the customers' desires for changes, new products, and new ways to solve their problems. For example, customers who are otherwise satisfied with a product's performance may be enraged at the difficulty of some particular items of maintenance or repair. With the close working relationships that normally develop between the engineering department and the factory, there is a tendency toward designs created for ease of manufacturing rather than regard to the customers' day-to-day problems. Witness, for example, the ratio of actual parts cost to labor cost in automobile repairs. An enterprising and creative engineer manager will help forge this chain of information.

11

DECISION SUPPORT, FEEDBACK, AND CONTROL

Gus W. Grammas, Greg Lewin and Suzanne P. DuMont Bays

GUS W. GRAMMAS is the Director of Technology & Company Programs, and an Adjunct Professor of Business at the Graduate School of Business at Columbia University. His degrees include: B.A., Columbia College, 1963; Ph.D., Department of Mathematical Statistics, Columbia University, 1972. Dr. Grammas has held the positions of: Associate Professor of Management, Graduate School of Business Administration, New York University (1977–1984); Assistant Professor of Management Science, Graduate School of Business, Columbia University (1970–1977); Assistant to the Director, the Executive Program in Business Administration (1968 & 1969), Instructor in statistics, School of Engineering, Columbia University (1966–1967). He currently lectures for Executive Programs. Dr. Grammas served as consultant to several major corporations and New York City agencies. Current publications and research concern such areas as the strategic management of high-technology companies, urban service systems, and mathematical models for energy and environmental planning. His articles have appeared in the *Journal of Environmental Economics and Management,* the *International Journal on Policy Analysis and Information Systems,* and the *Handbook of Operations Research.*

GREG LEWIN attended Northwestern University where he graduated in June 1979 with a Bachelor of Science degree in Industrial Engineering. Upon graduation he worked for IBM as a systems engineer. While at IBM he attended New York University's Graduate School of Business Administration where, in June 1982 he received his Masters of Business Administration in Finance. He then began a position with Merrill Lynch, Pierce, Fenner & Smith as a security analyst. His particular expertise was the analysis of companies involved with the technologies of local area networking. In April 1984 he joined the firm of Neuberger & Berman where he is presently responsible for the sale of securities to institutional investors.

SUZANNE P. DuMONT BAYS received a Master of Business Administration from the University of Western Ontario in 1980. She is currently employed in the Product Development department, Corporate & Government Banking for the Bank of Montreal. This group acts as a liaison between the user community and the development area of the bank, and is presently developing a system that will enable account managers to access current and historical information on a customer and product basis. Previous to this position, she worked in financial analysis for the controller's department.

11.1   INTRODUCTION

The ability of an organization to make good decisions is directly dependent on its effectiveness in gathering, analyzing, and reporting information. In today's complex business environment, the demand for specialization has emphasized the importance of dependent relationships. To solve problems efficiently a structure must exist to coordinate the flow of data between various departments or groups from the point of problem recognition, through problem resolution and subsequent evaluation.

In the field of engineering, constant feedback and control is required from the various design teams involved in major projects. Each aspect of the project requires careful coordination; otherwise the various teams may work at cross-purposes. Most of the time a project leader is not in a position to manage subprojects in sequence. Rather the teams may have to work in parallel, which requires constant communication, feedback, control, and analysis.

From a sales aspect, technologically advanced products cannot be designed and marketed successfully without a close relationship with the customer. For example, applications engineering requires open lines of communication between the customer and the engineering, production, and marketing departments. Customer understanding of the company's present and future products is essential. Of equal importance is the company's ability to understand customer needs and how a product, either current or envisaged, will fulfill these needs. This communications effort is essential in order to market the product from the research stage through to ultimate sales. This effort continues to be vital in maintaining sales and preparing for product extension, improvement, and, ultimately, replacement.[1]

With respect to problem solving, group decision making has many advantages over individual decision making. Groups have the ability to incorporate many more inputs concerning specific alternative courses of action. Moreover a group can utilize the wisdom and experience of individuals in various areas of expertise in order to evaluate these alternatives. In most cases participation in the process increases understanding and acceptance of the final decision. In addition, working with others has a synergistic effect in that it increases all levels of creativity. By having a group of people working on the same problem, individual biases may be offset and the potential unreliability of individual decisions may be reduced.[2]

The application of computer technology to the decision-making process has recently received a great deal of attention. A major conclusion is that efficient problem solving is contingent upon strong data-base management and reliable, adaptable information services. Such is not the case in many firms today. Although many managers would like to turn to their systems department for assistance in making key business decisions, often the information is not reliable or readily available. Cheney and Dickson's[3] empirical testing shows that the use of computer-based information systems improves the decision-making process and increases the stability of the decision-making environment. They also show that users' information level and job satisfaction increases with the introduction of a new system, most notably in companies whose information systems departments were well-managed. Their research supports the proposition that the degree of success in data-base management systems is not based upon the level of technological sophistication that an organization employs but rather on how the technology is initially implemented and supported on an ongoing basis.

In the context of business strategy, Lucas and Turner[4] argue that one of the keys to successful business management is the integration of information technology with strategy formulation. They identify three types of relationships that support this integration: independent information systems that create operational efficiencies; policy support systems that aid the planning process; and fully integrated systems that help define the range of possible strategic choices.

The issue of data processing and its effectiveness in meeting the existing gap between major business goals and the problems currently addressed by the typical operations system is discussed by Vierck.[5] The major concern is how to fit data-base management systems into an organization's information resource architecture to maximize the effectiveness of the people and the organization. Vierck concludes that the data processing industry should focus on information-resources management; that is, to help achieve the stated goals of the organization rather than simply processing data by computer.

The use of computers in business communications and the decision-making process has given rise to two pieces of terminology: Management Information Systems (MIS) and Decision Support Systems (DSS). A Management Information Systems is the combination of human- and computer-based resources that results in the collection, storage, retrieval, and communication of data for the purpose of efficient management of operations and for business planning. In order to enhance the interface between human and computer, a Decision Support System is one based on the design and implementation of a structure that controls the storage, access, and distribution of information throughout the organization. These systems are oriented towards upper-level management in an effort to improve the decision-making process.

The way in which an organization regulates the gathering and dissemination of information effectively, defines its managerial structure and its problem-solving process. In this Chapter the tools and techniques used in the design and implementation of an organizational communication system are examined. First of all, some of the traditional managerial techniques (schedules, meetings, group decisions, and so on) for regulating communication flow and enhancing decision making are discussed and their limitations analyzed. Secondly, the advantages and disadvantages of computer-based techniques are examined. The basic theme is that a comprehensive approach (human and computer) must be taken in implementing a computer-based system to support effective decision making.

11.2. TRADITIONAL TECHNIQUES

In the traditional business environment, the design of a structure to enhance communication in the decision-making process is based upon the implementation of an assortment of tools and techniques, each of which contributes to the problem-solving structure in a unique but complimentary fashion. However, policies must be instituted and followed in order to balance this structure; goals and objectives must be established in order to direct the process. Table 11.1 describes some of these techniques.

Each of these traditional techniques supports activities that contribute to the decision-making process. Effective implementation of the techniques can form a useful communication structure. As the complexity of the business environment grows, so will the number of problems and volume of information needed to solve them. This growth implies an increasing number of problem solvers, a larger number of departments involved, and further distance between problem solvers and information. As the dependencies become more complex, the timeliness and accuracy of the information is jeopardized. Under these circumstances the traditional methods used to

TABLE 11.1. Traditional Techniques Used In The Decision-Making Process

Technique	Description	Requirements	Advantages	Disadvantages
Schedules	To synchronize and document activities and resources	Must identify: Objectives Activities Resources Environment	Improve timing Reduce duplication	Encourage overdependence and inflexibility
Meetings	To evaluate past performance and predict future conditions	Must identify: Prior activities and conditions Scheduled activities Forecasted conditions	Complete information exchange	Substitute for task directed activities
Standard operating procedures	To balance & control formal and informal mechanisms of decision making	Document rules to balance formal and informal decision making	Expedite use of appropriate decision mechanism	Require review and update to avoid negative impact
Informal changing of communication flow	To define unstructured network of corporate resources and information	Document skills and responsibilities of employees	Identifies assistance, increases organizational flexibility, expedites crises problem solving	Alters critical relationships due to isolated decision making
Reporting	To review business information through a structured format	Must specify timely acquisition, layout, and distribution of information	Records business activities	Varies with quality of information

manage information are inadequate. As the participants, departments, and distance grow:

It becomes difficult to maintain comprehensive schedules that are up-to-date.

Meetings become increasingly difficult to schedule.

The problem solvers may be confronted by conflicting standard operating procedures that in turn may lead to poor process control.

The use of informal channels of communication to support information transfer is prevented by departmental boundaries.

The timely distribution of reports becomes increasingly difficult.

11.3. GROUP DECISION MAKING

As previously stated, a number of managerial tools and practices are available to aid decision making. Organized teams have the potential for making better decisions than an individual because expertise can be gathered and utilized from specialized fields. Conclusions are often reached in sequential order, with the findings of one group becoming the premise for the next. For example, the marketing department may assess the sales potential of a new product and in turn pass along their sales projections to the engineering department, which in turn assesses the firm's ability to produce such a product and develops requirements for start-up machinery. If a company wants to be confident that the final decision is optimal, careful attention must be paid to the basic assumptions. Personal bias and faulty communication can introduce substantial errors in the process.

Group decisions differ in the number of people involved and the degree of special effort required. As the complexity of a problem grows, so does the necessity for pooling more diverse managerial talents. A manager guiding the decision-making process must assess who will:

1. Identify and formulate opportunities
2. Arrive at creative alternatives
3. Produce projections of probable results
4. Evaluate separate projections and select one alternative course of action[6]

In order to achieve the first step, relevant information must reach a decision maker who perceives its significance. Provisions should be made to examine both the external environment and internal organization to detect future opportunities and problems. Generally speaking, information does not flow easily because people are poor transmitters of information. As information passes through additional intermediaries, transmission errors are compounded. Moreover, when a message passes up or down a chain of command, "protective screening" intervenes between subordinate and superiors as people selectively disseminate data. The combined effect is referred to as "organizational distance."

Creative alternative courses of action can usually be found through research and development teams, or simply a "think tank" group that is assigned the task of generating fresh ideas. This step of the decision-making process is particularly important in state-of-the-art technological firms. The basis for new ideas must be an organizational culture that fosters open communications.

The next stage is to produce projections of probable results. Because all the estimates used in decision making deal with the future, they involve some degree of risk and uncertainty. Many quantitative and qualitative analytical tools are available for dealing with complexity and uncertainty in decision making. For example, simulation and risk analysis can be particularly helpful.[7]

Estimates of future events can be either optimistic or pessimistic, depending on personal biases of the people making the projections. Some of the more common reasons why an individual's assessment may differ from that of the decision maker are: a person's or group's ability to persuade or sell other factions on their solution; the influence of societal pressures; personal aspirations; and different perceptions of the objectives. For these reasons, it is imperative that constant communication occur between all groups involved to minimize these effects. We inevitably must face uncertainty, absorption, and biased estimates, but the undesirable effects of these factors can be reduced by the appropriate utilization of a communications structure, the creation of a stable setting for projected activities, and the efficient use of a reliable data base.

11.4. COMPUTER BASED TECHNIQUES

The structure of a decision support system can be based on both manual and mechanized methods. Information is dependent on time: It must be collected, synthesized, stored, retrieved, communicated, and presented in a time frame appropriate to the decision-making task. A computer-based solution can address this requirement.

The computer can facilitate the decision-making process by:

1. Tracking and reporting problem incidents
2. Prioritizing, scheduling, monitoring and documenting problem solving activities
3. Analyzing and reporting performance
4. Integrating internal and external data to define the range of potential strategic alternatives

Once the computer has been properly programmed, its capacity to improve business activity can be significant. However, its ability to evaluate and process information is dependent on the ability of the decision makers to provide the computer with the proper questions.

The primary functional components of a computer-based decision support system are the data base, report generation, and inquiry processing. An analysis of the capabilities of these components and their application to the problem-solving process will assist in judging their value.

11.4.1. Data Base

The use of a central, integrated data base provides an organization with the ability to store all of its information in a single file. Traditionally, whether using manual or mechanized methods, information was stored in many separate files with the same information often common to more than one file. This redundancy required additional file space and a great deal of extra time and effort to manage the added workload. The use of separate files implies the potential use of separate file structures. This segmentation could place boundaries on the problem solver's ability to access information—boundaries that could defeat the fundamental objective of an information system, that of obtaining the proper information for the proper individual(s) at the proper time. In addition higher costs can be the result of the greater number of inaccuracies caused by a complex, multifile structure.

The use of an integrated data base eliminates this complexity by incorporating multiple files into a single hierarchical structure. A data base provides low-cost storage of billions of characters of data that can be accessed in seconds. The retrieval of any particular combination of datum is possible even with a data-base design that restricts access of data to selected individuals through the use of data protection schemes. The implementation of a single file structure can significantly reduce information costs and eliminate boundaries that impact problem solving.

11.4.2. Report Generation

The capability of the computer to generate reports can be used to produce three basic report types: performance analysis reports, exception reports, and special-analysis reports.

Computer-based performance reporting provides for the automatic access, processing and distribution of business information on a regularly scheduled basis. The computer allows a flexible presentation of requested information. Critical relationships can be emphasized by using the extensive graphics capabilities of the computer.

Exception reporting is used to emphasize unusual situations. It can be implemented by highlighting data on standard performance reports or triggering the generation and distribution of special exception reports. Exception reporting provides a mechanism for alerting problem solvers to the presence of potential crises situations. For example, a program could be designed that would record scheduled activities into a data base and would trigger the generation and distribution of exception reports if milestones and deadlines are missed.

The special analysis reporting function provides problem solvers with the ability to generate reports relevant to their problem-solving tasks. It allows them to examine unique relationships using information contained in the data base. This flexibility requires additional time and special programming skills to interface properly with the computer.

11.4.3. Inquiry Processing

Inquiry processing provides problem solvers with the ability to interact directly with the computer in a familiar language. The value of inquiry processing to the problem-solving process is dependent on the number and efficiency of computer programs dedicated to problem-solving tasks. Programs are logical collections of mathematical operations that are designed to interact with the data base to process user requests. A large number of computer programs, capable of processing a wide variety of problem solving requests, are currently available. The increased flexibility, improved usability, and reduced costs of these programs is rapidly bringing the approach of inquiry processing to the entire business environment.

As the size and complexity of the problem-solving structure grow, the need for a central facility to manage the flow of information becomes evident. A facility, commonly called the control center, can act as the information manager in a problem solving network.[8] It is typically staffed by one or more administrative-level employees, who serve to:

1. Record project information
2. Answer standard business questions
3. Speed complex requests to appropriate sources

Contact with the center is typically accomplished through a telephone call and entry into the data base serves as the permanent report record providing enforcement of project reporting standards.

The control center is responsible for maintaining a comprehensive record of currently scheduled activities and accepted standard operating procedures. Using this information the center provides employees with an identified service for answering questions, such as project status, documentation requirements, schedule changes, and usage of operating procedures. The use of sophisticated inquiry processing programs by requestors of information can assist them to directly access, retrieve, monitor, track, and report project information. This format would eliminate many of the routine requests handled by the control center.

Requestors can use the center to quickly locate sources for all business and engineering questions. The center will guide the requestor directly to a problem solver or to a business-functions representative. The principal business areas of the organization (engineering, finance, marketing, production, accounting, data processing, and so on) should be represented by an identified business contact. Typically, this contact function is a part-time position assigned to a manager in the business area. The representative serves as the primary interface between the business function and the entire organization. He or she is responsible for:

1. Disseminating information about the skills, responsibilities, and resources of the group represented
2. Monitoring organizational needs
3. Assisting in the allocation of resources
4. Monitoring the accomplishment of business objectives

The control center is a simple and efficient method for managing the information needs of the business. It expedites the controlled and coordinated flow of information in the problem-solving process. In addition, it serves to enhance productivity by minimizing the time needed to search for information and resources. The responsibilities of the control center can even be decentralized through the use of currently available computer programs. However, until an organization is prepared for this level of computer sophistication, the control center provides an effective way of managing information in the problem-solving process.

COMPUTER BASED TECHNIQUES

The introduction of an information network gives an organization the ability to centralize authority in the hands of top-level management. Although greater centralized control can be supported by a computer-based system, greater control may or may not be desirable. The degree of centralization is, of course, based on an organization's managerial philosophy and judgment.

There is no "best" way to implement the tools and techniques discussed into an integrated decision support system. Each organization adheres to unique management beliefs that demand unique strategies for achieving efficient problem solving. Regardless of the style and/or strategy chosen, however, the use of these tools and techniques will contribute significantly to an organization's ability to solve problems. As an illustration, an implementation strategy is described in Table 11.2 that merges the people and computer processes discussed previously into an integrated decision-making process.

From the initial point of problem definition, at which time functional management must determine objectives and available resources, through problem resolution and subsequent evaluation, a key component is the ability of control-center staff to work with functional management groups. It is essential that coordination of data to and from the systems department is properly directed and forms a cornerstone of management policy.

Common business practice employs crisis management to handle business problems. Typically, this practice is invoked at the expense of daily business functions, engaging too many resources, and resulting in decisions based on inadequate information. It is our contention that a decision support system based on the concepts and strategies discussed in this chapter will lead to:

1. Faster problem resolution
2. Less costly resolution
3. The anticipation and subsequent prevention of future problems

As the system evolves it should support the solution of known problems and prevent the development of potential problems so that the process ultimately goes beyond problem solving to problem management.

A computer-aided process enables the organization to develop a single problem-solving environment, allowing decision makers to:

Pool all problem-related information for the entire company and the industry as a whole into one data base to help define a range of possible alternatives

TABLE 11.2. Implementation Strategy For Decision-Making Process

Phase	Description	Requirements	Responsibility
I. Problem definition	Initial stage in the design and implementation of system	Construct organization matrix to define available resources Develop service level agreements to define objectives and subobjectives Select business function representative	Functional management groups
	Problem observed; information recorded by center staff and entered into data base	Should include: Definition of present condition Impact on the desired condition Problem observer identification Time and date of problem and reporting To service requests, center should have: Current set of standard operating procedure Automated schedule Directory listing current responsibilities and phone numbers of problem solvers and business function representative	Control center staff

II. Problem resolution	Daily inquiry into data base to see if any new problems entered	Representative should contact problem observer to clarify issue and prioritize impact on business objectives. Meet with managers of relevant business function. Resources allocated for problem solving task using schedules and organization matrix. All schedule changes entered into data base by business function representative.	Business-function representatives
	Weekly meetings	Examine current problem-solving activities and anticipate potential problem areas through use of schedules and performance reports. Exception reports produced automatically to highlight missed deadlines and milestones. Problem solvers use inquiry processing programs to examine unique data relationships.	Business-function representatives to chair; includes function managers and problem solvers
III. Problem evaluation	Meeting scheduled to evaluate solution and its alternatives Final review upon completion	Devise implementation plan. Allocate resources, adjust schedule. Document and distribute information detailing any operational changes.	Problem solvers, business-function representatives, function management, user group involved

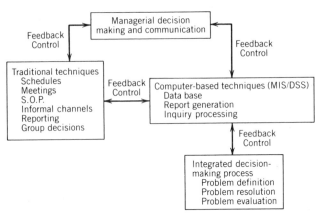

FIGURE 11.1.   Decision support, feedback, and control systems.

Store historical information that would be useful for generating solutions Access information quickly at remote locations, effectively eliminating the distance constraint.

Thus, the integration of information systems with decision making permits the identification of potential strategies to which a firm can adhere. The proper implementation, follow-up, and control of a fully integrated system is important to its success. Figure 11.1 depicts such a fully integrated system.

11.5. MANAGEMENT INFORMATION SYSTEMS VERSUS DECISION SUPPORT SYSTEMS

The concept of a computer-based management system is generally viewed as a centralized, manageable repository of the organization's data resources such as hardware, software, procedures, and specialized personnel. Management Information Systems (MIS) focus on generating better solutions for structured problems, as well as improving efficiency in dealing with structured tasks. On the other hand a Decision Support System (DSS) is interactive and provides the user with easy access to decision models and data in order to support semistructured and unstructured decision-making tasks. It improves effectiveness in making decisions where a manager's judgment is still essential. Sprague[9] has suggested the following DSS characteristics:

They tend to be aimed at the less well structured, underspecified problems that upper level managers typically face;

They attempt to combine the use of models or analytic techniques with traditional data access and retrieval functions;

They specifically focus on features that make them easy to use by non-computer people in an interactive mode; and

They emphasize flexibility and adaptability to accommodate changes in the environment and the decision-making approach of the user.

Waston and Hill[10] add that DSS also organize data and models around the decision(s) and are user initiated and controlled. DSS is relevant to situations in which a final system can be developed through an adaptive process of learning and evolution. This continuing adaptive nature of DSS contributes to providing current, nonobsolete management support.

The rise of the DSS movement can be traced over the past decade as computing services in most organizations have been transformed from a protected monopoly to a free market. The result has been a latent demand for certain applications that had been obscured by the supply-driven monopoly of a systems department but are now being satisfied in the demand-driven market characterized by DSS.

Debate continues regarding the validity of the DSS movement. Many critics, such as Naylor,[11] assert that DSS is neither new nor unique, and contend that there is no need for a cliché to describe a subset of management science. Watson and Hill (see note 10) have challenged this claim by showing that DSS is indeed different from other types of computer-based information systems, due to the focus on interactive, adaptive decision making, and will continue to play an increasingly important role in organizations.

Secondly, Naylor states that because DSS is not based on a well-defined conceptual framework, it is extremely difficult to compare or evaluate alternative systems. Recently, Blanning[12] has responded by discussing 5 principal areas of state-of-the-art DSS research, notably:

1. The construction of knowledge-based interactive systems
2. The development of frameworks for model management systems similar to those for data-base management systems (DBMS)
3. The integration of data management and model management to produce an emerging science of information management
4. The enhancement of current concepts in information economics to provide relevant criteria for evaluating proposed or existing DSS
5. The investigation of behavioral issues relevant to the design and implementation of DSS

Blanning concludes that the uncertainty and ambiguity in DSS should not be discouraging. Rather, he sees this stage as inevitable in the growth of the movement as concepts are challenged.

Thirdly, a question has been raised concerning the likelihood that managers will choose also to become computer programmers. Proponents state that the intent of DSS is to support semistructured and unstructured decision-making tasks at all levels of an organization.

Finally, the concept of the "office of the future" has been cast aside as merely a myth by Naylor, who states that the idea exists "primarily in the minds of academic visionaries and overly aggressive sales and marketing people" (see note 11). Office capabilities such as word processing, electronic mail, facsimile, teleconferencing, personal computing, and electronic filing are becoming increasingly common in a growing number of firms today. The debate will undoubtedly continue as this movement continues to grow and evolve.

11.6.   DATA-BASE MANAGEMENT

The traditional approach to data-base management has been dependent upon applications program development. Areas within the overall data network typically evolve independently from each other in response to specific needs for data support. Each program defines its own data requirements which are then gathered and stored with little or no external coordination with other groups. The resulting data system is typically a fragmented, overlapping collection of files, largely reliant on the originating program for definition, storage, and access procedures. Redundancy inevitably results and the data, once stored, cannot subsequently be revised without changing the programs. More importantly, the potential for future integration is limited by the relatively high conversion costs.

The data-base management concept includes centralization of the various data resources in singular files that are independent from applications programs. In this configuration two or more programs may execute independently while accessing the same data file. The updating process operates on a single file and avoids the problem of simultaneously updating duplicate files that may exist in other parts of the company's data network [Roberts and Boone[13]] The scale of resource commitment and the pervasiveness of data-base management systems' impact on the organization demands coordination of tasks and priorities effected by a routinely reviewed and updated master development plan.

It is clear that the integration of data management and model selection to produce a comprehensive framework to aid decision making marks a major step in systems development. The key managerial issue is to insure that a combination of people, data, models, and technical tools can interact and provide optimal answers in a convenient, timely, and cost-effective manner.

NOTES

1. S. Ramo, *The Management of Innovative Technological Corporations* (New York: Wiley, 1980), p. 264.

2. R.J. Aldag and A.P. Brief, *Managing Organizational Behavior* (St. Paul: West, 1981), Ch. 18.

3. P.H. Cheney and G.W. Dickson, "Organizational Characteristics and Information Systems: An Exploratory Investigation," *Academy of Management Journal* 25(1) (1982): 170–184.

4. H.C. Lucas and J.A. Turner, "A Corporate Strategy for the Control of Information Processing," *Sloan Management Review,* 23(3) (1982): 25–36.

5. R.K. Vierck, "Decision Support Systems: An MIS Manager's Perspective," *MIS Quarterly* 5(4) (1981): 35–48.

6. W.H. Newman and E.K. Warren, *The Process of Management,* 4th ed. (Englewood Cliffs, N.J.: Prentice-Hall, 1977), Ch. 15.

7. G.W. Grammas, "Quantitative Tools for Strategic Decision Making," in *Handbook of Business Strategy,* W.D. Guth (ed.); (Boston: Warren, Gorham & Lamont, 1985).

8. R.A. Bird and C.A. Hoffmann, "Systems Management," *IBM Systems Journal* 19(1) (1980): 140.

9. R.H. Sprague, Jr., "A Framework for the Development of Decision Support Systems," *MIS Quarterly* 4(4) (1980): 2.

10. H.J. Watson and M.M. Hill, "Decision Support Systems or What Didn't Happen with MIS," *Interfaces* 13(5) (1983): 81–88.

11. T.H. Naylor, "Decision Support Systems or Whatever Happened to MIS?," *Interfaces* 12(4) (1982): 92–94.

12. R.W. Blanning, "What is Happening in DSS?", *Interfaces* 13(5) (1983): 71.

13. C.R. Roberts and L.E. Boone, "MIS Development in American Industry," *Journal of Business Strategy* 3(4) (1983): 108.

12

THE IMPLEMENTATION OF MANAGEMENT BY OBJECTIVES

Michael K. Badawy

MICHAEL K. BADAWY is a professor of Management and Applied Behavioral Sciences at the College of Business, Virginia Polytechnic Institute, Falls Church, Virginia. He holds a doctorate with honors in Business Administration from New York University. He specializes in the management of technical professionals, human resource development, managerial skill development, and the strategic management of technology.

Dr. Badawy consults widely to business and industry. His action-oriented views make him a speaker in much demand and a national seminar leader to executive groups. He is also a trainer, author, and editor. Based on his experience, Dr. Badawy has written extensively in the areas of engineering, R & D management, and the effective management of technical professionals.

Dr. Badawy is the author of *Developing Managerial Skills in Engineers and Scientists* (New York: Van Nostrand Reinhold). His book was featured in the national news, including *The Best of Business, Research Management, Forbes,* and *Industry Week* magazines. It was also chosen as the main selection by the MacMillan and McGraw-Hill book clubs. The Japanese edition of his book has just been published.

In addition, he is the author of the Machine Design Magazine Series on "How to Succeed as a Technical Manager." He is also a contributing editor and the author of the Machine Design Series on "Your Toughest Tests as a Manager."

Dr. Badawy is editor-in-chief for a book series on Managerial Skills in Engineering and Science for Van Nostrand Reinhold Company. He is also the national president of the Management Education and Development Division, The Academy of Management.

12.1. WHAT IS MANAGEMENT BY OBJECTIVES (MBO)?

This chapter discusses the role, nature, and implementation of management by objectives as a total management planning and control system. Individual as well as organizational effectiveness measured by any criterion is contingent on several factors including that employees have an awareness of what is expected of them on the job, are given the opportunity to participate and be involved in their work, and get feedback on how they are performing their jobs.[1] These are precisely the factors that MBO attempts to satisfy. It is therefore a logical management system universally applicable to all organizations as long as it is properly implemented.

The application of MBO in engineering means applying it to a service organization, but such applications, as opposed to those in production-oriented organizations, have been relatively limited due to claims that because of the unique characteristics of these organizations, the system is inapplicable.[2] Research however, shows that service organizations do lend themselves to the application of MBO, and that applications are well underway in several types of service organizations—profit and nonprofit in both the United States and overseas.

Service organizations are of two types: professional or nonprofessional. Any organization whose essential job is to sell or render the knowledge of its people is characterized here as a professional services enterprise.[3] The organization can be a profit-making commercial business or firm, nonprofit enterprise, government agency, or an operating unit of any of these. In addition to engineering firms and staff departments, examples include R & D laboratories, advertising agencies, consulting firms, investment banking, educational institutions, hospitals, museums, religious institutions, and social service agencies. On the other hand nonprofessional services and enterprises include insurance companies, finance and real estate firms, transportation and public utilities, government agencies, and wholesale and retail trade services. Several of these encompass technical services at various levels. Since there are probably some conceptual difficulties and definitional gray areas in differentiating between professional and nonprofessional service activities based on the amount of sophisticated knowledge and specialized expertise they provide, it might be appropriate here to combine service organizations into one class for convenience purposes, without disregarding the important differences between both groups.

Management by objectives (MBO) is a total management system and a process whereby the superiors and subordinates of an organization jointly identify its common goals, define each individual's areas of responsibility in terms of results expected of him or her, and use these measures as guides for operating each organizational unit and assessing the contribution of its members.[4] Viewed in this perspective MBO is certainly a results-oriented approach, and emphasis is always on achievements rather than on means or efforts expended.[5] It is noteworthy that MBO is one of the most popular management systems in American industry.[6]

12.2.   WHY MBO?

As a total system of managing, MBO focuses on objectives to be defined, plans to be developed, a structure to enable management to achieve the organization's mission, a reward system linking rewards to performance, and an evaluation system based more on results and achievements and less on subjective and nonperformance related variables (interpersonal relations, personality characterization of subordinates, and so on).

A well-designed and well-implemented MBO system can provide several benefits:

1.   Gives the organization a sense of mission.
2.   Provides a high degree of coordination and direction.
3.   A total management system of planning, organizing, and controlling.
4.   Maximum utilization of individual talents.
5.   Provides a better approach for more effective performance appraisal and review.
6.   A basis for more equitable salary administration.
7.   An effective basis for coaching and developing subordinates.
8.   A high degree of motivation.
9.   Better communication and cooperation.

An important characteristic of MBO is that it is a total system of managing as it has implications for all phases of the management process. This interrelationship is shown in Figure 12.1.

12.3.   HOW MBO WORKS

MBO—or Work Performance and Review (WPR) as it is sometimes called—is a total system for managing an organization—any organization—that focuses on output and end-results rather than inputs, activities, and

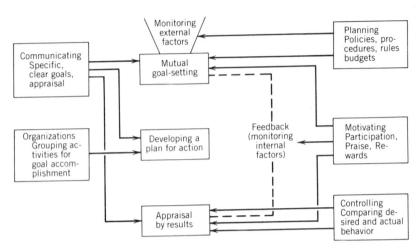

FIGURE 12.1.   MBO and management functions. Adapted from W. J. Duncan, *Management by Objectives* (unpublished manuscript).

means. If properly implemented MBO can help the manager in planning and goal determination, coordination of tasks and activities, and finally in effectively evaluating individual as well as organizational performance. Since MBO is a results-oriented approach it reduces the emphasis on personal and subjective factors in evaluation and sharply focuses on individual achievement and performance. In summary the theoretical and empirical evidence on the functionality of MBO as a management planning, motivation, and control system and its positive impacts on overall performance in numerous organizations is vast indeed.[7]

The logic of MBO is deceptively simple. Implementation is its true test. Experience shows that the way MBO has been implemented in organizations determined its chances of success. Basically installing an MBO system requires three interrelated steps or stages:[8]

Goal Setting.   This phase of MBO encompasses the determination of two types of objectives—overall organizational objectives and departmental or unit objectives. The overall objectives of the organization are broken down into subobjectives for each unit and in every functional area. The net result is a hierarchy of objectives or a mean-end chain showing the overall organizational, departmental, and task objectives for each employee. In addition to task or performance goals, each subordinate, jointly with his supervisor, will determine his personal development goals over a future period of time. Setting goals in MBO is certainly the crux of the system and does provide the organization with a deep sense of direction essential for effective management planning and control.

Intermediate Goal Review. This phase of MBO implementation consists of periodic reviews to check the progress and extent to which the preestablished organizational and unit goals have been achieved. If needed, appropriate adjustments in the objectives, means, and performance evaluation standards are made at this time. Although scheduling these reviews depends on several factors, including nature of tasks, markets, budgetary system, and organizational operating cycle, experience shows that it is reasonable to undertake these reviews on a quarterly basis.

Final Review and Objective Evaluation. The purpose of the final review is to determine final goal accomplishment. This review is undertaken at the very end of the period covered by the MBO system. Since the emphasis in MBO is on results and individual contributions to the organization, accurate documentation of results and achievements is an integral part of the system. Consequently, a set of operating procedures for MBO implementation, a time table, along with standard MBO forms must be developed and used throughout the organization. A data bank and an effective information system must also be established.

The MBO is essentially a participative management system in which the superior and the subordinate will jointly determine the subordinate's areas of responsibility and the overall plan for achieving the desired results. It is also crucial to emphasize that while they both agree on the "what to be achieved," the means for achieving it—"the how" is usually left to the subordinate's discretion.

The essential elements of MBO can be illustrated by the assumptions, concepts, and techniques upon which the system is based.[9] As shown in Table 12.1, there are four psychological principles underlying management by objectives:

1. People want to know what is expected of them.
2. People want to participate in decisions that will influence their lives and career development.
3. People want to know how they are doing.
4. People would like to obtain coaching and assistance as needed.

12.4. NEGATIVE VIEWS OF MBO

While applications of MBO have been successfully made in different types of production-oriented organizations—profit and nonprofit, large and small in the private and public sectors—a relatively smaller number of service organizations has attempted to do so. Effective implementation of MBO is

TABLE 12.1.   Essential Elements of MBO

Assumptions	Concepts	Techniques
People want to know what is expected of them.	Provide people with information about priorities, resources available, and expected results.	Establish clear, specific, and concise goals.
People want to participate in decisions that will influence their lives and career development.	Let people participate in important decisions.	Provide participative climate for goal-gathering.
People want to receive feedback about how they are doing.	Provide appropriate and timely feedback.	Provide frequent performance appraisal.
People would like to obtain coaching and assistance as needed.	Provide coaching and counseling relating to job performance and career development.	Substitute effective plans for crisis management and act as a helper rather than a judge.

based on several assumptions; for example, that organizational and individual goals can be clearly defined and that performance is easily measurable using tangible yardsticks. However, many managers in organizations, such as R & D and engineering departments, faced with the difficult task of defining clear and concise goals, of establishing feasible means, and of developing objective criteria for measuring intangible output often tend to reject MBO and dismiss the possibility of its successful implementation. The system, to them, is inapplicable in service organizations for one or more of the following reasons:

1. *Difficulty in defining goals.* In profit-making organizations—both production or service-oriented—the overall objectives and subobjectives are easily measurable using appropriate accounting and financial techniques like sales, return on investment, cost/benefit analysis, efficiency measures, and so on.

2. *Nature of tasks and activities of service organizations.* The second impediment, according to advocates of this view, concerns the nature of tasks, activities, and output of these organizations. Because of their intangible nature, defining clear goals and devising appropriate measuring devices could be a problem. Moreover, the final output is highly interdependent because many divisions have shared in pro-

ducing it. Therefore, no single department can get the full credit or is really able to see a tangible final result or product. In short, task forces and project teams usually replace functional task units, and group goals replace individual task goals—which makes identifying and measuring goals a rather complex task.

3. *Standards for performance measurement and evaluation.* Closely related to the above two problems is the difficulty of assessing the performance of service organizations and their members. While the efficiency of the operation can perhaps be measured, evaluating its effectiveness is still a very difficult matter. Typical of such difficulties are the problems faced in attempts to evaluate R & D activities and engineering designs. The effectiveness of these activities can be monitored at best, but not precisely measured or evaluated. Measuring the effectiveness of such activities, it is said, is still highly arbitrary and largely depends on the manager's background, tools he or she possesses, and the type of information at his or her disposal.

Just as there are success stories reported by many organizations implementing MBO, there naturally have been some failures.[10] Like any other management system, poor implementation of MBO will produce nothing but poor results. In addition, lack of appropriate top-management support of the system, inappropriate training of the individuals involved, creation of unneeded mountains of paper work, failure to prevent the system from becoming a statistical game, and too much emphasis on procedures but not the overall system were all reported as causes of MBO failure.

It follows that an incorrectly applied MBO can have unintended and highly negative results.[11] If it is based on a power-packed, reward-punishment psychology, the result can be damaging to employee self-esteem. If the establishment of subordinates' goals is intended to make the manager or unit look good rather than to contribute to the overall objectives of the organization, the effectiveness of the program can be reduced. If tangible measurements are overemphasized, they can have serious negative consequences for long-term goals and for employee competence. The key point to be remembered is that MBO is not a cure-all system or a panacea; and for the system to be effective, several conditions must exist—as will be discussed later in this Chapter.

12.5. MBO IN PRACTICE

Despite the aforementioned difficulties in establishing concise objectives, devising tasks, and employing appropriate standards and techniques for evaluating performance in service organizations, the allegations that MBO as a system of managing is inapplicable in such organizations are unwarranted. Experience shows—and research demonstrates—that MBO is a highly logical system universally applicable to different degrees in most organizations. Most of the problems facing MBO implementation in service organizations are more imagined or perceived than real.

First of all, if there is a good reason for an organization to exist, then there are certain functions to be performed and, thus, there is a certain mission to be accomplished. If the organization does not have a mission, it immediately loses its reason for existence and, thus, it must diminish and disappear. Service organizations, like production-oriented organizations, must therefore establish goals which are measurable. If the goals are definable, and if they are somehow measurable, then MBO is applicable.

Secondly, no organization has a right to be either inefficient or ineffective. In a sense, the profit motive exists in both private- and public-service organizations. Managers are charged with the responsibility of running the organization most efficiently and effectively so it can meet its accountability toward its publics, whether those publics are the stockholders, customers, clients, or the taxpayers upon whom the organization depends for funds and support.

Thirdly, there is no question that defining concise goals, means, and evaluation standards to measure performance of service organizations, is probably more difficult than in production-oriented organizations. In addition, what constitutes performance in service organizations is vague and difficult to define—let alone measure—because of the difficulty of defining goals in the first place. This however, does not necessarily mean that the goals are undefinable or unmeasurable; rather, it means that the goals are simply more difficult to define and measure. The difference, therefore, between service and production-oriented organizations in this area is one of degree.

It is also important to note here that qualitative measures can be used along with quantitative measures of performance, particularly in view of the fact that not all organizational variables, such as human and behavioral factors, are quantifiable. Nor is quantification of goals the most desirable way to measure the degree of their achievement. Quantitative figures, to be sure, will probably measure the operation's efficiency but not necessarily its effectiveness. For example, this is true when sales are high but mostly from relatively inexpensive and unimportant items or when the number of people trained and placed by a training agency is high, but the job turnover rate among them is excessive. While efficiency is doing things right, effectiveness is doing the right thing.[12] Without effectiveness, efficiency is useless. It does not do us any good to do the wrong thing very well.

As to the applicability of MBO in the profit-making sector of service organizations, the evidence of its effectiveness is mounting. The system has been successfully implemented in R & D and engineering activities in numerous companies including, for example, Honeywell, Inc., Diamond Shamrock Corporation and CIBA Pharmaceutical Company. Installation of MBO in R & D at these organizations has led to clarification of goals and better planning and evaluation of R & D results.[13] Typical of positive company reaction to MBO are these remarks made by an executive at Honeywell. "There are two things that might almost be considered fundamental needs at Honeywell: Decentralized management is needed to make Honeywell work and MBO is needed to make decentralization work." Another area in which MBO has been effectively implemented with positive results is the transportation industry. Examples include American Airlines, United Airlines and K L M (Royal Dutch Airlines) which introduced MBO into its world-wide field organization with a work force of over 13,000 people operating in 70 countries.[14]

12.6. HOW TO ESTABLISH OBJECTIVES IN AN MBO SYSTEM

In installing an MBO system, objectives are established on three levels:

1. Organizational objectives
2. Departmental or divisional objectives
3. Individual job objectives (task objectives)

12.6.1. Organizational Objectives

A statement of organizational objectives will define the areas, long- and short-term goals, the constraints, and the resources available for achieving corporate objectives. This is a broad statement that provides the framework within which divisional and individual task objectives will be identified.

There must be close links and interrelationships between objectives and goals at different organizational levels. This results in a mean-end chain since departmental objectives, while ends for these individual departments, are actually means for achieving the goals and the ultimate purpose of the organization. Table 12.2 illustrates a statement of corporate objectives.[15]

12.6.2. Departmental Objectives

How do I go about translating corporate objectives into objectives for my division? The essence of this question is what does the statement of corporate objectives mean to me in the engineering or R & D division? What is

TABLE 12.2.   Statement of Corporate Objectives

Profitability (stated level, growth, and stability)
Market position (stated share)
Product leadership (policy statement)
Productivity (stated levels and policy statement)
Balanced short- and long-range goals (policy statement)
Personnel development (policy statement)
Employee attitudes (policy statement)
Public responsibility (policy statement)

my department supposed to contribute? How can my division help in achieving the overall organization's mission?

As shown in Table 12.3, setting departmental objectives is a process consisting of several steps.[16]

Prepare a statement of the over-all mission of the department.
The first step for establishing departmental objectives is to define the goals and the overall mission of the engineering or R & D department.[17] The goals of these departments should be related to the overall purpose of the organization they serve. For example, in a manufacturing firm the purpose of the engineering group is the design and supervision of the manufacture of products that can be sold at a profit. Other secondary goals are updating the design of existing successful products, redesign of unsuccessful products, and planning and putting into action a reliable program for servicing, repairing, and overhauling the products manufactured by the firm. Thus, you can see that the engineering department or group has an important function in a manufacturing plant.

TABLE 12.3.   Steps in Establishing Departmental Objectives

Prepare a statement of the overall mission of the department
Define areas of responsibility
Provide specific quantifiable measures for conditions that will exist when each area of responsibility is met
Prepare an analysis of the present status of the operation as it relates to the major areas of responsibility
Pick out key areas where you would like to see improvement made and write specific objectives
Develop plans and programs designed to achieve the desired results

Source. Adapted from B. Scanlan, *Management 18: A Short Course for Managers* (New York: Wiley, 1974) p. 145.

HOW TO ESTABLISH OBJECTIVES IN AN MBO SYSTEM

The goals of the engineering group are equally important in other types of profit-making and non-profit-making organizations. Thus in consulting engineering, the main "products" for sale are design and construction experience. Without an engineering group, such a firm could not exist. The goals of the group are superior design at a minimum cost.

In plant engineering—the operation and maintenance of all types of manufacturing and service plants—the engineering group is directly responsible for an important part of the total cost of the service provided. Thus, the major goal of an engineering group at a typical plant is production of energy at minimum cost.

Even in non-profit-making organizations the engineering group has specific and important goals. Thus, in government—city, county, state, and federal—the engineering group must provide services at a cost acceptable to the taxpayer.

It follows that to define the goals of any engineering group, one must first identify the goals of the overall organization that this group serves. Once you know the overall goals, you can pinpoint the goals of the engineering group by using the guidelines shown in Table 12.4.

What makes up a sound objective? There are three criteria:

1. *Measurability.* A good objective should be measurable. However, wherever some creative effort in engineering or science is involved, measurement can be difficult. There are three ways for a measurable result descending from the very specific to the less specific:

    a. Measuring numerical results. This is the best measuring device such as by some number, cost, time, percentage, ratio, and so on.

    b. Measuring symptom results, such as survey data, results pertaining to attitudes, morale, safety records, and so on.

    c. Measuring effort results. These include things such as "give safety talks," "visit customers," "visit staff departments," and so on.

TABLE 12.4. Goals of the Engineering Department by Type of Organization: Some Examples

Type of Organization	Usual Goals of Engineering Group
Manufacturing	Profitable product design, updating, servicing
Advisory or consulting	Modern, economical design, surveys, reports
Plant engineering	Economical supply of energy services
Government	Economical design, advice, supervision, servicing

Source. Adapted from T. G. Hicks, *Successful Engineering Management* (New York: McGraw-Hill, 1966), p. 61.

The idea behind the last two measures is that through the measure of effort or symptom results, you might eventually pinpoint the controlling numerical variable.

2. *Time orientation.* A sound objective should have a time frame set up for its completion such as deadlines and time tables. Because of the importance of time on most projects, the incompetent manager usually uses time to cover the faults of poorly worded or vague objectives, or of just poor planning.

3. *Controllability.* The least understood criterion for a good objective is controllability. In discussing and agreeing on objectives, they must be within the manager's and the subordinate's area of responsibility so that he will have sufficient control over the resources (for example, people, equipment, budget) needed to accomplish the objectives.

Because it is usually difficult to give single and absolute control, especially in project engineering work, the degree of control should be discussed by both the superior and subordinate. Where responsibility interfaces must be crossed, help can be given. At least the recognition that this is not a one-man objective and that the job is therefore more difficult must be acknowledged.

Applying these criteria to engineering and R & D, there are four major components of engineering and R & D objectives:

1. Time—using time-oriented methods of planning like PERT (see Chapter 17)
2. Cost (an important project control mechanism)
3. Specifications (what is to be furnished in the way of quality, materials, dimensions, and so on)
4. Administration (responsibilities of the project manager relating to the managerial functions of planning, organizing, directing, and controlling)

Define Specific Areas of Responsibility.   As shown in Table 12.3 the second step in establishing departmental objectives is to identify the specific areas of responsibility. To do this, one should ask the question: "If we are to accomplish our general mission, what are all the specific functions this department must engage in?" Answering this question would require that you set forth each area of responsibility or accountability of your department and then delineate the specific activities which are or should be carried out as they relate to that area.

Generally, the engineering and R & D management function has the responsibility to provide for:[18]

1. Engineering and research objectives complementary to and supporting those of the firm
2. A sound plan of organization based on the objectives
3. Qualified personnel in all key positions
4. Effective means of control that will permit top executives to delegate wide responsibility and authority, thus freeing themselves to concentrate on broad planning and direction

Provide Specific Measures of Performance.   For each major area of responsibility, specific measures of successful performance should be established (see Table 12.3). As previously mentioned an important characteristic of a sound objective is its measurability. In terms of the quality of the measure, it ranges from the most specific (quantitative numerical results), to the less specific (symptom or effort results).

The point is that when quantitative measures are not available for evaluating performance you should develop a series of statements which precisely describe the conditions that will exist when that area of responsibility is adequately performed. Table 12.5 provides examples of quantified measures of performance.

However, examples of nonquantified measures of performance are presented in Table 12.6. Examples in this table describe a series of statements showing conditions desired as a result of improved performance. It is clear from this table that the technical training and development area might seem to defy measurement at first. However the statements described in the table will show the conditions that will exist when this function is adequately performed.

TABLE 12.5.   Quantified Measures of Performance in a Manufacturing Department

Cost	Production budget of standard cost data (stated as dollars)
Quantity	Production schedule or specific output standards for various machines and/or jobs
Quality	Rejection rate, tolerance standards, or scrap
Safety	Accident frequency and severity rates
Individual job performance	Standard output data

Source.   Adapted from B. K. Scanlan, *Management 18: A Short Course for Managers* (New York: Wiley, 1974), 150.

TABLE 12.6.   Nonquantitative Measures of the Technical Training and Development Function

New employees are able to progress to jobs requiring a higher degree of skill within certain specified periods of time (to be determined for each situation).

There are at least two people in the department capable of performing any given job operation.

All employees are producing at a level consistent with the minimum standard required.

New employees are able to reach standard performance in the normal expected period of time. These times would be based on past experience for the average employee.

Qualified candidates for new or upgraded jobs can be chosen from the ranks.

Source. Adapted from B. K. Scanlan, *Management 18: A Short Course for Managers* (New York: Wiley, 1974), pp. 150 and 152.

Prepare an Analysis of the Present Status of the Operation as It Relates to the Major Areas of Responsibility.   This is the fourth step in establishing departmental objectives (see Table 12.3). Once the conditions for ideal accomplishment have been described, you must determine exactly where your department stands. This is an inventory of the "status quo."

Using the worksheet shown in Figure 12.2, you can identify the present conditions in every activity or area of responsibility falling within your jurisdiction, and then describe the conditions desired as a result of improved performance. Through a comparison between actual and desired levels of performance, you can identify the current problem areas as well as the potential sources of troubles. You can also detect where the biggest payoff would be if improvement could be effected.

Pick Out Key Areas Where You Would Like to See Improvement Made and Write Specific Objectives.   In writing these objectives, remember the three criteria for sound objectives discussed above:

Measurability
Time orientation
Controllability

Examples of two different types of objectives are provided in Table 12.7.

Conditions now	Activity of area of respon-sibility	Conditions desired as a result of improved per-formance
1. _____		
2. _____		
3. _____		
4. _____		
5. _____		
6. _____		
7. _____		
8. _____		
9. _____		
10. _____		
Comments		

FIGURE 12.2.   Worksheet. Statement of Conditions. Adapted from R. K. Scanlan, *Management 18: A Short Course for Managers* (New York: Wiley, 1974), p. 151.

Develop Plans and Programs Designed to Achieve the Desired Results.   This is the last step in establishing departmental objectives (see Table 12.3). After specific objectives have been set, appropriate plans and programs should be developed to achieve the desired results. Undertaking this task requires defining the problems or areas of difficulty which must be overcome for better achievement (that is, causes of delays in test reports or of high rate of accident frequency). Alternative courses of action to deal with these problems can then be analyzed and appropriate solutions will be chosen. Plans can be developed to implement these solutions.

12.6.3.   Individual Job Objectives

The third, and last level on which objectives will be established in an MBO system is the task's or job's level. As shown in Table 12.8 there are several guidelines that would help you in joint objective setting with employees.

TABLE 12.7.   Performance Objectives: Some Examples

Quantifiable objectives	Conditions to be met when objectives are achieved
To reduce the average amount of time lapse between testing and the test engineers' written reports from six weeks to three weeks during the next three months.	To improve the work of the design engineering department in estimates of cost savings and the amount of time various projects will take. This objective is operative for the next six months and will be accomplished when the following conditions are met:
To reduce the accident frequency rate in department X from a level of 5.31 to a level of 4.25 during the next six months.	a. Cost savings estimates are within 7 percent of actual savings for 90 percent of the projects worked on.
	b. The actual time spent on projects is within 3 days of the estimated time in at least 95 percent of the cases.

Source. Adapted from B. K. Scanlan, *Management 18: A Short Course for Managers* (New York: Wiley, 1974), p. 153.

Limit the Number of Objectives.   The fewer the number of objectives on which the employee should concentrate in a given period, the better. This would not only help the employee focus on achieving these objectives, but help the supervisor to make a more efficient follow-up as well as measure and evaluate employee performance.

Set Specific Objectives.   These should meet the previously discussed criteria of sound objectives. In case of difficulty of quantitative measure-

TABLE 12.8.   Guidelines for Establishing Individual Job Objectives

Limit the number of objectives
Set specific objectives:
   Quantifiable terms
   Specific measures of actual performance
   Specific time period
Make certain that individual objectives blend with departmental objectives.
Maintain a balance
Make sure that objectives are reasonable and yet offer a challenge.
Do not avoid areas where employee does not have complete control.
Consider setting different levels of accomplishment.

Source. Adapted from B. K. Scanlan, *Management 18: A Short Course for Managers* (New York: Wiley, 1974), p. 156–157.

ment of the objective's degree of achievement, the conditions that would exist when objectives are met should be specified. Make certain that individual objectives blend with the objectives of the department during a given period and the employee's situation. This will guarantee a unified effort and ensure that individual objectives are in tune with the overall department mission (that is, reducing overtime and improving safety record). This would also reduce the potential conflict between the individual and the organization, and thus enhance opportunities for cooperation.

Maintain a Balance.   Achievement of results in one area should not be at the expense of performance in other areas. Increasing quantity, for example, at the expense of quality is obviously not a desirable objective. Balance should also be maintained between short-range objectives and long-range purposes.

Make Sure That Objectives Are Reasonable and Yet Offer a Challenge.   Objectives should be set at a reasonable level taking into account the nature of the task, technology used, employees' skills and backgrounds, and past experience. It is desirable that challenging objectives are set. Remember that setting objectives at very low or very high levels is equally undesirable. Excessively low objectives lack the necessary challenge and motivation, while excessively high objectives generate a lot of pressure leading to sheer frustration and possibly lower productivity!

Do Not Avoid an Area Even Though an Employee or Department Does Not Have Complete Control Over That Area.   It is important to recognize these areas so subordinates will realize that you understand that this is not a one-man objective and, thus, the variables are not totally under the employee control. This recognition also tends to discourage "buck passing" through putting the blame on "them not us." The subordinate is also "forced" to seek the necessary integration of effort with people inside and outside your department.

Consider Setting Different Levels of Accomplishment.   The conditions describing satisfactory, above average, and exceptional performance should be outlined. This would help differentiate between performance for people in similar jobs, and would also ensure tying the reward system—to a large degree—with actual performance.

To summarize: Installing an MBO system requires setting objectives on three levels: the overall organization, departments or units, and individual tasks. Along with defining objectives, standards of performance will be formulated. Establishing objectives is only the first step in the process of implementing an MBO system. The other two steps are the intermediate goal review and the final performance review and objective evaluation—as previously discussed.

12.7.   PERFORMANCE PLANNING AND REVIEW

The above discussion clearly shows that MBO is a results-oriented system of management. It is not activity- or effort-oriented since results speak louder than activities. Using mutually negotiated objectives agreed upon by both superiors and subordinates, the system demands both the identification of job-related goals and the measurement of results against shared expectations and performance.

The system provides for intermediate reviews where superiors will continually be apprised of progress toward goal achievement and developments taking place between the time of initial goal setting and final performance review. Appropriate goal modifications will be made based on these developments. Finally, at the end of the MBO time period (usually one year), the actual results of subordinates are jointly reviewed against the previously agreed upon goals. Based on this review, an appraisal of the employee is then made and the process is repeated again. Recommendations will then be made for salary adjustment and promotability decisions, if any.

As you can see, management by objectives provides you with an integrated performance planning and review system. As noted in our previous discussion in this Chapter, the entire MBO system consists of five interrelated steps.[19]

1. Identify key results areas
2. Set objectives for key results areas
3. Establish priorities
4. Establish performance standards
5. Develop action plans

These five steps are illustrated in Figure 12.3 which presents a Performance Planning and Review Form that can be used in the goal-setting process. As you will note the first element on the worksheet concerns areas of responsibility. This should be of interest since each major area of responsibility—as just discussed—must have an objective. Of course, the areas of responsibility listed on the form should be the same as those identified in the job description. Typical areas might be supervision, profitability, staff development, policies, and operations. The priority of these areas of responsibility should also be noted as well as the steps to be taken in achieving stated

	Date of
Name: _____	Performance Planning: _____
	Date of
Department: _____	Performance Review: _____
	Length of
Supervisor: _____	Planning Period: _____

Performance Planning

Results to be Achieved A specific statement of the major goals the employee is expected to achieve in the period.	Action Steps to Achieve Objective	Priority of Objective 1 = primary 2 = secondary 3 = least important
Area of Responsibility: (Use key word(s)) Objective:		1 2 3 (circle one)
Area of Responsibility: Objective:		1 2 3
Area of Responsibility: Objective:		1 2 3
Area of Responsibility: Objective:		1 2 3
Area of Responsibility: Objective:		1 2 3

FIGURE 12.3. Suggested form for performance planning and review. Adapted from P. P. Schoderbek, Management by Objectives in M. G. Newport (ed.), *Supervisory Management: Tools and Techniques* (St. Paul: West, 1976), p. 85–88.

Actual Achievements	Level of Goal Achievement				Continuing Responsibilities (Responsibilities not covered in objectives, to be considered only when they have a significant effect on performance.)
	exceeded	fully met	not fully met	unsatisfactory	
					_____
					_____
					_____
					_____
					Unprogrammed Accomplishments
					_____
					_____
					_____
					_____
					_____
					_____
					_____
					Overall Rating
					☐ Consistenty exceeds requirements of job.
					☐ Fully meets the requirements of job.
					☐ Does not fully meet requirements of job.
					☐ Unsatisfactory performance.

FIGURE 12.3. *(Continued)*

Suggested Training _____

_____

_____

Interview Comments (Significant items discussed in the interview but not recorded
elsewhere on this form)

_____

_____

_____

I certify that this report has been discussed with me. I understand that my signature
does not necessarily indicate agreement. Please indicate any substantial disagreements with the evaluation.

Employee Signature	Date
Reviewer's Signature	Date

MANAGEMENT REVIEW (Optional comments by reviewer concerning goals/per-
formance review)

_____

_____

Employee Review (Optional comments by employee concerning/goals performance
review)

_____

_____

FIGURE 12.3. (*Continued*)

objectives. When performance is reviewed, the actual results will be noted and a suitable rating given.

12.8. MAKING MBO WORK

Many an organization has tried to install an MBO system hastily without a keen understanding of its logic, mechanics, and limitations, and as a result the system did not work. Failure in such situations is not due to weaknesses in the system itself, but rather to its poor implementation. Put differently, the proper climate for MBO implementation is extremely vital for its effectiveness.

The following conditions are some "prerequisites" for effective MBO functioning in all organizations.[20] These conditions revolve around establishing an adequate attitudinal, structural, and behavioral climate conducive to MBO functioning. These conditions are summarized below, describing the five phases of implementating an MBO system (see Table 12.9).

12.8.1. Pre-MBO Installation Phase

For MBO to be operational, the organization must perceive a real need for the system based on complete understanding of its logic, strengths, and limitations. Otherwise, the organization will not be able to cope with the various ramifications in organizational structure, management philosophy, and methods of operations resulting from MBO installation. In short unless MBO is perceived by managers at all levels to be of benefit to themselves, considerable resentment against its development will be created. It is crucial, then, that the system be seen as a process of self-control rather than a process of imposed control.

The organization must be psychologically ready for the system. MBO will not work with the same degree of effectiveness in all organizations. One of the most important psychological criteria for MBO implementation is that the system must have the full and total commitment of top management, their enthusiastic backing, active involvement, and acceptance of ground rules. Also, management must understand the full implications of MBO before the organization is fully committed to the system.

Management must recognize that MBO is not a panacea. It is not a management tool; nor is it all there is to managing. Rather it is a total system of managing with many facets and applications. It is not, however, a substitute for good management. Moreover MBO is not a trouble-free system. There are some inherent problems or built-in weaknesses such as: a tendency of

TABLE 12.9. Making MBO Work

1. Pre-MBO Installation Phase
   a. A real "need" for the system must be perceived by the organization
   b. The organization must be "psychologically ready" for the system
   c. Management must recognize what MBO is not
2. The Learning Phase
   a. Staff should be well trained
   b. Managers should be trained in participative management techniques
   c. The use of a consultant or a team of consultants would be helpful
3. Acceptance Phase
   a. An appropriate behavioral climate is needed
   b. The system should fit appropriate psychological and social needs of organizational members
   c. Designing a proper climate supportive in nature to the kind of attitudes desired and the desired changes in performance behavior
4. Implementation Phase
   a. Goals, means, and evaluation criteria should be identified at all organizational levels and in each functional area
   b. Intermediate and final performance reviews should be undertaken regularly, and not as an annual "one-shot deal"
   c. Superiors and subordinates should recognize the limits and constraints of participation in mutual goal setting
   d. MBO must be fully integrated within the formal organizational structure in such a way to reinforce desired behavior
5. Post-MBO Installation Phase
   a. Flexibility should be an integral part of the system
   b. Adequate review, counseling, and monitoring of subordinates' performance and achievement should be constantly undertaken. Follow-up and feedbacks should keep flowing in all directions.

goals to be short-run, too much paper work, danger in overemphasizing objectives, and overinsistence on numbers and statistics.[21]

12.8.2. Learning Phase

Once a need for the system has been recognized, the entire staff who in any way will be involved in MBO implementation must be well-trained in the philosophy, procedure, and implementation of the system.

MBO is a participative management system. Managers must be made fully aware of the effects MBO will have on the authority structure of the organization. Participation is meaningful only when a change in the distribution of influence occurs. This is possible only when superiors assume less control. Managers at all levels, therefore, must be well trained in participative management techniques. They must be taught joint goal setting. Otherwise the superiors will impose their will on subordinates.[22]

MBO is not a carefree system. Successful installation of the system requires the use of a consultant or team of consultants working closely with top management and other executive managers to get the project moving. The consultant will play multiple roles including those of the teacher, facilitator, and change agent. Because of the significant impact the consultant will have on the overall implementation and functioning of the system, caution must be shown in selecting the right individual for this capacity.

12.8.3. Acceptance Phase

An appropriate behavioral climate for MBO implementation and functioning must be created. An important condition is that the system must be initiated and used at the very top of the organization in order to get reinforcement and acceptance at lower levels.

The system must fit appropriate psychological and social needs of organizational members. It must not be forced on people, rather they must buy it. For example, policies must be established and procedures developed that support the system and facilitate participation of all managerial levels. It is also crucial that reward systems of the organization must reinforce MBO implementation by relating them to performance and achievement.

A climate must be created that is supportive of MBO attitudes and the desired changes in performance behavior. This climate facilitates acceptance of the system by subordinates. Examples include opportunities to use learned skills on the job, participation in goal-setting, appropriate feedbacks and appraisal systems, and consistency with informal group rules, norms, and standards.

12.8.4. Implementation Phase

Goals must be clearly identified at all organizational levels and in each functional area. Goals must be seen as an interlocked system with each area as a subsystem where realistic, attainable, and verifiable goals will be set. Flexibility, however, in attaining or changing goals must be maintained. Means to achieve goals must also be identified and sound criteria for evaluation must be developed based on factors and characteristics which are closely related to task performance requirements.

Goal setting and intermediate and final reviews must be jointly conducted and mutually accepted by both superiors and subordinates. Naturally, the superior must engage in appraisal of the subordinate's actions consistent with the changes desired in performance behavior. Moreover, regular reviews of progress toward objectives must be undertaken on a continuous basis, and not as an annual "one-shot deal." Experience with the latter method at many organizations, including General Electric, has been rather disappointing as it violates one of the basic tenets of MBO success.

Superiors and subordinates must recognize the limits and constraints of participation in mutual goal-setting. As previously mentioned, participation is power redistribution, and superiors must be willing to relinquish some influence by giving subordinates the opportunity to participate.[23]

MBO must be fully integrated within the formal organizational structure. Managers must be given proper tools to achieve objectives in the sense of clear organizational relationships, adequate delegation of authority, planning premises, and clear understanding of company objectives and strategies. Moreover, formal aspects of the structure (communication systems, reward systems, operating cycle and budgetary system, decision-making, coordination, performance evaluation and manpower planning, and so on) must be consistent with desired changes in employees' job attitude and performance behavior.

12.8.5. Post-MBO Installation Phase

Managing by objectives is a continuing process. Flexibility is an integral part of the system since adequate changes in goals, means, and evaluation standards must be undertaken whenever needed.

Adequate review, counselling, and monitoring of subordinates' performance and achievement must be constantly undertaken. Follow-up and feedbacks must keep flowing in all directions so that top management as well as organizational members are constantly appraised of the status of the system, problems that arise, and appropriate strategies to deal with them.

12.9. BEATING MBO: HOW TO PUT MBO OUT OF ORDER

To make MBO work, a five-phase system was described in the previous Section. By way of putting things in the right perspective, discussion should focus on both the positive and negative sides of each issue. In line with this, what if you want to beat MBO? How would you go about it?

That is easy! Here is a prescription of twenty ways to kill MBO.[24] As shown in Table 12.10 they all are self-explanatory; even a small selection of them should do the job!

TABLE 12.10.   Twenty Ways to Kill Management by Objectives

Consider MBO a panacea.
Tell people what their objectives should be.
Leave out staff managers.
Delegate executive direction.
Create a paper mill.
Ignore feedback.
Emphasize the techniques.
Implement overnight.
Fail to reward.
Have objectives but no plans.
Stick with original program.
Be impatient.
Quantify everything.
Stress objectives, not the system.
Dramatize short-term objectives.
Omit periodic reviews.
Omit refresher training.
Don't blend objectives.
Be a coward.
Refuse to delegate.

Source.   Adapted from D. McConkey, "20 Ways to Kill Management by Objectives,"
Management Review (October 1972): 4–13.

For the reader who is "sold" on MBO, the message is to watch for these pitfalls and stay away from them for a well-implemented and smoothly functioning system.

12.10.   PRACTICAL IMPLICATIONS FOR THE MANAGER

Table 12.11 presents a summary of the major points made in this chapter along with their practical implications for you.

Discussion and analysis in this chapter clearly shows that the system of management by objectives is applicable to both service and production-oriented organizations as long as it is properly implemented. Research suggests that the problems facing MBO implementation in service organizations (such as engineering and R & D operations) are not insurmountable.

Effective implementation of the system requires that appropriate modifications in the organizational structure, planning, control, reward systems, and procedures be undertaken so MBO would be fully integrated with the formal system.

**TABLE 12.11.   Implementing Management by Objectives in Service
Organizations**

1. MBO is a total management system and a process whereby the superior and subordinates of an organization jointly identify its common goals, define each individual's areas of responsibility in terms of results expected of him or her, and use
   these measures as guides for operating each organizational unit and assessing the
   contribution of its members.
2. Installing an MBO system requires three interrelated steps:
   a. Goal setting
   b. Intermediate goal review
   c. Final review and objective evaluation
3. Although MBO has been widely implemented at many types of organizations,
   there have been some failures caused, primarily by poor implementation of the
   system.
4. Establishing objectives in an MBO system involves setting these objectives on
   three levels:
   a. Organizational objectives
   b. Departmental or divisional objectives
   c. Individual job objectives (task objectives)
5. Sound objectives should meet three criteria:
   a. Measurability
   b. Time orientation
   c. Controllability
6. MBO provides you with an integrated performance planning and review system
   consisting of five steps:
   a. Identify key results area
   b. Set objectives for key results areas
   c. Establish priorities
   d. Establish performance standards
   e. Develop action plans
7. In order to make MBO work for you, installing the system must go through the
   following five phases:
   a. Pre-MBO installation
   b. Learning
   c. Acceptance
   d. Implementation
   e. Post-MBO installation
8. If you want to "kill" MBO and put it out of order, simply use some of the prescriptions offered in Table 12.9.

NOTES

1. Some parts of this chapter are based on M.K. Badawy, "Applying Management by Objectives to R & D Labs," *Research Management* (November 1976): 35–40.

2. See, for example, S. Sloan and D. Schrieber, "What We Need to Know About Management by Objectives," *Personnel Journal* (March 1970): 206–208; W. Wickstrom, "Management by Objectives or Appraisal by Results," *The Conference Board Record* (July, 3, 1966): 27–31; T. Kleber, "The Six Hardest Areas to Manage by Objectives," *Personnel Journal* (August 1972): 571–575; H. Levinson, "Management by Whose Objectives?," *Harvard Business Review* (July–August 1970): 125–134.

3. R.E. Sibson, *Managing Professional Services Enterprises: The Neglected Business Frontier* (New York: Pitman, 1971), p. 7.

4. S. Odiorne, *Management by Objectives* (New York: Pitman, 1965), pp. 55–56.

5. For a good discussion of MBO and its strengths and weaknesses, see, for example, P.F. Drucker, *The Practice of Management* (New York: Harper & Row, 1954) and A.P. Raia, *Managing by Objectives* (Glenview: Scott Foresman, 1974).

6. F. Schuster and A.F. Kindall, "Management by Objectives—Where We Stand—A Survey of the Fortune 500," *Human Resources Management* (Spring 1974): 8–11.

7. See, for example, R. Howell, "A Fresh Look at Management by Objectives," *Business Horizons* (Fall 1967): 51–58; A. Raia, "A Second Look at Management Goals and Controls," *California Management Review* (Summer 1966); 49–58; H. Tosi and S. Carroll, "Managerial Reactions to Management by Objectives," *Academy of Management Journal* (December 1968): 415–426; H. Meyer, E. Kay and F. French, "Split Roles in Performance Appraisal," *Harvard Business Review* 42 (1965): 123–129; and J. Ivanevich, J. Donnelly, and H. Lyon, "A Study of the Impact of Management by Objectives on Perceived Need Satisfaction," *Personnel Psychology* (1970): 139–151.

8. For a detailed discussion of these steps see, for example, Odiorne, *op. cit.,* 68–79; Raia, *op. cit.,* Ch. 3–6; H. Tosi and S. Carroll, *Management by Objectives* (New York: Macmillan, 1973): Ch. 3–5.

9. W. Duncan, "Essential Concepts and Elements of Management by Objectives" (unpublished manuscript, August 1977).

10. See, for example, J.P. Muczyk, "Dynamics and Hazards of MBO Application," *Personnel Administrator* (May 1979): 51–62; J.P. Muczyk, "A Controlled Field Experiment Measuring the Impact of MBO on Performance Data," *Journal of Management Studies* (October 15, 1978): 318–329; S. Singular, "Has MBO Failed?" *MBA* (October 1975): 47–50; H. Tosi, J. Hunter, R. Chesser, J. Tarter, and S. Carroll, "How Real Are Changes Induced by Management by Objectives?," *Administrative Science Quarterly* (June 21, 1976).

11. See E.F. Huse, *The Modern Manager* (St. Paul: West, 1979), p. 149; and J. Bucalo, "Personnel Directors . . . What You Should Know Before Recommending MBO," *Personnel Journal* 56 (April 1977): 176–178.

12. Drucker, *op. cit.*

13. R. Henderson, "MBO: How it Works in a Sales Force," *Supervisory Management,* 20, 4 (April 1975): 9–14.

14. J. Humble, "Avoiding the Pitfalls of the MBO Trap," *European Business* (Autumn 1970): 13–20.

15. For a detailed statement of the areas in which business should establish objectives, see, for example, Drucker, *op. cit.*, Ch. 2.

16. This discussion is partly based on B.K. Scanlan, *Management 18: A Short Course for Managers* (New York: Wiley, 1974), pp. 144–154.

17. T.C. Hicks, *Successful Engineering Management* (New York: McGraw-Hill, 1966), pp. 59–61.

18. D.W. Kargar and R.G. Murdick, *Managing Engineering and Research* (New York: Industrial Press, 1969), p. 38.

19. This discussion is based on P.P. Schoderbek, "Management by Objectives" in Newport, *op. cit.*, pp. 83–84.

20. For more discussion, see P.F. Drucker, "What Results Should You Expect—A User's Guide to MBO," *Public Administration Review* (January–February 1976): 12–19.

21. Koontz, *op. cit.*, p. 8.

22. Levinson, *op. cit.*

23. Tosi and Carroll, "Managerial Reactions" *op. cit.*, p. 425.

24. D. McConkey, "20 Ways to Kill Management by Objectives," *Management Review* (October 1972): 4–13.

13

FACILITIES FOR THE ENGINEERING FUNCTION

Adolph M. Orlando

ADOLPH M. ORLANDO is a registered architect in practice in New York City. Since he graduated from Pratt Institute in 1977 with a Bachelor of Architecture degree, he has been involved in facilities design for domestic and international companies. Mr. Orlando is licensed in New York and California, and is certified by the National Council of Architectural Registration Boards.

13.1. Office Environments and Places to Work

The amounts and quality of space and the requirements of equipment, storage, and ancillary services are the basic determinants of design in the facilities used for engineering purposes. This may seem obvious, but actually a well-designed engineering office requires considerable effort and thinking to become a reality.

Again, there are basic differences in the approach, depending on whether it is desired to house an independent engineering office or the engineering department of a firm essentially engaged in something else, like manufacturing. In the latter case, there may be substantial undifferentiated space available, whether as part of the typical light manufacturing facilities in open bays in the virtually standard buildings of industrial parks or similar developments, or in the street-end of such buildings which are typically intended to serve office functions. Elsewhere surplus manufacturing space is pressed into action, perhaps even in old buildings where ceiling heights or floor loads are inadequate for the modern needs of manufacturing. Whatever the details, the need for good space for the engineering function is often slighted in favor of other planning concerns.

In many modern industries, engineering and administrative functions have greatly increased in comparison with other activities. This is especially so in military production with a high technical content. One such plant has 7000 employees, of whom only 1200 are production workers, and this proportion is not unusual in such work. Offices, and engineering offices, in particular, then inevitably become the central issues in the overall design of the plant, where in other cases they must take second place to the need of direct production.

The central importance of engineering space is also self-evident for firms of engineers, architects, or other designers. However, many of them are required, by their nature and location, to secure their space in standard and undifferentiated office buildings, which may impose major constraints on the scope and variation of possible space design. Perhaps most crucial of all, office rents in metropolitan areas, central cities, as well as the more desirable suburbs have reached levels where economy in space is essential if the firm is not to have excessive overhead. Under such conditions good office design becomes even more important.

A further problem for engineering firms may be posed if their business is of the feast-or-famine type, as so often happens. In that case, a real space shortage may develop, calling perhaps for emergency space with hastily arranged appointments that may be very much less than optimal, or for overcrowding in the existing office. It is highly advisable even under such conditions, however, not to lose sight of the advantages of a good physical work environment, which can be quite decisive.

13.2 GENERAL LAYOUT

Until recently the layout of engineering offices was more or less set. With remarkable unanimity engineers and architects describe places of work similar to that shown in Figure 13.1. They are seemingly well-ordered places with the rationality of a factory layout and comforts of a distinctly spartan sort. Engineers and drafters sit in regular rows and can examine the back of their neighbors' necks whenever they are not bent over their own work tables. In a way this resembles the process layout of the job shop; all like activities are grouped together in neat patterns, but do not conform to product needs. Such a layout is inevitable in much of mechanical and other manu-

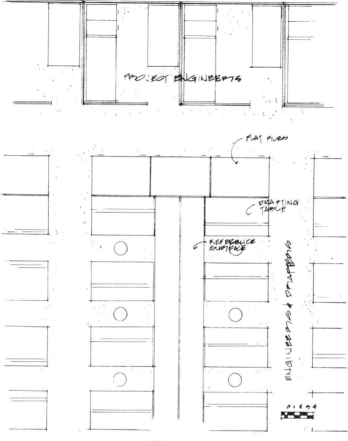

FIGURE 13.1.

facturing; but in the environment of an engineering office, it comes into conflict first of all with the need for close contact among members of a design or project team; a more flexible layout is clearly useful. Furthermore, engineering is creative work and is difficult to carry out where there are major distractions in the form of noise, snatches of discussion about other and unrelated work, movement of people, and so on. A degree of confidentiality may also be advisable in some projects and this may not be easy to assure in the old standard open plan.

A different arrangement, which allows both communication with others and privacy, is shown in Figure 13.2. Project engineers and job captains are given small partitioned offices and drafters are likewise put into small groups, rather than in the traditional "bullpens."

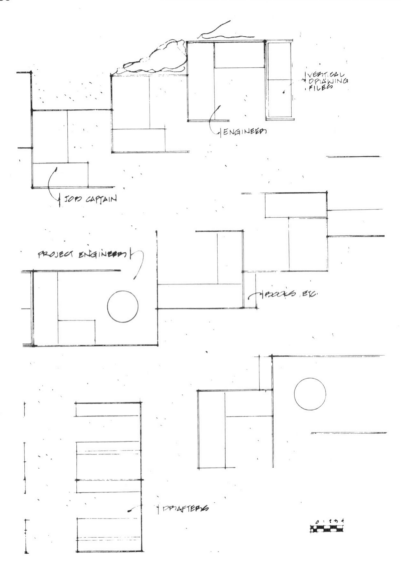

One common factor in layouts is some sort of standardized approach, as it often is in general office design. This need not take the excessively formal character of the pecking order in design of large corporations, in which one's progress up the pyramid of line management is carefully synchronized with a parallel change from simple office furniture made of metal to the wooden variety and on to the chairman's antiques. Rather our discussion here focuses on a determination of physical needs for the job and on a brief review of what is now not only available but, necessary, as the profession evolves. This is important not only in a degree of standardization in the amount of space allotted to each member of the staff but, most closely, in the design of the work stations themselves.

Another major aspect is a recognition that just as a degree of privacy is important, so is communication among team members. While this can clearly be overdone, old managerial attitudes, that is, engineers should be chained to their drawing boards and only speak when talking to their superiors, are generally accepted as outdated or unenforceable under present conditions. They are also functionally absurd; a successful project is the result of successful personal interactions as well as professional skills, as is noted in several other chapters in this volume.

13.3   WORK STATION DESIGN

13.3.1   Basic Elements and Space Requirements

Work-station design must therefore respond to the need for privacy, flexibility, accommodation of varying work habits within reasonably standardized guidelines, and a way of providing for team communication. It is difficult to imagine a team approach occurring in the kind of traditional office layout shown in Figure 13.1. The flow of information must include not only the logical line sequence from project engineers to assistant project engineers, other supervisors, staff engineers, designers, and drafters, but must also include coordination at all levels so that different aspects of the job do not disastrously converge from or conflict with each other. The skills, structure, and methods of management are obviously crucial here, but even on a simple functional level, traditional office furniture is usually too cumbersome and varied in size to accommodate the kind of flexibility needed.

A more suitable plan might be that shown in Figure 13.3 which shows a typical work space for an engineer. It is based on a generic furniture system with modular panels and components. The relative ease with which these can be changed around is a physical embodiment of a team approach. An office such as this can be enlarged, rearranged, or eliminated as team requirements and membership change, jobs progress, and the resources of the firm must be deployed in differing ways.

Engineers are essentially engaged in preparing instruments of communication in the form of drawings, specifications, technical manuals and research reports, designs and correspondence. The basic instrument of communication is still the drawing board and it still requires a basic 30

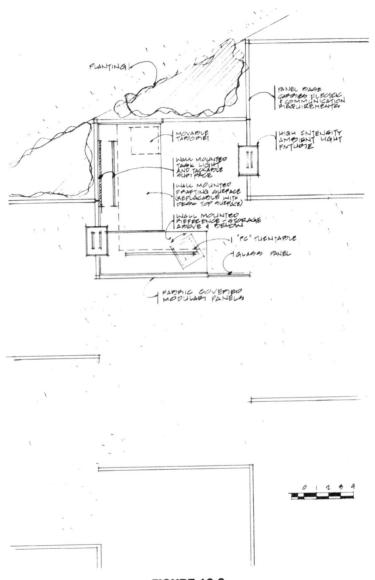

PLANTING

PANEL BASE
CARRIES ELECTRIC
& COMMUNICATION
REQUIREMENTS

MOVABLE
TABOORET

HIGH ENTENSITY
AMBIENT LIGHT
FIXTURE

WALL MOUNTED
TASK LIGHT
AND TACKABLE
SURFACE

WALL MOUNTED
DRAFTING SURFACE
(REPLACABLE WITH
DESK TOP SURFACE)

WALL MOUNTED
REFERENCE STORAGE
ABOVE & BELOW

"PC" TURNTABLE

GLASS PANEL

FABRIC COVERED
MODULAR PANELS

0 1 2 3 4

FIGURE 13.3.

square feet of which 15 square feet are for the board itself and the other 15 square feet for the person working at it. This is a practical minimum. It allows for a board used in part for actual drawing and in part for reference drawings; this would be needed, for instance, for anyone making detail drawings from general arrangement drawings or basic layouts. The paper sizes used for these determine the actual space needs. In certain industries some 40 square feet might be needed.

Note that in the arrangement of Figure 13.1, reference tables are arranged at right angles to the drafting table, thus saving some floor space, while providing extra reference space. In Figure 13.2, a similar arrangement might be followed, although it is not shown. Alternatively, as in the detail shown in Figure 13.3, a large tackable panel may be provided which affords space for reference materials and, incidentally, a modest amount of space for personal items like photographs, samples, newspaper clippings, and so on. Such items may relieve stress, inform and comfort, as well as provide a degree of individuality. Such a panel, as shown, is part of a semiprivate office and would have to be part of an internal partition or external wall.

By and large there have been few changes over the years in the basic design and layout of a drafting station. There are vertical, easel-type drafting machines, but these have not been widely adopted. If they had been they might, under some conditions, lead to a reduction in floor space required, especially when combined with a tackable reference board.

One addition which is bound to increase in importance is the computer terminal or personal computer. It is part of the equipment shown in Figure 13.3. What purpose it is to serve depends on the work to be done. Even if computer-aided design (CAD) is not or not yet used in the office in question, engineering analyses requiring the solution of multivariate equation systems, for example, would find access to a machine with suitable software quite essential. For many of the tasks however, a good hand calculator is amply sufficient. The issue of computers in engineering is discussed in detail in Chapter 14. For the present it need only be noted that as the sample layout shows, the space required is quite small and might become even less, once monitors are replaced more widely by liquid crystal displays which are flat. One could envisage a future engineering office with networks of personal computers in which design is done on the monitor or other display, and the printing is done at some central station, perhaps generating copies at the same time. In that case the work station would begin to lose its present specialized appearance and look more like a usual clerical or general office work place, including the computer, of course.

13.3.2. Furniture and Furniture Systems

A considerable variety of furnishings for engineering or drafting stations is available. There are drafting tables with a number of configurations, providing drawing and book storage, file drawers and integral lighting. While the basic 30 square feet per station are still used, their utility is enhanced. However, the greatest potential for utility and efficiency is in the use of the vertical plane. Again, as Figure 13.3 shows, storage and reference space is available above as well as below the drafting surface and the benefit of privacy is added.

The design of engineering facilities today relies heavily on systems furniture and modular components, and it is useful to discuss these briefly. The ancestor of systems furniture was a partition standing about 48 inches high; the bottom two-thirds were metal and the balance was of clear or frosted glass. This was the so-called "bank-screen" partition. It surrounded standard, that is, self-supporting furniture and did little except provide a barrier. The most advanced design element incorporated in such a unit was the electrical outlet, also of the standard type.

The availability of lighter and stronger materials, largely in metals and plastics, allowed the concept of a moveable screen to progress to flexible systems in which partitions are combined with desk tops, shelves, drawers, and so on. Lighting and electrical outlets are included. Such systems are modular in design, meaning that the components have standardized connections which allow them to be arranged in a variety of ways, depending on need. The only major restriction may be that once a certain system has been selected, it is not compatible with another one, but a good system does, as noted, leave a lot of choices. For example, desk surfaces are available in increments of one foot in length and six inches in depth, accommodating almost every design task and style of working. There is similar flexibility in bookshelves, task lights, tackable surfaces, and drawer components. A wide assortment of colors and finishes is available, thus leaving ample creative scope for interior designers of such offices.

Finally, a major advantage of the modular or systems approach is the flexibility afforded to management in making changes in the work stations; it is easy to order new components wherever needed, change the internal configurations of the units and thus responding to the changing needs of the organization without managers having to spend excessive time on such tasks.

13.3.3 Lighting and Cable Management

Lighting is clearly very important in engineering work and so its provision is an important issue in facilities design. The focus of interior lighting design has recently changed from quantity of light to its quality, that is, the contrast it creates at the work surface. This change has been influenced in part by rising energy costs, but also by better understanding of what is needed. There is, therefore, an alternative to the traditional grid of fluorescent ceiling fixtures, providing uniformly intense light all over the office.

A new approach is called task/ambient lighting. In it, an intense light source is directed toward a reflective surface, usually the ceiling, developing a fairly low but even light level. This basal light level is supplemented by lighting at the work surface, or task lighting, which is made adjustable to individual needs. The problem of ceiling reflectance, especially as it affects CRT screens is greatly reduced by this combination of ambient and task lighting.

Much as in systems or modular furniture, the major early work on ambient and task lighting was done by European designers. Some of them have expanded the concept to heating and climate control in which a base standard is provided, with individual services added by way of raised floor-type ducts and outlets in desks or other parts of the furniture system.

There is an additional problem, resulting in part from the needs of modern lighting systems but also from modern communication needs. It is that of cable management. Obviously nobody would be happy with cables all over the floor, such as extension cords, telephone wires and so on, creating tripping hazards and possible fire dangers.

An understanding of the scope of the problem is of fairly recent origins. Lighting is still the most obvious first concern but telephones run a close second. Yet not so long ago not every engineer was given one as a matter of course. Today it is a necessity, because not only are projects often built almost simultaneously with their design, but engineering changes, feedback from the field or from laboratories, and the gathering of technical information from outside library sources or product suppliers all require instant access to the telephone. As computer networking becomes commonplace, its cables will likewise have to be accommodated.

As previously noted, system furniture may provide some cable management in the electrical services it provides, but its capacity is limited. The clear space above suspended ceilings may provide a design option for cable runs, but the accessibility of such areas is poor and special protection is needed for possible fire hazards from power cables. Raised floor panel systems have long been used in computer rooms and are quickly finding appli-

cations in general office design. The convenience of floor outlets can quickly
be extended to telephone or data communication (and, as noted previously,
to climate control).

13.3.4. Acoustics

For a long time durable hard surfaces have been espoused for office furniture because of their obvious maintenance advantages. However, they have
also contributed decisively to the distinctive and distracting din of offices
because they have so little sound absorption. By introducing softer materials
like freestanding modular panels or by applying suitable materials to existing conventionally constructed surfaces, sound can be managed effectively.
Synthetic materials ease the problem of durability and maintenance while
still providing a great variety of decorative effects.

Lack of carpeting can contribute to a large portion of the noise within an
office. Even carpets glued directly to the structural slab (that is, without
padding), add a degree of noise absorption. An equally satisfactory solution
is using carpet tiles which are a variant of the glued-down carpet. Heavy
drapery or upholstery are not appropriate to an office environment because
they require a lot of maintenance (for example, in cleaning) and tend to appear worn even when still serviceable.

In large installations where privacy is deemed critical, there are more sophisticated approaches to noise management. One such method is generating white noise, a random jumble of sound frequencies, to effectively mask
speech and thus restrict its reception to the immediate areas. However, such
methods are not usually needed in the normal engineering office.

13.4. ANCILLARY FACILITIES

An engineering office must also determine its needs in other than direct engineering areas. In what follows, wide discretion is possible and wide variation is found due to the diversity of engineering work. However, the
following items should always be considered.

13.4.1. General Offices

Some type of general offices are always needed, comprising executive and
clerical facilities as well as reception areas. The amount of office space
needed depends on the character of the business. Obviously an engineering
facility that generates mainly technical reports has much greater need for

general office space than one that mainly turns out plans and drawings. The general principles of office design are beyond our scope in this chapter, but certain guidelines are quickly established.

First, the space per clerical worker is also of the order of 30 square feet per person. In engineering, this is just about the minimum required, but for clerical work, a little downward adjustment might be possible in practice. However, the clutter and crowding one often observes in offices does not contribute to efficiency. Moreover, as computers are added to clerical work stations, whether as full PC's or as word processors, the space requirements for office workers are rapidly tending to the stage where 30 square feet per capita is also becoming something of a minimum.

The amount of space to be provided for reception also varies a great deal, depending on the nature of the traffic. There may be minimal space in a hall for 2 people or so, or there may be a waiting room for 6 to 12 people. The question to be asked is whether the costs are justifiable by the degree of utilization.

Executive offices are even more a matter of discretion. They tend to be rather austere in many engineering offices, at least relative to other businesses or professions, but they need not be so. Some principals of engineering firms carefully arrange their offices to look like work stations for engineers, thus conveying an impression of direct involvement in the technical as well as the commercial aspects of the work. The final decision again turns on costs and, in this case certainly, on image.

13.4.2. Conference Rooms

Conference rooms may or may not be needed, depending on the nature of the linkages of the firm. If meetings are generally held only in connection with presentation to the clients, there is little need for having such facilities. If there are regular staff meetings that require such earmarked space, or if there are many meetings of any sort involving more people than will readily fit into an executive office, separate meeting facilities will be needed. A space of about 12 by 15 is enough for 8 to 10, possibly 12 people. A well-designed conference room should have enough space for slide projectors, overhead projectors, screens, blackboards, or similar equipment.

13.4.3. Reproduction Services

In all engineering offices there is need for a copying machine and, in those turning out plans, for a duplicating or print machine. About 80 square feet is required for the latter, including storage for paper and work in progress.

13.4.4.   Print Storage

Space for print storage is an old problem in engineering offices. Obviously a firm that decided to keep all its plans forever would incur an increasingly unmanageable storage problem. This is resolved in part by the renting of low-cost warehouse space somewhere and using it for materials no longer needed or ever thought likely to be needed. There is often still a need for keeping them, in the event that revisions must be made to buildings, or spare parts reproduced, and so on.

Here too there has been some change in methods and styles. The large, heavy drawer cabinet for drawings has given way in many places to vertical hanging files. Microfilm has also been utilized, especially in industrial plants. To improve the resolution possible with compact storage, the so-called "blowback negative" method has been introduced in which drawings are reduced to 8½ × 11 inch negatives which can be kept in binders and speedily enlarged in case of need. For simple consultation they are large enough to be viewed without elaborate equipment such as would be needed for microfilm. Except in the largest firms, such work is usually sent out to be done.

13.5.   FUTURE POSSIBILITIES

This chapter has described a variety of approaches to designing work space for engineers; some of them are only now emerging as useful methods, and their increasing application is thus itself an indication of where the future of engineering office design lies.

However, one should note that advances in communications, computer systems, facsimile transmission, and similar methods make it possible that at least some kinds of engineering will become cottage industries, carried out at home by engineers who will depend on electronic media for their communication and interaction with the outside world. If that day comes, our previous discussion will, of course, become academic.

14

COMPUTERS AND THE ENGINEERING FUNCTION

Joel N. Orr

JOEL N. ORR is a CADD/CAM and computer graphics consultant. He is chairman of Orr Associates, Inc. (OAI), one of the most active consulting firms of its type in the world. OAI counts among its clients IBM, Burroughs, the US Air Force and Navy, Hasbro-Bradley, Xerox, Applicon, TA Associates, the Government of Israel, and many others. OAI provides both technical and marketing counsel to users and vendors of CADD/CAM and computer graphics equipment and services, as well as to investors. Dr. Orr also founded the CADD/CAM Institute, a seminar and publishing firm which he directs.

Dr. Orr is CADD/CAM editor for *Computer Graphics Today, Computer Graphics World,* and *Hardcopy.* He is a founding member and elected director of the National Computer Graphics Association. He holds a Ph.D. in mathematics and computer science.

14.1. ENGINEERING IN THE COMMUNICATION AGE

For centuries engineers have required and exercised conceiving, planning, modeling, and testing skills. Their work has depended on good drawings and models and the ability to observe and record accurate measurements. But an age in which things appear to be different is upon us. How has this happened?

Evocative terms have recently come into use as names for three periods of history: the Agricultural Age, the Industrial Age, and the Post-Industrial, or Communication Age. The end of each of the two prior ages was marked by cascading cataclysmic events—the Industrial Revolution and the Electronic Revolution. The former, announced by the advent of the mechanized loom and steam power, was characterized by the rapid emergence of nonanimal power sources—the replacement of sweat with steam and streams. The harbinger of the latter was the transistor (invented in 1948 by Bardeen, Shockley, and Brattain, at Bell Labs). Like the first explosion in a popcorn popper, that development was followed by numerous others. These quantum leaps of technological progress have brought us quickly to a time in which hundreds of thousands of transistors are routinely produced in pieces so small and inexpensive that multifunction calculator wristwatches can be profitably sold for $25. The same developments have brought about the astounding profusion of cheap computers that are changing the work of engineers.

Traditionally, engineers design, measure, evaluate, plan, and construct. Each of these tasks has certain skills and types of knowledge associated with it. Designing has necessitated drafting, measurement, and the use of surveying and measuring instruments. Evaluation has required a knowledge of accounting convention, planning, mastery of scheduling methodologies such as PERT and CPM. Construction requires familiarity with materials and assembly techniques. Most of these activities are not germane to the central task of the engineer—solving problems through design.

What does an engineer do nowadays? The answer to this question is becoming stranger every day. The activities that used to occupy most of an engineer's time no longer do; motor skills essential to such activities are not necessary anymore. Instead of drafting, negotiating the construction of models and prototypes and testing them, and attending meetings and performing lengthy calculations, a growing number of engineers now spend over half their workday designing. Computers are making it possible for the process that precedes construction or manufacturing to be devoted almost entirely to examining design alternatives. This age contrasts with the Industrial Age, in which engineers spent less than 30% of their time designing.

The Bureau of Labor Statistics expects the number of engineers in the United States to grow in the 1980s from 1.5 million to over 2.1 million. So while engineering is definitely not disappearing, the activities of an engineer of the late 1980s might not be identifiable as such to a practitioner trained in the 1960s. Let's take a look at the engineer of this decade and his or her relationship to the revolutionary tool—the computer.

14.2. THE COMPUTER TRADITION

The two ways in which people interact with computers are called *batch* and *interactive*. In batch computing jobs are written out on forms which are then keypunched. The punched cards are submitted to the computer; hours or days later the user receives the cards back with the results of the job "run." In interactive computing the user works at a terminal, interacting directly with the computer.

Computers were primarily used by engineers in the 1960s in the batch mode—and infrequently, at that. They were used as analytical tools for simulation and modeling. However, users had to understand many things about computers that had nothing to do with engineering. This discouraged many engineers from approaching these awesome marvels. Engineering therefore developed a "tradition" based on the inaccessibility of computers—an attitude that was a mixture of awe and scorn, characterized by the following exchange:

"We'll have to use the computer to analyze this problem."

"—Has it come to that?"

Computers today are more accessible to engineers than ever before for two reasons: First, their programs have been designed to "reach out" to the user who has no knowledge of computers; second, in their basic professional training engineers are learning more about computers and their use. In "end user" form computers are actually systems, consisting of hardware (computers and peripheral equipment), software (computer programs), and operational methodologies. Let us examine the basic building blocks of computers: hardware and software.

14.3. HARDWARE

The electronic and mechanical portions of computing systems are collectively called *hardware*. Computer hardware has developed at a dizzying rate. From the huge electrical calculators of the 1940s, based on solenoids, to the plethora of solid-state devices available today, computers have grown smaller, cheaper, faster, and more powerful by several orders of magnitude—and the rate of change is accelerating.

There are three sizes of computers that are important to engineers: large computers (often called mainframes), minicomputers, and microcomputers. Large computers are expensive; entire systems frequently cost more than $1 million. They are used principally for data processing (accounting and administrative tasks) and batch analysis. In some large firms they are shared by engineers.

Computers are also distinguished by "word" length—the number of bits (binary digits) they can handle at one time. Smaller computers used to have shorter word lengths; today there is little correlation between overall computer size and word length. Some microcomputers, the Apple "Macintosh," for example, have 32-bit words, while older minicomputers, like the PDP/11 series, have 16-bit words.

Minicomputers became widely available in the 1970s; their relatively low prices ($25,000–$300,000) placed them within reach of a large number of engineering organizations. Using them usually required some computer knowledge on the part of the engineer. Most of these systems can support multiple users simultaneously.

The low cost of minicomputers made it possible for them to be used in "turnkey" CADD systems—computing systems for design and drafting that did not require computer knowledge.

In the last five years, microcomputers—devices whose computational functions reside in a single integrated circuit—have become very popular among engineers. Ranging in cost from a few hundred dollars to about $20,000, many of these systems have as much power as minicomputers. While most are designed for individual users, some are powerful enough to support multiple terminals.

Microcomputers are divided into several size categories—desktop, transportable, and laptop. Desktop units are typified by the IBM PC; transportables by Compaq; and laptop by the Radio Shack Model 100. Desktop units cannot usually be conveniently moved, but their power and graphics capabilities can approach those of minicomputers or even mainframes. Transportables are simply repackaged desktop units; by and large they can usually be carried (by a fairly strong user) aboard airplanes. However, they require an ac connection for operation. Typical units weigh 20 to 30 pounds.

Laptop units are the most rapidly growing segment of the microcomputer population. These devices are battery-operated, and as a result must use low-power displays and mass storage devices; LCD displays and cassettes are the norm, at present. As bubble memory, a nonvolatile storage medium, becomes less expensive, it is sure to find its way into these units.

Common among laptops is software that is "frozen" in ROM (read-only-memory). It is thus quickly accessed, and does not use up user memory.

Laptops often contain built-in telephone communication capabilities. Many laptop computer users also have desktop units.

Of the different types of mass storage devices in use today, most important are magnetic tape and magnetic disks. Both store data by magnetizing particles in a ferrous coating. Tape is the principal archival medium for computer systems; it is inexpensive and has a long "shelf life." Data are stored sequentially on tape and must be accessed that way.

Magnetic disks are used for random access. There is a great variety of disk types, ranging from small, inexpensive "floppies" that hold tens of thousands of characters, to large "Winchesters"—sealed packs containing multiplatter packs and read/write electronics that can store over a billion characters.

Disks are generally more fragile than tape; that is why tape is used as a back-up and archival medium. Tape and disk drives, being delicate electro-mechanical machines, are the costliest part of most computer systems—especially small ones.

While many have predicted the disappearance of paper from engineering, the rate of growth of the computer output device industry gives no hint that such an event is imminent. Printers, plotters, and film output units of all sorts are proliferating and becoming cheaper than ever before. Recent years have witnessed an explosion in color output devices and units with very high resolution.

Networks are now available that make it possible for a wide variety of different devices to communicate among themselves at high speeds and thus share data rapidly over long distances. ETHERNET, for example, is a pro-tocol developed by Xerox for sharing data over coaxial cables at rates of about 1 million characters per second. However, network hardware must be accompanied by network software, which has taken longer to develop and is less standardized than the hardware.

14.4.   SOFTWARE

In the experience of the classical engineer, one particular aspect of computers makes them different from most other machines: Software. This ineffable essential of computerdom is the biggest conceptual obstacle to the engineer's comprehension of computers. Machines operate according to certain principles, largely physical; computers' principles are changeable and are dynamically controlled by computer programs—software.

The two major types of software are operating systems and applications programs. An operating system is the "autonomous nervous system" of the computer; it keeps the basic system functions, such as the organization of information in memory and on disk and tape, under control. Transfer of data among these devices is regulated by the operating system so that the mechanical aspects of this activity need not concern the programmer.

Application programs cause the computer to perform functions having to do with the interests of the user rather than the operating of the computer.

Examples include programs for thermal analysis, cost accounting, or scheduling.

Standards have become very important in software, due to the accelerating development of hardware. Software is, and probably always will be, far more labor-intensive than hardware; consequently, it is important that its value to the user be protected over successive generations of computers. Some standards, such as NAPLPS (North American Presentation Level Protocol System—a graphics standard), are being brought about by consensus; others, like the UNIX operating system, come into being because of the popularity of a product.

Compliance with standards should be a significant selection criterion for the engineer considering the purchase of a computer system. If software does not comply with standards, it might not survive the current generation of hardware, whose "lifetime" is very short.

A fundamental understanding of the distinction between hardware and software, as well as an essential knowledge of the functioning of the computer, are needed. Programming is a helpful skill, but is not required for every type of engineer.

14.5. COMPUTER PACKAGES

Computers can be bought with a wide range of hardware options; software can be purchased with the computer or separately. There are three ranges of computer systems, differentiated by size: large ("mainframes"), medium ("minicomputers"), and small (microcomputers). The technology underlying these devices is changing too rapidly to admit of clear distinctions; the important trends are *smaller* and *cheaper*. All computers of a given computing power are smaller and cheaper today than they were three years ago and these trends are showing no sign of slowing down.

When a computer is packaged with devices and programs for performing a certain function so that the purchaser need only "plug it in and turn it on" to make it work, the system is called a turnkey. Organizations that do not have computer experts should buy only turnkey systems; the problems of integrating software and hardware from different vendors can be terrible.

14.6. ENGINEERING APPLICATIONS

The best known use of computers by engineers is for analysis. In the recent past engineering designs were represented as mathematical equations in standard forms; loading of these models was effected by providing the computer with the appropriate numbers. The results were usually returned in numerical form. Special knowledge was required in both preparation of the input to the program and interpretation of the output. Many analytical programs today allow data to be entered in graphical or simplified numerical form so that the analytical power of the computer can be used by engineers with limited mathematical understanding.

With the computer some design problems can be parametrized, that is, they can be defined as equations relating independent variables. By entering values for these variables a new design can be produced. One large manufacturer of television sets has found a way to characterize its cabinets using no more than fifteen numbers to specify component size, shape, angles, and so on; full implementation of this process will cut design time for a new cabinet in half.

CADD—Computer-Aided Design and Drafting—is the fastest-growing area of engineering automation. CADD systems provide automated tools for the drafter, reducing by a large factor the time and effort required to produce accurate and aesthetically pleasing drawings. The greatest productivity gains achieved through the use of CADD are found in situations where there are many changes to drawings; productivity improvement factors can be as high as twenty or thirty.

Development of tools for direct input of graphical information, obviating the need to reduce images and models to mathematical representations, has revolutionized the design process. Models for analysis can now be generated by engineers who have little or no understanding of the computer and its operation. Powerful analytical programs that have existed for years are now widely accessible as a result.

Once the geometry of a design is in the computer, it can be used to control machine tools for its manufacture. This process is called *numerical control*—an important part of CAM (Computer-Aided Manufacturing).

Storing the geometry in the computer does not enable it to characterize the part; *group technology* is a coding system for classifying parts so that they can be retrieved by characteristics, for example, "show me all parts with three or more holes and two threaded protrusions." Adding manufacturing methods, for example, drilling, reaming, annealing, and so on, to the code makes it possible to use the computer to select an appropriate order of operations based on available resources; this is called CAPP—Computer-Aided Process Planning.

In project-oriented engineering the computer is frequently used for planning and control. PERT and CPM software, which prompt the user for appropriate input (for example, early/late finish dates, and so on), can generate graphical output in the form of precedence network charts. Reports of various types can also be printed out at will.

In civil engineering computers are used for cut-and-fill, road design, structural analysis, surveying and automated mapping, and many other functions.

In sum, the key factor in the usefulness of computers in any type of engineering design work is their ability to store geometry. Once the shape of the design is recorded, categorization and analysis can be performed and drawings of various sorts can be produced. CADD is the first step.

14.7. THE PROMISE OF CADD

CADD claims it will be able to raise engineering productivity in many ways; here are some of them:

More efficient drafting, leading to shorter turnaround times and more design iterations

Fewer errors

Greater dimensional accuracy

Multiple views, with little additional effort

Tie-ins to bill-of-material processing, analysis programs, and computer-aided manufacturing

Improved communications with engineering and with construction or manufacturing

Many users believe CADD will, as a side benefit, lead to paperless engineering; let's examine this concept.

Engineers have always used drawings as the principal means of communication between the designer and the constructor. Today, as in the distant past, in most engineering facilities, the designing agent creates a drawing and passes it along to the constructing agent for implementation. This holds true to architecture, civil engineering, manufacturing—in fact, in all engineering disciplines.

With the accelerating growth of Western society, more and more paper has been created in this way: A glut of engineering drawings! Through the use of microfilm, attempts have been made to reduce the total amount of paper that must be kept on hand. But, even with microfilming, we're still using a lot of paper.

Computer-aided design and drafting offer us an opportunity to get rid of paper entirely. It is now possible to provide both the designer and the constructor with a graphics terminal connected to the same computer system, so that graphical information can be transmitted in the form of electro-magnetic impulses, instead of as pencil or ink on paper.

In fact, multiaxis NC (numerical control) machines can themselves be directly connected to the computer (DNC—direct numerical control) so that the designer, in effect, actually constructs the output.

14.8.   THE FOUNDATION OF CADD

Computer graphics is the basis of CADD. What is computer graphics? It is the use of computer systems for creating and manipulating pictures. The main hardware components of a CADD system are input devices, displays, and output devices. Each of these is discussed in its own subsection. They fall into two broad categories corresponding to the two basic systems for encoding pictorial information for computer manipulation: vector and raster; these are described below.

14.8.1.   How Are Pictures Stored and Retrieved?

Vector Graphics.   With vector encoding an image is represented in terms of the lines it comprises. Each line is defined by the coordinates of its end points; curves are described by using very short straight line segments, or can be represented by polynomial approximations.

A vector input device typically looks like a large drafting table with a movable part called a "cursor." The cursor is placed over the point whose coordinates the operator wishes to determine; the operator than presses a button, causing the coordinates of the location indicated by the cursor to be stored. This process is called "digitizing." The operator moves the cursor to another position and digitizes that point, indicating (by means of a button) whether or not it is connected to the previous point.

Most vector graphics displays are cathode ray tubes. The displays are similar to plotters except that an electron beam is the writing device (tracing the outline of the drawing), and the phosphor-coated front of the CRT is the drawing medium.

A typical vector output device is a pen plotter, in which a pen is moved from point to point on a piece of drafting medium, under computer control. The plotter moves the pen just as a person would draw lines on paper, with the pen in contact with the drafting medium as it moves to draw the line. The pen lifts and moves to a new point (without touching the medium) to start a new line.

Raster Graphics.   In the raster method, the entire picture area is divided into small homogeneous cells as if by a grid. Each cell is colored in some way, like a tile in a mosaic, with all tiles having the same size and shape. In the simplest form each cell contains either a color or the absence of color.

In more complex systems each cell contains one of a range of colors and shades. These cells are called "picture elements," or "pixels."

The quality of a raster picture depends on two parameters: the number of pixels per unit of area (resolution) and the range of colors and shades permitted each pixel. Of course, the smaller the pixel and the greater the variety of available colors and shades, the better the picture will look.

Raster input devices do not distinguish individual lines, but take in the entire image pixel by pixel. Typical devices are TV cameras (with digitized output) and digital facsimile machines.

In raster CRT displays the electron beam always traverses the same path. The only thing that changes from one pass (or frame) to the next is the beam intensity which changes on a pixel-by-pixel basis. An image is thus created by light and dark dots evenly sized and spaced.

Raster output devices generate images either one line of pixels at a time (as do electrostatic plotters) or a single pixel at a time (as do digital facsimile receivers).

It is, of course, possible to use a vector input device, such as a digitizing tablet, with a raster display or hard-copy output unit. Off-the-shelf hardware and software can take care of the conversion from one format to the other, albeit at some cost.

14.8.2. Parts of a CADD System

All CADD systems are made up of the same types of components, regardless of their size; these components are listed below.

Computer and Peripherals
Graphics Input Device
Graphics Display
Graphics Output Device
CADD Software

The graphics display and input device are sometimes called a *workstation*. The workstation is the portion of the CADD system with which the operator interacts. The nature of the workstation is a very important system characteristic because it is the part of the system with which the user has direct contact.

Sometimes it is difficult to distinguish between workstation characteristics and software qualities; in other words, it is not always possible to know where to place the blame or credit for a particular system phenomenon. In examining systems it is helpful to make such distinctions carefully; often apparently serious deficiencies can be resolved by simply acquiring a different workstation.

CADD systems may be acquired in turnkey form or as separate components. In a turnkey arrangement the system vendor supplies the entire system—hardware, software, training, and service. This is the recommended arrangement for the new or small user of CADD. Buying separate components can be (but is not necessarily) more frugal; however, it saddles the user with the responsibility for interfacing the various parts and maintaining them after they are installed.

The various CADD systems on the market today can be grouped in many ways; one convenient grouping is by size. While the boundaries among computer size classes have become somewhat indistinct (e.g., is a VAX 11/780 a "mini"? Is an IBM 4361 a "mainframe?"), we can still classify CADD systems by this parameter if we also consider the number of terminals that can be supported and price of the system. Together these factors facilitate the characterization of a CADD system as small, medium, or large.

There is a wide variety of CADD systems available to the purchaser today, distinguished primarily by cost and number of workstations and secondarily by capability. These systems are usually acquired as turnkeys, in which the vendor supplies all the components, as well as the system maintenance; it is also possible, however, to buy "separates": a computer from one vendor, displays, input and output devices from other vendors, and software from yet another. Buyers of "separates" take on the responsibility of interfacing all the parts—not a desirable burden for any CADD user let alone a beginner!

Input Devices: Functions.   Three major functions are served by graphic input devices: First, they are used to get pictures into the computer—such as maps, seismograms, engineering drawings, and so on—as well as continuous-tone imagery—such as remotely sensed data and advertising art.

Second, the devices are used in conjunction with interactive graphics displays to control a cursor that moves on the face of the display. The area of a small digitizing tablet is "mapped" to the face of the display in such a way that the motion of a stylus or other type of cursor on the tablet controls the corresponding motion of a cross or other marker on the graphics display.

A third use of graphic input devices is for the purpose of function selection; this happens in three ways. First, there are menus that are specially-designated areas of digitizer boards that are laid out similarly to keyboards. They have function names and symbols placed inside of grid cells. The computer system associated with the digitizer "knows" that when a point is digitized within that area, it is to activate the selected function rather than simply enter the XY coordinates of the point into the database.

The second way is with a screen menu whereby instructions are selected on a graphic display by using the graphic input device as a cursor control.

In the third way the operator draws simple characters or symbols within the stylus of the digitizing tablet; these are recognized by the associated computing system as cues invoking certain commands.

Features.   The main operational features of input devices are listed below.

1. *Size of Input Document Allowed.*   Where applicable this varies widely; digitizers as small as 8 inches square and as large as 4 feet by 5 feet, are available.

2. *Resolution.*   Resolution refers to how close together two points can be and still be recognized as two separate points.

3. *Accuracy* in the plane. This is difficult to define, but basically refers to the correlation between measurements determined by the graphic input device and "real world" measurements. For example, if a line is known to be a foot long, and the digitizer says it is 1.001 feet long, then that is a measure of the accuracy of the digitizer.

4. *Repeatability.*   This refers to how small the range of values will be if a point is digitized and redigitized many times. These terms are frequently used very loosely and should be carefully defined within the context of each individual piece of equipment under consideration, before making a decision based on them.

5. *Device Intelligence.*   Many input devices contain powerful microcomputer systems. A user must determine how much intelligence is required for tasks such as rotation, translation, and other forms of encoding in relation to the system into which the input device is to be integrated.

6. *Raster or Vector Format.*   If the data are ultimately going to appear in vector format, it is likely that a vector input device is needed. If the data are to be manipulated in raster format, a raster input device is required. It is possible to gather in raster format and to convert it to vector format; however, this requires highly specialized and expensive software. Similarly, vector data can be converted within a computer into raster format, but slightly more easily than the conversion in the other direction. However, these considerations must be carefully examined as conversions are typically computer-consuming tasks.

7. *Color Discrimination.*   Some input devices, raster scanners in particular, are capable of discriminating among different colors in the source materials.

8. *Operational Environment.* It is important to specify the conditions under which the device is expected to operate—temperature, humidity, dirt, and so on.

9. *Computational Environment and Interfaces.* Will the device have to operate with existing systems or software? It is important to specify precisely what systems and software the device will have to support.

Types. The varieties devices include digitizer tablets; joysticks; trackballs; "touch" screens; mouses; light pens; potentiometers; raster scanners; and TV cameras.

Displays: Functions. Graphics displays provide the CADD system user with a "window" into the "mind" of the system. Models that exist only in mathematical form can be viewed on the graphics display.

If the display is large enough, as in the case of projection displays, it can be used for sharing the model with other users, as in meetings.

Features. Graphics displays are characterized by the following features:

1. *Resolution.* As in input devices this refers to how close two points can be together and still be separable. The range of resolutions for commonly-used graphics displays is from 200×300 points or pixels to 4000×4000 points or pixels.

2. *Display Size.* This may vary from about 2 inches diagonally up to 24 feet diagonally, for projection displays.

3. *Speed.* The ability of the display to show dynamic images depends on the speed at which it can redraw the image on the display. This is usually measured in microseconds or nanoseconds per pixel.

4. *Number of Vectors.* Applies primarily to vector refresh displays in which the number of vectors—expressed sometimes in vector/inches—is limited.

5. *Color.* The number of colors that can be shown simultaneously varies generally with the price of the unit. Displays may have from as few as two colors up to as many as millions of possible colors.

6. *Physical Operating Environment.* Some displays are more sensitive to temperature, humidity, and vibration than others. In some situations the display itself may affect the environment, as in the case of high security installations, which require displays that are shielded so that they do not radiate information into the environment.

7. *Interfaces.* Displays vary widely by available interface. Standards—RS232C, GPIB, and so on,—are usually supported; specialized parallel and serial interfaces can often be ordered.

8. *Software Support.* The GKS, NAPLPS, and CORE graphics software standards are supported by some display vendors. There are also some de facto standards in graphics display command sets, such as those of the Tektronix 4010 and DEC VT 100. A display manufacturer will often say that his display is "Tektronix compatible."

9. *Light Output.* Varies from dim, requiring a partially darkened room, to very bright. While spec sheets will sometimes express this factor in candelas or lumens, it is generally more useful for the user to judge by demonstration.

10. *Hardware Function Generation.* Some displays have local graphic "intelligence" and are able to generate various geometric shapes, such as circles, ellipses, curves, and rectangles, based on simple commands. Another common feature is "panel fill" or "polygon fill," in which an enclosed area can be "filled" or "flooded" with a specified color. The speed of the "fill" function varies considerably from one unit to another. Local text and line font generation are also sometimes available. Other functions include alternate displays, meaning sufficient memory for the terminal to locally store more than one full screen. Panning, zooming, drafting, and 3D rotation are other functions that are sometimes implemented locally.

11. *Programmability and Local Memory.* Many displays are actually complete graphics computers, and can store data and be programmed.

12. *Hard Copy Interface.* Many graphics displays offer simple interfaces to common hard-copy output units, such as the Tektronix 4631 or the Versatec V-80, or thermal, impact or ink-jet units.

13. *Input and Interaction Devices.* Graphics displays generally come with some form of graphic input device. Other displays might include a tablet or a mouse. Not all displays offer all types of input devices.

Types. There are several different types of graphics displays; CRTs (Cathode Ray Tubes) are by far the most common. Various "flat panel" technologies are under development; they will weigh less, use less power, and be less harmful to people than CRTs.

Output Devices: Functions. Hard-copy output is the interface between the CADD system and the nonautomated world. Output devices are primarily used to generate images on human-readable media. The output can be used in the same way as manually created drawings, thereby minimizing the trauma of automation for the user organization—at least temporarily.

Features.   Output devices have the following features.

1. *Media*—Standard output media include paper, mylar, kronar, linen, vellum, and photosensitive film. Some types of output devices, for example, pen plotters, can handle a wider range of media than others, for example, electrostatic devices. Devices also vary widely as to the dimensions of the output they provide.

2. *Speed and Throughput*—Do not confuse these two parameters. The time required to draw a particular type of line may be irrelevant to "start-to-finish" time in your application. Both parameters must be considered in the context of what you will be doing with the hard-copy device. For example, a flatbed pen plotter may have a pen motion speed far in excess of that of a given drum plotter; however, in an application requiring 25 plots at a time, the drum plotter could show greater throughput. Requirements should be defined in terms of "plots per day" rather than "inches per second."

    In a related vein, pen plotter speed figures can be deceiving in another way, unless the device's acceleration rates are known. For example, a "40 inch per second" pen plotter that needs to accelerate in a straight line for 35 inches to attain that speed is going to be spending most of its time plotting at considerably lower speeds. Pen plotter accelerations currently range from under 0.5g to 6g; speeds range from about 3 IPS (inches per second) to over 40 IPS.

    With regard to electrostatic and other raster plotters, while output rate is usually constant, the computer time required to rasterize a vector plot isn't! This parameter should be checked carefully.

3. *Resolution, Accuracy and Repeatability*—Similar to the terms used for input devices.

    a. "Resolution" usually refers to how close together two points can be plotted and still be two separate points. Resolution may also determine how smoothly curves are plotted.

    b. "Accuracy" has a variety of common definitions. Users should define it for their own use, then let the suppliers know their definition. One definition: "How different from one inch is a plotted one-inch line?"

    c. "Repeatability" refers to how close a second set of points will be to a first set of points if four points are plotted in the corners of a plotting area and then plotted again.

    Remember that in the case of graphic COM it may be necessary to consider the parameters of the "blowback" unit (the apparatus used to enlarge the COM output in normal office use) as well as those of the COM unit.

4. *Character, Curve and Pattern Generation*—Many graphic hard-copy output devices today contain hardware character, curve (usually conic sections) and pattern generators. These features can greatly increase productivity—if the application is capable of taking advantage of them. They do tasks in hardware that were heretofore accomplished in software—namely the drawing of lines that make up the characters, curves, and patterns of output. The controlling device has only to issue an appropriate code, for the letter "A," for example; the character generator produces the lines that make up the letter.

   Similarly, the hard-copy device generates patterns or curves based on succinct control codes from the controlling device rather than having to receive commands defining all the points in a pattern or curve. Before deciding to require any of these features, make sure they will be useful in the intended application(s).

5. *Raster or Vector Format*—Raster devices "paint" pictures by letting the point of image-creation cover the area of the drawing in a line-by-line, area-filling pattern, much like a commercial television set. Vector devices "paint" from point to point, with the point of image creation traversing the component lines of the drawing. Electrostatic plotters are examples of raster devices, while pen plotters are vector devices.

   While there are other considerations, the format in which data is to be used should have some effect on the selection of a raster or vector hard-copy device. Vector devices need data in vector format, while raster devices are happiest with data in raster format. It is certainly possible to convert from one format to the other; however, such a conversion is typically a computer-gobbling task. Since most graphic data at this time is stored in vector format, raster device vendors offer rasterizing processors as options with their gear.

   Vector devices, until recently, had a clear advantage in resolution over raster devices; raster technology has now almost caught up. However, vector devices still have a clear lead in resolution per dollar.

6. *Color*—Color is useful but can be expensive. If just colored lines are needed, with only small areas in color, pen plotters will do. If you need color-filled areas in quantity, you can select from electrostatic color plotters, color ink jet plotters, the various color matrix (impact) printers, color thermal printers, or color COF units. Remember that for small quantities, the cost of color copies is almost negligible; it is well worth your while to use color if you can.

7. *Operating Conditions*—What type of environment will the machine work in? Not all hard-copy output devices can survive dirty environments, extremes of temperature, humidity, or line current fluctuations. Who will run it? Will it have a dedicated operator or will the night watchman be instructed to push the red button when the paper box is empty? These considerations should be expressed in the statement of requirements.

8. *Software*—With the device, some device suppliers provide certain programs free and charge for others.

Types. Devices include pen plotters, electrostatic printers/plotters, laser printers, ink-jet printers, film output devices, thermal printer/plotters, impact printer plotters, electroerosion printer/plotters, dry silver screen copiers, and others. These units vary widely in price and performance.

CADD Software. Usually a form of application software, CADD comes "bundled" with computer and workstations in turnkey systems or "unbundled." The latter is often more flexible than the former but requires considerably more computer knowledge to use to best effect. Turnkeys have the advantage that the user need call only one party if anything is not working. When the user is the system integrator, much time and energy is often lost in "fingerpointing matches"—occasions when all vendors disclaim responsibility for the problem and blame one another.

Mechanical and architectural CADD systems can be grouped by the sophistication of their data structures. The simplest systems have a 2D database. The next level is a 3D "wireframe" database, which accommodates 3D data but only for points and lines. Next there are systems that support surface data; these provide volumetric analysis and in some cases color-shaded imagery.

The acme of sophistication is the solids system. In it, "building blocks" have full volumetric definition. In addition to much more through volumetric analysis than is possible in a surface system, a solids system permits realistic sectioning of models.

Each successive level of sophistication requires more computing power and usually more money. It is important that the user have a clear understanding of his requirements so that the proper system is selected.

14.9. OTHER USES

Large organizations, including Xerox and IBM, have found that their engineers spend less than 30% of their workday performing engineering design functions; the rest of their time is taken up with meetings and administrative activities. Computers can increase the amount of time available for design by providing a matrix within which the engineer's time can be better organized as well as by automating many administrative functions.

The best example of such help is the electronic spreadsheet. This type of software (there are now many commercial models available) reduces by a factor of 100 the effort in examining economic and schedule trade-offs. Here's how it works: A spreadsheet—a matrix of cells normally used for financial calculations—is created within the computer. In contrast to paper spreadsheets, the computer version allows each cell to contain formulas as well as numbers and words. The formulas can refer to the contents of other spreadsheet cells; they can contain logical or arithmetic expressions. When the spreadsheet is "calculated," the indicated operations are performed. The results can be displayed on a screen or printed on paper; some programs can create charts and graphs using the spreadsheet data.

The power of the spreadsheet lies in its flexibility. For example, a cash-flow analysis for a proposed capital acquisition can be set up in a spreadsheet. Testing a variety of assumptions—regarding, for example inflation or productivity—is simply a matter of modifying the appropriate variable and recalculating. This provides a "what if" capability hitherto unavailable to the average engineer.

Electronic mail enables the engineer to send messages to superiors, peers, and subordinates for instantaneous delivery or for "store and forward" if the party is not there to receive the message. Electronically created documents can be shared by a number of viewers and edited "in concert" with ease. On many systems messages can even contain color graphics images.

Word processing puts the ability to produce a legible document at the disposal of even the poorest typist. Spelling programs can go a long way toward improving the quality of the document; soon intelligent editors will also be available to flag jargon, bad usage, obfuscatory phraseology, prolixity, and other common maladies of the engineering writer.

The output of word processing programs can be directed over telephone lines to typesetting services. This makes it easy to quickly publish manuals and reports.

Calendar programs can prevent impossible scheduling and can serve as an "alarm clock" for specified events.

Data libraries can be accessed via telephone lines, making it possible to search enormous quantities of information automatically (e.g., "find all references containing 'fusion' and 'superconducting' but not 'Russia,' 'Russian,' or 'USSR' ").

14.10.   PUTTING COMPUTERS TO WORK

"A risky affair!" That's what engineering automation implementation has been called by those who have tried it. The way has been poorly mapped and there are many obstacles. Are you prepared for them? Avoiding these roadblocks may not make you a hero, but it can help you keep your job. Here are descriptions of the major ones and of detours around them.

Financial roadblocks are generally more complex than they appear. In most organizations cost justification of automation is a formality or a veil for more arcane decision-making reasoning. In general there must be a predisposition on the part of management toward doing it or it won't be approved. Careful inquiry and informal discussions with decision-makers will help you circumvent this problem.

"We just can't afford it" is seldom a valid objection anymore, especially in computer graphics. Prices of hardware and software are dropping at an astonishing rate; for example, "starter" CADD (Computer-Aided Design and Drafting) systems have gone from $75,000 to $7,500 within the last 18 months. Similar price drops have occurred in almost all computer-based gear. Business graphing programs are available for under $100 and plotters can be bought for $200.

Technological roadblocks are usually considered to be the main and sometimes only obstacles. Lack of familiarity with the implements and benefits of automation is a source of technophobia for many engineering workers and managers. It takes more than general education in the technology to overcome this fear; "demystification" must point out "what's in it for me."

The more obvious side of the technological roadblock is the challenge of converting from the old way of doing things to the new, and the implications of such conversion for activities that are not being automated at the same time. This is a formidable obstacle, but not a tricky one; it can be overcome through diligence. Some guidelines:

Include those who will be most affected by the automation project in the planning and implementation, evaluation, and selection; involve drafters in CADD system selection, for example.

Set clearly-defined and measurable short- and long-term objectives and publish them; then make sure you meet them! But condition expectations by "underpromising and overproducing."

Consider the entire work cycle, even if only a portion of it is being automated. Automating a procedure that will be rendered irrelevant in the next phase of mechanization is usually unwise.

Think through the automated process carefully and model it if possible. A pilot project, limited in terms of resources and objectives, can be very helpful. Use it to provide material for the preparation of standards and procedures.

Organizational issues are often challenging—the question of who will do what after the change must be carefully researched and resolved. The potential for a bona fide roadblock, however, is greatest in the politics of the change—the actual and perceived shifts in control. There is no simple formula for overcoming this problem; just make sure you don't ignore it because that won't make it go away.

The most common roadblocks are veritable social diseases in American industry today: Lack of willingness to lead and lack of discipline. It is unpopular to tell people what to do—yet everyone loves to know what to do and hates not knowing! Taking responsibility for finding out and accepting the authority of leadership are essential to successful automation. Likewise, following the agreed-upon (or imposed) plan is so rudimentary a requirement it shouldn't have to be mentioned—yet it must because so few do it.

Computer graphics, in particular, is one of the most wonderful currents in the wave of technological advances that is sweeping engineering. Its dazzling nature blinds many to the need for caution in its implementation. Don't be overwhelmed—be thoughtful and deliberate, and get the full benefit of the new tools.

How can an engineering organization best structure its computing resources? The answer is easy to state, but can be difficult to implement: The computing facilities must mirror the working structure of the organization. Highly structured organizations work best with highly structured computer facilities; loosely structured groups require decentralized, somewhat eclectic facilities. Selecting the appropriate structure requires management to discern and acknowledge the real modus operandi—not the "wished-for" one.

Are computer professionals necessary? Not usually, in small installations; however, having one or more internal consultants can enable an organization to derive much greater benefits from its computers. Such consultants should be engineers who have become "computer-knowledgeable"; they must be able to communicate well with users.

14.11. WHAT NEXT?

Today's engineering computer byword is "micro"; tomorrow's are "workstation" and "network." There is no technical reason for an engineer to use more than one work station for all computation, communication, and data organization tasks. Early versions of such thoroughly integrated systems have already appeared; better ones are just around the corner. Networks, of course, make it possible to share data and resources—why buy more than one expensive, seldom used microfilm plotter, for example?

The prudent engineer will learn to touch-type. This skill is helpful now and will be essential tomorrow.

FIGURE 14.1. Turnkey minicomputer-based CADD systems are changing Engineering. (Applicon photo)

FIGURE 14.2. Electronic and mechanical design can be integrated on a CADD system. (Intergraph photo)

FIGURE 14.3.   Computer can be accessed via telephone lines. (3M photo)

FIGURE 14.4.   A mainframe computer. Control Data Cyber 170/85 computer system. (CDC photo)

FIGURE 14.5.   A minicomputer system. (Vectron photo)

FIGURE 14.6.   Microcomputers——desktop and laptop. (H-P photo)

FIGURE 14.7. Computer graphics simplifies engineering analyses. Control Data Corporation offers the industry's broadest range of delivery systems for its **ICEM** (Integrated Computer-Aided Engineering and Manufacturing) program. The options allow users to process jobs from a single to 40 or more workstations. (CDC photo)

FIGURE 14.8. A low-cost turnkey CADD system. The producer electronic drafting system by Bausch & Lomb.

A Raster Picture

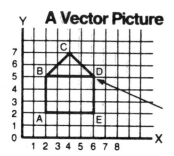

A Vector Picture

PIXELS

FIGURE 14.9. Vector and raster pictures.

FIGURE 14.10.   Workstation with digitizing tablet. (Calma photo)

FIGURE 14.11.   Microcomputer with mouse. (Mindset photo)

317

FIGURE 14.12. Touch-screen terminal. PLATO computer-based training (CBT) delivers the In-Plant Maintenance Training program that helps employers protect their investment in new electronic equipment and other sophisticated manufacturing hardware. (CDC photo)

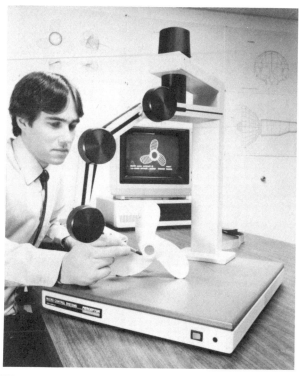

FIGURE 14.13.   A 3D digitizer. (Micro Control Systems photo)

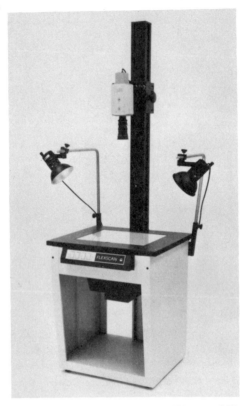

FIGURE 14.14.   TV camera for digital input. Dicomed Flexscan black and white scanner. (Dicomed photo)

FIGURE 14.15.   A raster graphics workstation. (Vectron photo)

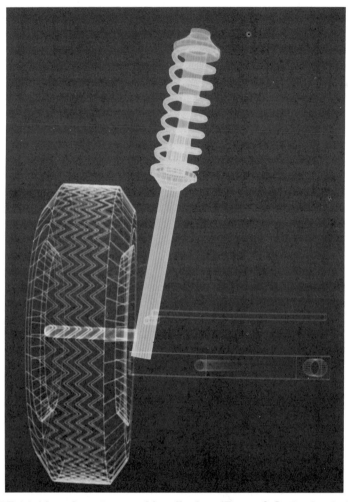

FIGURE 14.16.   A vector graphics display. (Evans & Sutherland photo)

FIGURE 14.17. Microcomputers can be turned into good graphics displays. (PCVision photo)

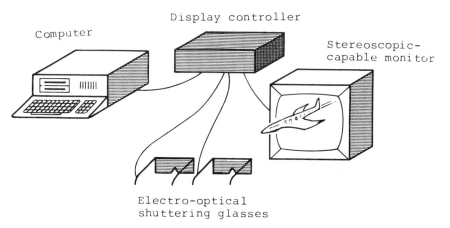

Computer

Display controller

Stereoscopic-capable monitor

Electro-optical shuttering glasses

FIGURE 14.18. A technique for showing stereo imagery. (Stereodimensional photo)

FIGURE 14.19.   A large-screen projector. (Inflight photo)

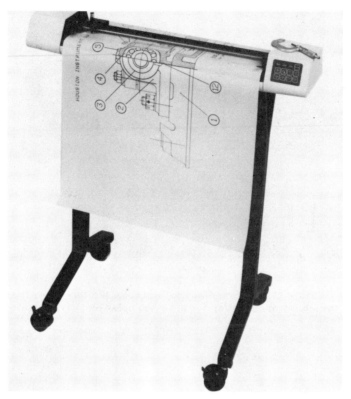

FIGURE 14.20.   A pen plotter. (Houston Instrument photo)

FIGURE 14.21.   A color thermal printer/plotter. (Seiko photo)

(a)

FIGURE 14.22. (a) The first electrostatic color plotter——only from Versatec, (b) Principles of operation for Versatec color plotter. (Versatec photo)

Color Plotter Diagram

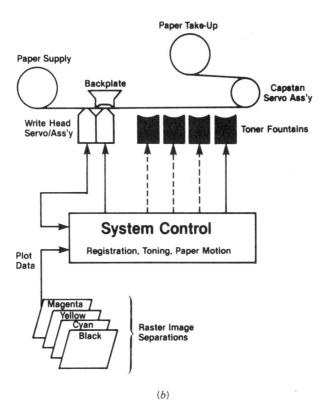

(b)

FIGURE 14.22. (*Continued*)

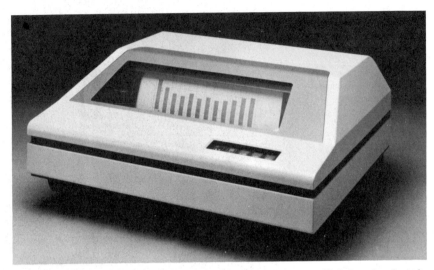

FIGURE 14.23. A dot-matrix impact printer/plotter. (Printronix photo)

FIGURE 14.24. An ink-jet plotter with automatic sheet feeding. (Joel Orr photo)

FIGURE 14.25. Good-quality slide and photo output from a microcomputer. (Polaroid photo)

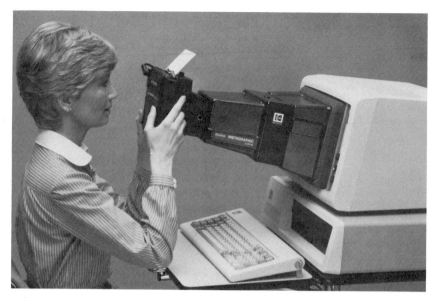

FIGURE 14.26. A simple device for glare-free, in-focus CRT pictures. (Kodak photo)

FIGURE 14.27.   Solid modeling on a microcomputer. (Cubicomp photo)

15

PLANNING AS A FUNCTION OF ENGINEERING

Robert R. Godfrey

ROBERT R. GODFREY is a generalist of broad experience in industry and government. He has served in management positions of responsibility that provide him with a real world base for understanding and responding to the needs of management.

He has worked in research and development, manufacturing, new product development, corporate planning, operations research, and business analysis. He has participated in assignments exploiting the most effective disciplines and techniques.

Mr. Godfrey entered the field of management consulting in 1970 after more than 20 years in general management and research. He has had overall responsibility for design and installation of management systems throughout the United States.

Mr. Godfrey attended the Rutgers Institute of Management. He received his B.S. from Indiana State University and his M.S. and Doctoral Studies from New York University.

We cannot cast out pain from the world, but needless suffering we can. Any calamity visited upon man could have been avoided or at least mitigated by a measure of thought. Whatever failure I have known, whatever errors I have committed, whatever follies I have witnessed in life have been the consequence of action without thought.

BERNARD BARUCH

15.1.　INTRODUCTION

If you want to turn a good idea into reality, then you must plan it. Without a logical coherent plan of action, one is a victim of fate. An effective plan of action and implementation forms the bridge over which an idea travels from your brain into being. An idea is useless until you have planned how to make it a fact.

The ability to plan multiplies your efficiency and effectiveness. With planning you can manage several operations in the same time frame and still give proper attention to each. You know the difference between the critical few activities that need the most attention as opposed to the trivial many that need little. You guard against uncoordinated activities. It keeps you from muddling and becoming a wall target for the unexpected.

15.2.　THE PLANNING FUNCTION

The concept of futurity is the essence of the philosophy of planning. It is unnecessary to accept one's fate in the future. Some thought about yesterday and today, mixed with the anticipation of tomorrow, offers an opportunity to manage the future with more certainty than it can be forecast. Effective planning can change tomorrow's assumed fate into a controlled benefit.

15.2.1.　The Benefits of Planning

Managers and organizations usually fail, not from technical incompetence, but from the lack of planning ability. Effective planning is the secret ingredient in any success. Its biggest enemy is the belief that it is a mysterious and complex activity of questionable value. It isn't! Planning is a straightforward methodical process that is fundamental in synchronizing a group of activities. The benefits of planning can be immediate and profound.

Good planning can promise many benefits. Some of the significant ones are listed below:

1.　Provide a road map or course of action and the optimum allocation of limited resources
2.　Reduces uncertainty and ambiguity by clearly defining the plan's objectives and how they are to be achieved
3.　By clearly showing how tasks are related to goals, guidelines are developed for day-to-day decision making

4. People participating in the planning have an opportunity to develop plan ownership and commitment

5. Allows for faster and smoother implementation and operating efficiency

6. Facilitates communication by supplying people with the vital information they need to get the job done

7. Because responsibility is clearly defined, delegation is easier and more certain

8. People are motivated and work harder because they know and understand what is expected of them

9. Becomes a means for monitoring progress and evaluating deviations or the standards for control

15.2.2. The Purpose of Planning

An important factor in the acceptance and appreciation of planning is the perception of its purpose—why plan? Planning involves perceptions of and projections into the future that when executed are converted into reality. It identifies stages of progress that are to be accomplished in a given time frame and provides a yardstick for measuring effectiveness.

Planning also provides guidance for the interaction of all of the other functions of management, keeping the organization focused on its objectives and direction it must follow. It is a unique function because of its integrative nature. Developing a good plan will help to ensure that a project will be analyzed, understood, and communicated. An effective plan serves three important purposes.

1. *Provides a Framework.*   A plan is a series of ideas that have been developed and organized into a series of parts. All of the plan's parts are interdependent, each one leading to the next and each influencing the other. In planning, many things must be taken into consideration. Therefore the planning process is a series of evolutionary steps that provide the necessary framework. The process links all of the parts to each other so they are all in their proper focus for development, understanding, and execution.

2. *Check Plan Feasibility.*   The product of the planning process is an integrated system of plan elements that fit together and complement each other. It shows the relationships between the objectives, activities, and strategies that represent the real world. To reduce risk and enhance success, the plan should be evaluated to see if it will work.

The feasibility of the plan can be checked as one would read a blueprint. A dry run can be made on paper to reveal problems and bottlenecks that can be corrected or planned for. If something doesn't work, it is easier to make adjustments on paper.

3. *Communication.* A well-developed plan is an instrument of quality communication. It provides a clear sense of direction by stating purpose, what is to be done, when, and by whom. It affords the basis for decisions, problem solving, resource utilization, and control. It is the foundation for coordination and cooperation. A plan that is effectively communicated invites participation, the antecedent to commitment, responsibility, and motivation. Good communication creates understanding which makes the job easier and the work more effective and rewarding.

15.2.3. Planning Fundamentals

Planning is the development of a scheme of action that defines what should be done and why; how it will be done and by whom; where it will be done and when; and how much it will cost. Therefore, it involves determining the results to be achieved, the resource requirements needed, and the required strategies and procedures for success. Planning is also critical in reducing the effects of unexpected events, failures, and human error. It is an organized process of deciding "now" what should be done tomorrow, of linking today with tomorrow through purpose.

A plan is a road map of the future, but it is not a final set of details. It should not be cast in bronze and never changed. More precisely it is a model for the future that must be tempered with judgment, management intuition, and the changing environment.

The following is a conceptual model of the fundamentals of planning. It presents a series of steps for plan development: What, why, when, how, where, and who. These elements are stated as questions with an answer and brief explanation. They provide the intellectual stimulant for developing the basic information necessary for an effective plan. The model outlines the concept of planning in clear, easy terms that moves progressively from top to bottom as a conceptual process which, in practice, is iterative. The final result can be either a simple plan or a very elaborate one depending on the detail desired. The model also serves as a guide for the development of a more explicit, tactical planning system adaptable to a wide range of planning needs.

CONCEPTUAL MODEL OF PLANNING FUNDAMENTALS

What	Objectives	End result
Why	Purpose	Rationale
How	Strategy	Approach
Who	Responsibility	Personnel
Where	Scope	Location
When	Schedule	Start, finish
How much	Budget	Resources

What is a statement of the objectives that define the end-result of the plan.

Why is the purpose of the plan or the rationale for selecting the objectives.

How is an explanation of the basic strategy or approach to be used in accomplishing the plan.

Who identifies the responsible lead and support personnel participating in the plan.

Where indicates the scope of the plan as designated by a geographical or organizational level.

When defines the time frame of the plan and presents a schedule of all activities.

How much is the plan budget that spells out the resource requirements necessary to accomplish the plan.

15.2.4.   The Characteristics of a Good Plan

A good plan of action is essential but what should it contain? There are plans and plans. Since they may be good, bad, or indifferent, special effort is needed for protection. We need the capability of developing a plan that reduces risk and focuses action on purpose, without waiting for the verdict of trial by experience. There are certain characteristics in plans that work well and set them apart. These factors are presented to guide you in developing an effective plan. Your plan:

1. *Must Clearly Define the Objectives' Plan.*   The objectives define the results to be achieved and are the heart of the planning process. They provide direction in developing strategy, assigning priorities, and making decisions.

2. *Must Identify a Course of Action.* The plan must detail the activities to be done, who is responsible for their execution, sequence, and resource requirements.

3. *Must Contain the Structural Dimension of Time.* Time is a structural dimension in all planning. The plan must present a schedule for starting and completing all plan activities. Sequence and timing minimize delays and confusion and form the fundamental basis for control.

4. *Must Specify Resource Requirements.* Resource requirements are the result of a trade-off between performance, cost, and schedule. The interaction has a major effect on a plan's needs. A negotiated schedule is a time-phased projection of how resources are integrated into the final plan and provides the rationale for budgeting.

5. *Must List the Planning Assumptions.* The assumptions made in the planning process must be clearly stated. Although assumptions are carefully considered, their validity may change when tested against reality and fact. If an assumption becomes unrealistic, the entire plan must be reassessed.

6. *Must Provide Contingency Planning.* Contingency planning (worst-case planning) seeks to identify or anticipate occurrences of less than desirable situations and minimize their effect. It is concerned with developing an alternative strategy should the primary one fail. It limits losses by providing a fall-back alternative.

7. *Must be Concurrent.* A plan should be flexible so that it can be revised in response to deviations in the operating environment. We live in a dynamic world so a change can occur. Be alert to it, expect it, and update the plan to reflect it. A plan must respond to change, otherwise it will become insensitive and ineffective as any kind of early warning system.

8. *Must be an Instrument of Communication.* Proper communication is vital to the success of any plan. A plan held secret is totally ineffective. It tells all concerned the strategy to be used to get the job done. A plan is a complete information system that clearly defines key areas of responsibility and authority, the relevance and priority of every activity involved in the success of the plan, and provides essential support and focus to the management decision process. It facilitates understanding and coordination.

15.2.5. Goals of the Planning Process

We have covered the characteristics of a good plan. However, there are other factors to consider. Successful plans have particular qualities that cause them to produce the best results. To give direction to your planning expertise, we have earmarked them. In developing effective plans, there are four major goals to achieve.

1. *Focus Attention on the Objectives.* The purpose of every plan is to facilitate objective achievement. It is the essence of the planning process. To be effective, objectives should be feasible, acceptable, flexible, measurable, and understandable. They should have the commitment of everyone. They target action, reduce confusion, and stimulate responsibility. They give a clue to the importance of productivity.

2. *Reduce Uncertainty.* Uncertainty makes planning a necessity. Planning detail defines the work required, uncovering areas of risk and uncertainty—the major causes of negative deviations. It exposes misunderstanding and the associated side effects and makes possible more intelligent measurements and feedback. Good planning information makes an important decision base for faster, easier implementation.

3. *Develop an Optimum Course of Action.* Planning generates efficiency by organizing and synchronizing action to achieve a common set of objectives. Plans focus action on purpose, yet maintain flexibility to cope with and adopt to change. Good planning reduces changes in resource requirements, scheduling, and priorities. Without plans, action would be random and chaotic.

4. *Facilitate Control.* Planning and control are almost inseparable. Unplanned action cannot be controlled; there is no course deviation—there is no course. Plans furnish information and standards for quantitative (objective) evaluation and control.

15.3. ELEMENTS OF THE PLANNING PROCESS

Regardless of your capabilities, if you are on the wrong road, it matters little the speed you travel. Whatever the activity, whatever the stake, there must be some thought given to developing a plan of action to enhance the odds of success.

Planning is the bridge over which an idea must travel from the brain to reality. An idea is useless until a plan has been developed to make it a fact. A plan assures logical development, orderly progress, and protection against uncoordinated action. If you don't plan effectively, you are likely to be forced behind an eightball of your own creation.

Before the planning process begins, it is necessary to assess the current environment and its future trends. There must be a realistic analysis of the opportunities to be exploited and weaknesses to be avoided. From this situational audit, guidelines are formulated to develop the most effective strategy for plan development and resource allocation. In developing a planning strategy, every effort must be made to reduce the elements of risk and chance.

Planning is an evolutionary activity in which the planning steps are interrelated—one step leads to the next with each influencing the other. They are all interdependent. As the plan develops there is an evolving understanding of reality and the actual environment. Gross assumptions are replaced with fact—enhancing the feasibility of the emerging plan.

There is some art to the planning process. It requires the development of a mental image of the proposed plan and then through a series of tools and techniques, it is brought to "hard copy." The planning process is a step-by-step evolutionary procedure divided into four distinct but interrelated steps. They are:

1. Defining
2. Programming
3. Scheduling
4. Controlling

15.3.1. Defining—The Development of Plan Objectives

After the decision has been made to develop a plan, the first step is the development of specific plan objectives. They are simple statements of the results to be achieved—the end-point of the plan. The purpose of the objectives is to give direction to where you want to be and what you want to accomplish. They define the target for focusing your efforts. A goal cannot be reached until it is identified. Identification of the end-point will tell us when the plan has been completed. It is a better game of golf if everyone tees off for the same green and has an idea of the fairway's condition.

In defining a plan we need to know what is to be done, what results are expected, when the work must be done, who will do the work, where the action will take place, and what resources are available and will be required. Here are some important questions to answer in developing a statement of plan objectives.

What: Write a detailed description of the work to be accomplished. Doing this brings the plan to life. Time spent on this step will avoid ambiguity and confusion later.

Final Result: Specifically what does the objective hope to achieve? What is the desired end-product?

Why: State rationale for selecting the objective. Why is it important? What is the end-purpose?

How: Describe the major tactics to be employed. What is the basic approach to be used in achieving the objectives?

Where: Define the geographic area or organizational level at which the objectives are directed.

When: What is the time frame for the plan? A gross schedule should be presented, also a major milestone schedule should be included.

How Much: At this point in the process only a tentative estimate can be made of the plan's resource requirements. Schedule drives budget so a final estimate cannot be made until all activities have been scheduled and the degree of acceptable performance determined. Budgets are often in error because it is assumed that a final budget can be prepared at this point.

Who: Who is responsible for doing the work? Plan direction is an extension of the planning process so all activities must show either an organizational or individual responsibility. Examine the plan and establish the categories of resources that will be needed. Within each category individuals should be identified and assigned responsibility for the appropriate work elements that match their skill and capability.

Assumptions: In planning there must be predictions and assumptions. Although carefully considered, they can be wrong or become irrelevant. As assumptions are tested against time and reality, their value changes. Should assumptions become unrealistic, it must be known and the plan reassessed.

15.3.2. Guidelines for Preparing Objectives

It is a challenge to define adequately the overall nature and scope of a plan prior to becoming involved in its detailed planning. To assist in writing the objectives, the following guidelines may be helpful.

1. Follow the KISS (Keep It Simple and Short) principle. Use simple terms to describe the results to be achieved.

2. Be sure that the objectives are achievable.

3. Write the objectives so they are measurable and verifiable. Spell them out in concrete terms that leave no doubts.

4. State the objectives so they are readily understandable by all who will be working with them.

5. Make certain the objectives are consistent with the resources available.

15.3.3. Programming—Develop Strategy

Following the development of the plan's objectives, the next step is to identify the work action, strategy, and resources required to achieve them. This step is primarily one of synthesis based on the technological aspects of the plan. To identify the work to be done, we list the major elements of work. These can be presented graphically against time with a Gantt (bar) chart. The next step is to define the work in greater detail for better definition. The major work elements are restructured to smaller units or activities. This makes the work more manageable for assigning resources and specific responsibility. It also makes the plan easier to control. This consists of listing the activities necessary to complete each of the major work elements.

A plan consists of separate but interrelated activities. An activity is an item of work. It may be a task, operation, waiting time, procurement time, processing time, and so on. An activity has a beginning and an end, and requires time and resources to do it. The activities represent a "to do" list for the plan that must now be sequenced. We now know what work is to be done but little else. We know where to go but we need a map to get there. In the programming phase we develop that map—a model of the work to be done.

A model is a scaled-down version of the "real" world—in this case, the work to be done. It will provide a framework for managing the plan. It affords an opportunity to present information graphically so we can analyze the important variables in the plan. The model we will use simplifies reality but closely approximates the important technical and logical relationships of the various activities. The product of the model will be the sequence and priority of the activities. It is a powerful tool that provides us the proper guidance for effective execution of the plan.

Following the development of the series of major work element models, they are interfaced to produce a complete plan. At this point we know the logic sequence of the work activities to which we must now add the element of time—the duration of each activity. Many rules have been developed for determining activity duration, but the wisest is to use the estimate from the person most familiar with the work. For many reasons his or her best estimate is the wisest choice. After time has been added to the model, it can be analyzed. This analysis will reveal how long it will take to complete the plan and also single out the critical few activities that actually determine the plan's duration. The isolation of the critical few is valuable information for focusing management attention.

In summary, the programming phase has generated the following information:

A Gantt chart of major work elements to grossly track the plan.
A list of the essential activities for achieving the plan's objectives.
A model of those activities showing priority sequence.
The duration and timing of each activity.
The critical activities which, if delayed, would delay the plan.

The mechanics of planning section presents the detail for developing and analyzing the models.

15.3.4. Scheduling—Allocation of Resources

In the programming phase activities are sequenced by logic and technical demands. It was assumed that the resources were available to perform the plan. The plan developed was an *ideal* or theoretical plan showing when an activity *could* be done. In developing the schedule we determine when it *should* be done. The activity schedule is a function of resource availability, time, and cost. It is concerned with the effects of activity duration and resources. To schedule requires a knowledge of the availability of resources that can be used optimally. In scheduling we introduce a new kind of start and finish. Now we have to consider scheduled start and scheduled finish.

In a plan, many activities can be done faster or slower, depending on the number of people assigned, kind of facilities and equipment used, and so on. Schedules are created out of negotiations based on commitments, resources, activity time requirements, priorities, and input dependencies from other activities. Since resources are always limiting, the schedule is identified as the negotiated plan. The difference between the ideal plan and the negotiated plan is the effect of resource availability.

After the plan schedule has been developed, a budget can then be prepared. The budget should be based on the work to be performed and the optimum schedule for the plan. This calls for the preparation of a specific budget that is different from conventional budgets. A specific budget covers the cost of implementing a specific action plan and is directly related to the actual work to be done. It recognizes the fact that the schedule drives the budget. Therefore, it is a realistic and useful guide for pricing out alternative courses of action. Specific budgets enable us to weigh the cost of an action plan against the results expected from that plan and to determine if it is worth doing. These budgets are also helpful in determining the priority of a plan—the larger the impact on profitability, the higher the priority.

Little recognition has been given to the importance of preparing a specific budget after an acceptable schedule has been done. All too frequently budgets are prepared very early in the plan and long before a schedule has been developed. This results from a disregard of the true meaning of what schedule entails and the effect of resources on schedules. In addition little consideration is given to the difference between a planned convenience date and a demand date. A planned convenience date gives maximum consideration to resource utilization, while demand dates usually ignore resource requirements and are thus very costly to meet.

When preparing a specific budget, the main concern is determining the cost of the work to be done in each activity and the cost of personnel covered by the schedule. Direct costs and the cost of time are of primary interest.

Care must be exercised in calculating the specific budget to be assured of its integrity. Be aware of a particular tendency to reduce the cost of a plan by simply changing the numbers and not reducing the plan effort. Each time a budget amount is changed, there must be an associated change in the plan's work. Once a plan's schedule and specific budget has been agreed to for implementation, it becomes the baseline budget for control. A budget underrun then becomes a suspect as a potential problem as an overrun.

The preparation of a specific budget should not be a function of the accounting department. They do not have first-hand knowledge of the plan's work and operations so cannot determine the specifics of the assumptions that must be made.

There are many products of the scheduling phase. Some are:

1. A Gantt chart presenting the major work elements of the plan.
2. A schedule for each activity and for the total plan.
3. A resource analysis showing the resource requirements for each activity.
4. An order and delivery schedule for needed equipment and services.
5. A specific budget of the cost of resources, in monetary units, necessary to carry out the plan within the scheduled limits. It is a useful guide for action, showing a detailed cost of the plan's objectives.

15.4. CONTROLLING—SUCCESS INSURANCE

In sailing, the rudder is as indispensable as the compass and chart. In management, control is as important as planning—they are integrated systems. The work of planning tells us where we want to go, how best to get there, and what it will cost. Control is the navigation system that keeps us on course and tells us when we have reached our destination. Control systems are information systems. Control is successful insurance that safeguards the work we do, the results we achieve, and protects our material and emotional investments.

The first purpose of control is verification to minimize any deviations from the plan. The second purpose of control is remedial action. It utilizes a feedback system of information on performance along with the objectives and plans as guidelines to spot potential deviations. Control focuses on deviations and these deviations recognize problems.

Controlling is measuring, comparing, and adjusting. A plan establishes a course of action, a series of activities to be completed, and then control guides it through implementation. If all of our plans were executed flawlessly and the unexpected never occurred, we could dispense with the whole concept of control. However, we live in a dynamic world of change—expect it and be ready to cope with it.

The actual progress of a plan may be difficult to measure. There is no one information sensor that will meet all the requirements. In a plan there are three key elements to be managed: schedule, cost, and technical performance.

1. *Schedule Control.* Schedule control involves integrating all of the schedules in the plan which the model does effectively. As implementation proceeds, data is collected for comparison, analysis, and schedule status.

2. *Cost Control.* Cost control deals with the organization, administration, and monitoring of all cost data in the plan. All actual costs are monitored and accumulated for comparison with original estimates. Cost data are also correlated with schedule accomplishment to equate cost with progress.

3. *Technical Control.* Technology is the most difficult parameter to define and control. Technology refers to the state of the product's performance characteristics which can be difficult to measure. Often the best sources of information are the technical people involved with the plan.

15.4.1. The Process of Control

Controlling is detecting and assessing what is being done and, if necessary, making corrections. The process can be fractioned into four steps.

1. *Monitor Progress*—What is happening? Monitor is to observe and watch, not only normal progress, but check for any unexpected progress and delays. It is important to sense any deviations before they become major problems too costly to resolve. The plan is the baseline that states how the future is to happen.

2. *Update*—Information processing to show completion. Periodically we freeze a point in time and record the state of the plan to indicate progress.

3. *Evaluate Performance*—How are we doing? Evaluation is assessing the progress of the plan. Using the concept of steering control for strategy, evaluation determines if the plan is on course. The plan itself is the baseline for checking against the update for any deviation. If the comparison shows any deviation, corrective action is initiated.

4. *Corrective Action*—Implement change. Taking corrective action is the principal reason for the entire controlling function. It is remedial in nature rectifying mistakes and improving performance to increase the potential of plan achievement.

15.4.2. Requirements of an Effective Control System

In controlling the operation phase of a project, there are some basic requirements to consider for a system that will deliver a successful project.

1. Must have a well-developed project plan and objectives
2. A capable information system
3. An organization with clearly defined responsibilities
4. Contingency plans for anticipated problems
5. A clearly understood change system
6. A control period that meets the needs of the project

15.5. PLAN VALIDATION

After the scheduling phase has been completed and the plan is ready for implementation, critique it for completeness. There is always the possibility that a critical step was skipped or something became irrelevant but was not dropped. Does the plan answer all of the questions of what, why, how, who, where, and how much? For whatever reason, the plan should be reviewed before a final commitment.

There are many ways the review process can take place, ranging from an individual assessment to a peer attack or a full team effort. Whatever approach is used, it is important to look at the entire plan. If during the review process any elements are found flawed or inadequate, there should be an immediate change to make the plan more effective. All personnel associated with the plan must understand it, have confidence in it, and commit to its success.

Below are some key questions to aid in the review process:

1. Are the plan's objectives consistent with those of the organization?
2. Are the plan's assumptions realistic?
3. Is the specific budget in balance with the expected benefits of the plan?
4. Has adequate time and resources been allocated to the plan?
5. Are the completion dates realistic?
6. Will the required resources be available when needed?
7. Are the responsible people capable of doing the work?
8. Do the plan personnel understand the work they are expected to do?
9. Do current policies and procedures meet the plan's requirements?
10. Are there contingency plans available if needed?
11. Have all alternatives been evaluated?
12. Does the plan have adequate check points?
13. Is the management information system adequate to assess plan performance?
14. Does the plan provide for unique operations?
15. Does the plan comply with government regulations?
16. Does the plan deal adequately with the termination phase?

15.5.1. The Mechanics of Planning

Following the development of the plan's objectives in the definition step, the major work segments are identified. This is followed by a further breakdown of segments to work packages or tasks and then to activities—the smallest element of work in a plan. If the complexity of the plan does not require it, the work package level of indenture can be deleted. Keep the plan simple.

Plan Breakdown Structure

Element	PLAN											Model
Segment	S			S			S					Bar Chart
Task	T		T		T		T		T		T	Network
Activity	A	A	A	A	A	A	A	A	A	A	A	Network

The major work segments of a plan are best presented in a bar chart format for fast, easy reading. Any further detail of the plan can be best modeled by precedence networks.

15.5.2.   How to Create a Precedence Network Plan

An activity list is a plan's description in tabular form, but it lacks clarity in showing technical logic and dependency. To clearly illustrate these relationships, precedence networks are the technique of choice. They show dependency clearly and succinctly. The activity list serves as a starting point for dependency analysis and networking.

Network Conventions

☐ A box is used to represent an activity—the smallest element of work. The description of the work is written in the box which is further identified by a number.

→ Arrows are used to show how activities are related. They may be drawn any way that is convenient to show work flow. Arrows may cross but that is not important so long as it is clear where each arrow leads.

A logic network for the whole plan is created by these two symbols. The term precedence network is used to identify this technique of planning.

Network Guidelines

Work flow in the network is from left to right.

A network should have one starting point—a start box and one finishing point—a finish box.

Use a list of plan activities as input for developing the network.

List each activity on a small card the size of the activity box.

Create the network plan by arranging the activity cards on a table top.

The logic of the plan is developed by continually asking the questions, "What must be done before I can start this job?" This is a rigid test of logic. Networking is an exercise in logical thinking.

No activity can begin until all activities leading to it are complete.

After the cards have been arranged, the network is then drawn.

Example: A Precedence Network

		Make A			
S	Design		Assemble	Test	F
		Make B			

Adding Time to the Network

The second step in preparing a network is to add the dimension of time. It begins with estimating the duration of each activity and placing the number in the associated box.

15.5.3. Analyzing the Network

The purpose of this analysis is to determine when each individual activity can be done. In a network there is a sequence of activities that determines the duration of the plan. This is the longest sequence through the network and is referred to as the critical path. All other activities have time to spare (slack).

Steps:

1. *Estimate Activity Duration.* Estimate the duration and add to the activity box.
2. *The Forward Pass.* Calculate the earliest start and finish times.
    ES = the earliest an activity can start.
    EF = the earliest an activity can finish.

    The earliest a job can start is when the preceding job has finished. The earliest a job can finish is the early start plus the activity's duration.
    (ES + duration = EF.)
3. *The Backward Pass.* Calculate the latest start and finish times.
    LS = the latest an activity can start.
    LF = the latest an activity can finish.

    Start at the end of the network and subtract durations.
4. Example: *Network Analysis*

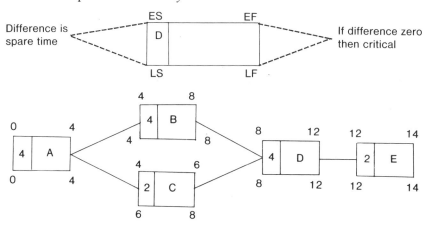

5. *Determination of Spare Time* (Slack). The forward and backward pass have been completed so it is easy to see by inspection the amount of spare time available for each activity. Spare time is calculated by the difference between ES and LS or EF and LF.

$$S = LF - LS$$
or
$$S = LS - ES$$

If the difference is zero, then the activity has no spare time and is critical. It is then part of the sequence that determines the plan's total time. If spare time does exist for an activity, it is utilized in resource scheduling.

15.5.3. Scheduling

Scheduling is the act of developing a timetable for the plan that shows when each activity begins and finishes. The critical activities schedule themselves.

Factors to Consider.   In many cases it is easy to decide when an activity should start and finish. The difficulty arises if there is a problem of resource availability. The essence of scheduling is to use spare time to match the resources required to those available.

The process begins with activities in the earliest start position. Where activities compete for resources, they are allocated according to the plan's system of priorities.

Example: Time Phased Bar Chart

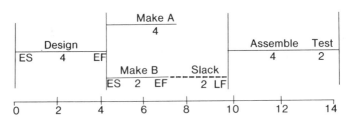

This time phased bar chart quickly identifies the critical activities which are represented by the solid path through the network. Note activity "Make B" has spare time and can be scheduled by sliding it into the slack area represented by the dotted line. The activity is moved to the right until it matches the time the needed resource is available.

Scheduling is the application of common sense within the framework of the available start and finish times. In scheduling resources there isn't always an optimum solution. However, an acceptable allocation can be done.

15.6.   WHY PLANS FAIL

Planning is a function that many think important but not totally necessary. They feel that time shouldn't be wasted on it. These people harbor doubts that planning can be very effective. They feel that planning can restrict independent action and deny one the freedom to make decisions and take action at a moment's notice. There seems to be more fascination for being knee-deep in a flurry of daily activities.

There are many obvious signs as evidence of why plans often fail. We see idle employees waiting, people not sure of what they are supposed to do, people doing unnecessary work, and all kinds of work being duplicated. We see these conditions further expressed in personnel inefficiencies, organizational ineffectiveness, uneven work loads, improper facilities and equipment, a plethora of problems and bottlenecks, and productivity going down the drain. The importance of planning has been forgotten.

In recognition of the fact that planning is not perfect, surveys have produced evidence that most planning failures can be traced to one or more of the following:

1.   Management assumes that if the work is planned, it will automatically happen (They plan the work but don't work the plan)
2.   The budgeting process controls planning
3.   It is felt that the only plan required is a budget to obtain resources
4.   There is a lack of understanding about the process of planning
5.   The planning process is recognized as being weak but control is expected to ward off failure
6.   There is a lack of accountability in the management process, so why bother to plan
7.   Planning does not follow a systematic process
8.   The plan's objectives are not clearly defined
9.   The plans are not effectively communicated
10.   Personnel is not adequately trained for the work to be done
11.   Information input to the plan is inadequate
12.   The planning process restricts personnel participation

15.7.   THE BEHAVIORAL ASPECTS OF PLANNING

Regardless of how thorough the planning effort, no plan is better than the willingness of people to do it. We have made important progress in the mechanics and technology of planning. However, a major problem today is our intense involvement in the physical aspects of planning and the disregard for the consideration of people. Plans can be technically complete, yet not recognize the needs of personnel that must implement them.

The heart of planning is the thinking and action that people do. The plans they produce will work only as well as people want them to. Successful planning is as much motivating people to perform as it is developing the plans to direct their work. Highly motivated personnel can do a better job of planning with crude methods than poorly motivated people with a super system.

Good planning procedures must encourage people to participate. There are four key words that must be understood and used in building consideration for people into any plan. They are motivation, participation, communication, and commitment. Unfortunately we assume that the activity implied by these terms will occur automatically. It won't!

Motivation is the engine that drives people to action and goal achievement. A good plan can be a powerful motivator if it seeks acceptance and understanding. It must solicit participation and consensus so that people can contribute their range of views and ideas. They must have an opportunity to make a contribution, to seek consensus, and to become involved.

Participation requires open communication, recognition, and mutual respect. People will develop a sense of belonging, of ownership and a sense of responsibility. When one feels a part of a group, a member of a team, then one becomes committed and develops a willingness to serve; he or she is inspired to strive for a common purpose. Participation requires patience and understanding and the reward for it is involvement, commitment, and a sense of responsibility.

16

BUDGETING FOR THE ENGINEERING MANAGER

Earl B. Hitchcock

EARL B. HITCHCOCK's background includes experience in marketing consulting, sales and marketing management, as well as general management.

He has held sales management positions with Hoke, Inc. (fluid controls), Potter Instrument Company (computer peripherals), and General Electric (industrial electronics). During that time he built two different manufacturers' representative and distributor organizations.

As marketing vice-president, Mr. Hitchcock established a world-wide sales and distribution group for Hoke. Later he became vice-president and general manager and guided the company through a growth period to where sales reached $40 million in 1978.

He has over six years experience in marketing consulting with emphasis on formulating successful marketing policies for companies with representative and distributor sales organizations.

He is the author of a complete guide to selling through representatives and distributors as well as a number of magazine articles on the same subject.

16.1   BUDGETING DEFINITIONS

The word budget conjures up so many images in the minds of those who hear and use it, depending on individual managerial backgrounds and goals, that a definition for the purposes of this chapter is almost mandatory. At the same time it will be instructive to consider definitions for those terms most frequently associated with the budgeting activity to insure that the discussion is as clear and uncomplicated as possible. In presenting these definitions some thought has been given to the relationship of each term to the other as well as to the function of budgets in a business organization, so that the importance of the engineering budget and its relationship to the business as a whole will become clear.

The engineering budget in its ideal form is an integral part of a larger corporate plan which includes the formulation of a strategy to achieve previously agreed-on objectives. Without such a foundation the engineering manager is either at a loss for directions to follow or in danger of striking off in a direction away from the rest of the company. The format of the plan may be implied rather than explicit, but must be real and practical, since few companies are able to achieve growth and profitability without some kind of direction from top management.

Nevertheless a clearly articulated, well-thought-out strategic plan containing short- and long-range elements will benefit the whole company and most particularly the engineering department in channelling available resources toward clearly specified goals. Since all aspects of the engineering budget are controlled by the demands of the strategic plan, the engineering budget may be thought of as an integral component of the overall plan.

That plan is composed of a series of discrete elements and starts with the formulation of corporate objectives to be achieved in a given period of time. Since they are generally major steps, often in new directions, the time periods may vary from 2 to 5 to 10 years or longer, depending on the urgency of the task, type of business, and the company's capabilities.

These objectives are arrived at after a careful review of the company's present situation, its competition, advances in technology, the business climate, and other factors affecting the company's ability to survive. They may include development of major new products, abandonment of older products or markets, diversification, and so on. Objectives lead to strategies for accomplishing them and strategies require implementation. Most well-thought-out plans provide for risk assessment and methods to control execution, costs, and adherence to schedule.

By most definitions budgets cover the execution of the strategic plan during a specific time period (usually a year), detailing the work to be accomplished, scheduling of each part of the work, personnel to accomplish the work, time required, plus facilities and cost. Generally, the budget also includes a method for reporting progress and controlling the results. The engineering expense budget covers the portion of the work, scheduling, personnel, facilities, costs, and controls to be carried out by the engineering department. If the budget is not driven by an overall strategic plan, it degenerates into a forecast of the people, time, and money expenditures for the oncoming time period and may leave management wondering what value was received from these expenditures.

16.2. THE ENGINEERING BUDGET STRUCTURE

The plan behind the engineering budget is similar in form to the corporate plan. For instance it begins with a statement of objectives. But in most cases these objectives are technological in nature and derived from corporate objectives. For instance, if the corporate plan calls for the development of a new line of products for a specified market within three years, the corresponding engineering objective will specify each product plus the time allotted for research (if required), development, and finished design.

Both the corporate objective and engineering objectives must be realistic and achievable. For that reason a screening process must be applied at both the corporate level and development level to insure that new product and business ideas are practical and have a reasonable chance for success in the marketplace. From an economic point of view, the importance of screening new product ideas at the engineering level is quickly demonstrated by the fact that only one out of three is ultimately successful, based on industry-wide averages.

The company's engineering objectives can be either short or long range in nature. Those associated with product improvement or redesign and requiring engineering facilities only can be more quickly and predictably achieved. Completely new products or product concepts take longer to achieve and the time required is less predictable. If basic research is required, the ultimate results are unknown and the difficulties of reaching the desired goals cannot be measured. Basic research is so difficult to budget accurately that most companies tend to underfund it. Applied research and development on the other hand are easier to measure, since the technological data is available to apply to specific design and performance problems.

In the typical engineering department activities are divided into three principal areas:

1. Basic and applied research—used to explore and refine new product ideas, related technologies, and concepts.

2. Development—takes new product ideas and concepts and reduces them to physical form by reviewing and analyzing alternative approaches and measuring performance and manufacturability.
3. Maintenance—covers product improvement, minor redesign, application engineering, value analysis, improving manufacturability, and so on.

Companies using advanced technology will tend to emphasize research and development and these activities may dominate engineering or be administered separately. In other companies, which manufacture products with less technological content, greater emphasis is placed on engineering and R & D is a small-scale activity or even nonexistent. Typically, companies devote up to 10% of their engineering budgets to basic and applied research, 15–40% to development, and the rest to maintenance.

In most engineering departments the potential workload in all three areas is greater than the manpower and facilities available to complete it. To insure optimal use of the workforce and facilities, the major projects resulting from the engineering objectives previously adopted must be prioritized. The establishment of these priorities depends on the results of the screening process, financial considerations, and some form of risk analysis.

The screening process is most successful when carried out by all those activities involved and includes attention to such factors as:

1. Marketability—customer need, market size, product compatibility, potential market share, growth potential, market risk, and so on
2. Technology—development cost and time, potential design success, resource and facility requirements, engineering capability, patentability, and so on
3. Manufacturing—compatibility with current facilities, new capital requirements, manufacturability, and so on
4. Financial—potential profitability, risk, return on investment, cash flow, and so on

Based on these factors a numerical value can be established to yield the relative importance of this project with respect to others and thus a basis for its priority. Other factors come into play in prioritizing engineering projects. For instance, corporate management may dictate the urgency of some work based on overall strategic considerations.

Next in line come those projects that appear strongest from a financial standpoint, for instance, those with the highest potential return on investment. Basic and applied research projects deserve attention at this point. Although difficult to plan and schedule, this type of work often contains opportunities for future growth. Maintenance projects represent the routine tasks that generally take up the bulk of the engineering department's time. Since the level of work is fairly constant from year to year, a fixed proportion of the budget can be allocated for the purpose.

16.3. BUDGET CHARACTERISTICS

In most companies budget preparation is divided into two principal parts: the operating (or expense) budget and the capital budget. As an engineering manager using your department's facilities and people to create new products, as well as improve or reduce the costs of the old, you will be primarily concerned with the operating budget. Only when the company's strategy demands a sizable increase in engineering effort or a new direction for the effort will your capital budget become of major importance. You may need additional office space, furniture, laboratory or shop facilities, or computer-aided design equipment.

In most cases capital budgets are based on capital requests submitted during budget preparation. These requests normally include an analysis and justification for the department's need based on corporate strategic requirements. The capital budget is administered and controlled in the same way as the operating budget with monthly and quarterly reports that compare actual expenses to budgeted figures.

16.4. THE BUDGETING PROCESS

Since budgeting has been defined as a planning process, it is helpful to outline a method for preparing and achieving this plan in order to optimize results and avoid as many barriers and missteps as possible.

The flowchart in Figure 16.1 depicts such a method. As has already been indicated, it begins with the definition of objectives. The general direction and established constraints flow to the engineering department from the corporate strategic plan. In its ideal form that plan presents the engineering manager with a variety of concepts, new product ideas, and areas for research sufficient to occupy the time and facilities of his or her department. However, there ought to be sufficient slack to take advantage of new ideas and opportunities which may come from random sources.

The maintenance phase of engineering work may originate in many areas within the company, mostly unrelated to corporate strategy. Product improvement requests may originate in marketing, cost cutting studies and value analysis, manufacturing departments, or competitive pressures that

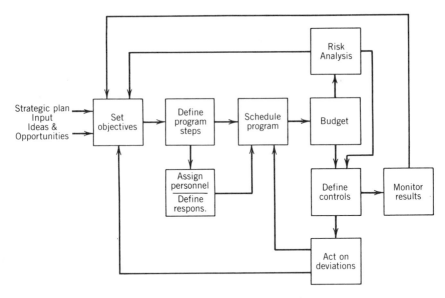

FIGURE 16.1.   The engineering planning process.

may make demands on engineering. The aforementioned activities are intended to keep products and services abreast of the market.

At this point it may be assumed that projects to be included in this process have survived screening and have a sufficiently high priority to be included in the budget under preparation. It should also be assumed that the projects being planned are sufficiently large in scale to merit a detailed plan, and that the logic of the plan is valid whether it is an R & D or maintenance project. Small-scale work is normally assigned as personnel become available and generally requires less planning and control.

The flow chart may be applied to major engineering projects regardless of nature, and the objectives to be established are project objectives. To be useful these should be expressed in specific terms, for example: "The company needs a new line of battery-powered tools for the home workshop consisting of a drill, saber saw and orbital sander to be available in the Fall of 1989."

After project objectives are established, it remains to define the steps necessary to achieve each objective. Once these major steps are identified (test concepts, design mechanical parts, develop materials, etc.), they need to be sequenced into the order in which they can be accomplished. For instance, mechanical parts cannot be designed until a concrete design concept has been completed.

These tasks are then assigned to individuals or functional groups within the organization. At this point it is vital to invite the participation of the individuals or groups involved to insure their understanding, agreement, and full cooperation. Also be sure you have the participants' approval of the sequencing of the program steps! If the assigned tasks are fashioned into work packages, each participant becomes fully aware of his or her responsibilities and a measure for the completion of the task is acquired. The work package is also useful in measuring progress, as will be seen later.

The next step is to schedule each of these activities with particular emphasis on the interrelationship of those dependent on the prior completion of other activities. A Gantt chart is helpful in visualizing the schedule and a simplified example is shown in Figure 16.2.

The chart is constructed from the following task information:

Activity	Time Required	Prerequisite	Number
Preliminary design	2 months	None	1
Performance calculations	1.5 months	Complete 1	2
Layouts	2 months	Complete 1	3
Part drawings	2.5 months	Complete 3	4
Cost estimate	1 month	Complete 4	5
Construct model	3 months	Complete 4	6
Test model	1 month	Complete 6	7

The symbol 0 designates the starting and stopping points for each activity. Each of these points is a milestone in the completion of the project. The triangle denotes the latest time by which an activity must be completed in order not to delay the succeeding dependent activity. Note that there is no

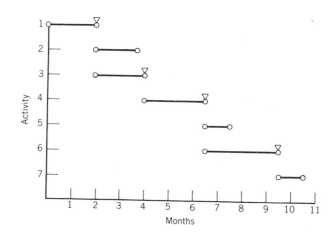

FIGURE 16.2. Simplified Gantt chart.

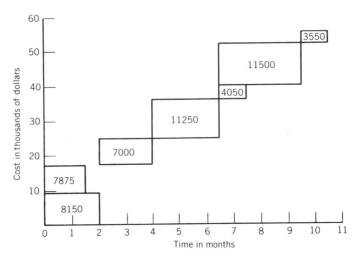

FIGURE 16.3.   **Budget chart.**

slack in the scheduling of the critical path activities. The critical path runs from 1 to 3 to 4 to 6 to 7. The chart can be used to monitor progress by filling in the actual times required to complete the activities alongside the planned times.

The project budget is then developed by preparing manpower cost estimates for each of the work packages. When these labor costs are added to material costs, the project budget is complete. If the incremental costs are plotted against time, a picture of the total budget as it develops results. Assuming that the material and labor costs for each activity listed above is as follows:

Activity	Labor	Material	Total Cost
1	$ 4,000	$ 150	$ 8,150
2	7,875	100	7,975
3	7,000	175	7,175
4	11,250	250	11,500
5	4,000	50	4,050
6	12,000	1,200	13,200
7	3,250	300	3,550

The graph in Figure 16.3 shows how costs accumulate to add up to the total budget.

For major projects it is prudent to perform a risk analysis during the planning process. There are three purposes for this type of analysis:

1. To identify and record risks that may affect progress or reaching the planned objectives successfully.
2. To assess the impact of these risks on schedule and objectives.
3. To identify ways of limiting risks, monitoring their occurrence, or modifying objectives to avoid them.

Use the combined resources of your task force to identify and assess the importance of potential risks. The questions to be answered at this stage and procedures required to complete the risk analysis are:

1. Determine what the risks are—invite suggestions and ideas from the group involved and list them without comment.
2. Assess how probable each risk is—after completing the list, evaluate the probability of each one by assigning a value to it (use high, medium or low or use a numerical scale).
3. Assess the importance of each one—evaluate in terms of its effect on the success of the project (same type of assessment as in No. 2).
4. Now combine these two factors and decide whether the risk is major or minor
5. Finally, determine what kinds of action can be taken to prevent the major risks from occurring as well as how to limit the damage if they do occur
6. Decide how to monitor the process to detect risks or what performance standards are required to avoid them—in some cases it may even be desirable to modify plan objectives to prevent major risks

Once the budgeting of resources is completed, the risk analysis is made, and the project is moving ahead, control steps are generally necessary to insure that:

1. The project goes according to schedule
2. It remains within budget
3. Results meet specifications established during the program definition step
4. Project objectives are ultimately met

Step 1 can be met by recording actual completion dates of project activities on the Gantt chart alongside the planned dates. A look at the chart will immediately reveal when additional manpower or overtime is needed to keep critical path activities on schedule. Step 2 can be reduced to visual form by charting actual completion dates and dollars spent on the chart in Figure 16.3. Again comparison of actual times and costs versus projected ones will be revealing. Step 3 requires monitoring the design and performance achieved versus project specifications. Step 4 demands a periodic review of the initial objectives measured against progress to date of the whole project.

It should be obvious that the flow chart as a whole is an iterative process and requires a return to the previous step whenever obstacles are encountered or previously agreed-on schedules or specifications cannot be met. In this way an overall awareness of the project's progress is maintained and the records reflect the actual state of affairs. There is nothing so demoralizing or damaging to the sense of urgency as schedules and charts so out of date as to be meaningless.

16.5. DEPARTMENTAL BUDGETS

The engineering department budget is a compilation of all the costs, regardless of source, accumulated by its managers and personnel in the execution of projects, R & D assignments and maintenance engineering. The only exception to this description is the case where an engineering department takes on contract work for outside parties, the costs of which are borne by them. In most companies costs are compiled on a direct and indirect (overhead) basis.

Direct expenses generally include all labor costs (salaries, wages, and employee benefits) plus those expenses under the direct control of the department: materials, office supplies, printing, copying and other reproduction, travel, education, and services. Hours chargeable to projects are calculated by assuming that each employee works a 5-day week, 40 hours per week and 52 weeks a year. Holidays, sick leave, vacations, and other nonproductive time are subtracted.

The calculations are started by determining the available chargeable hours for each employee. An example for one engineering department is shown in Table 16.1. Total hours per employee are based on a 52-week year, 40 hours per week less 8 holidays and 5 days of sick leave. If the total accumulation of direct expenses for drafting are added up and divided by 14,152, the result is the chargeable direct rate for drafting.

Indirect expenses include:

1. The portion of company-wide fixed charges allocated to engineering
2. Engineering overhead including administrative expense, clerical costs, and miscellaneous departmental costs.

TABLE 16.1. Drafting Hours

Employee	Total Hours	Vacation Time	Administrative Time	Available Project Time
Abel	2176	80	40	2,056
Delatour		80	0	2,096
Francesco		240	0	1,936
Gilbert		80	60	2,036
Johnson		240	100	1,836
Karten		80	0	2,096
Monkmeyer		80	0	2,096
Total available hours				14,152

The combination of direct and indirect charges for a typical engineering department are shown in Table 16.2. The direct and indirect hourly rates are used to establish project costs in terms of time consumed. Adding the two rates yields the total loaded cost per hour for each department. Dividing total department hours into each total cost figure shows the department rate in each category.

16.6. MANAGING THE BUDGET

The first step in budget management is to insure that it is prepared on time, all the participants are informed and cooperative, and the preparation steps are completed in proper sequence. To understand the relationship of the engineering budget to the corporate whole, it may be helpful to describe the process that takes place in a typical, well-managed company.

First, all departments concerned must start with complete information on such matters as planned average compensation rates for employees, projectable economic information (expected rate of inflation, for example) and corporate profit goals. This is to supplement the previously published strategic planning information. In addition each department head must know when each phase of his budget must be ready for review by management as well as when the final budget must be submitted. In companies with corporate budget staffs, these are usually responsible for this internal scheduling. Most companies begin budget planning a month or so after midyear by preparing a sales forecast for the next year. Except for companies in highly volatile businesses (some consumer products, electronics, and military, for example) the sales forecast is generally fairly accurate. Preparation and review take about a month.

TABLE 16.2. Combined Direct and Indirect Departmental Charges

(1) Activity	(2) Direct Costs	(3) Indirect Costs Overhead	(4) Allocations	(3 + 4) Total Indirect	(2 + 3 + 4) Total Charges	(5) Available Hours	(3 + 4/5) Rates Indirect	(2/5) Direct
Drafting	$ 235,820	$21,500	$7,500	$ 29,000	$ 264,820	14,152	$2.05	$16.66
Laboratory	172,560	18,600	6,200	24,800	197,360	11,825	2.10	14.60
Testing	77,800	9,400	3,700	13,100	90,900	6,220	2.11	12.51
Model shop	56,210	7,800	2,500	10,300	66,510	5,421	1.90	10.36
Current engineering	375,810	31,000	9,400	40,400	416,210	19,756	2.04	19.02
Prototype design	127,340	15,200	7,200	22,400	149,740	10,956	2.04	11.62
Application engineering	82,160	12,400	4,400	16,800	98,960	7,896	2.13	10.41
Departmental								
Totals	$1,127,700			$156,800	$1,284,500	76,226	$2.06	$14.79

The forecast is then analyzed by manufacturing to establish how well it can produce to the forecast, as well as the associated costs. From this data a gross profit forecast is prepared and completed by the end of September. At this point each operating department can start work on its own budget. The final corporate budget must be completed (including profit figures) in time for review during November. Often it must be revised after the review, but in any case detailed figures must be ready for budget versus actual comparisons when the new year starts.

In many cases the budget review is done on a monthly and quarterly basis. Today most financial departments can provide computer-based budget comparisons very shortly after the close of the month. The format generally provides overall financial figures for the month including sales results, profits, comparisons with budget and last year's performance for the equivalent period.

In most businesses operating budgets are tracked by comparing actual figures spent with the budget previously prepared. The accounting department prepares both combined budget figures for each department as well as a detailed breakdown for each activity most often monthly, sometimes quarterly. These figures are for the guidance of department heads and supervisors in monitoring the performance of their activities against budget standards. An example of a departmental budget report is shown in Table 16.3. Usually the report compares the month just completed and the year-to-date figures.

The primary requirement for these budget figures is prompt delivery. Budget management is almost impossible to achieve if budget comparisons are received three or four months after the close of the period being measured. It is also imperative that the reports be accurate. Mistakes in allocations or assigning expenses to the wrong account number guarantee that the numbers will be useless for control purposes.

In dividing up the annual budget into monthly periods, the temptation to divide the total by 12 should be carefully avoided unless in fact, those expenses occur at such an even month-by-month rate. Since most activities vary depending on vacations, travel, seasonal demand, and so on, these monthly variations should be estimated as well as possible when the budget is first calculated. Otherwise the manager is confronted by inexplicable variances which again complicate the control task. In Table 16.3 note that actual expenses for May are roughly 2.5 % over budget and that the year-to-date compliance is within less than 1 %. At this juncture the budget is in excellent shape.

On a smaller scale the budgets of individual activities can be tracked the same way (see Table 16.4).

Note that on a year-to-date basis, drafting is almost 6% over budget, a trend that must be changed before year-end.

TABLE 16.3. Monthly Engineering Department Budget Report

Activity	Month of May Budget	Month of May Actual	Year-to-Date Budget	Year-to-Date Actual	Annual Total
Drafting	22,000	21,750	111,300	110,750	264,820
Laboratory	17,000[a]	18,100	85,000[a]	86,220	197,360
Testing	8,200[a]	8,700	41,000[a]	40,510	90,900
Model shop	5,200[b]	5,000	25,500[b]	23,750	66,510
Current engineering	35,000	36,500	174,720	177,050	416,210
Prototype design	12,450	12,310	63,000	64,765	149,740
Application engineering	8,200	8,350	41,600	42,510	98,960
	108,050	110,720	542,120	545,555	1,284,500

[a] These activities are busier during the first 6 months.
[b] The model shop is more active in the last quarter.

TABLE 16.4. Monthly Budget Report for Drafting

Account	Month of May Budget	Month of May Actual	Year-to-Date Budget	Year-to-Date Actual	Annual Total
Salaries	11,828	12,020	59,375	63,720	142,500
Fringe benefits	3,884	3,960	19,500	20,464	46,800
Drafting supplies	2,620	1,350	6,333	7,455	15,200
Reproduction	411	397	2,062	2,184	4,950
Field work	1,426	1,485	7,158	7,475	17,180
Materials	763	540	3,829	2,970	9,190
Indirect costs	2,407	2,407	12,083	12,083	29,000
	23,333	22,159	110,340	116,532	264,820

Although this kind of track is useful for keeping routine expenses in line, it provides no information on what results are being achieved. To monitor real results some measure of what has been accomplished to-date is needed in addition to the actual versus budgeted figures just discussed. The work packages described earlier are useful for this purpose if they are planned properly. In essence the work must be divided up in such a way that progress can be easily measured at any time. For practical purposes this means making small, well-defined work-packages so they can be easily characterized as to status. For instance, the design of a key component can be described as 10% or 50% complete with fair accuracy. Larger-scale work-packages on the other hand are more difficult to evaluate so judgments become more subjective.

By charting the value of work packages completed on a regular basis and comparing it with costs incurred to date, an informed judgment on project status can be made. Here is a simplified example of this work package concept. This particular project consists of four packages with the work values shown when each step is completed:

Work Package	Value	Time Required
1. Complete design concept	$13,500	2 months
2. Prepare detail drawings	22,400	1½ months
3. Build model	12,650	1½ months
4. Test model	8,715	1 month

The project budget calls for expenditures of $9,500 per month. At the end of four months, Step 1 is completed and Step 2 is 50% complete. Thus the value of the work completed is $24,700. Although the work is slightly behind schedule, the project is under budget so additional people can be assigned to the work or overtime scheduled to speed it up.

For more complex projects the data can be put in the form of a chart similar to the one in Figure 16.4.

16.7. BUDGET FLEXIBILITY

The ideal budget is one so carefully planned that results are predictable to within a few percentage points and that expectations are met. In reality, such results are seldom attained. It is particularly difficult to measure productivity in the engineering department where innovation is the primary objective. Finding new ideas and evaluating their practicality is an inherent part of engineering design and development. But the very act of putting time

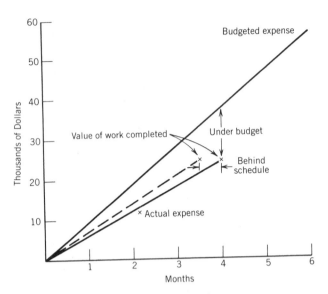

FIGURE 16.4.   Budget curve.

limits on design work tends to limit creativity. In addition, delays in material delivery, backlogs in the model shop, problems in testing, and many similar events play havoc with project scheduling. Thus the prudent manager leaves slack in the schedule to take potential problems into account. At the same time some projects encounter few holdups and consequently finish well ahead of schedule with a resultant scramble to reassign the personnel involved.

To deal with this lack of exact predictability a number of companies ask their operating departments to prepare budget alternatives for expected, better than expected, and worse than expected conditions. This type of budget planning offers a number of benefits:

On a company-wide basis it permits top management to recognize the profit consequences of less than optimum results

Within each department managers have the flexibility to deploy their resources to counteract problem situations when they arise

It provides a visible incentive to aim for the better than planned results displayed in the budget

Lastly the flexible budget provides an opportunity (as the budget is being prepared) for realistic contingency planning if it becomes necessary during the planned period. Chapter 17 discusses the procedures in detail.

16.8.   PROBLEMS IN BUDGETING

Although budgeting, properly used, is a prime management tool, it is often the source of discord, basis for bad management decisions, and even excuse for poor performance. Most of the problems involved in budgeting are rooted in a lack of understanding of its purposes and a failure on the part of management to grasp how it works. There are, first of all, corporate difficulties with budgeting. Many executives fail to recognize the budget's direct relationship to the implementation of corporate strategy.

Management's basic task is to plan, execute, and control. A well-conceived budget is said to be the detailed exposition of the expected results from a project or a plan. In the successful company the plans flow from the strategic plan and the budget is presented as the annual business plan. Budgeting is almost never successful when it is considered only as a way to financially discipline a manager.

The fact that a company does not have an explicit strategic plan does not free its managers from the responsibility of preparing a budget which forwards the company's implied goals and objectives. Obviously it is very difficult to prepare a budget (or annual business plan) which optimizes progress toward those goals and objectives if they're not clearly specified and thoroughly understood by all concerned. Thus the logic of the strategic plan.

Assuming then that the departmental budget is tailored to the objectives of the corporate strategic plan or that it complies with agreed-on policies and goals leading toward an improvement of the company's market position, what are management's expectations from it?

First of all it provides the ability to forecast performance, at least during the time period it covers. By the same token it provides management with the capability of controlling performance. Secondly, perhaps the budget's strongest benefit is to provide information which will assist management in allocating limited resources for the company's optimum performance. If well-prepared, a budget offers an analytical tool that is timely enough to be immediately beneficial. For instance, it quickly reveals variations from forecast and points to developing problems or sudden opportunities. Third, it provides past performance data as a yardstick to measure progress. Lastly it puts the company's performance data into clear and understandable form. This fosters interdepartmental cooperation since each group's contribution is visible and its value apparent.

Unfortunately corporate affairs seldom run so smoothly. Problems crop up in unexpected fashion, disrupting the smooth flow of business and causing managers to resort to crisis management, reacting to rather than anticipating developments. Events such as strikes, recessions, and fires cannot normally be anticipated in strategic planning.

There are other reasons for unsuccessful budgeting. Lack of support from the top is the most frequent disrupter of budgets. If top management is not committed to the budget objectives or goes off suddenly in a new policy direction, the budget loses its importance as a guide and its reality as a standard.

At the other extreme is the managerial tendency to substitute the budget for good judgment. If the budget acts as a straitjacket, it can prevent managers from responding to opportunities or sudden dangers in the marketplace.

The budget review which precedes the adoption of the budget in most companies, often places great stress on the management group. It is here that top management decides where to allocate limited resources and where pet projects can be ruthlessly rejected. Trade-offs are the order of the day and if the individual manager cannot subordinate his or her department's needs to the overall demands of the business, results can be chaotic. A delicate balance is required since the manager must defend his or her budget request, particularly if he or she supports the need with strong evidence that the return on investment will match or exceed the others under consideration.

The task of preparing and managing an engineering budget has its particular problems even within the department itself. If, for instance, the budget is shaped and imposed by an ambitious engineering manager without the help and cooperation of those who must make it work, it is doomed to failure. The tendency will be to regard it as unworkable (even if it is quite reasonably constructed). Employees will show up for work daily and carry out their assignments faithfully for 8 hours. In the end when projects are hopelessly behind schedule, there will be fingerpointing and head shaking but no one will feel responsible.

If, on the other hand, the workforce is fully involved in planning, scheduling, cost estimating, and risk analysis, people will work harder and longer, ideas for shortcuts and improvements will sprout, and the project may well be completed ahead of schedule.

Budget controls are particularly difficult to administer in an engineering environment because engineers typically resent constraints on their freedom of action. Since they tend to dodge procedures, schedules often get out of hand. The emergency efforts to repair schedules imposed on top of normal controls frequently become unwieldly. This is how departments sometimes lose their best engineers.

It is particularly necessary in engineering administration to plan thoroughly and employ sound control methods. In many engineering activities personnel are assigned to component design without prior system analysis, project work is assigned without regard to the interdependency of work packages, and assignments are made without a clearcut definition of required results. This planning deficiency is complicated by a lack of controls to pinpoint obstacles. Needless to say successful budgeting is impossible in such an environment.

These conditions generally arise out of the engineers' discomfort with schedules and controls. They do not make the life of the engineering administrator any easier, but once again it appears that the most practical solution is to fully engage the engineering staff in the planning and scheduling task. Once they consider the plan theirs, the necessity for coordination and control seems perfectly natural.

One of the conflicts faced by the engineering manager arises out of the dual nature of his work. Because of his or her engineering background he or she may tend to overemphasize engineering responsibilities and thus shortchange administrative obligations.

Administrative obligations include:

project and overall planning and scheduling

developing project monitoring and departmental cost controls

responsibility for foreseeing, planning, and completion of facilities and equipment

developing the basis for standard cost systems

translating corporate strategic plans into engineering objectives

supervising and training the engineering staff

Note how many of these tasks are budget-related! The manager who neglects them often winds up underestimating costs, offering inaccurate project schedules, over- or understaffing and slighting control systems.

17

WORK SCHEDULING

Jack Yurkiewicz

JACK YURKIEWICZ is Associate Professor in the Management Science Department, Lubin Graduate School of Business at Pace University. Dr. Yurkiewicz earned his Bachelor's degree in Electrical Engineering from The City College of New York. His Master's degree in Electrical Engineering is from Columbia University, and he later graduated from the Courant Institute of the Mathematical Sciences with a master's in Mathematics. Professor Yurkiewicz received his master of Philosophy and his Ph.D. in Operations Research from Yale University.

A licensed professional engineer, Dr. Yurkiewicz has worked for major firms as an electronics-communications engineer and a nuclear power-electrical engineer.

Although Dr. Yurkiewicz's specialty is optimization theory, he has many interests in operations research including applied probability models, stochastic processes such as Markov Processes, optimal control theory, and statistical methods. Three articles written by Dr. Yurkiewicz on optimization models have been published in the *Encyclopedia of Management,* 3rd edition. Other articles have appeared in *Networks.*

17.1. INTRODUCTION

Most engineering tasks consist of a large number of interrelated activities, both at the stages of design and of execution. The required planning, scheduling, and control can be very complicated, especially since frequently the activities must be done in a certain sequence in order to complete the task. Over the years many methods have been proposed that facilitate the management of complex projects; the purpose of this chapter is to explore some of these procedures, examining both their merits and shortcomings. This chapter thus extends previous discussions of project management.

A *project* is a task which consists of activities, usually interrelated, which must be performed in some certain order. An *activity* is some job that is part of the project and requires time and resources for its completion. During World War I, Henry L. Gantt developed a graphical approach to solving scheduling problems having many activities. The so-called *Gantt chart* simply displays the start and finish times of each activity on a time scale as a horizontal bar. The length of time required to complete all the activities of the project is called the *makespan*. Inspection of the Gantt chart allows the schedule to be evaluated for waiting time and makespan. However, it has a serious weakness as a scheduling technique in that the logical connections between activities (which really control the progress of the project) cannot be seen. It thus does not provide any structured approach to schedule improvement. However, Gantt charts are used effectively as visual aids for the presentation of carefully prepared plans and schedules.

Example 1

Consider the project of Buying a New Car. There are many activities involved with this undertaking, including doing research by reading car magazines, visiting dealers, shopping for car loans at various financial institutions, and so on. A *partial* Gantt chart of this project might look like Figure 17.1. (Note that not all the activities of this project are shown.)

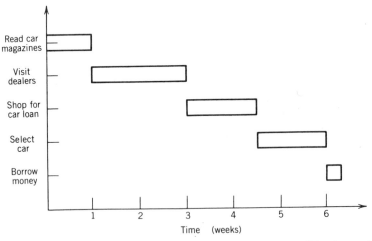

FIGURE 17.1. A portion of Gantt chart for the project "Buying a New Car."

The Gantt chart indicates how long each activity will require. But it does not show that shopping for a car loan can be done simultaneously with reading car magazines, for instance.

17.2. PERT/CPM—A BRIEF HISTORY

The PERT (Program Evaluation and Review Technique) system was originated in the late 1950s as an aid to plan the Navy Polaris ballistic missile. At the time, various management tools available for the planning of the Polaris program did not provide sufficient information necessary to evaluate how the program was progressing. In particular, information about the project's progress vis-a-vis the program objectives, as well as an analysis of the chances of meeting those objectives, was lacking. A better management system was needed. PERT was the planning and scheduling tool developed to aid in the development of the missile. Many of the activities of the project had uncertain completion times and also had to be completed in some set order. The PERT model proved to be useful in scheduling this and other large projects which had such time-described characteristics.

The Critical Path Method (CPM) was developed by E. I. du Pont de Nemours slightly earlier for the purpose of utilizing a computer in scheduling construction projects. The approach needed the activity sequences, duration times, and starting times for a minimum number of activities. With this, the computer, a UNIVAC, could work out, via a diagramming method, a project schedule. CPM and PERT were later recognized to be quite similar, but with important differences. Under the CPM method the amount of time necessary to perform the activities is known with certainty; under PERT it is not. In addition, the relation between the amount of resources employed and the time necessary to perform the activity is also known under CPM. Thus CPM deals with time-cost trade-offs while PERT is concerned with uncertain activity times. Because of this PERT is used more in research and development projects where there is little or no experience in performing the various activities necessary for the completion of the project. CPM is used more in construction projects where there has been prior experience, which allows rather clear estimates of activity times. However, there are so many similarities in the two methods that they are almost always referred to in the same breath, PERT/CPM, rather than separately as two distinct procedures.

The purpose of this chapter is to introduce the basic concepts of PERT and CPM. These two methods are subsets of the larger subject of network models which are used in sequencing problems and resource allocation problems. Both the positive as well as the negative attributes of PERT/CPM will be discussed and examples will be given to demonstrate how the methods work.

17.3. NETWORKS

A *network* consists of a series of junction points called *nodes,* with certain pairs of these nodes connected by lines called *branches.* The branches are shown as arrows and represent the activities. The tail of the arrow is the start and the head of the arrow is the completion of the activity. It is usually not drawn to scale, nor is it a vector. When drawing activities, the arrows may be bent or curved. An *event* is defined as a point in time designating the completion of one or more activities and the beginning of new activities. It is represented by a node on the network. See Figure 17.2.

This is an activities-on-arcs, (AOA) network representation. A much less widely used approach is the activities-on-nodes (AON) representation. Here the activities are nodes and the arrows show only the precedence relationships between the activities.

Activities are usually labeled by capital letters. Figure 17.3 shows how some logic relations are shown via network representation.

The situation in Case (5) requires some explanation. In order to get the correct logical representation, a fictitious activity called a *dummy,* of zero time duration shown by a dotted line must be incorporated into the network. Without dummy activities it may at times be impossible to establish the correct logical relationship between activities. When to have dummy activities in the network is one of bigger problems facing beginning practitioners. In general dummy activities are required when the project has groups of two or more jobs having some, but not all, of their immediate predecessor activities in common. If an activity has more than one predecessor and one or more of these predecessors is also a predecessor for some other activity, then a dummy activity is needed to draw a logical network. Thus, in drawing a network for each activity one needs to know: Which activities precede, follow, or may be concurrent with it?

Finally, a few other tips in drawing a network would be: The network should have both a unique starting event and a unique completion event, with nodes being numbered for convenience. A completion node for any activity should have a higher number than its starting node, and no activity

FIGURE 17.2.

Representation

Network

1. Activity B starts at the conclusion of activity A

(a)

2. Activity C must be completed before activities D or B can start

(b)

3. Activity F cannot start until both activities D and E are completed

(c)

4. Activity F or G can be started once both activities D and E are completed

(d)

5. Both activities F and G must be completed before activity H can start. Also activity J needs only the completion of activity G before it can start

(e)

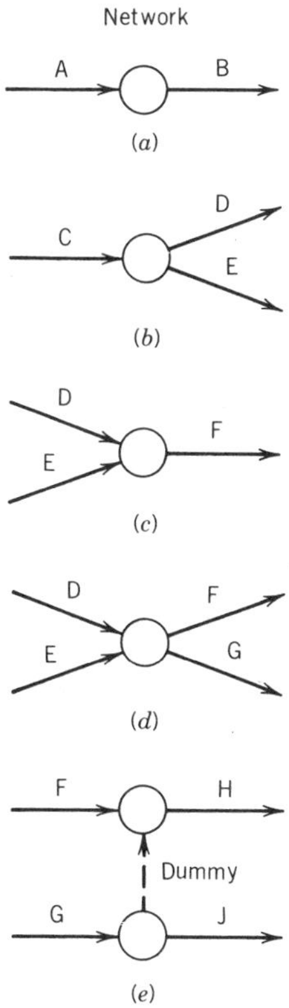

FIGURE 17.3.   Network logic. (a) activity B starts at the conclusion of activity A; (b) activity C must be completed before activities D or E can start; (c) activity F cannot start until both activities D and E are completed; (d) activity F or G can be started once both activities D and E are completed; (e) both activities F and G must be completed before activity H can start. Also activity J needs only the completion of activity G before it can start.

should be represented by more than one arrow in the network. Lastly, a commonly but not universally followed rule is that no two activities should share the same starting node and completion node. See Figure 17.4.

When there are two or more concurrent activities this last rule may be violated and the inclusion of a dummy activity avoids the violation. However, the inclusion of many dummies makes the network complicated and unwieldy. Therefore Figure 17.4a is considered by many to be just as correct as Figure 17.4b. Most texts prefer Figure 17.4b to Figure 17.4a, and therefore that convention will be used here.

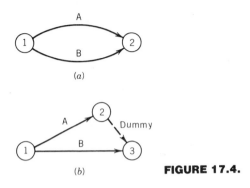

FIGURE 17.4.

17.4. ANALYZING ACTIVITY NETWORKS

In analyzing a project, the most important question is, "When will it be completed?" Related to this is, "Which activities contribute directly to the duration of the project?" In order to answer these questions the activity duration times must be known. Under CPM these are known with certainty; under PERT they are expected times, as the time needed to perform any activity is a random variable. For the moment it can be assumed that the times are deterministic. Later the probability aspects will be considered. Rather than refer to the network as a PERT or CPM network, it will be referred to as an *activity* network since that is exactly what it consists of, a series of activities.

Associated with each event are two time points: the *earliest start time,* which is the earliest point in time at which the event could possibly occur, and the *latest finish time,* which is the latest point in time at which the event could occur without delaying the completion of the project. In general a network will have n nodes. Since each node is numbered, with the beginning of the project being node 1, and the completion of the project node n, let $T_E(i)$ denote the earliest start time of node i, and $T_L(i)$ be the latest finish time for node i, where $i = 1, 2, \ldots, n$. Arbitrarily set $T_E(1) = 0$; that is, the project starts at time zero.

Associated with each activity is a duration time (assumed for now to be deterministic). Let $t_e(i,j)$ denote the duration of the activity connecting node i and node j. (Based on this, it must be assumed that there is only one activity between nodes i and j. If Figure 17.4a is permissible, then the activities can just be numbered and their durations denoted by $t_e(j)$, the jth activity.) In order to find the earliest start time of each event, a *forward pass* is made through the network, meaning calculations begin with node 1, the start node, and proceed to note n, the conclusion node. The following recursive formula is used:

$$T_E(j) = \max_{i<j}\{T_E(i) + t_e(i,j)\}$$ (1)

That is, to the earliest start of each directly preceding event node, add the duration of the connecting activity and find the maximum of all such sums found. This is done for all event nodes.

To find the latest finish time for each node, a *backward pass* is performed. This means calculations begin with the completion node n, and move backward ending with the start node 1. The terminal event is assigned a latest finish time equal to the project due date, and practice has it that it is set to the earliest possible completion time. That is, $T_L(n) = T_E(n)$. To find the latest finish of every other node, the recursive formula

$$T_L(j) = \min_{j>i}\{T_L(j) - t_e(i,j)\}$$ (2)

is used. This means that for the latest finish of each directly succeeding event, the duration of the connecting activity is subtracted, with the minimum of all such differences then found. This is done for all event nodes.

Regarding the activities, there are four times (other than the duration times) that must be calculated.

1. The *earliest start time* for an activity is the earliest time the activity can be started. It is designated as $ES(i,j)$, meaning the earliest time the activity connecting nodes i and j can be started. Since an activity cannot start until its predecessor event has occurred, its earliest start is the earliest start time of the predecessor node.

$$ES(i,j) = T_E(i)$$ (3)

2. The *earliest finish time* for an activity is the earliest time that activity could possibly be completed. Labeled $EF(i,j)$ it is easy to see that

$$EF(i,j) = ES(i,j) + t_e(i,j)$$

or

$$EF(i,j) = T_E(i) + t_e(i,j)$$ (4)

3. The latest finish time of an activity, designated $LF(i,j)$, is the latest time an activity can be completed and still not delay the completion time of

the project. Since an activity can be finished no later than the latest permissible time of the successor event, then the latest finish time of an activity is just the latest finish of the activity's completion event node.

$$LF(i,j) = T_L(j) \qquad (5)$$

4. The *latest start time* for an activity, written $LS(i,j)$, is the latest time the activity can be started without delaying the completion time of the project. It is thus the difference between the latest finish time for the activity and its duration.

$$LS(i,j) = LF(i,j) - t_e(i,j)$$

or

$$LS(i,j) = T_L(j) - t_e(i,j) \qquad (6)$$

Example 2

To illustrate these ideas, consider some hypothetical project consisting of seventeen activities. Table 17.1 gives the activities, their duration times in

TABLE 17.1. Information About the Activities of the Hypothetical Project of Example 2

Activity	Predecessor Activity	Expected Time (days)
A	—	6
B	A	6
C	A	10
D	A	16
E	B	12
F	C	5
G	C	16
H	B,F	30
I	E	5
J	I	5
K	G,H,I	4
L	C,D	14
M	D	16
N	J,K	12
O	L,M,N	8
P	L,M,N	11
Q	O,P	7

days, as well as their predecessor activities. The latter are those, judged by a manager, project leader, and so on, that must be completed immediately prior to a given activity can begin. Hence they describe the set of event nodes. The resulting network is shown in Figure 17.5. The nodes are numbered consecutively, with node 1 designating the start of the project and node 15 is the completion of the project. Notice that dummy activities are necessary in order to get the correct logical precedence. For instance, the completion of activity C permits the work on activities F or G; the completion of D allows M to begin, but both C and D must be done before L can start. Thus the dummy activity emanating from node 4 is necessary. In keeping with the general rule of not having two activities share the same beginning and ending node, another dummy activity is employed between nodes 13 and 14.

A forward pass is made, setting the first event, the start of the project, to time zero. Equation (1) is used to find the earliest start times for the various nodes. The results are:

$$T_E(1) = 0$$
$$T_E(2) = T_E(1) + t_e(1,2) = 0 + 6 = 6$$
$$T_E(3) = T_E(2) + t_e(2,3) = 6 + 6 = 12$$
$$T_E(4) = T_E(2) + t_e(2,4) = 6 + 16 = 22$$
$$T_E(5) = T_E(2) + t_e(2,5) = 6 + 10 = 16$$
$$T_E(6) = T_E(3) + t_e(3,6) = 12 + 12 = 24$$

$$T_E(7) = \max \left\{ \begin{array}{l} T_E(3) + t_e(3,7) = 12 + 0 = 12 \\ T_E(5) + t_e(5,7) = 16 + 5 = 21 \end{array} \right\} = 21$$

$$T_E(8) = T_E(6) + t_e(6,8) = 24 + 5 = 29$$

$$T_E(9) = \max \left\{ \begin{array}{l} T_E(5) + t_e(5,9) = 16 + 0 = 16 \\ T_E(4) + t_e(4,9) = 22 + 0 = 22 \end{array} \right\} = 22$$

$$T_E(10) = \max \left\{ \begin{array}{l} T_E(8) + t_e(8,10) = 29 + 0 = 29 \\ T_E(7) + t_e(7,10) = 21 + 30 = 51 \\ T_E(5) + t_e(5,10) = 16 + 16 = 32 \end{array} \right\} = 51$$

$$T_E(11) = \max \left\{ \begin{array}{l} T_E(8) + t_e(8,11) = 29 + 5 = 34 \\ T_E(10) + t_e(10,11) = 51 + 4 = 55 \end{array} \right\} = 55$$

$$T_E(12) = \max \left\{ \begin{array}{l} T_E(11) + t_e(11,12) = 55 + 12 = 67 \\ T_E(9) + t_e(9,12) = 22 + 14 = 36 \\ T_E(4) + t_e(4,12) = 22 + 16 = 38 \end{array} \right\} = 67$$

$$T_E(13) = T_E(12) + t_e(12,13) = 67 + 8 = 75$$

$$T_E(14) = \max \begin{cases} T_E(12) + t_e(12,14) = 67 + 11 = 78 \\ T_E(13) + t_e(13,14) = 75 + 0 = 75 \end{cases} = 78$$

$$T_E(15) = T_E(14) + t_e(14,15) = 78 + 7 = 85$$

These earliest event times are usually shown on the network to facilitate other future calculations.

In order to calculate the latest completion times for each event a backward pass is made. Since there is no indicated due date for the completion of the project, the latest completion time is set equal to the earliest completion time. Using Equation (2), the latest completion times can be found.

$$T_L(15) = 85$$
$$T_L(14) = T_L(15) - t_e(14,15) = 85 - 7 = 78$$
$$T_L(13) = T_L(14) - t_e(13,14) = 78 - 0 = 78$$

$$T_L(12) = \min \begin{cases} T_L(14) - t_e(12,14) = 78 - 11 = 67 \\ T_L(13) - t_e(13,14) = 78 - 8 = 70 \end{cases} = 67$$

$$T_L(11) = T_L(12) - t_e(11,12) = 67 - 12 = 55$$
$$T_L(10) = T_L(11) - t_e(10,11) = 55 - 4 = 51$$
$$T_L(9) = T_L(12) - t_e(9,12) = 67 - 14 = 53$$

$$T_L(8) = \min \begin{cases} T_L(11) - t_e(8,11) = 55 - 5 = 50 \\ T_L(10) - t_e(8,10) = 51 - 0 = 51 \end{cases} = 50$$

$$T_L(7) = T_L(10) - t_e(7,10) = 51 - 30 = 21$$
$$T_L(6) = T_L(8) - t_e(6,8) = 50 - 5 = 45$$

$$T_L(5) = \min \begin{cases} T_L(7) - t_e(5,7) = 21 - 5 = 16 \\ T_L(10) - t_e(5,10) = 51 - 16 = 35 \\ T_L(9) - t_e(5,9) = 53 - 0 = 53 \end{cases} = 16$$

$$T_L(4) = \min \begin{cases} T_L(9) - t_e(4,9) = 53 - 0 = 53 \\ T_L(12) - t_e(4,12) = 67 - 16 = 51 \end{cases} = 51$$

$$T_L(3) = \min \begin{cases} T_L(6) - t_e(3,6) = 45 - 12 = 33 \\ T_L(7) - t_e(3,7) = 21 - 0 = 21 \end{cases} = 21$$

$$T_L(2) = \min \begin{cases} T_L(3) - t_e(3,6) = 21 - 6 = 15 \\ T_L(5) - t_e(2,5) = 16 - 10 = 6 \\ T_L(4) - t_e(2,4) = 51 - 16 = 35 \end{cases} = 6$$

$$T_L(1) = T_L(2) - t_e(1,2) = 6 - 6 = 0$$

These are also put directly on the network diagram.

In a network, a *path* between two nodes i and j is defined as a sequence

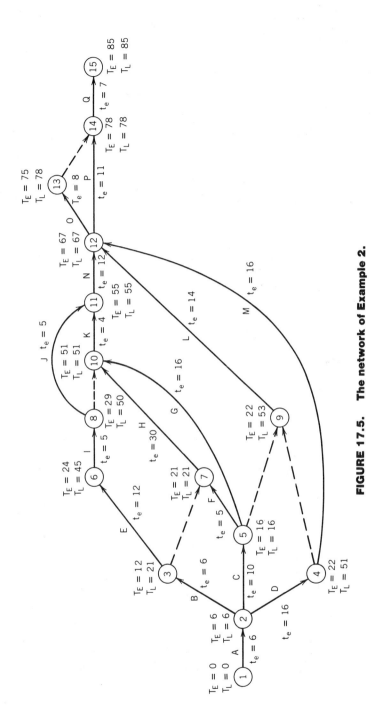

FIGURE 17.5. The network of Example 2.

383

arcs or branches connecting these two nodes, using the correct orientation. For example, a path from node 1 to node 15 (the start and end nodes of the previous example) is A–D–M–P–Q (or written by labeling the nodes as 1–2–4–12–14–15). Another is A–B–E–I–J–N–P–Q (or equivalently 1–2–3–6–8–11–12–14–15). Each path has a certain duration or length found by adding the durations of the activities that comprise it. Thus path A–B–E–I–J–N–P–Q has duration 6+6+12+5+5+12+11+7 = 64 days. There are obviously many paths connecting the start event and the completion event. The path with the longest duration is called the *critical path* and the activities that comprise it are called the *critical activities,* because any increase or delay in their start times will lengthen the entire project. The critical path is important because it indicates how long, at a minimum, the project will take. There can be no delay in either the start or the finish of the critical activities if the project is to be completed on time; each such activity must be on time. Thus there is no *slack* in either the earliest possible start time and the latest allowable start time for each of the critical activities; they must coincide. There ideas result in an easy way to find the critical path. It is the path through the network such that all the event nodes on this path have zero slack. All events having zero slack must lie on the critical path, but no others can. There may be more than one critical path in a network, as ties in length may exist. The slack of an event thus shows how much the event can be delayed and still not postpone the completion of the entire project.

Returning once again to Example 2, in order to find the critical path it is necessary to find $T_L(i) - T_E(i)$ for nodes $i = 1, 2, \ldots, 15$. If this difference is zero, then there is no slack for that node, and it is on the critical path. From the previous calculations it can be seen that the critical path is 1–2–5–7–10–11–12–14–15 or, equivalently, activities A–C–F–H–K–N–P–Q. The duration or length of this path is 85 days.

Closely related to the slack time of an activity is the notion of *float times,* which are measures of scheduling flexibility of a project. Certain activities in projects should be done during key time periods so as to even out the demand for resources, as frequently one group is responsible for performing more than one activity. Float times yield information which makes activity scheduling easier and more efficient. There are four types of float times.

1. *Total float* of an activity (i,j), labeled $TF(i,j)$, is the maximum time that activity can be delayed without delaying completion of the entire project. That is, it is the difference between the maximum time available to perform the activity (which is $LF(i,j) - ES(i,j)$) and its duration (which is $t_e(i,j)$). Hence

$$TF(i,j) = LF(i,j) - ES(i,j) - t_e(i,j)$$
$$= LF(i,j) - t_e(i,j) - ES_e(i,j)$$
$$= LS(i,j) - ES(i,j)$$

or $\qquad\qquad TF(i,j) = T_L(j) - T_e(i) \qquad\qquad\qquad\qquad (7)$

Thus the total float of an activity is just its slack as defined previously, and is helpful in determining the critical path. Hence it is useful for scheduling the overall project, but helps little in scheduling individual activities.

It should be clear that if a noncritical activity is delayed, frequently the start of succeeding activities are delayed also. For instance, if activity B in Example 2 is delayed and will not begin until day 10, then activity E cannot begin until day 16 at the earliest (it takes 6 days to perform activity B). So the total float of activity E is reduced to 17 days (33–16), which in turn reduces the total float of activity I to 17 days (45–28). Thus the total float in succeeding activities may be affected. This may of course not always be the case, so the next float time is useful.

2. *Free float* for activity (i,j), written $FF(i,j)$, measures the maximum time activity (i,j) can be delayed and still not affect the start of the succeeding activities. In order to calculate the free float of an activity it must be assumed that prior activities are started as early as possible. Then the free float for activity (i,j) is the difference between its earliest finish time and the earliest of the early start times of its immediate successors. That is,

$$FF(i,j) = T_E(j) - EF(i,j)$$

or $\qquad\qquad FF(i,j) = T_E(j) - T_E(i) - t_e(i,j) \qquad\qquad (8)$

Free float is not larger than the total float for an activity. If the total float is zero, then the free float must be zero also; the converse is not true. There may be some nonzero total float for a noncritical activity and yet the activity has zero free float. Activity B in Example 2, for instance, has a total float of nine days but has no free float.

3. The *independent float* of activity (i,j), written $IF(i,j)$, is the most this activity can be delayed from starting without delaying the succeeding activities, assuming all the prior activities have been completed as late as possible. It is found by

$$IF(i,j) = \max \begin{cases} 0 \\ T_E(j) - T_L(i) - t_e(i,j) \end{cases} \qquad (9)$$

Independent float is perhaps the most useful of the activity float times since it gives the delay in start time that can be absorbed by an activity independent of any other scheduling decisions made elsewhere. In Example 2, it can be seen that activity G has an independent float of 19 days.

4. *Safety float* of activity (i,j), designated $SF(i,j)$, is the maximum time it can be delayed from starting without affecting the completion of the project, assuming all the prior activities are completed as late as possible. Thus only some succeeding activities may be delayed but not the project itself. It is given by

$$SF(i,j) = T_L(j) - T_L(i) - t_e(i,j) \qquad (10)$$

Activity B in Example 2 has a safety float of 9 days.

Table 17.2 summarizes all the pertinent activity time calculations of Example 2. Calculations are rather easy, and most computer software presents the results in a table such as this one in order to facilitate the understanding of the project and its scheduling aspects.

17.4.   PROBABILISTIC NETWORK ANALYSIS (PERT)

In industries using new or radical technology to develop special or nonstandard products, many projects are undertaken involving the performance of numerous activities. There is thus a fair amount of uncertainty in the duration times of these activities, which may involve research in development, engineering design, and the final construction of some component. There is little past history to fall back on to determine the duration times of such activities. The network ideas of the previous section are usually applied to construction projects which involve activities by-and-large unchanged from earlier similar projects. Thus it is not too difficult to make reasonable estimates for the duration times for the various activities; that is, it can be assumed that they are known with certainty, or are *deterministic*.

PERT recognizes in its methodology that activity times are uncertain; in other words, they are random variables. However, the activities and their network relations are themselves well-defined. Thus the PERT model assumes that the activity duration times, while random variables, have a known probability distribution. Knowing this distribution, one can calculate the *expected value* or mean duration time for each activity as well as its *standard deviation*. Thus the activity duration time $t_e(i,j)$ is now actually an *expected duration time*. The actual time is unknown and all that can be estimated is its mean and its standard deviation. So all the calculations involving the network proceed exactly as before (when the times were assumed deterministic) only now some further probabilistic calculations can be made. One might typically attempt to answer the question, "What are the chances

TABLE 17.2. A Summary for Example 2

Activity	Predecessor Node i	Succeeding Node j	Duration $t_e(i,j)$	Earliest Start $ES(i,j)$	Earliest Finish $EF(i,j)$	Latest Start $LS(i,j)$	Latest Finish $LF(i,j)$	Total Float $TF(i,j)$	Free Float $FF(i,j)$	Independent Float $IF(i,j)$	Safety Float $SF(i,j)$	Critical Path
A	1	2	6	0	6	0	6	0	0	0	0	Yes
B	2	3	6	6	12	15	21	9	0	0	9	No
C	2	5	10	6	16	6	16	0	0	0	0	Yes
D	2	4	16	6	22	35	51	29	0	0	29	No
E	3	6	12	12	24	33	45	21	0	0	12	No
F	5	7	5	16	21	16	21	0	0	0	0	Yes
G	5	10	16	16	32	35	51	19	19	19	19	No
H	7	10	30	21	51	21	51	0	0	0	0	Yes
I	6	8	5	24	29	45	50	21	0	0	0	No
J	8	11	5	29	34	50	55	21	21	0	0	No
K	10	11	4	51	55	51	55	0	0	0	0	Yes
L	9	12	14	22	36	53	67	31	31	0	0	No
M	4	12	16	22	38	51	67	29	29	0	0	No
N	11	12	12	55	67	55	67	0	0	0	0	Yes
O	12	13	8	67	75	70	77	3	0	0	0	No
P	12	14	11	67	78	67	78	0	0	0	0	Yes
Q	14	15	7	78	85	78	85	0	0	0	0	Yes

the project will be completed by a certain date?" There are also other questions that could be posed at this point.

The PERT model has two basic assumptions that were not made earlier:

A1. The activities are all statistically independent. *Very loosely* put this means the activities do not interact with each other.

A2. The critical path contains a "large" number of activities. What this means is that the number of activities is sufficiently large enough to make the Central Limit Theorem from statistics hold. (The Central Limit Theorem says that the sum of a large number of independently distributed random variables has a distribution looking more and more like a normal distribution as the number of activities increases).

Since the activity times are random variables, an estimate must be made of their expected duration time. Because there is frequently no previous data to draw upon, PERT assumes that it is possible to make "educated guesses" about three specific activity times:

1. An optimistic estimate of the duration of the activity, labeled *"a"*
2. A pessimistic estimate of its duration, labeled *"b"*
3. The duration time most likely to occur, labeled *"m"*

It is frequently assumed that a and b are equally likely to occur, and m is assumed to be four times more likely to occur than either of the other two, so an estimate of the expected duration time of activity (i,j) can be found from

$$t_e(i,j) = \frac{a + 4m + b}{6} \tag{11}$$

In addition to estimating the mean duration time of an activity, it is also necessary to have its standard deviation. For many probability distributions most of the observations are within three standard deviations of the mean, implying the spread of the distribution is about six times its standard deviation. Based on this an estimate of the standard deviation of the duration of activity (i,j), written $\sigma_{i,j}$, can be found from

$$\sigma_{i,j} = \frac{b - a}{6} \tag{12}$$

As far as what the actual distribution of the duration time for the activities is, the originators of PERT assumed the *beta* distribution. This distribution is continuous, unimodal, has finite nonnegative end points, and can assume various shape configurations based on different parameters. (The probability density function of the beta distribution is a function of four constants, two of which are "location" parameters, while the other two are "shape" parameters.) Figure 17.6 shows three beta distributions with different parameters. Because of these attributes the beta distribution is frequently used as the assumed activity duration time distribution. Its choice is not based on empirical data, as most activities in these projects occur just once, but is based on its supposed "flexibility" as far as its shape is concerned.

With the three time estimates for each activity known, Equations (11) and (12) are used to find the expected duration time and its standard deviation. The duration times are then treated as deterministic and the calculations are made as before. The critical path is found and the (expected) duration time of the project is the sum of the (expected) duration times of the activities that comprise the critical path. However, since the activity times are random variables, so is the project duration. It thus has some probability distribution

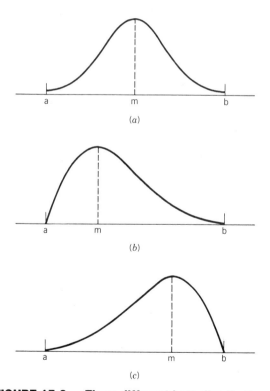

(a)

(b)

(c)

FIGURE 17.6.   Three different beta distributions.

TABLE 17.3.   Information About the Activities of Example 3

Activity	Nodes	Predecessor Activity	Optimistic Time a	Most Probable Time m	Pessimistic Time b	Expected Time $t_e(i,j)$	Standard Deviation $\sigma_{i,j}$
A	(1,2)	—	3	6	9	6	1
B	(2,3)	A	4	6	8	6	0.67
C	(2,5)	A	4	9	20	10	2.67
D	(2,4)	A	8	12	40	16	5.33
E	(3,6)	B	8	10	24	12	2.67
F	(5,7)	C	3	4	11	5	1.33
G	(5,10)	C	6	18	18	16	2
H	(7,10)	B,F	17	32	35	30	3
I	(6,8)	E	2	5	8	5	1
J	(8,11)	I	4	5	6	5	0.33
K	(10,11)	G,H,I	2	4	6	4	0.67
L	(9,12)	C,D	8	15	16	14	1.33
M	(4,12)	D	10	17	18	16	1.33
N	(11,12)	J,K	8	10	24	12	2.67
O	(12,13)	L,M,N	2	9	10	8	1.33
P	(12,14)	L,M,N	6	9	24	11	3
Q	(14,15)	O,P	4	7	10	7	1

with a mean and a standard deviation. The two assumptions A1 and A2 of PERT are employed to yield the result that the distribution of the critical path duration time is normally distributed with mean equal to the sum of the expected duration times of the activities on the critical path, and variance equal to the sum of the variances of the durations of those activities. Hence it makes sense to make probabilistic statements about the duration of the critical path and thus the project's length.

Example 3

Consider Example 2 one last time. Table 17.3 lists the activities, their predecessors, the optimistic, most likely, and pessimistic estimates of the duration times. The resulting network is identical to the one found earlier and the reader should once again refer to Figure 17.5. To illustrate the method consider Activity A for instance. Using Equations (11) and (12) for this activity (between nodes 1 and 2),

$$t_e(1,2) = \frac{3 + 4(6) + 9}{6} = 6 \text{ days}$$

$$\sigma_{1,2} = \frac{9 - 3}{6} = 1 \text{ day}$$

The calculations for the other activities are similar. The earliest start and latest completion time for the events are found as before, as well as the earliest start, earliest finish, latest start, and latest finish times for the activities. Once again the critical path is A–C–F–H–K–N–P–Q. Let T_{cp} denote the duration of the critical path. Then the expected duration of the project is $E(T_{cp})$ which is the sum of the expected duration times of those activities on the critical path.

$$E(T_{cp}) = t_e(1,2) + t_e(2,5) + t_e(5,7) + t_e(7,10) + t_e(10,11) + t_e(11,12)$$
$$+ t_e(12,14) + t_e(14,15)$$
$$E(T_{cp}) = 6 + 10 + 5 + 30 + 4 + 12 + 11 + 7$$
$$E(T_{cp}) = 85 \text{ days}$$

If $(\sigma_{cp})^2$ denotes the variance of the duration of the critical path, then $(\sigma_{cp})^2$ is the sum of the variances of the durations of those activities that comprise it.

$$(\sigma_{cp})^2 = (\sigma_{1,2})^2 + (\sigma_{2,5})^2 + (\sigma_{5,7})^2 + (\sigma_{7,10})^2 + (\sigma_{10,11})^2 + (\sigma_{11,12})^2$$
$$+ (\sigma_{12,14})^2 + (\sigma_{14,15})^2$$
$$(\sigma_{cp})^2 = (1)^2 + (2.67)^2 + (1.33)^2 + (3)^2 + (.67)^2 + (2.67)^2 + (3)^2 \ (1)^2$$
$$(\sigma_{cp})^2 = 36.44$$
or $\quad \sigma_{cp} = 6.04 \text{ days}$

Because of Assumptions A1 and A2, the duration of the project T_{cp} is approximately normally distributed with mean 85 days and standard deviation 6.04 days. A manager can then easily calculate the probability that the project will be completed by, say, day 92.

$$\Pr\{T_{cp} \leq 92\} = \Pr\left\{\frac{T_{cp} - E(T_{cp})}{\sigma_{cp}} \leq \frac{92 - 85}{6.04}\right\} = \Pr\{Z \leq 1.16\} = 0.8770$$

where Z represents the standard normal random variable. Thus the probability is just under 88% that the project will be done by day 92.

The probability it can be completed three days sooner than expected is

$$\Pr\{T_{cp} \leq 82\} = \Pr\left\{\frac{T_{cp} - E(T_{cp})}{\sigma_{cp}} \leq \frac{82 - 85}{6.04}\right\} = \Pr\{Z \leq -0.50\} = 0.3085$$

So the probability is about 31% that the job will be completed three days sooner than expected.

It is also possible to find an approximate $100(1 - \alpha)$ percent confidence interval for the project duration time. Letting $(T_{cp})^*$ denote the true project completion time, this confidence interval is found from

$$E(T_{cp}) - z_{\alpha/2}\sigma_{cp} \leq (T_{cp})^* \leq E(T_{cp}) + z_{\alpha/2}\sigma_{cp}$$

where $z_{\alpha/2}$ is a percentage point of the standard normal distribution such that $\Pr\{Z > z_{\alpha/2}\} = \alpha/2$.

For example, a 95% confidence interval estimate for the true completion time of the project is

$$85 - (1.96)(6.04) \leq (T_{cp})^* \leq 85 + (1.96)(6.04)$$

or
$$73.16 \leq (T_{cp})^* \leq 96.84$$

or approximately between 73 and 97 days.

In addition, it is possible to find the probability of any event having positive slack. If an event has a small probability of having positive slack then it may become critical; perhaps management should pay close attention to it as the project progresses. Assuming that there are many activities in the network so the Central Limit Theorem can be invoked, the distribution of the slack $(T_L - T_E)$ of event i is normally distributed with mean $E[T_L(i) - T_E(i)] = E[T_L(i)] - E[T_E(i)]$ and variance equal to the total of the sum of the variance of the activities along the backward path to even i, written $\sigma_L^2(i)$, and the sum of the variances of the activities on the forward path to event i, written $\sigma_E^2(i)$.

To illustrate, in Example 3 what is the probability event 3 has positive slack? Now $E[T_L(3)] = 21$ days and $E[T_E(3)] = 12$ days. Also

$$\sigma^2_L(3) = (\sigma_{14,15})^2 + (\sigma_{12,14})^2 + (\sigma_{11,12})^2 + (\sigma_{8,11})^2 + (\sigma_{6,8})^2 + (\sigma_{3,6})^2$$
$$= (1)^2 + (3)^2 + (2.67)^2 + (0.33)^2 + (1)^2 + (2.67)^2$$
$$\sigma^2_L(3) = 24.34$$
$$\sigma^2_E(3) = (\sigma_{1,2})^2 + (\sigma_{2,3})^2 = (1)^2 + (0.67)^2$$
$$\sigma^2_L(3) = 1.44$$

Thus

$$\sqrt{\sigma^2_L(3) + \sigma^2_E(3)} = \sqrt{25.78} = 5.08 \text{ days}$$

The probability that event 3 has a positive slack translates to

$$\Pr\left\{[T_L(3) - T_E(3)] > 0\right\}$$

$$= \Pr\left\{Z > \frac{-[E[T_L(3)] - E[T_E(3)]]}{\sqrt{\sigma^2_L(3) + \sigma^2_E(3)}}\right\} = \Pr\left\{Z > \frac{-(21 - 12)}{5.08}\right\} = \Pr\{Z > -1.77\}$$
$$= .9616$$

This says the probability is over 96% that event 3 has positive slack, implying that it is highly unlikely event 3 will become critical (be on the critical path).

17.5. CRITICISMS OF PERT

Practitioners and researchers have pointed out that many of the assumptions of PERT are faulty, leading to incorrect, or at best, overly optimistic results. Because of this PERT's usefulness has been questioned. For a time some practitioners abandoned it completely. Today PERT is widely used, but the successful manager must be made aware of the limitations of the method and judge the results with a cautious eye.

The critical path is defined as the longest path in the network, based on *deterministic* activity times (which under PERT is not the case). The actual length of a path is a random variable. If the different paths in the network are assumed independent, the critical path is found from the maximum of a finite set of random variables. The distribution of the maximum of a sum of random variables is not normal, as PERT assumes. Even if it can be approximated by a normal distribution, the mean and variance of the duration of the critical path are not those given by PERT (which says just add the expected durations of the activities on the critical path to get the mean, and add the variances of those activities to get the variance). Errors can be as high as 30% when using the methods indicated by PERT

The problem is further compounded because activities are frequently not statistically independent (Assumption A1). Activities may interact because they compete for scarce resources (manpower, materials, and so on), especially when the project seems behind schedule. Also paths are not statistically independent as they more than likely share activities. Hence the Central Limit Theorem cannot be invoked, which implies that the distribution of the duration of the paths is not normal. Even if the activities were "approximately" independent, their duration need not be normally distributed. Even if the duration of each path was "approximately" normal, the distribution of the critical path duration is not normal, as has been previously explained. Finally, even if the critical path duration was "approximately" normal, its mean and variance is not what PERT says they are.

Advocates of PERT point out that the errors due to faulty statistics are minor compared to the errors made in getting the time estimates of the activities. The latter may be several times more than the former in practice, making them almost irrelevant. Hence the criticism, though valid, is not of major consequence.

A second major criticism is concerned with the distribution of the activity times themselves, which is assumed to be a beta, and with its four parameters, permitting "any" desired shape and location. Thus this distribution supposedly allows great flexibility in possible "fits." However, because Equations (11) and (12) are used to find the mean and standard deviation of the activity duration times, the shape of the beta distribution can be shown, in fact, to be restricted to only *three* different possible shapes and not the infinite number thought. Hence the "flexibility" assumption of the beta distribution is totally uncalled for, and there is doubt that this distribution should be used at all. Proponents concede this point, but counter by saying that the choice of the beta distribution is really unimportant, as the Central Limit Theorem yields the desired normal distribution regardless of the distribution of the activity time.

Another frequently raised objection is the notion of a "near-critical" path. The activity times are uncertain, all having some standard deviation. It is quite possible that some other path, while having a slightly smaller expected duration than that of the assumed "critical path," but with a higher standard deviation, might really take a longer period of time. Hence this "near critical" path might actually be the "official" critical path. Practitioners should be aware of this possibility and also know that PERT generally yields optimistic results. The calculations made earlier, concerned with the probability of the project being completed by a certain date, are thus in error. The probabilities calculated are higher than they are in actuality, and managers should thus plan cautiously. One potential solution would be to apply simulation techniques to the network, although such methods become impractical for larger networks because of complexity and cost.

17.6. INCORPORATING COSTS IN PROJECT SCHEDULING—CPM

While the techniques presented to this point emphasize the *time* necessary to complete a project, managers recognize that the costs needed to do so should obviously be considered also. Frequently an activity duration time can be decreased if extra resources (such as people, extra and better machines, and so on), are assigned to them. These involve higher costs, but if certain gains could be made (such as incentives or rewards) if the project is completed early, then it may pay to expedite the activity or *crash* it. However, if reducing the duration time of an activity does not reduce the project length, or if its crashing cost is too high, the duration time should not be reduced but rather the activity should proceed at its regular or *normal* pace with the usual resources allocated to it to insure its completion.

There are two types of costs associated with a project. *Direct costs* are related to the activities and the resources they consume as they are being completed and are inversely proportional to the duration of the activity. *Indirect costs,* such as managerial services, equipment rental, security, and so on, are usually directly related to the length of the project itself. Reducing certain activity times will increase direct costs but may decrease indirect costs if the project is completed earlier. There are thus some time-cost trade-offs to be considered, and this is done with the aid of CPM methods.

It is assumed that each activity has a cost-duration relationship. The activity can be completed in some *normal* time at some normal cost (This is *not* to be confused with the normal distribution. Here "normal" means "usual.") and completed in some crash time (the shortest possible duration time) at some higher crash cost (because of the necessary extra resources). Introducing any further resources to perform this activity will not decrease its duration below the crash time. This time-cost duration is assumed for convenience to be linear, as shown in Figure 17.7, but it need not be so.

Referring to Figure 17.7, let

$$D_n(i,j) = \text{the normal duration of activity } (i,j)$$
$$D_c(i,j) = \text{the crash duration of activity } (i,j)$$
$$C_n(i,j) = \text{the normal direct cost of activity } (i,j)$$
$$C_c(i,j) = \text{the crash direct cost of activity } (i,j)$$

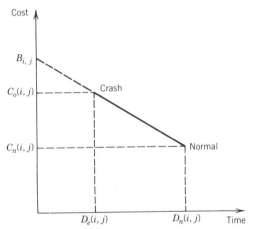

FIGURE 17.7.   The cost-time relationship for Activity (*i, j*).

The slope of this line thus gives some indication of the time versus cost relationship for the activity. Denote

$$C_{i,j} = \frac{C_c(i,j) - C_n(i,j)}{D_n(i,j) - D_c(i,j)}$$

as the incremental direct cost for activity (i,j) for a unit decrease in the time of its duration. These are computed for all the activities in the project. The activity times are at first assumed to be the normal values, the network is drawn, and the event and activity earliest start and latest completion times are calculated as before. All times are assumed deterministic. The critical path is found and so is its associated normal cost, which is the sum of the normal costs of the activities that comprise it. Then an attempt is made to reduce the length of the project, that is, the critical path. Only those activities on this path need be considered, as noncritical activities that are shortened will not reduce the overall critical path. A critical activity is chosen and an attempt is made to "crash" it; that is, to reduce its duration to its crash time, with an appropriate increase in overall cost of the project. Which activity to choose is determined by the one with the smallest incremental direct cost. By how much the activity can be reduced in duration depends upon other activity durations in the project; it may not be possible to compress the activity to its actual crash point. For example, it is possible that by such compression an additional or new critical path is found. The procedure then repeats with the selection of a critical activity and an attempt is made to crash it, and so forth. The method ends when no further crashing can be done; that is, any further attempt to reduce the duration of any critical activity will not decrease the projects' overall length.

Example 4

Consider a simple project consisting of seven activities labeled A through F. Each has a normal duration time (all times are assumed in days) and cost as well as a crash duration time and cost, as indicated in Table 17.4. Figure 17.8a shows the network for the project. Table 17.4 also shows the calculated incremental direct costs for each activity. It is further assumed that indirect daily costs are $225.

The method starts with all activities performed at their normal duration times. The earliest start times $T_E(j)$ for all events $j = 1,2,3,4,5$ are calculated, yielding a project length of 36 days along critical path 1–2–3–5 (or equivalently A–D–F). An attempt is now made to reduce the project length by reducing the duration of some activity on the critical path. Activity D has the smallest incremental cost of the three critical activities, so it is compressed. However D cannot be crashed to its crash limit (10 days) because another path, 1–3–5 (or B-F), becomes critical prior to that. In fact D is crashed only 4 days (from 16 to 12 days) because at that point path B–F becomes critical. See Figure 17.8b. The project length is now 32 days.

From Figure 17.8b, it can be seen that along critical paths B–F and A–D–F reducing the duration of activity F reduces the project length. Reducing A still leaves B–F critical; reducing B leaves A–D–F critical. Activity F can be crashed just one day (from 10 to 9) because at that point A–D–F is

TABLE 17.4.   The Normal and Crash Information for Example 4[a]

| Activity | Nodes | Normal | | Crash | | Incremental cost |
		Duration $D_n(i,j)$	Cost $C_n(i,j)$	Duration $D_c(i,j)$	Cost $C_c(i,j)$	$C_{i,j}$
A	1,2	10	200	7	500	100
B	1,3	22	300	18	900	150
C	2,4	15	150	9	750	100
D	2,3	16	300	10	750	75
E	3,4	4	200	2	600	125
F	3,5	10	500	6	1000	125
G	4,5	5	375	2	750	125
			2025		5250	

[a] Note that all time units are days and all cost values are in dollars.

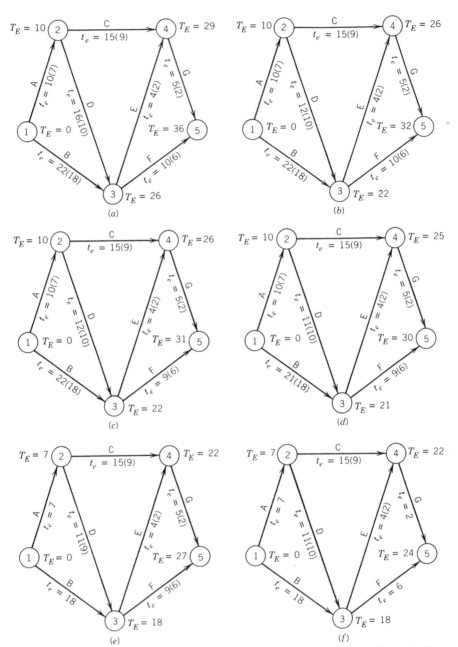

FIGURE 17.8. Networks illustrating crashing method. Note the notation is t = normal time (crash time).

no longer critical, but A–D–E–G and B–F and B–E–G are the new critical paths. Figure 17.8c shows the result. The project length is now 31 days.

Along critical paths B–F, A–D–E–G, and B–E–G, the crashing of one activity alone does not reduce the overall project length. Hence two activities must be reduced simultaneously. Choices are B and D, A and B, and F and G with B and D being the cheapest. Reduce B and D each by one day because at that point all paths become critical (A–C–G, A–D–F, A–D–E–G, B–F, and B–E–G). The project duration is now 30 days and the network at this point is shown in Figure 17.8d.

In order to reduce the project length further, either A and B or F and G must be compressed simultaneously. If both have the same combined incremental cost here ($250), choose the pair which reduces the project by the greater amount. In this example, once again it is a tie, and both pairs can be crashed just three days to their *crash limits*. At this point it is irrelevant which pair to choose, so arbitrarily choose one. Let A and B be compressed three days, the project then has a duration of 27 days and the network is in Figure 17.8e.

Once again all paths are critical, so pair F and G are compressed because they have the least expensive sum of incremental costs of any pair in the network that will reduce the project's duration. Hence F and G are crashed three days each, so that they are now at their crash limits. The project's length is 24 days and the network at this point is shown in Figure 17.8f. There are still several activities that can be crashed at this point. However, it can be seen that no other activities should be crashed because the project's length would not be decreased by doing so. The problem is finished and the steps can be summarized in a table. See Table 17.5.

Figure 17.9a and 17.9b show plots of the indirect, direct, and total costs as a function of time. It is frequently true that crashing all activities is rarely the best advice. Even crashing those activities necessary to get the project done in the minimum time is rarely the optimal solution. The total cost curve (here a piecewise-linear one) is drawn and the minimum cost is found. In Example (4) it is $9425 and occurs at either 30 or 31 days. The optimum policy would then be: crash activity D four days and F one day (or D four days, F one day, and B and D one day each) to get a project duration of 31 (or 30) days at an optimal total cost of $9425.

The total cost versus time plot usually yields a convex function, thus assuring a minimum cost will occur at some project length. (Example 4 has two distinct project lengths yielding an equal minimum cost.) The astute manager need not always choose a plan yielding a minimum cost; there may be other factors that should be considered, such as safety and risk to workers for example. These qualitative considerations must be examined in addition to the quantitative solutions that CPM yields.

TABLE 17.5. Summarizing the Crashing Process from Example 4[a]

Step	Compress Activity (i,j)	Amount Crashed	Crash Cost	Critical Path	Critical Path Duration	Indirect Cost	Direct Cost	Total Cost
1	—	—	—	A–D–F	36	36(225) = 8100	2025	10125
2	D(2,3)	4	4(75) = 300	A–D–F B–F	32	32(225) = 7200	2325	9525
3	F(3,5)	1	1(125) = 125	A–D–E–G B–E–G B–F	31	31(225) = 6975	2450	9425
4	B(1,3) D(2,3)	1 1	1(150) = 150 1(75) = 75 225	All	30	30(225) = 6750	2675	9425
5	A(1,2) B(1,3)	3 3	3(100) = 300 3(150) = 450 750	All	27	27(225) = 6075	3425	9500
6	F(3,5) G(4,5)	3 3	3(125) = 375 3(125) = 375	All	24	24(225) = 5400	4175	9575

[a] Note that all time units are days and all cost values are in dollars.

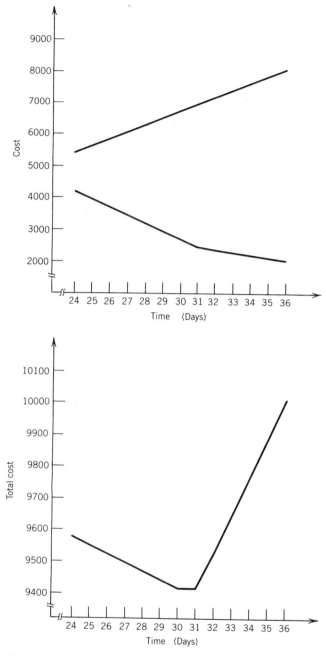

FIGURE 17.9. **(a) A plot of indirect and direct cost as a function of time for example 4; (b) A plot of the total cost as a function of time for example 4.**

17.6.   LINEAR PROGRAMMING AND CPM

For small problems, solving a CPM problem by "hand" is relatively simple. But for projects containing many activities, yielding numerous paths, such a solution procedure is very difficult if not impossible. Other approaches, in particular mathematical programming, are used to find the minimum cost of the project. A solution is found with the aid of a computer. How to formulate such a problem as a mathematical program will be considered here. The reader not familiar with linear programming may omit reading this.

The cost–time relationship for each of the activities will once again be assumed to be linear (though this need not be so) so the resulting problem of finding the optimal cost and hence the optimal duration of each activity becomes a linear program. Looking once again at Figure 17.7, let $B_{i,j}$ represent the projected intercept of the cost–time line for activity (i,j). If t_{ij} represents the duration of the activity (the e subscript used earlier will be suppressed for convenience and the (i,j) will be relegated to a subscript) the direct cost this activity (i,j) is

$$B_{i,j} - C_{i,j}t_{i,j}$$

where once again $C_{i,j}$ is the incremental direct cost per unit decrease in time for activity (i,j). The total direct cost of the project is then

$$\sum_{i,j} B_{i,j} - C_{i,j}t_{i,j}$$

where $\sum_{i,j}$ means to sum over all the activities in the network. The goal is to minimize the total project cost (direct plus indirect costs). Since the indirect cost is a linear function of the project's duration, it can be written as $KT_E(n)$, where K is some constant (the per-unit indirect cost), n is the ending node in the network and $T_E(j)$ is as before the earliest completion time of node (event) j. The objective function is thus

$$\text{minimize} \sum_{i,j} B_{i,j} - C_{i,j}t_{i,j} + KT_E(n)$$

or

$$\text{maximize} \sum_{i,j} C_{i,j}t_{i,j} - KT_E(n)$$

since $\sum_{i,j} B_{i,j}$ is just some constant, and thus may be ignored.

Each activity's duration is constrained to be between its normal and fully crashed times, so

$$D_n(i,j) \leq t_{i,j} \leq D_c(i,j)$$

Also the difference between the earliest occurrence times of two connected events must be at least as great as the duration of the activity joining them, or

$$T_E(i) + t_{i,j} - T_E(j) \leq 0$$

The linear program formulation is thus

$$\max \sum_{i,j} C_{i,j} t_{i,j} - K T_E(n)$$

$$
\begin{aligned}
\text{s.t.} \quad & t_{i,j} \leq D_n(i,j) && \text{for all } i,j \\
& t_{i,j} \geq D_c(i,j) && \text{for all } i,j \\
& T_E(i) + t_{i,j} - T_E(j) \leq 0 && \text{for all } i,j \\
& T_E(j) \geq 0 && \text{for } j = 1,2,...,n
\end{aligned}
$$

This formulation can be changed if nonnegativity constraints are desired for the $t_{i,j}$ decision variables by writing

$$t_{i,j} = D_c(i,j) + t'_{i,j} \tag{13}$$

everywhere in the model, thus yielding $t'_{i,j} \geq 0$. Notice that decision variables $T_E(j)$ are automatically nonnegative because $T_E(1) = 0$ (the start of the project) and $t'_{i,j} \geq 0$, so the third set of functional constraints is really $T_E(j) \geq T_E(i) + D_c(i,j) + t'_{i,j}$ which is nonnegative.

Example 5

Consider once again Example 4. This problem can be formulated as a linear program. Remember that the objective function value found will not be the correct answer but will be off by an additive constant.

$$
\begin{aligned}
\max \quad & 100t_{1,2} + 150t_{1,3} + 100t_{2,4} + 75t_{2,3} + 200t_{3,4} + 125t_{3,5} \\
& + 125t_{4,5} - 225T_E(5) \\
\text{s.t.} \quad & 7 \leq t_{1,2} \leq 10 \\
& 18 \leq t_{1,3} \leq 22 \\
& 9 \leq t_{2,4} \leq 15 \\
& 10 \leq t_{2,3} \leq 16 \\
& 2 \leq t_{3,4} \leq 4 \\
& 6 \leq t_{3,5} \leq 10
\end{aligned}
$$

$$2 \leq t_{4,5} \leq 5$$
$$t_{1,2} - T_E(2) \leq 0$$
$$t_{1,3} - T_E(3) \leq 0$$
$$T_E(2) + t_{2,4} - T_E(4) \leq 0$$
$$T_E(2) + t_{2,3} - T_E(3) \leq 0$$
$$T_E(3) + t_{3,4} - T_E(4) \leq 0$$
$$T_E(3) + t_{3,5} - T_E(5) \leq 0$$
$$T_E(4) + t_{4,5} - T_E(5) \leq 0$$
$$T_E(j) \leq 0 \qquad \text{for } j = 1,2,3,4,5$$

While this formulation is correct, in order to use the simplex algorithm to solve the linear program, it is necessary to have all variables be nonnegative. Equation (13) is thus utilized, resulting in another linear program. The reader must remember that this new formulation will yield an objective function value off by an additive constant from the above formulation, which in turn differs from the actual total cost optimum value by an additive constant. The details are straightforward and will be omitted here.

Alternatively, the problem could be formulated by minimizing only the sum of the direct costs of the activities in the network. The indirect cost term is omitted from the objective function, and instead a given maximum project completion target time, written T, is incorporated into the model. This is done via the addition of a new constraint

$$T_E(n) \leq T$$

T is really a parameter that may be varied (if a specified deadline for the project has not been given) to see the effect of project length and cost. Parametric programming can be used to solve this problem. Easier yet would be working with the dual of this problem, which can be shown to be a variant of a problem seeking the maximum flow through a network, a classic network problem for which computer codes are readily available.

From Example 5, it is clearly seen that even moderately sized projects involve formulating linear programs with many variables and constraints, requiring much computer time to solve. (Working with the dual as mentioned above is the answer to this problem). However, solutions can frequently be found and managers can thus use CPM techniques effectively to plan projects.

Finally, if the cost–time curve for an activity cannot be assumed to be linear, but is at least convex and continuous, a piece-wise linear approximation can be used and a linear program can still be formulated to solve the problem. If it is neither convex nor continuous, integer programming techniques may be incorporated, using piece-wise linear approximations once again.

17.7. OTHER CONSIDERATIONS—LIMITED RESOURCES

The procedures considered thus far assumed that the resources needed to perform the activities were always available. This may of course not always be the case. Manpower, raw materials, cash flow, and so on may be the resources not in sufficient supply (temporarily) to perform the activity. The ideas of CPM may be extended to handle situations like these. With the aid of Gantt charts, special diagrams called resource loading diagrams, can be constructed to give management the profile of resource usage as a function of time. Hence the planner can make adjustments to the schedule of the project in order to insure that all activities are finally done. Such adjustments are called resource leveling and involve the float calculations for the various activities.

The resource loading diagrams give a picture of the use of the resources to perform the activities. If the amounts available are constrained, the situation may become more complicated, and perhaps some rescheduling must be done in order to avoid the use of resources beyond that which is on hand. Procedures that do this are called constrained-resource scheduling. Most of these methods are heuristic in nature, yielding approximate schedules that are resource feasible. Other approaches involve mathematical programming, but because of the complexity of the problem such methods work quickly for only relatively small problems.

17.8. CONCLUSION

The ideas of PERT and CPM have been around for close to three decades. The first decade saw their use extensively in many areas. But as the limitations of the models were pointed out, their popularity declined in the second decade. In fact, the methods were actually shunned for a period of time. Lately managers have recognized the merits and value of these models and once again they have found their way into the planner's bag of tools to expedite scheduling. They are relatively simple to understand and rather easy to program on the computer. Thus many PERT/CPM computer codes are available, allowing the solution of even large projects to be found quickly and inexpensively. Assuming the manager is aware of their limitations and weaknesses and that they are used with caution and care, PERT and CPM are valuable tools in scheduling work problems.

While this chapter delved into PERT and CPM fairly extensively, the coverage was not exhaustive. The interested reader wanting to learn more should consult the bibliography, especially Note 3 for some of the more advanced notions referred to in Section 17.7.

NOTES

1. Baker, K. R., *Introduction to Sequencing and Scheduling* (New York: Wiley, 1974).
2. Elmaghraby, S. E., *Activity Networks: Project Planning and Control by Network Models*, (New York: Wiley, 1977).
3. Moder, J. J. and C. R. Phillips, *Project Management with CPM and PERT*, 2nd ed. (New York: Van Nostrand, 1970).
4. O'Brien, J. J., *Scheduling Handbook* (New York: McGraw-Hill, 1969).
5. Phillips, D. T. and A. Garcia-Diaz, *Fundamentals of Network Analysis* (Englewood Cliffs, N.J.: Prentice-Hall, 1981).
6. Taha, H. A., *Operations Research*, 3rd ed. (New York, Macmillan, 1982).
7. Wiest, J. D. and F. K. Levy, *A Management Guide to PERT/CPM*, (Englewood Cliffs, N.J.: Prentice-Hall, 1977).

PART

THE ENGINEER
AS MANAGER

18

SUPERVISING ENGINEERING AND TECHNICAL PERSONNEL

Leonard J. Smith

LEONARD J. SMITH, executive director of Training Services, Inc., Rutherford, New Jersey, has taught on the Engineering and Science faculties of Rutgers and Fairleigh Dickinson universities and has lectured at various universities. Mr. Smith has held technical management positions with Lightolier, Inc., Conmar Products Corporation, Arvey Corporation, and General Magnaplate Corporation. He has also been a consultant and trainer to leading industrial and pharmaceutical companies and government agencies. Mr. Smith is the author of several widely used textbooks and articles on industrial management. He received the M.B.A. and B.S. degrees from New York University in Personnel and Industrial Management. He is an accredited personnel diplomate (APD).

18.1. INTRODUCTION

The ability to supervise engineers is the key to successful engineering management. There are several principles and techniques which need to be known and be used in supervising or directing an engineering staff.

The primary supervisory principle is the understanding of the individual members of an engineering staff and their behavior patterns. Although every engineer is different, behavior and performance can be predicted and controlled to a significant extent. To accomplish this, we must properly utilize the Law of Human Behavior. It is a supervisory tool which states that a stimulus applied to an individual results in a response or reaction which we call behavior. Thus to achieve a desired behavior or performance an engineering manager needs only to determine, and then apply, the proper stimulus. Stimuli are also known as supervisory techniques. They include communications, motivation, contacts, and similar managerial actions. These influence an engineer in his or her behavior or performance. The proper stimulus will enable a manager to obtain desired results from individual subordinates.

18.2.   PERSONALITY FACTORS

We know from experience that engineers are not alike. However, there are several principles of human nature which are common to all.

The first of these is that each of them is the center of his or her own world. Each is inherently self-centered. This is known as the self-concept. Engineers, like other human beings, tend to look at work and life in terms of how they are affected. Thus when a stimulus is applied by an engineering manager it must be in terms of his or her personal interests if it is to properly influence behavior or performance. In reviewing work interests of engineering personnel, there is a wide range of different interests. Assigning individuals to work that is in their area of interest will increase productivity and performance. Figure 18.1 lists the different areas of interest of engineers.

Second, engineers are conditioned to behave as they do by various influences in their lives. These include their cultural upbringing, education, and experience. This conditioning is reflected in their value system, attitudes, perception of work and assignments, expectations, and interests. In order to predict their behavior or performance we need to know more about their backgrounds.

Third, engineers have basic psychological needs similar to other human beings. However their priorities tend to differ when translated into job needs, desires, or wants. Challenge, opportunity, and achievement tend to be more important than other needs.

Resistance to change is the fourth fundamental of human nature. This also is known as inertia. In most people it is fear of the unknown. In engineers, this shows up in their tendency to accept established laws, techniques, principles and practices. It is an unwillingness to introduce new things or ideas.

As with most people, engineers tend to be creatures of habit. This fifth

1. Ideas—planning, concepts, processes, advertisement, sales promotion, methods of industrial engineering
2. Things—products, equipment, packages, materials
3. Numbers/figures—mathematics, statistics, calculations
4. People—management, sales purchasing, customer services
5. Creative, innovative and imaginative—design, new products, and services
6. Diversity versus repetitive activities
7. Complexity versus simplicity of activities, sophisticated versus basic
8. Indoor versus outdoor
9. Clerical—recordkeeping, being detail-oriented
10. Analytical—inductive & deductive reasoning
11. Problem solving and decision making—trouble shooting
12. Specialist versus generalist
13. Literary—writing reports & papers, preparing proposals, advertising copy-writing, preparing instruction manuals
14. Graphic—art, photography, illustration, drafting
15. Research versus applications
16. Legalistic—patents, liability
17. Financial—accounting, budgets, and estimating
18. Manual versus mental
19. Conservative versus speculative—playing it safe" versus gambling
20. Academic versus practical—theory versus practice
21. Sociological—improving society
22. Environmental—protecting environment
23. Linguistic—translating interpreting

Figure 18.1   Interest areas of engineers and scientists.

factor is based on the fact that people develop habitual patterns of behavior whenever they are involved in repetitive acts. A habit is a learned pattern of behavior in which thought is not involved. Most engineering knowledge tends to be learned by rote. This creates mental blocks or mind-set which is in reality a habit of thinking in a predetermined manner. This is valuable in training individuals to perform repetitive tasks, to work safely, or to follow a prescribed procedure. On the other hand, it inhibits creativity, originality or innovation.

A sixth fundamental involves engineers as social beings. Although most people are influenced in their behavior by society and group dynamics, the nature of engineers' education and work assignments tend to make them more individualistic and less influenced socially. They are the rules of the work place. They will be influenced by the technical competence of their managers rather than normal leadership.

Engineers will be found to be emotional beings. They will be influenced by emotional stimuli or situations. They will react to stress, anxiety, tension, pressure, and frustration. These are work related due dates, deadlines, interruptions, reports, travel, and so on. Their reactions to these emotional situations are known as defense mechanisms. As managers we need to recognize this seventh fundamental. That engineers when stimulated emotionally may behave in an irrational or unthinking manner. The existence of an unusual behavior or performance may be an indication of an individual being influenced emotionally.

An eighth fundamental of human nature which applies to engineers and technicians is that they are *total beings*. That is, their behavior is influenced by a multiplicity of factors both on-the-job and off-the-job. To supervise them effectively, predict their behavior, and influence their performance, we need to know nonjob related factors that affect engineers and technicians as well as those whom they control.

18.3. THE ASSIGNMENT OF WORK

Another fundamental in supervising engineers involves the assigning of available personnel. Whether we are assigning tasks or functions to be performed or staffing projects, our successful accomplishment of results involves placing the right people in the right jobs.

We already have discussed some salient characteristics of engineers and their behavior. Our next task is to develop a *skills inventory*. This is an evaluation of the knowledge, skills, and abilities of each of our engineers. It should be a quantitative and qualitative analysis of what they know and can do about the work in our department.

Ideally, we should not assign any of our engineers to perform any work for which they are not qualified. To the degree that we can assign properly qualified individuals to all assignments and projects, we can be assured of achieving desired results. In the absence of a sufficient number of qualified engineers to staff all assignments, we may find it necessary to recruit qualified individuals, train available personnel to become qualified, or be satisfied with less than desired results.

The skills inventory permits a manager to match the requirements of regular department activities and special project needs with the talents of the engineering staff. Priorities must be established for each function and project. Available skills should be assigned to those jobs with the highest priorities. Lower priority work should be assigned to those who have available time and unassigned skills.

The assignment of work is a form of *delegation*. It is the process of making an engineer responsible for a particular activity or task and providing the individual with the degree of authority with which to act. There are several methods by which engineering managers can assign work in addition to qualifications for the job. When two or more engineers are relatively equal in qualifications, we can rotate the assignment or make the assignment by having the individuals draw lots to determine who is assigned. Assigning on the basis of nepotism, that is family or personal relationships, or politically so as to impress or influence higher management, are not acceptable methods although they may be used by engineering managers. When an assignment is difficult, dangerous or otherwise unpleasant a manager might request a volunteer. A similar method is employed in companies which encourage promotion from within. A new assignment might be posted for engineers to bid for consideration as candidates to handle the work. This method permits engineers to identify their job interests and enables managers to satisfy them whenever possible.

In assigning work or delegating there are several *fundamentals* which are considered important. These include:

1. Determine and explain the goals—the results to be achieved
2. Explain the background of the task or project
3. Establish time limits
4. Inform individual of your plan to follow-up on progress and performance
5. Communicate the assignment to those involved or concerned
6. Make sure the individual understands the assignment and the extent of authority delegated—what's expected
7. Provide for recognition of performance

The *extent or degree of authority* delegated defines the action which an engineer can and should take in the course of performing. It is rare that blanket authority would be delegated for the performance of an assignment or project. Usually engineers are given several different levels of authority. These may range from acting only on orders, whether oral or written, to acting entirely on their own. The latter, discretionary authority, would be delegated to engineers in whom we had complete confidence of their knowledge and ability in regard to the assignment. The former, reflected in standard practices and procedures, would be used for new engineers or those in training. These prescribe the what, when, how, where, and why of their assignments. They are predetermined courses of action established by us as managers to guide them in their work performance.

An effective tool for identifying the degree or extent of authority for tasks or functions delegated can be found in Figure 18.2, which presents a scheme for organizational clarification.

Degrees of Authority: (A) Act only on orders
(B) Recommend action for superior's approval
(C) Plan action in name of superior or to be countersigned
(D) Notify before taking action
(E) Act and then report or inform superior
(F) Act entirely on own
(G) Not responsible
Responsibilities of Scientists and Engineers

Materials and Supplies
1. Requisitioning
2. Evaluation
3. Bills of materials
4. Quality control and inspection
5. Substitution
6. Testing and analyzing
7. Procurement specifications

Machinery and Equipment
1. Design
2. Specification
3. Maintenance
4. Tooling, jigs, and fixtures
5. Installation
6. Inspection
7. Layout

Methods
1. Operation sheets
2. Operating procedures/processes
3. Performance standards
4. Quality standards
5. Waste treatment
6. Waste disposal
7. Methods improvement and work simplification
8. Pilot plant
9. Instruction sheets and operating manuals
10. Start-up

Financial
1. Budgets
2. Capital equipment
3. Cost estimating
4. Cost reduction

Facilities
1. Space utilization
2. Utilities
3. Pollution control

Figure 18.2   Organizational clarification.

4.   Noise
5.   Vibration
6.   Heating, ventilating, air conditioning
7.   Steam
8.   Air pressure
9.   Construction
10. Plant engineering

Record Keeping—Reporting
1.   Research studies and reports
2.   Documentation
3.   Literature studies
4.   Costs and expenses
5.   Tests
6.   Inspections
7.   Patents
8.   Drafting
9.   Presentations

Products/Services
1.   Origination
2.   Design
3.   Improvements
4.   Standardization
5.   Specifications
6.   Applications
7.   Evaluations
8.   Reliability
9.   Packaging
10. Product liability

Customer Relations
1.   Customer services
2.   Customer contact
3.   Troubleshooting and problem diagnosis
4.   Advisory
5.   Applications engineering
6.   Support engineering
7.   Proposals and bids

Figure 18.2   (continued)

18.4   SETTING STANDARDS OF PERFORMANCE

The assignment of work to an engineer also requires setting standards of performance. When we set standards, we are establishing our expectations for our staff. The standards can be set for individual tasks, special assignments, regular work, or projects.

Experience has found that an engineer's level of performance is related to the standards which have been established. Generally the higher the standards, the higher the level of performance.

In the absence of established managerial standards, an engineer will set his or her own standards. While they might be satisfied with their own standards, we as managers may find that we have abdicated our authority and diminished our supervisory responsibilities. All of our subordinates need to know what is expected of them in the performance of their work.

There are a number of factors involved in the *performance of an engineer's work*. While these vary with different assignments, the common factors include:

1. Time—progress or achievement of specific results in a given period of time or meeting a deadline or due date

2. Results—completion of the assignment or project or achievement of the pre-established goals, targets, objectives and commitments

3. Budgetary—control of expenditures within predetermined amounts, cost effectiveness in performance, or accurate estimating of project costs

4. Cost reduction and savings—Value analysis applied to work or project

5. Quality or accuracy of work performance

6. Productivity—number of activities or assignments completed within a period of time

7. Creativity—degree of innovation, originality, and imaginative thinking applied to work or project

8. Documentation of assignment and projects—completeness, accuracy, and up-to-date

9. Effective utilization of available or assigned resources

10. Problem identification and resolution

11. Exercise of judgment in decision making

12. Supplementary contributions—byproducts of assignments which are above and beyond those anticipated

In setting *standards of performance* we should keep in mind several essential considerations. These include:

1. The standards should be measurable

2. They should be adequately communicated to all concerned

3. They should be capable of being achieved by the individuals

4. They should be related to the work being performed
5. They should be controlled by the engineer or project team performing the work
6. Each standard should be weighted to indicate its relative value or importance to the work

It is considered a sound managerial practice to have the engineers involved participate in setting the standards after a discussion of them between the individuals and the manager. In either situation there is a better understanding between the individuals involved and the manager as to what is expected.

18.5. GIVING ORDERS TO ACHIEVE RESULTS

Another fundamental of supervising engineers is that of giving orders. These are directions to guide engineers in their conduct, behavior and performance while employed. They should be stated in positive terms. This will enable our staff to know what it is that we expect of them. The more results oriented the more likely we will achieve the performance desires.

As has been stated, the assignment of work or projects provides the major method for giving orders. This involves communicating what is to be done by the engineer or team and when, where, how and why it should be done. The more detailed these orders, the greater the need to reduce them to writing. Written orders or instructions become ready references to ensure that all engineers involved will know what the engineering manager wants them to know and do.

When these are standard and uniform for a department, they could be issued as standard operating procedures, commonly call SOPs. These usually are included in a manual in a loose-leaf format. The manual becomes a reference source for engineers in need of guidance as to what actions they are expected to take in the performance of their work.

For standard and uniform personal conduct and behavior at work managers usually issue rules and regulations or codes of conduct. Most frequently these are developed and issued by an organization for all employees, not necessarily for engineers.

When orders are reduced to writing but are not to be considered as SOPs, they are frequently issued as memorandum or special instructions. These apply to specific assignments or projects. They are valid only for the duration of those assignments or for a specified period of time.

18.6. CONTROLLING ENGINEERS AND THEIR PERFORMANCE

The one technique that has frequently been considered in the act of supervising is that of control, specifically of control by an engineering manager involving the follow-up and checking of the work of the engineers being supervised.

This is the facet of a manager's work that enables him or her to keep abreast of what each engineer and project team is doing. It involves measuring progress and performance against plans and programs and includes checking of resources that have been and are being utilized. It takes into consideration time and money.

The controls could be the manager's observations, including a review of progress reports; examination of Pert charts, milestone charts, and similar planning tools; performance evaluations of individual engineers and their participation in project meetings. All serve to keep the manager informed of the staff's activities.

On the basis of this information, the manager could motivate those engineers and teams who were not performing as desired, restrain those who were working too fast, and redirect those who were moving in the wrong direction. Thus the manager could regulate all work being performed to ensure achievement of overall goals and objectives. He or she could coordinate the work of the individuals on his staff, as well as the direction of all projects.

The more sensitive an engineering manager is to his or her established controls, the more effective his or her supervision will be. Since communications skills and interpersonal relationships serve as control tools, a manager can detect misunderstandings, reassure those in need of guidance, and encourage those who are performing properly. A manager's controls permit the manager to keep his or her fingers on the pulse of the staff's work.

18.7. RESOLVING ENGINEERING PERSONNEL PROBLEMS

An extremely important fundamental of supervising engineers is that of resolving their personnel problems. This involves a recognition of the fact that engineers, like all other personnel, could have problems or could cause them.

Studies of engineers and their behavior indicate that there are a number of typical personnel problems which are experienced by engineers and their managers. These include:

1. *Dissatisfaction, Complaints and Grievances.* These involve their feelings and experiences about work or conditions under which they

work. Included are dissatisfaction with compensation, opportunities for personal and professional growth, work assignments, being kept informed, and a lack of managerial appreciation.

2. *Technical Incompetence/Obsolescence.* These are the result of technological changes taking place in many industries without a corresponding development of professional staff members. Many engineers have been found to be involved in little or no continuing professional education since being graduated.

3. *Disciplinary and Behavioral Problems.* These arise when engineers are unwilling to comply with a company's rules and regulations or codes of behavior and conduct. They also include problems which could be considered attitudinal. That is, engineers who are nonconformists working in an atmosphere where conformity is required. It also has appeared in companies where engineers as activists resist a company's efforts in working on military products or engaging in activities which have been deemed antienvironmental.

4. *Personality Conflicts.* These include engineers who are unable to work cooperatively with other engineers, other departments, or with their managers. The diversity of engineering personalities has resulted in a larger number of personality conflicts. The demand for interpersonal work relationships has increased, creating more situations in which conflicts can arise.

5. *Personal Problems.* Engineers, like others, experience nonjob-related problems. These may influence the performance of engineers on the job. Engineering managers now recognize the importance of getting these problems resolved to minimize their adverse affect on performance.

6. *Emotional Problems.* These are the problems which are frequently called stress. Both work and nonwork factors cause stress in engineers. Like personal problems, the effect of stress or emotional situations may be detrimental to performance.

In order to deal effectively with personnel problems, the engineering manager must develop sensitivity to these problems. A manager must anticipate these problems and attempt to prevent them. However, prevention is not infallible. It is necessary for a manager to develop empathy with his staff. This permits him to become aware of early indicators of problems. Any changes in the behavior pattern or performance of an engineer are possible indicators of problems. They reflect the existence of problems but not the causes.

An effective supervisory technique is to encourage engineers to communicate with you as the manager whenever they have a problem—Whether you use the "open door" policy, an informal complaint procedure, a formal complaint or grievance procedure, or a periodic manager-engineer performance review. The engineers need encouragement to communicate their problems. Conversely a manager should not delay sitting down with any engineer whose performance is regressing. Facilitating the identification of the cause of an engineer's problem is a major step in resolving the problem.

Not all personnel problems can be resolved by an engineering manager. Many have to be referred to professionals either within or outside the company for resolution.

18.8. BUILDING THE ENGINEERING TEAM

The essence of supervising engineers is to build an engineering team. This requires the engineering manager to provide positive leadership. He or she must direct the efforts of the staff to the achievement of common department goals. He must build an esprit de corps within the department. Every engineer must be proud to work for the manager, department, and company. Each must be willing to subordinate individuality for the good of the group.

To accomplish this an engineering manager must stimulate feelings of belonging within the staff. All engineers must be encouraged to cooperate with each other. All developments must be publicized as team products. A "we" attitude must be present in all relationships.

A technique frequently used to build an engineering team is to conduct staff meetings. At these meetings special assignments and projects are discussed, problems are introduced for group solution, and results are announced for all to share. This is a team-building communication technique. (See Chapter 4.)

Another technique which has been successfully used by engineering managers is to create a climate or environment in which individual engineers and project teams are encouraged to achieve *self-accomplishment*. Personal projects or assignments are recommended for the manager's review and approval to stimulate individual creativity. Individual initiative is encouraged. Individuals are permitted to learn from their own mistakes without fear of reprimand or discipline.

When the engineering staff looks to the engineering manager for advice and guidance, leadership is being established and the team being formed. Supervision of a team will be found to be more effective in achieving results than supervision of a group of individual engineers.

19

SCREENING AND SELECTION

William C. Byham

WILLIAM C. BYHAM has been President and Chief Executive Officer, Development Dimensions International (DDI), Pittsburgh, Pennsylvania since 1970. Dr. Byham obtained a B.S. and M.S. from Ohio University and a Ph.D. in industrial psychology from Purdue University. He has been a faculty member at the Graduate School of Business of the University of Pittsburgh since 1970, having previously taught at the Baruch Graduate School of Business in New York. Dr. Byham is author of 72 articles and book chapters and 14 books. He is a Certified Psychologist in the states of New York and Pennsylvania, a Diplomate in Industrial and Organizational Psychology of the American Board of Professional Psychology, and is a member of several major professional societies.

19.1. INTRODUCTION

Probably the least expensive way an organization can increase productivity is through improved employee selection. If an organization can hire better people:

1. Turnover is decreased since people are successful and satisfied on the job
2. Initial training and development expenditures are decreased
3. Management can devote time to making good employees better rather than focusing the majority of time on a few subpar employees

4. Organizations will have a larger proportion of high-producing employees.

The alternative to good selection is training and development. Few organizations have such faith in their training and development competencies that they can afford to minimize their efforts toward selection. Even if an organization has ultimate faith in its training and development programs, it is still better off starting out with good people. Successful training and development can make adequate performers better more successfully than it can make poor performers adequate.

Yet management often does not realize the importance of employee selection. Organizations seldom have organized, thought-out selection systems to fill positions. They often use methods or instruments that are inappropriate, misleading, or ineffective. Managers do not spend enough time making selection decisions, yet they often spend a great deal of time with an employee correcting mistakes or working out problems. If a small amount of that time had been spent in better selection, the problems might well have been avoided and the manager could spend his or her time in areas that would have a more valuable impact on the organization.

In addition to the obvious financial rewards for improved employee selection, there is another compelling reason in the United States why managers should be concerned—the Equal Employment Opportunity Commission (EEOC). The federal government and many state and local governments have focused on selection decisions as the primary culprit in creating race and sex differences in job level and pay. Penalties in the tens of millions of dollars have been levied against organizations. Even organizations that have won cases have found the victory expensive. In addition to legal fees, organizations charged with violations of EEOC regulations have spent large sums of money putting together documentations required in court cases. Compliance with EEOC regulations produces economic as well as social good.

This chapter will take the reader through a step-by-step approach to devising a selection system for any type of engineer. Various alternative selection procedures that might comprise a system will be described and their values and limitations will be discussed. Finally, methods for integrating data obtained about applicants will be described. Throughout the chapter, the EEO implications of each step or procedure will be highlighted. The chapter will end with a brief description of selection trends that a manager might look for in the late 1980s and 1990s.

19.2. WHAT IS A SELECTION SYSTEM?

Hiring decisions are based on an evaluation of application information accumulated from a variety of sources. These sources may include interviews, reference checks, psychological tests, medical examinations, assessment centers, and so on. The combination and order of these sources of information, along with the method of integrating the data obtained from the sources, comprise the selection system for a particular position. In this Chapter the term "system" will refer to a uniformly applied, step-by-step procedure for collecting applicant information and making hiring decisions.

Using a selection system has many practical advantages. It provides a more efficient method of collecting the necessary applicant information and at the same time ensures fair hiring and rejection decisions for all applicants. Figure 19.1 illustrates a typical selection system for the selection of an electrical engineer. The arrows represent points where decisions must be

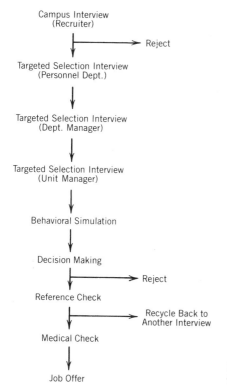

FIGURE 19.1. Selection system for electrical engineer.

made—either to reject a candidate or to continue processing him or her through the system. These stages are called "decision points."

There are many advantages to having a formalized selection system.

1. Selection systems are standardized. Every individual who goes through the selection system is exposed to the same number and type of interviews and other selection elements.

2. Selection systems present selection elements in an efficient sequence. More expensive selection elements such as medical checks or interviews by senior managers occur at the end of an effective system. These expensive resources are economized because they are used only with the most promising applicants.

3. Selection systems provide defined decision points. Good selection systems contain several clearly defined decision points. Managers have common standards as to applicant qualifications needed to pass the decision point and move on in the system. Decision points placed early in the selection system allow for the rejection of less qualified applicants. As a result more time can be spent on candidates who have a better chance of getting the job.

4. Selection systems ensure full coverage of applicant information (dimensions) required for job success. A good selection system makes sure that none of the job dimensions defined for the position to be filled are overlooked. It also insures that sufficient information is collected from each applicant on all dimensional areas.

5. Selection systems prevent unplanned overlap in coverage of an applicant's background. In an effective selection system each participating manager knows his or her responsibility for collecting dimensional information. This prevents two interviewers from unknowingly covering the same dimension in separate interviews or perhaps completely skipping over other dimensions that are important.

6. Selection systems allow more emphasis on the most important dimensions of a person's background. An effective selection system is designed so that important dimensions are covered as many times as necessary. Particularly important dimensions may be double checked or even triple checked if necessary.

7. Selection systems eliminate needless overlap in providing information about the job and organization. A good selection system identifies each interviewer's responsibility for giving information about the organization and the position. Overlapping of effort is eliminated when managers know their responsibilities.

19.3.   SELECTION SYSTEMS SHOULD BE BUILT AROUND JOB-RELATED DIMENSIONS

Figure 19.2 shows how a selection system facilitates gathering information on dimensions identified as important to the target position. An "X" indicates that information on the dimension is obtained by the selection element. First, applicants meeting the general requirements of desired employees are identified through screening application forms and one-to-one screening interviews. More information is then obtained as later system elements are used. More and more dimensions are evaluated as the selection process continues. There is also increased repetition in the coverage of dimensions.

Industrial Engineer	Phone Screen	Chief Industrial Engineer Interview	Department Head Interview	Manufacturing Manager Interview	Reference Check
Technical proficiency	X	X	X		X
Decision making	X	X		X	X
Organizational awareness			X	X	X
Organizational sensitivity			X	X	X
Adaptability		X	X		X
Independence		X	X		X
Tolerance for stress		X	X		X
Job motivation	X	X			
Energy		X	X		X
Written communication				X	X

FIGURE 19.2.   Selection elements used to cover dimensions.

19.4.   WHO IS RESPONSIBLE FOR ESTABLISHING A SELECTION SYSTEM?

In most organizations the responsibility for establishing an appropriate selection system for any position rests with the personnel department. The personnel department may not be responsible for making the hiring decision, but logically should play the key role in making sure the hiring managers consider all the information necessary to make an appropriate decision and have all the resources necessary to aid them in making the best decision.

19.5. DIMENSIONS—THE KEY TO AN EFFECTIVE SELECTION SYSTEM

The primary responsibility of a selection system is to collect the information about an applicant's past experiences. However, a manager should not be interested in all past information that he or she might be able to collect. The manager should collect only information that is relevant to the position the applicant is seeking. For example, an efficient interviewer will ask only questions that encourage the applicant to discuss job-related topics and will direct the applicant back to job-related areas when the conversation strays to irrelevant topics.

In an effective selection system, background interviews and all parts of the system are guided to job-related topics by "dimensions." A dimension is simply a label that represents a group of behaviors exhibited (or not exhibited) by successful people in a position. Some common dimensions and their definitions are provided in the accompanying box.

Common Dimensions

Analysis—Identifying problems, securing relevant information, relating data from different sources, and identifying possible causes of problems

Judgment—Developing alternative courses of action and making decisions that are based on logical assumptions and reflect factual information

Decisiveness—Readiness to make decisions, render judgments, take action, or commit oneself

Planning and Organizing—Establishing a course of action for self and/or others to accomplish a specific goal; planning proper assignments of personnel and appropriate allocation of resources

Delegation—Utilizing subordinates effectively; allocating decision making and other responsibilities to the appropriate subordinates

Independence—Taking actions in which the dominant influence is one's own convictions rather than the influence of others' opinions

Tolerance for Stress—Performing stably under pressure and/or opposition

Leadership—Utilizing appropriate interpersonal styles and methods in guiding individuals (subordinates, peers, superiors) or groups toward task accomplishment

---

Initiative—Making active attempts to influence events to secure organizational supplies, resources, or information to achieve goals. Actions taken to achieve goals must be beyond normal performance expectations of position

Work Standards—Setting high goals or standards of performance for self, subordinates, others, and organization; dissatisfied with average performance

Energy—Maintaining a high activity level

Job Motivation—The extent to which activities and responsibilities available in the job overlap with activities and responsibilities that result in personal satisfaction

Written Communication—Clear expression of ideas in writing and in good grammatical form

Oral Presentation—Effective expression when presenting ideas or tasks to an individual or to a group when given time for preparation (includes gestures and nonverbal communication)

Oral Communication—Effective expression in individual or group situations (includes gestures and nonverbal communication)

Of course the responsibilities and skills needed to perform well differ widely from job to job. The use of dimensions focuses selection elements on the most relevant and important topic areas for each specific target position.

19.6.   ELEMENTS OF SELECTION SYSTEMS

On the following pages the most common elements of selection systems are reviewed. They are presented roughly in the order in which they are normally found in selection systems. However, there are many possible variations, depending on organizational needs, dimensions sought, and other considerations.

19.6.1.   Recruiting Activities

Strictly speaking, recruiting activities are not "selection elements," yet the quality of individuals hired by the use of any selection system is determined by the quality of the applicant who is first attracted to apply for a position. It is always to the organization's advantage to have a large number of applicants from which to choose. Recruiting is expensive and time consuming, but it is important. Most organizations must compete vigorously for talented applicants.

There are several methods of attracting applicants to an organization.

1. Advertisements in newspapers, industry-specific newsletters, and on radio
2. Visits to schools and other training institutions by organization representatives
3. Referrals by organization members
4. Signs outside the plant or office
5. Use of private employment agencies and executive search firms
6. Use of government-sponsored employment agencies or referral services

19.6.2. Screening Application Forms or Resumes

Applicants will often send a resume or fill out an application blank at an office. In this way the organization obtains a "paper portrait" of the applicant. The initial review and evaluation of these application forms and resumes is an important part of a selection system. The purpose of this review is to select applicants who seem promising and from whom details will be gathered during the remaining selection elements. This process is called "screening." Screening applicants for the remaining steps in the selection system helps the system operate more efficiently by quickly eliminating obviously unqualified applicants from the applicant pool.

Screening application forms is always difficult and can be somewhat unreliable due to variations in information provided. This is particularly true when one must screen resumes that are voluntarily submitted by applicants. A wide range of detail is encountered. Even when all applicants are asked to complete a company-designed application form, some individuals provide a great deal of information while others use only one or two words to answer a question.

Dimensions to be considered on an application form or resume screen should be chosen based on their importance and how well they can be evaluated. It is fairly easy to determine if applicants meet certain requirements, such as the need to hold a professional license or certificate. General technical skill areas can often be accurately evaluated.

Organizations should check the reliability of their resume screening processes to be sure that all people involved are using the same standards and tending to the same clues in making dimensional evaluations.

19.6.3.   Targeted Selection Interviews

The objective of a targeted selection interview is to provide specific behavioral information on dimensions defined as important to a specific job or family of jobs. Many selection systems have two targeted selection interviews. Some have three or more. The number of interviews depends on the number of individuals responsible for evaluating applicants for a position, importance of the job filled, and difficulty in obtaining accurate evaluations of applicants. A typical selection system involves a representative of the personnel department and the hiring manager.

An effective interview targeted to specific dimensions has the following characteristics:

1. *Use Past Behavior to Predict Future Behavior.*   Managers recognize the importance of evaluating an individual's past behavior when making personnel decisions. Deciding who can handle a problem today is a matter of recalling who was successful in solving a similar problem last week, last month, or time and time again during the past few years. Managers conclude that the individual who solved the problem or carried out the work assignment well in the past can do it again. They are using *past* behavior to predict *future* behavior. They may not always be right, but the odds are certainly in their favor!

   Finding out what an applicant has done in the past is the heart of an effective interview. Once the interviewer knows what an applicant has done on the job, he or she can accurately predict the behaviors, skills, and decisions the applicant will probably repeat in the future. With this information, the applicant(s) with the best chance of being successful can be offered the position(s).

   The term "behavior" is used to describe a person's past actions and accomplishments as well as his or her actions and reactions during the interview. A complete description of behavior includes the situation under which an action occurred, the action itself, and the result of that action.

2. *Apply Effective Interviewing Skills and Techniques.*   In the day-to-day operation of most selection systems, the in-depth interview provides the key opportunity for the organization to collect applicants' past behaviors. This means the interviewer has a very important responsibility to the overall success of the system. The interviewer must use the time allotted for the interview to explore applicants' unique past experiences and collect specific behavioral information

according to the list of target dimensions. To do this the interviewer must work efficiently and effectively. This is accomplished by using proven, well-worded interview questions designed to bring out relevant past behaviors.

19.6.4.   Behavioral Simulations and Job Replica Tests

A behavioral simulation is a controlled situation in which applicants display behavior relative to target dimensions. For example, if negotiating behavior is sought, an applicant might be given an "assignment" to negotiate with a "vendor." Information about the product and the vendor would be provided. The applicant's behavior would be directly observed by an organization representative playing the role of the vendor, or by a third party. Behavior in the exercise would then be categorized under the appropriate target dimensions.

Behavioral simulations are particularly appropriate in situations in which it is difficult to evaluate an applicant in certain dimensions because he or she has had little or no opportunity to demonstrate behavior within those dimensions. For example, it is often difficult to evaluate the planning skills of an individual who has never had a job that required planning.

19.6.5.   Tests

No selection device has come under closer government scrutiny than the paper-and-pencil psychological test. This high degree of attention stems from the many misuses of tests, which have brought about intended or unintended discrimination against protected classes, and from the fact that tests are easily quantifiable, thus making them an easy portion of the selection system to attack. The vulnerability of the selection test (quantitative scores) is also its strength and the source of its great appeal. For the last 80 years organizations have sought an easy solution to selection problems and many have felt that the solution was a paper-and-pencil test. Most of the tests used in industry are relatively easy to administer and can be scored by a clerk. They produce information with very little expenditure of management time or energy. If they can accurately predict success on the job, these written tests are a cheap solution to many problems of personnel managers.

A major difficulty, however, is that most paper-and-pencil tests used in employee selection produce differential results for black and white applicants taking the test; therefore, different distributions of scores for black and white applicants are obtained. Figure 19.3 shows the typical national distribution of test scores for a commonly used selection test. It is quite evident that the average white person does better than the average black person, but

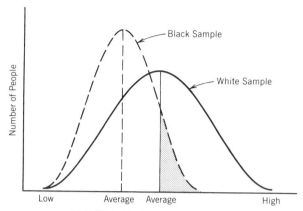

FIGURE 19.3. Test scores.

there are blacks who do far better on the test than the average white, and whites who do far worse on the test than the average black. The point is that this selection instrument, if used, would produce a disproportionately low number of blacks in the work force.

Because this problem arises with the use of most paper-and-pencil tests, various government agencies have focused their attention on psychological testing. Unless an organization can specifically show that such tests are necessary, the use of test data probably works against employment of blacks.

To state the situation in another way, if an organization were to administer a paper-and-pencil test which had absolutely no relationship to the target job for which individuals were being considered and used a fixed cutoff on that test to make selection decisions, the general result would be discriminatory toward blacks. Because the distribution of scores is different, any arbitrary cutoff test score would tend to select a greater percentage of whites. Obviously a very high cutoff score will produce more of a disproportionate effect than a low cutoff score.

The reason for the poor performance of blacks on most paper-and-pencil tests can be traced to many factors, including early cultural deprivation, unfamiliarity with test-taking techniques, and possible unintended cultural biases built into certain tests by the white middle-class psychologists who develop them. Whatever the reason, the problem exists and has resulted in the high rate of test scrutiny.

If an organization is considering a new test, it must be deemed practical for the situation and meet the highest standards dictated by law. This should be done by a recognized test expert who can guide the organization in every step of the process, from the initial selection of the test, through validation, to implementation. The investment in a testing expert's consulting time is a small price to pay relative to the potential hundreds of thousands or millions of dollars worth of damages that have been paid by organizations who have been found to be using invalid selection techniques.

19.6.6. Use of Psychologists

Many organizations use professional, licensed psychologists ("shrinks") to evaluate prospective employees. The psychologist usually administers a battery of tests and conducts one or more in-depth interviews with the applicant. A written profile of the applicant's strengths and weaknesses results.

In-depth, clinical evaluations of applicants can be a valuable contribution to a selection process. The psychologist can add excellent insights into the motivation and life goals of the individual.

However, one should always remember when dealing with psychological evaluations that they are "clinical" judgments based on the psychologist's experience in giving tests and interviewing many applicants for jobs. They are not based on impersonal, criterion-related research. For this reason an organization should be especially concerned with possible overreliance on the evaluation results. Clinical evaluations should be given no more weight than a manager's interview results. Psychological evaluation findings should be compared and contrasted with other data on an applicant and weighted appropriately. Negative data should be checked out in subsequent interviews—not just accepted.

A psychologist cannot aid your organization unless he or she understands your organization and the jobs for which the individuals are to be selected. Time spent in orientation and briefings is well invested. It is advisable to bring the psychologist to your organization in order to provide him or her with a "feel" for it, to furnish organizational charts and job descriptions, and thoroughly define the dimensions wanted.

19.6.7. Reference Checking

This Chapter has consistently emphasized the need to obtain examples of past behavior that can be used to predict future behavior. Much time has been spent on getting "behavior" from the interview, and yet the information obtained during the interview is not really "behavior." It is a self-report of behavior that needs to be confirmed, authenticated, and validated. An effective means of such validation is reference checks. They provide a means of obtaining additional information about an applicant that is unavailable in other parts of a selection system.

Reference checks are a vital component of all selection systems. Only when dealing with applicants who have no previous work history can they be satisfactorily bypassed. Even then one would usually expect to check educational background.

There are two purposes of reference checks: checking facts and obtaining data. The first is the easier and more common. Information about critical dates (length of employment with an organization), critical accomplishments (achievement of a college degree or salary level), and critical experiences (type of products handled) obtained in interviews or from the application form is checked for accuracy. Validation of factual information is easy and can usually be obtained from the personnel department of an organization or from a school registrar. Such information is readily given. This type of reference checking can be assigned to a personnel department specialist.

Obtaining new data and confirming suspicions comprise the second goal of reference checking. This is much more important and more difficult to accomplish. However, the potential payoff, in terms of aiding the accuracy of the selection process, is enormous. This is the kind of reference checking that is considered here. One should still check facts, but because this is an easy task it is not emphasized here. Moreover, checking facts can usually be accomplished while obtaining new data, making it easy to do both at the same time.

1. *What to Check.* In addition to confirming facts obtained in the interview and from the application form, a reference check can accomplish many other goals such as: (1) following up leads from interviews or other parts of the selection system, (2) testing "gut" feelings regarding the applicant, (3) seeking examples of the applicant's behavior in tasks required in a new job, and (4) checking for common failings, for example, drinking and absenteeism.

Probably the most important function of reference checks is to follow up leads or hunches. For this reason reference checks are usually scheduled near the end of the selection process. When reviewing information obtained in interviews or from other sources, the interviewer may often find certain areas that were not fully explored or he or she may not be satisfied with the answers obtained. A reference check is the place to investigate such areas.

An interviewer might have had a feeling that an applicant had trouble dealing with time pressure, but was not able to pin down a specific example during an interview. Such examples could be sought in a reference check. Confirmation of leads or hunches is extremely important because it is not fair to deprive a person of employment without facts. An organization gives the person the benefit of the doubt by further checking.

2. *Whom to Check.* Current behavior is more predictive of the future

than past behavior. Thus a reference from an individual who has observed the applicant in recent jobs is more valuable than one obtained from someone in the applicant's more remote past. Also a reference from an individual who has observed the applicant in situations similar to the one in which the applicant may soon find himself or herself is more valuable than one from an individual who has observed him or her in markedly different situations. For these reasons references such as high school coaches are seldom useful, since exposure to the applicant was typically at some time in the past and the situation may have been considerably different from the one in which the applicant would now be operating.

The number of individuals from whom it would be appropriate to obtain reference checks is limited only by the creativity of the reference checker. Appropriateness of references depends on the particular applicant being checked, his or her background, importance of the job, and other similar considerations.

The truism in reference checking is "avoid the personnel department if possible." The personnel department will usually give only the barest factual information about an applicant.

High school and college administrators used to be very good sources of information on individuals. However, as school systems have increased in size and become more "gun-shy" about giving information, this source has nearly dried up. If the reference checker knows an official at a university, it is still possible to get useful information, but in general he or she must be satisfied with confirmation of attendance and grades.

Former supervisors are likely candidates for most reference checks. They have been in a position to observe and evaluate the applicant's work. If a former supervisor has moved on to another position or organization, do not let this inhibit your reference checking. Obtain the individual's new phone number and address and follow up. He or she may feel even more free to talk under these circumstances.

Although it is possible to do reference checking with past supervisors without asking for names from the applicant, this is inadvisable. It is good policy to tell the applicant the names of the people with whom you will be speaking. A good applicant will be impressed by the thoroughness of your efforts in making a selection decision. You may also be able to get some insight into the individual and his or her relationship with the person providing the reference. This will help you interpret the obtained data.

A common problem in reference checking is the candidate whose only significant work has been at an organization where he or she is currently employed. Assuming that the applicant has not told his or her supervisor that he or she is looking for a job, the supervisor cannot be approached. An alternative open in this situation is to check information from the following:

(1) former employees of the organization, (2) clients and business associates in other firms, and (3) co-workers in the organization, if they are known to the manager seeking information and can be trusted not to jeopardize the applicant's job situation.

The easiest source of names for reference checking is the applicant. Most personnel forms ask for business references. One might assume that because the applicant has volunteered a name, the person specified will almost certainly give the applicant a good reference. This is not necessarily so. Many applicants with a poor work history take the chance that their references will never be checked. Others do not realize that a former boss may provide negative data. Furthermore, applicants will sometimes interpret information they know a former boss will provide as being positive, while the hiring manager would put the opposite interpretation on the same data.

On the other hand, the names of personal references provided by applicants on resumes are usually very poor sources of data. They are more in the nature of character references, and ministers and physicians are frequently listed. For most positions, any effort to follow up these references is not worthwhile.

If names of references are not requested on the application blank but contacting references is the organization's custom, the usual alternative is to ask applicants to suggest appropriate supervisors, clients, or previous co-workers. It is very difficult to obtain names in any other manner.

A seldom used but excellent source of names for reference checking is individuals suggested by the references the applicant has listed. Many experienced reference checkers end an interview with the question: "Is there anyone else I can talk to who would be familiar with our applicant's performance?"

The best reference is a person known to the reference checker. Thus a good stratagem is to develop a contact with some of the better sources of applicants. When a personal relationship has developed and the natural fears associated with reference checking and giving reference information have diminished, the whole process is materially eased.

3. *Who Should Check.* It is possible to train subordinates or to use specialists from the personnel department to take care of less important reference checks and to authenticate facts obtained in interviews and on the application blank. If reference checking is to be delegated, then a form should be developed to guide the person given this responsibility.

The majority of reference checks should be taken care of by the managers who will supervise the individual if hired, or by a key manager in the selection process.

Most managers will accept the fact that they should perform these functions. The time required is the only factor that keeps them from taking on the assignment. Yet, like all parts of the selection system, it is critical. Because reference checking usually comes near the end of the selection system, a manager needs to concern himself or herself only with the final applicants, thus cutting down considerably on the time devoted to this element.

4. *How to Check.* Phone calls are the most popular method of reference checking. They are definitely worth the money no matter how great the distance. In fact, reference calls of unusually long distance (such as overseas calls) are often exceptionally effective because they show the person giving the reference the importance of the conversation. The main element of success is to keep the call from being rushed. Therefore, it is very important to check the availability of the reference before getting into the substance of the reference check.

5. *When to Check.* Ordinarily, reference checks should be made near the end of the selection system, but an opportunity should always be open for scheduling an additional interview to check out leads from the reference check should a poor report be obtained.

Most organizations schedule their reference checks immediately before or after the final selection decision (not the job offer). The concerned managers make up their minds contingent on good references. If good references are obtained, the decision holds and an offer is made. If the references are poor or indicate an area of concern, an additional interview is scheduled and the decision delayed.

6. *Interpretation of Information Obtained.* While reference information is extremely important, it must never be accepted without further checking. There are many reasons why an employer may be angry with a former employee. Even if the employee did an extremely good job, the employer may feel resentment because the person left the company at a busy time; perhaps the employer is slightly jealous. Also it is possible to obtain inaccurate reference data because the source often tries to recall events from several years past and may naturally make mistakes. Therefore, dates and information obtained must be double checked, unless the reference is reading dates from an official form.

Negative information obtained from a reference is not sufficient to eliminate from consideration an otherwise good candidate. This candidate deserves the opportunity of another interview. Since the manager must abide by the "off-the-record" promise made to the reference, this interview should not be a confrontation of the applicant with the reference data. It should be a thorough attempt by the manager to check out the data received from the reference. Collaborating or conflicting data should be pursued subtly through planned, nonleading, behavioral questions. This interview should be exceptionally well-planned, as some very sensitive issues usually must be considered. Often the manager simply must decide whom to believe, the applicant or reference. This decision should be based on as many facts as possible.

19.6.8. Medical Examinations and Health-Related Questions

A forgotten part of most selection systems is the medical examination. It is viewed by most managers as a last hurdle forced on the organization by its insurance carriers. It is something for a candidate to pass, not a source of information about the candidate. However, as in most areas related to selection, this attitude is rapidly changing, principally because of increasing government concern and legislation. The United States federal government is increasingly forcing organizations to obtain certain medical information, while at the same time promulgating new restrictions as to what can and cannot be asked of an applicant.

The principal legal requirement for preemployment physicals comes from the need to establish preexisting conditions, for example, hernia, hearing loss, back injury, spot on lung, foot trouble, poor eyesight, and so on, which if not identified upon employment, can later become the responsibility of the employer. Monitoring of intake condition has always been important relative to avoiding compensation claims. It has become more important as rulings and guidelines stemming from the Occupational Safety and Health Act have extended the list of potentially harmful physical working conditions from which an organization is responsible for protecting its employees.

At one time Worker's Compensation claims were strictly the prerogative of nonmanagement employees. This is no longer the case. Increasingly, executives and the widows or widowers of executives are being compensated for heart and other conditions brought on by stress, frequent traveling, and long hours. Thus, for all levels in an organization, a physical examination may be a good investment.

Medical information can best be obtained by the organization's medical department or by designated physicians. Questions regarding health and medical status should not be asked in the selection interview or on an application blank. Under United States law individuals who have diabetes or a terminal illness cannot be turned down unless a professional medical examination shows they are incapable of performing the job.

19.6.9. Explaining the Job and Selling the Organization

Effective selection interviewers fully explain both the positives and negatives of the job, using as many specific examples as possible. Problems such as high turnover can be eliminated by a complete and truthful description of job responsibilities. However, this is difficult for an interviewer who is very anxious to hire an employee. He or she might tend to overlook some of the negatives of the job, while emphasizing the positives. The interviewer should never "sell" the job to the individual. However, the organization can be "sold" to the individual. It is definitely appropriate to point out all the benefits of working for the organization, including information on salary and benefit plans, special organization features, and less tangible advantages. For example, the interviewer could relate data on important clients and the organization's public image.

The process of giving job and organization information is planned in a selection system. This task is divided among interviewers, as is the information-seeking function. Usually a selection system is set up so that individuals receive fundamental information about the job and the organization early in the process. This information is expanded as the process continues. In the latter stages of the process, interviewers often not only give information but also check an applicant's understanding of the information provided by previous interviewers. They also answer questions about the organization the individual hasn't had a chance to ask or questions that have developed as a result of the interviewing process. These questions are often about working conditions, vacation, pay, benefits, necessary personal tools and equipment, dress codes, and work schedules. The task of explaining the job and selling the organization generally does not show up as a distinct step in an organization's selection system. It is part of each interview situation.

19.6.10. Decision-Making Meeting and Hiring Decisions

Obviously one of the last elements of an employee selection system is the final go/no-go decision on an applicant. This decision is made by the key organization members involved in the selection process.

The data integration session is a meeting of all individuals who gathered behavioral example information about an applicant. They share all available behavioral and other information for each target dimension. Covering one dimension at a time, participants discuss behavior within a given dimension and develop a consensus rating for that dimension. After this process is carried out for every target dimension, the interviewers consider the applicant's profile of strengths and weaknesses across all the target dimensions and make the final hiring decision.

The exchange of information in an effective data integration session is in many ways very much like giving evidence in court. Nothing is taken at face value. All statements must be supported by evidence, usually behavioral examples. The data integration session is also similar to court proceedings in that both rely on the judgments of more than one person in making decisions. A jury system helps assure a fair trial and verdict by providing various views and perspectives on the evidence. In the same way, the data integration session is conducted so the best judgments of several managers are combined to produce fair and accurate decisions.

During the data integration session the discussion among interviewers centers around the dimension rating grid. The grid is a chart with dimensions listed vertically on the left side and each interviewer's name (and other sources of data, if used) heading a separate column. The squares on the grid are used for recording dimension ratings. A separate grid is completed for each applicant.

19.6.11. Making a Hiring Decision

Once a consensus rating is reached for all target dimensions, each interviewer reviews this "dimension profile" and makes an overall summary rating of the applicant. The formats used to make the ratings differ from one organization to another. For some organizations the summary rating is simply expressed in a hire/no-hire decision. Others rate the applicant using a five-point scale similar to the one used in rating behavior within a dimension. Still others devise their own scales. Some kind of multiple-step rating scale is most advantageous when comparing several candidates for a position.

19.7. WHAT TO EXPECT IN THE FUTURE

Major advances have occurred over the last 10 years in the area of interviewing. They have occurred principally as a result of the dissection of the interview process by a number of researchers to determine what makes interviewing successful or unsuccessful. Up to that time researchers had taken a global "Does it work or doesn't it?" approach. Generally they found that interviews didn't work. As we've learned more about the process of interviewing, the dimensionally based, behaviorally oriented interview process has developed, and great strides have been made in both the efficiency and the accuracy of the process.

The assessment-center technology, that is, using behavioral simulations to produce behavior, has also played a significant role. The continuing research by Dr. Douglas Bray, AT&T, and by many other individuals has shown that simulations can play an important and unique role in making selection decisions. Thus, one could anticipate a steady growth of behaviorally oriented interviewing systems, assessment-center methodologies, and a welding of the two. Rather than a formalized assessment center, many organizations are opting for merely adding assessment-center exercises (behavioral simulations) to their interview process.

Organizations are beginning to use videotape technology to capture the applicant's behavioral simulation. They are making arrangements for applicants to go through simulations and have those simulations recorded on videotape. The videotapes can be reviewed at a more convenient time for the manager involved in the selection decision.

Another innovation to look for is the use of computer-managed selection exercises. Many of the typical behavioral simulations used in selection systems can now be adapted so that they can be administered by a minicomputer. Sometimes it is necessary for a videotape playback component to be attached to the computer. For example, research is currently under way to evaluate a behavioral simulation exercise where the applicant looks at 20 minutes of a group discussion and then answers a series of questions about the discussion and how he or she would go about changing the views of the various participants in the group. If these types of simulations prove to be as effective in predicting future behavior as the more standard performance simulations, they will make an attractive economic alternative.

Reference checking seems to be coming out of the bad grace in which it was held for many years. More and more organizations are realizing that very important data can be obtained through reference checks and are training key individuals to conduct effective checks—particularly by using specially designed interview forms. Moreover, there is increasing clarity of the legal status of reference giving. It appears from review of the law, that references (even very negative references) can be given as long as the reference givers understand certain easy points in regard to limiting their comments to the questions asked. Finally, the EEOC has pushed reference checking as a means of obtaining applicant information.

20

THE PERSONNEL DEPARTMENT AND THE ENGINEERING MANAGER

Edward Alexander Tomeski and Michael Klahr

EDWARD ALEXANDER TOMESKI, Professor of Management, Barry University, is the author of five business and computer books and about one hundred published articles, monographs, and chapters in books. Dr. Tomeski has consulted with such organizations as Western Electric, the United Nations, the City of New York, the Rockefeller Family, and the Commonwealth of Puerto Rico, among others. He has held management- and staff-level positions with ITT Corporation, W. R. Grace & Co., Bankers Trust Company, and Mobil Oil Corporation. In addition, Dr. Tomeski has held national offices in The Institute of Management Sciences, the Association for Systems Management, and the American National Standards Institute. He has lectured at several universities, including New York University and City College of New York, and is a frequent speaker at international conferences.

MICHAEL KLAHR is an Associate Professor of Marketing at Barry University. He also serves as a marketing consultant and lectures frequently in the United States and Europe. He has an M.S. from Massachusetts Institute of Technology and a Ph.D. from Columbia University.

20.1. INTRODUCTION

This chapter focuses on the nature, organization, and function of personnel departments. If such departments are large, they are found most often in large corporations whose engineering departments must work with personnel in the processing of their employees and rely on it for services rendered to all employees. The small firms, which predominate in engineering, usually do not have personnel departments; however the functions described here and the principles governing them are essential even in small firms. In such cases these functions may form a part of the work load of principals or other delegated individuals, and the clerical work must be done by the office staff; it may also involve the accountant who comes periodically to audit the books. For large firms, however, as is shown below, personnel departments may be a large function, encompassing specialists of various kinds, elaborate data bases, and associated wide-ranging staff activities.

20.2. THE IMPACT OF GOVERNMENT REGULATION AND LEGISLATION

Government regulation over the past 20 years or so has transformed the personnel function. First, there is need for more highly skilled specialists. Second, the importance of the personnel function in most organizations has increased. Third, there has been a trend toward more centralization to deal with the new constraints when hiring, firing, disciplining, directing, training, promoting, and compensating employees. In sum, this means the emergence of a central personnel staff as a primary unit for overall organizational planning and control to advise top management on matters of governmental regulations. Combined with the earlier emergence of collective bargaining, the above trends have largely molded the characteristics of the personnel function.

The following descriptions are only intended as brief summaries. Special chapters in this volume discuss the areas concerned in detail.

20.2.1. Nondiscrimination in Employment

The Civil Rights Act of 1964 was directed at eliminating discrimination against individuals on the basis of race, color, religion, sex, or national origin. Title VII of the Act has the greatest impact on industry because it prohibits discrimination in employment policies and practices.

The law established the Equal Employment Opportunity Commission (EEOC). The EEOC's main powers were to seek voluntary conciliation between the parties in disputes brought under the Act. The EEOC also issued guidelines on matters of compliance under Title VI in order to help employers know what changes they should make in their personnel policies and practices. The main responsibility for interpreting and enforcing the Act has fallen to the federal courts.

In 1972, Title VII of the EEO Act amended the Civil Rights Act of 1964, expanding the powers of the EEOC to file civil suits in federal court against parties violating the provisions of Title VII.

In 1965, President Johnson issued Executive Order 11246 making it the policy of the government to provide equal opportunity in Federal employment for all qualified persons. E.O. 11246 established the Office of Federal Contract Compliance (OFCC). Under the order a company that was found to be discriminating could be deprived of all its government contracts. Government contractors are required to maintain a written affirmative action compliance program.

The Equal Pay Act of 1963 provides that there should not be discrimination between employees on the basis of sex for equal work that requires equal skill, effort, and responsibility, and that is performed under similar working conditions.

The Age Discrimination in Employment Act, passed in 1967, prohibits discrimination against individuals aged 40–65 on the basis of age.

20.2.2. Employment and Training Programs

The employment and training programs have their roots in the Full Employment Act of 1946 which was designed to try to alleviate cyclical unemployment. Fiscal and monetary policies are the primary tools that government uses to try to stabilize the economy and employment. However new attention was devoted to undesirable features of the existing supply and demand equation for labor.

The Area Redevelopment Act of 1961 made federal aid available for depressed areas of the country. In 1962 Congress passed the Manpower Development and Training Act aimed at unemployment due to technological displacement. MDTA was designed to give displaced workers new skills through training programs. One of the programs has been the National Alliance of Businessmen Jobs (JOBS) which aimed at getting large numbers of employers to hire and train people who, because of poverty, race, age, sex, mental or health handicaps, become marginal to the work force.

The Economic Opportunity Act of 1964 was aimed at involving the poor and otherwise disadvantaged in various training programs.

The Comprehensive Employment and Training Act (CETA) of 1973 was intended to integrate the various manpower programs of the government.

20.2.3.   Occupational Safety and Health

OSHA brought three new federal agencies into being—the Occupational Safety and Health Administration in the Department of Labor; the National Institute of Occupational Safety and Health (NIOSH) in the Department of Health, Education, and Welfare; and an independent Occupational Safety and Health Review Commission. The first and second of these agencies have authority to conduct inspections and investigations under the Act, while the Review Commission handles the appeals of employers.

20.2.4.   Employee Benefits

ERISA, the Employee Retirement Income Security Act of 1974, attempts to protect employees' retirement income. It prescribes standards of administration for pension funds, requires employer insurance against pension fund failure, limits a variety of pension fund transactions, and in general prescribes rules governing pension plan liquidation, mergers, consolidations, and amendments.

20.3.   CHANGING BUSINESS ENVIRONMENT

Developments in the business environment have had an impact on the approach to managing human resources.

The day of the patriarchal owner-manager who personally presided over the welfare of his workers is continuing to disappear. Increasingly organizations of any size have professional managers. They tend to recognize that the vitality of the organization is dependent on the character and motivation of the work force, quality of work relationships, and capacity of employees to respond to or initiate change.

Government regulations and demographic changes have resulted in a generally more diverse work force. Managing diversity makes the personnel job more difficult.

A few years ago many organizations were diversifying their products and markets. This change gave rise to more complex structures, usually called divisionalization. More recently, in response to economic pressures, there has been some movement toward divestiture and reduction in product lines and markets. Both the earlier trend and its reversal were accompanied by a rising trend in technological resources required by the organization. More directly important to the structure of the personnel function, divisionalization tends to generate more staff units at lower levels of the organization. There is also a headquarters staff which is often the major means by which top management carries out the overall job of planning and control.

20.4. THE WIDENING OF THE PERSONNEL FUNCTION

The concept of the personnel function has always covered more than the narrow purview of the personnel staff. The management of people has always been at the center of the practice of management. Business and societal changes and turmoil have involved a wider group of managers directly in the formulation of personnel policy.

Increasingly the personnel staff has come to deal with the long-term future effectiveness of the company. A major focus has been on integrating the company's personnel plans with the business plans of the divisions of the company as a whole. Of course, there is growing preoccupation with assuring that the company's policies are in accord with the laws and values of the countries and communities where the company does business.

20.5. THE CHANGING EMPHASIS OF PERSONNEL STAFFS

The traditional role of personnel often focused on industrial relations, and a preoccupation with hourly workers and their collective bargaining units. Other activities that were sometimes present included medical, safety, employment, and cafeteria services—and the leaders in personnel management introduced wage and salary administration, employee benefits, training, and employee publications.

More recently leadership organizations are focusing their efforts in such areas as manpower planning, staffing, organization development, and personnel management information systems. The new government compliance units inside personnel, such as equal employment opportunity, are also part of the change.

A key word for personnel executives has become "motivate." This calls for managing people so their imagination and initiative, as well as their hands, are directed to the task of company survival and growth. Accompanying this emphasis on motivation has been a greater emphasis on planning and control activities including setting standards, measuring performance, turnover analysis, and evaluation of operations. Most large personnel departments are now headed by vice-presidents, and many report to the president or chief executive officer of the firm.

20.5.1. The Array of Personnel Activities

Reflecting the changing objectives is the long list of activities that can occupy a personnel staff. They include: labor relations, recruitment, selection, and employment of new employees, equal employment opportunity, compensation, benefit programs, occupational safety and health, communications, personnel planning and research, organization development, organization structure and employee services (e.g., recreation, food services, medical). Some personnel departments are also focusing on out-placement for employees who are departing from the organization.

20.6. ORGANIZATION STRUCTURE OF THE PERSONNEL FUNCTION

The number and location of personnel units within any given company can vary considerably. One unit is almost always found on the headquarters staff. But in addition to this unit, personnel staffs with varying responsibilities, competence, and numbers of personnel employees, are often also found reporting to operating and other staff units. Most arrangements conform to one of four models.

20.6.1 Headquarters Personnel Unit Only

This model has one unit which provides all specialized personnel services for all units of the company. See Figure 20.1. Smaller organizations and those which are physically centralized in one location are most likely to use such a model.

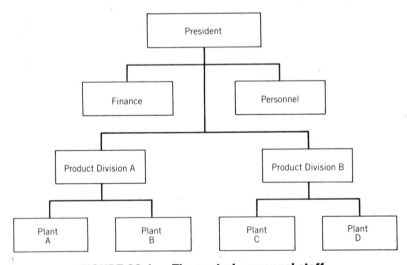

FIGURE 20.1.   The central personnel staff.

20.6.2.   Headquarters and Plant Level Units

In addition to a headquarters personnel unit, some companies have personnel units in one or more plants or similar installations. Typically, the personnel units in these installations report to the installation manager, although they usually must adhere to the personnel policies set by the headquarters unit. See Figure 20.2.

20.6.3.   Headquarters and Division or Functional Operating Units

In such organizations there is a headquarters personnel staff and also personnel staffs that report to the general managers of the units at the operating level. The personnel units in divisionalized companies report to the heads of the operating divisions. In companies with operations specialized along functional lines, decentralized personnel units may report to the head of one or more of the functional units, for example, the head of manufacturing. See Figure 20.3.

20.6.4.   Headquarters, Division or Functional Operating Units, and Plant Level Units

In such cases, personnel staffs exist at headquarters, divisional, and plant levels. See Figure 20.4.

20.7.   DECENTRALIZATION IN THE PERSONNEL FUNCTION

Personnel functions carried out by personnel units on several levels are often dubbed decentralized, due to the fact that many organizations introduced the multilevel personnel units to make profit-center concepts work more effectively. The profit-center manager gains a greater feeling of authority over the services he or she needs if there is a personnel staff under his or her direction. Also placing a personnel unit organizationally within an operating or staff unit, and not merely geographically close to it, makes for better service. It should aid coordination and greater cooperation between members of the served and servicing units, and hence foster greater and more efficient utilization of the services offered by the personnel unit.

Typically, the headquarters personnel unit has responsibility for overall policy, planning, and control; the decentralized units are in charge of local implementation and routine administration.

Headquarters planning and control relative to those activities where in-

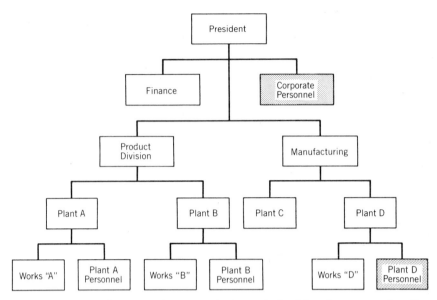

FIGURE 20.2.   Company has corporate and plant-level personnel units.

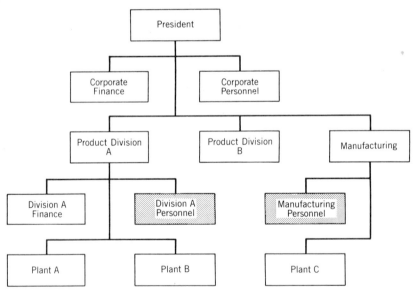

FIGURE 20.3.   Company has corporate, division, or functional operating personnel units.

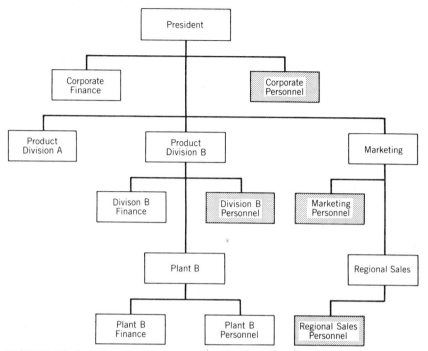

FIGURE 20.4.   Company has corporate, division, or functional operating personnel and plant-level personnel units.

terdivisional consistency is required are generally assigned to and become the most important tasks of the headquarters personnel staff. In addition, services affecting future organizational vitality, such as management development, are usually retained at the central personnel unit. The local units usually have an input role in such activities, but the central personnel staff has the overall coordinating responsibility.

20.8.   STAFF–LINE RELATIONSHIPS

The establishment of personnel staff reporting to the various levels of the organization has solved one set of problems: the local units can provide better services. But it has given rise to a whole new set of problems surrounding the relationships of lower-level staff to headquarters personnel. Essentially the problem for the lower-level unit comes down to "who is boss"—the headquarters personnel executive or the decentralized unit head? Usually the resolution of this problem is at best imperfect. This is partly resolved by spelling out more clearly the distinct and different relationships that derive from the advisory, service, and control roles of staff. In order to clarify the responsibilities and authority of personnel at the various levels of organization, managements have sought to reduce confusion by establishing clear headquarters personnel policies and objectives from which the differing roles and responsibilities of the headquarters personnel staff and the decentralized units may be derived and delineated.

20.9.   THE HEADQUARTERS PERSONNEL UNIT

The size of the headquarters unit depends on the size of the organization and the extent to which personnel units exist at other levels of the organization. The activities which are most frequently handled by the headquarters unit include compensation, equal employment opportunity, benefits, recruitment, selection and employment, contract negotiations, training and development, compensation of managers, labor relations, and recruitment of managers.

The personnel unit can be organized in different ways. Figure 20-5 shows an arrangement with a labor-relations emphasis, that has two subunits; labor relations and compensation. Figure 20.6 shows a compensation-oriented structure which has subunits for salary administration and benefits. The unit in Figure 20.7 is occupied mainly with personnel planning activities such as personnel development, organization planning, and executive recruitment. In Figure 20.8 the departmentalization is more extensive; there are subunits for labor relations, training, medical services, safety, wages, and benefits. In contrast, Figure 20.9 shows an organization that stresses management appraisal and development, manpower planning and recruitment, and organizational planning. The needs of the organization will dictate the actual combination of specialized subunits which will be part of the personnel function.

20.10.   PERSONNEL IN THE OPERATING UNITS

The operating manager has increased need for personnel service and higher-level expertise in personnel work. The core responsibilities, especially at factory levels, are to deal directly with employees.

The branch-level personnel unit is still heavily involved in generalized activities of an administrative nature. The unit often has the grievance procedure as a major activity.

Division personnel staffs tend to be characterized by the service activities

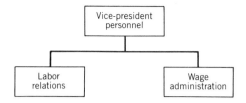

FIGURE 20.5.   A labor relations-oriented staff.

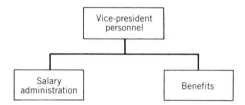

FIGURE 20.6.   A compensation-oriented staff.

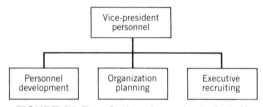

FIGURE 20.7.   A planning-oriented staff.

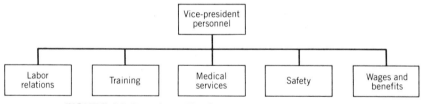

FIGURE 20.8.   A staff oriented to hourly workers.

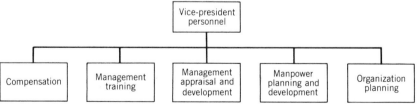

FIGURE 20.9.   A staff oriented to the work of top management.

they carry out in support of organization-wide programs. Much of the paperwork and administration of salary structures, management development programs, and nonroutine grievance procedures tend to be centered in division-level personnel staffs in those companies that have them.

Below are some specific activities and short descriptions of the personnel function.

20.11. HUMAN RESOURCES PLANNING

During the last decade the best-run large companies have been using human resources planning methods. Such planning deals with the following questions: What are our human resource goals? What is our current human resource position? What is the best strategy to reach our desired goals? Did we accomplish our objectives?

This long-range planning approach deals with the human resources inventory and need of the organization which is factored into the enterprise's overall organizational plans and goals.

20.12. JOB ANALYSIS AND DESIGN

Jobs are dynamic and are subject to change and even elimination as advances are made in technology and systems. Work content, the qualifications required to perform it, and returns and rewards for performing it, constitute the attributes of work. Job analysis provides input for a variety of activities including job descriptions, developing training programs, recruitment and strategies, and compensation packages.

Job design is concerned with alternative arrangement of job content, rewards, and qualifications. The techniques involved include simplification, rotation, enlargement, and enrichment.

20.13. RECRUITMENT AND JOB SEARCH

The recruitment process for the organization begins with employment planning which determines the numbers and kinds of people the organization needs to recruit. Afterwards both internal and external sources may be used. Internal sources have the advantage of knowing the organization, are of a proven quality, and it generally is a morale builder to promote from within. Among external sources, walk-ins, referrals, and respondents to newspaper advertisements are most common. College recruiting is the most expensive and time consuming.

20.14. SELECTION

Selection takes place after recruitment has generated a pool of available people. Some screening goes on during the recruiting process itself. Since it determines who gets the job offers, selection plays a critical role in attempting to match the nature of the work, the supervisor, and work group with the person's qualifications and interests. Selection processes include such steps as ranking candidates, comparing candidates' qualifications against job specifications, interviewing, checking of references, assessing prior work experience, and testing, among others.

20.15. ORIENTATION

Orientation programs to indoctrinate new employees and to acclimate them to the organization and job assignments are a critical part of the employment process.

20.16. CAREER PLANNING AND DEVELOPMENT

Career planning and development is often part of the annual evaluation process. By making it a joint activity, the individual can be made aware of future possibilities within the firm. Properly used, such a program can help motivate and develop employees' talents. It helps both the employer and employee set up career goals and work together in facilitating the employee to progress towards those goals.

20.17. PERFORMANCE EVALUATION

Some evaluation tools are widely applicable (e.g., essay, critical incident, graphic rating scale, MBO, ranking, paired comparison, and forced distribution), while others have fewer applications (performance test, field review, and forced choice). Firms usually evaluate employees at least once a year and some standardized form is used. If applicable, merit increases may be tied in with the level of performance. A constructive system involves employees in the evaluation process and encourages them to improve areas of weakness while rewarding areas of strength.

It is usually advisable to have a review process so that an employee who is negatively evaluated by a supervisor can appeal to a third party, for example, the next higher level of management.

20.18.   EMPLOYEE TRAINING AND DEVELOPMENT

Training and development is vital for maintaining a skilled labor force and to provide incentive for advancement. Training needs and objectives must be defined and translated into programs, including participant selection and program evaluation.

20.19.   COMPENSATION

Compensation must be structured so as to have a ladder within the firm for its job classifications, as well as for steps within any one job. In addition, the market situation must be considered; typically, firms want to be within a certain range of the highs and lows for similar job in the local job market. Obviously, the supply of persons for a particular job and the demand for those skills will greatly influence the salary needed to hire talent.

The major tool used to help employers set the pay level is the pay survey. The most frequently used tool in structuring pay within the firm is the job evaluation. After jobs are evaluated, pay policy lines, rate ranges, and pay classes are developed. These then become the basis for individual pay determination and establishing the actual pay for each individual.

20.20.   EMPLOYEE BENEFITS

Benefit programs include such things as pension plans, educational benefits, disability programs, life and medical insurance, among others. Some benefit programs are mandated by law, such as: unemployment compensation, social security, etc. Others such as vacation program are set by the firm. It is not unusual for benefit programs to be equal to 15 to 20% of the worker's gross pay.

REFERENCES

D. Yoder and H. Heneman, Jr. *ASPA Handbook for PAIR* (Washington, DC.: Bureau of National Affairs, 1979).

J. Kerr and J. Rosow, eds., *Work in America* (New York: Van Nostrand Reinhold, 1979).

James Walker, *Human Resource Planning* (New York: McGraw-Hill, 1980).

Barbara Schlei and Paul Grossman, *Employment Discrimination Laws* (Washington, DC: Bureau of National Affairs, 1976).

William F. Glueck, *Personnel* (Plano, TX: Business Publication, Inc., 1982).

American Society of Personnel Administration. *The Personnel Executive's Job* (Englewood Cliffs, NJ: Prentice-Hall, 1977).

Gary Latham and Kenneth Wexley, *Increasing Productivity Through Performance Appraisal* (Reading, MA.: Addison Wesley, 1981).

Edward A. Tomeski, Harold Lazarus, and Kurt Sadek, *People-Oriented Computer Systems* (Melbourne, FL: Krieger, 1983).

Thomas Kochan, *Collective Bargaining and Industrial Relations* (Homewood, IL: Richard D. Irwin, 1980).

21

THE SELECTION INTERVIEW

Ronald B. Morgan

RONALD B. MORGAN is an industrial/organizational psychologist specializing in employee selection and placement, employee development, and organization development. He holds a B.S. from Michigan State University, and a Ph.D. from Ohio State University. He taught personnel interviewing and industrial psychology at Ohio State and served as a management consultant to both private and public sector clients. Much of his consulting work has involved development of employee selection systems and the training of personnel interviewers. He is currently in the corporate organization development department at Burroughs Corporation.

21.1. INTRODUCTION

The interview is the most widely used employee selection technique. According to most estimates, American industry spends over one billion dollars a year interviewing potential employees. Organizations have made this commitment to the interview for a variety of reasons. First, both applicant and company use it to collect information needed to "size up" each other. Second, both use the interview to "sell" each other. Few applicants or few companies would make their choices "sight unseen."

It is clear that the interview does provide the "setting" in which both applicant and company can meet, get to know one another, and make their respective choices. The important question, and the one which forms the basis for this chapter, is how well does the interview really serve the purpose of evaluating and selecting employees? Although few companies ask that question of their interview process, the fact is that generally the interview does not serve as an effective way of choosing new employees. Studies conducted over the past 25 years have found that the interview *frequently* lacks those qualities we expect of a good employee selection procedure.[1] However, there is hope! While the "typical" interview may not serve its purpose well, through training, planning, and practice, the interview can be improved and can become a useful tool in the selection of employees.[2]

The goals of this and the following chapter are: First, to give the reader a better feel for both the strengths and limitations of the selection interview. What makes for an effective interview? What kinds of things inherent in the players and in the process of interviewing limit its effectiveness? What are the legal issues involved in interviewing? Through a greater awareness of the interview *process,* the reader will be more able to capitalize on its strengths, and minimize some of the risks inherent in its limitations. The second goal is to then familiarize the reader with techniques used by *effective* interviewers. Many specific recommendations will be made regarding how you should prepare for and conduct the interview. An example of an interview format as well as a detailed list of suggestions for conducting the interview will also be presented. As an organizing scheme, this Chapter deals with concepts which will help the reader understand the interview process and become a more effective interviewer, while the next Chapter focuses more on specific techniques for conducting the interview.

21.2. STANDARDS FOR THE SELECTION INTERVIEW. WHEN IS THE INTERVIEW EFFECTIVE?

All methods of the hiring process can, and should be, evaluated in terms of their effectiveness. There are four standards by which hiring procedures are evaluated. Together, these qualities serve as the definition of an effective tool for selecting new employees. The four standards are consistency, validity, fairness and utility. The discussion of each of the four standards is organized in the following way. First, the standard is defined. Then it is described in terms of how well the "typical" interview meets the standard. Finally, techniques for improving how well the interview meets the standard are introduced. These techniques are then elaborated on and illustrated in the next Chapter.

21.2.1. The Effective Interview Is Consistent

All methods used to screen and select employees (for example, tests and interviews) are in a sense measurement instruments. As such they must be consistent in the measurements they provide. Consider the analogy of a ruler. A ruler is a measuring device and we expect it to be consistent in the measurements it provides. We expect two different rulers to agree in the measurement of a given object. Similarly, we expect the same ruler to give consistent measurements of different objects having identical dimensions. Finally, we expect the ruler to give an identical reading of the same object measured at two different times. If we used an elastic ruler to measure an object, repeated measurements would produce different results. It is clear that we could not depend on a ruler that lacked consistency.

Similarly, we can't depend on an "inconsistent" interview process. If two interviewers disagree about the information they have collected or in their evaluation, clearly one must be wrong. Or if the same interviewer were to reach different conclusions based on his or her mood at the time of interviewing, or on the questions he or she happened to think of while interviewing the second time, the process could not be considered consistent.

How Consistent Is the Interview? Studies tend to show that the *typical* interview lacks consistency. That is, two interviewers often fail to agree on the strengths and weaknesses of a particular applicant. Further, the same interviewer may apply different standards to different applicants.

Finally, although the "typical" interviewer lacks consistency, some *are* highly consistent. The next section discusses the way in which interview consistency can be improved.

Improving Consistency of Interviewer Judgments. It has been shown that the consistency of the interviewer can be improved through use of a "structured interview." A structured interview is one that sets a predetermined pattern and follows that pattern for all interviews. The same information is asked of all applicants. (An example of a structured interview appears in Chapter 22).

One study, for example, had professional interviewers conduct interviews under one of three conditions. One group conducted a structured interview. A second group conducted a semistructured interview. A third group conducted an unstructured interview. The results of the study indicate that the degree of inter-interviewer consistency was directly related to the amount of structure. Those who conducted structured interviews produced more reliable judgments (e.g., there was greater inter-interviewer agreement), while those who conducted unstructured interviews showed little consistency across interviewers.[3]

By using a structured interview guide, the interviewer is more likely to be consistent in the way he or she evaluates different applicants. In addition multiple interviewers in the organization are more likely to be using a similar frame of reference.

21.2.2.   Validity: Do Interview Results Reflect True Job Potential?

The second standard or requirement of a measurement is that it be valid. To be valid a selection device must have two characteristics. First, it must actually measure what it is intended to measure. For example, a test of mechanical aptitude must be able to separate those of high mechanical aptitude from those of low mechanical aptitude. It must accurately categorize people. Second, a valid selection tool must measure characteristics which are important for successful performance of the job. We expect the interview to meet *both* of these conditions. If the interviewer evaluates the applicant's "supervisory knowledge and skill," there should be a high relationship between the interviewer's evaluations and the *true* ability of the applicant. Also the characteristic (e.g., supervisory knowledge) must be important for successful performance of the job. To restate this point, for any selection tool which is used to make predictions about a person's knowledge, skills, and abilities, we expect those predictions to be accurate and to be relevant to the actual job-performance potential of the person.

How Valid Is the Typical Interview?   Many interviewers express an almost absolute faith in the quality of their judgments. The bad news is that the research does not support their optimism! Studies show that judgments made on the basis of the *typical* interview are often invalid. In fact, the prevailing attitude among those who research the interview is that most of the characteristics measured in the interview are more accurately measured by other "tests." However, like consistency, the accuracy of individual interviewers differs. Some make highly accurate predictions of job performance, while others do not.

An additional fact of interest is that some characteristics seem more effectively measured by the interview than others. For example, verbal communication skills, sociability, and motivation seem among those most accurately assessed in an interview.

Improving Interview Validity.   Although the general conclusion is that the validity of judgments made in the interview are typically low, they can be improved. This can be accomplished by two practices:

1. Use of a structured interview focused on collecting information which is *relevant to actual job performance*
2. Interviewer skill in building rapport with the applicant

The first practice requires careful consideration of exactly what kinds of things should be asked, and resistance of the temptation to be misled by characteristics irrelevant to the true capability of the applicant. The second requires that the interviewer relax the applicant, build trust, and get the applicant to open up. Only by doing this will the interviewer get the applicant to reveal an accurate picture of him or herself.

21.2.3. Fairness

Fairness in employment has become a national policy. Therefore, the third expectation that we have of the interview is that it be "fair." That is, it must not systematically overestimate or underestimate any groups ability to perform the job. For example, if the interview consistently determines that women can't perform the job, but in fact they can, the interview is unfair.

How Fair Is the Interview?   Clearly, it is difficult to generalize on the fairness of the interview. However, because of the *subjectivity* of the typical interview process, it is highly susceptible to the personal biases of the interviewer. If one wishes to discriminate against a group, the interview is one of the easier ways to do so! This is because so much is typically left up to the interviewer, (for example, what questions to ask, which to probe into more deeply, as well as the evaluation and often the eventual decision). This subjective basis makes the interview a ready vehicle for unfair discrimination. Because of the importance of this topic to the interview, these issues are discussed in more depth in the section titled "EEO and the Interview."

Improving Interview Fairness.   Fairness in the interview can be achieved by:

1. Making sure that the interview is focused on questions which are *relevant* to the job, and that irrelevant questions related to the person's race or sex are eliminated.
2. Through personal recognition on the part of the interviewer of the biases which influence his or her perceptions of other people.

The first is accomplished through the use of a structured interview guide focused on collecting information clearly related to performance of the job. The second is accomplished only through the willingness and motivation of the interviewer to face up to his or her own biases and eliminate them as a basis for making decisions about applicants.

21.2.4. Utility: Does the Interview Justify Its Costs

The fourth standard by which we judge the interview is the degree to which it adds useful information to the selection decision. Is it worth its cost? Research has shown that, in many cases, the interview fails to add any improvement in prediction of subsequent job performance over that provided by well-developed tests. In particular, most skills can be assessed more accurately by test than by interview. For example, you would not ask the secretarial applicant if he or she could type. Instead, you would give the applicant a typing test.

However, the usefulness of the interview is based to some degree on factors other than information used to make the selection decision (e.g., its public relations value). For this reason it *will* persist as the most widely used selection technique. The challenge is to make it an effective one!

Improving Interview Usefulness.   The utility of the interview can be improved by focusing interviewer effort on getting *accurate* information. Increasing the accuracy of the information obtained in the interview improves the quality of the applicant(s) chosen. This, in turn, makes the investment of time and energy more productive (e.g., more useful). Two practices improve the utility of the interview:

1. Ensuring that the interview questions focus on job-related information. Otherwise, the information may decrease the accuracy of the selection choice; and

2. Conducting the interview in a relaxed, friendly manner. Doing so builds good rapport with the applicant and this encourages candor. An open, candid applicant is more likely to make an accurate (and useful) choice.

21.2.5.   Summary

To summarize, we ask four qualities of the interview and interviewer. First, we ask that decisions and judgments made are consistent over both different interviewers and over time. Second, we ask that those decisions and judgments (e.g., predictions) be valid, or in a sense, correct and accurate. Third, we require that the results of the interview are not unfair to any group. That is, we must assure that the interview is not systematically overestimating or underestimating the performance potential of individuals, based on their race, sex, national origin, religion, or age. Finally, the interview should be cost effective. It should add important information to the selection process, and the cost of that information should be in line with its usefulness.

21.3. FACTORS WHICH REDUCE INTERVIEW EFFECTIVENESS

Before we elaborate on some of the "how to's" of interviewing, it is important that we deal with some of the reasons why interviewers often make poor judgments. In particular, the following discussion focuses on some of the factors which distort our judgments of others and consequently reduce the consistency, validity, fairness, and usefulness of the interview.

21.3.1. The Challenge of the Task

The first limitation to effective interviewing is the difficulty of the task. The task facing the interviewer is extremely complex. He or she must maintain a smooth conversation while attending to relevant and important information. Then, the interviewer must use the information to make evaluations and important decisions. "Does the applicant have sufficient knowledge to perform the job? . . . Does he or she have sufficient skill? . . . Will he or she be highly motivated?"

The objectives of the interviewer are made more difficult because both the interviewer and applicant often hold incompatible goals. That is, they are simultaneously trying to attract and to evaluate one another. Each is putting his or her best foot forward while trying to collect as much information as possible about the other. As the interviewer is trying to learn what the applicant is "really like," the applicant is trying to appear as "ideal" as possible.

The result of this dilemma is that neither party has an easy time getting accurate information. The applicant doesn't find out enough about what it is really like to work at the organization, and the organization doesn't get as true a picture of the applicant's strengths. The role of the interviewer is to get the information needed to make an accurate and fair choice among applicants. An important step to doing so is to conduct the interview in a manner which reduces the defensiveness and apprehension of the applicant. By building and maintaining a sense of "rapport" with the applicant, the interviewer is more likely to get at the true potential of the applicant. The section entitled "Building Rapport with the Applicant," which appears in the next chapter, identifies specific things that you can do to help get at the "truth" behind the impression the applicant is trying to build.

21.3.2. The Interviewer—Applicant Mix

The interview is an interpersonal process. Because of this, the course of the interview and the effectiveness of the judgments made from it are very much dependent on the "mix" between the interviewer and applicant. Each person affects, and is affected by, the other.[4]

The appearance, mannerisms, and behaviors of the interviewer affect the applicant's performance. For example, an inattentive interviewer can demoralize an otherwise highly motivated applicant. Similarly, an overly talkative interviewer may diminish the motivation of the applicant. Finally, the formality of the impression presented by the interviewer affects the applicant's degree of comfort. The important point is that who you are and how you conduct yourself influences the performance of the applicant.

Given that your appearance, mood, and how you conduct yourself (e.g., eye contact and smiles) affect the applicant, it is very important for you to ask yourself whether you behave *similarly* with all applicants? To the extent that you don't, you may be inducing errors in your judgments of people. Be sensitive to this fact and ensure that your own presentation is nonthreatening, friendly, and most of all, *consistent* with all candidates. The use of a structured interview format for all applicants will aid in this process.

21.3.3. Errors in the Evaluation of Applicants

The most basic factor which acts to limit the effectiveness of the interview stems from the fact that it's difficult for people to make objective evaluations of one another. A number of biases influence how each of us evaluates others. The following discussion focuses on some of these. Through awareness of these biases the interviewer can keep track of his or her own impressions and thereby minimize the effects of these interpersonal biases on the judgments he or she makes.

First Impressions: The Four-Minute Interviewer.   Generally when we meet someone for the first time, our initial encounter plays an important role in our overall opinion of the person. Psychologists have a simple experiment which demonstrates this.

Individuals are asked to read a two paragraph description of a fictional person named "Jim." One paragraph describes Jim in ways which would characterize an introverted person. The second describes him in ways which would characterize an extroverted person. Although all participants in the experiment read the same information, the order in which it is presented is varied. Half of the participants read the description of the "introvert" first, while the other half reads the description of the "extrovert" first. The participants are then asked to rate Jim on various characteristics, including "introversion-extroversion." Usually participants rate Jim according to the first information they read of him. This phenomenon is termed "first impression," and casual observation will confirm that it operates in our everyday interactions with people. In situations of short duration, we simply have a natural tendency to latch on to the first information that we get about a person and give it undue importance in our overall impression of the person.

The phenomenon of first impressions has been demonstrated in the employment interview. In a classic study done years ago, a team of researchers found that the impressions formed early in the interview tend to play a prominent role in the decision made by the interviewer. In fact these studies found that the typical interviewer had often made up his or her mind after only four minutes of interviewing![5] Although there may be some uncertainty as to how applicable this finding is, it does alert the interviewer to the natural tendency he or she may have to make abrupt judgments about people.

A final interesting note about first impressions is that they do tend to diminish over time. That is, if you continue to interact with a person, your initial impression becomes less stable. However, because of the short duration of the interview, first impressions can mislead the interviewer before he or she has had the opportunity to collect an accurate impression of the person. Consequently, the best advice to the interviewer is to resist making that *initial* judgment of the person. Instead use the *full* interview period to collect information and make judgments *only* after the interview is completed. When judgments are made, use information from the entire interview to challenge and substantiate the decisions that you have made.

Stereotypes.   We all develop and carry around ideas and beliefs about what characteristics go together in people. We believe that people who have certain attributes (such as being part of a racial, ethnic, religious, or age group) have some characteristic pattern of attitudes, beliefs, habits, values, or capabilities. In short, we hold a stereotype of the group. Stereotypes often influence our opinions and evaluations of those that we meet. Because we lack information about what they are really like, our stereotypes seem to "fill in" the details. For example, we know that all red-heads are hot-tempered and all "engineers" are introverted!

Because interviewers are human, they hold stereotypes. These may mislead the interviewer into making determinations about a person, without sufficient information. Because the stereotype fills in the gaps of missing information, no in-depth probing for information is needed. Clearly, the dangerous consequence of this is an inaccurate evaluation of the applicant by the interviewer.

Similarity Bias.   There is some evidence to suggest that we evaluate those who are similar to us more favorably than we do those who are dissimilar. For example, studies from social psychology suggest that we like those who are similar to us in attitudes, values, and characteristics more than we like those who are dissimilar. One study found that students serving as interviewers rated similar applicants as more competent and worthy of a higher salary than dissimilar applicants. However, they did not recommend hiring the similar applicant more often. See Note 3. Be alert to the fact that you may quite naturally favor those who are similar to you in some way.

Rating Errors.   One of the most pervasive errors in applicant evaluation is termed "halo." It is a phenomenon which occurs when we are asked to simultaneously evaluate a person on a number of different characteristics. What happens is that, because the applicant is outstanding on one of the relevant job characteristics, we develop an overall favorable attitude about him or her. This overall impression diminishes the chance that we will spot the person's weaknesses as well as his or her strengths. For example, you are to rate applicant "Joan" on the following characteristics: judgment, knowledge of engineering nomenclature, and interpersonal skill. However, because of her pleasant disposition and relaxed manner (for example, her effective interpersonal skills), you fail to see the weaknesses in her judgment. The overall positive impression you have of Joan misleads you. We see evidence of halo when we look at the pattern of ratings across judgments of a person's characteristics and find that all of the ratings are high (or all are low). The truth is that very few people excel in all areas. Most applicants have both strengths and weaknesses. Therefore when you do evaluate a person, force yourself to consider each characteristic separately. If all of your ratings of the applicant are either high or low, look again at your assessment of the person.

21.3.4. Human Memory: A Good Reason to Take Notes

An additional source of bias in the interview is the result of imperfections in memory. Psychologists have demonstrated systematic distortions in memory. We occasionally remember what we wish, rather than what we experienced. More important, the errors we make are often associated with personal biases we hold.

Because so much information and so many impressions are collected during the interview, memory distortions tend to limit the accuracy of judgments made by the interviewer. The solution is simply to rely less on memory and more on *interview notes.*

An experiment, described by Carlson, documents the need for note-taking in the interview (see Note 3). Forty managers viewed a 20-minute tape of an interview. Before doing so they were informed that they would be asked to make judgments about the applicant at the end of the film. At the end of the film, they were given a 20-question test on factual information revealed in the tape. The results were quite discouraging. The 40 managers averaged 10 errors! Some in fact missed up to 15 of the 20 questions.

Effective note taking is among the most difficult interviewing tasks. The challenge is to take notes without allowing the task to infringe on your interaction with the applicant. Let the applicant know in advance that you are going to be taking notes and explain that the reason for doing so is to assure that all of the facts are recorded accurately. Then minimize the degree to which you let your note taking interfere with eye contact and smooth conversation flow. A structured interview guide aids in this regard.

21.3.5. Situational Influences on the Interviewers Judgments

A number of factors in the situation affect how favorably the interviewer evaluates the applicant. These influences are independent of the true qualities of the applicant. Consequently, they lead the interviewer to inaccurate judgments.

Time Pressures.   Carlson (see Note 3) describes research which suggests that interviewers make more favorable judgments of applicants when they are under time pressure, or trying to meet some quotas. The need to have the hiring process complete by "this Friday" may result in different evaluations than if the process occurs over a longer period of time. Although the best advice is to avoid hiring under time pressures, it may not always be under the reader's control. To minimize distortion under these conditions, the interviewer must force himself or herself to stick closely to the applicant standards that have been identified as important.

Applicant Contrasts.   A second situational factor which affects evaluation of an applicant is the quality of the preceding applicants. We tend to be misled by contrasts in applicants. For example, one study asked interviewers to evaluate a series of applicants. The applicants had been prerated as "good," "average," or "poor." The prerating was unknown to the interviewers. The results revealed that the evaluation assigned to the "average" applicant was *higher* when he or she followed a "poor" applicant. Further, it was *lower* when he or she followed a "good" applicant. This has been termed the "contrast effect," and it seems most severe when the interviewer does not have a clear standard of what constitutes a good or a poor applicant. Lacking this, he or she tends to rely upon contrasts in applicants.

21.3.6. Summary

Three general factors which contribute to interviewer error in evaluations
have been discussed. First, the complexity of the interview process often exceeds the interviewer's capability. To deal with this the interviewer should
plan and outline the interview in advance. This will help assure that he or
she maintains control and gets the needed information. Taking notes during
the interview also helps the interviewer overcome errors due to faulty memory. Second, the goals of the applicant are often to conceal or inflate, rather
than to open-up. This can be overcome by creating an atmosphere in which
the applicant is comfortable, open and trusting. Finally, the interview is an
interpersonal process. The evaluations which are made are, therefore, subject to errors originating from bias the interviewer holds. To deal with this,
the interviewer must acknowledge his or her biases, and confront them when
evaluating an applicant.

21.4. EEO AND THE INTERVIEW

21.4.1. Historical Perspective on the Interview

Since 1964 and the passage of the *Civil Rights Act,* public policy regarding
personnel practices such as hiring, transfer, training, and termination have
gone through many changes.

Government surveillance of personnel practices has had interesting effects on the interview. With passage of Title VII of the Civil Rights Act and
the subsequent Court decisions upholding its legitimacy, many employers,
fearful of suits, concluded that because their tests had not been shown to be
job-related, they should abandon testing. Consequently, the late 1960s and
early 1970s saw a sharp decrease in the use of written tests. As tests were
abandoned, employers became increasingly reliant on the interview. The
irony is that, as the law has developed and become more focused, *it is clear
that the interview itself is a "test" and is subject to the same requirements as
any other selection procedure.* The questions asked, and the judgments made
on the basis of them, must show clear relation to the characteristics needed
to perform the job.

21.4.2. The Interview in the Courts

A number of cases have either dealt directly with, or had implications for,
the use of the interview as a hiring tool.[6] Generally the interview has lost
court cases when it has been shown to disproportionately reject minorities or

women *and* when the basis for those rejections did not show a clear relationship to important characteristics needed to perform the job. In two cases, *Hester v. Southern Railway*[7] and *United States v. Hazelwood School District*,[8] the courts criticized the subjective nature of the interview. In *Hester* the district court ruled that the interview process was faulty because of its subjectivity. In particular, the process was criticized because it lacked "formal guidelines, standards and instructions." Although a higher court reversed the ruling on the basis that adverse impact was not clearly demonstrated, the subjectivity of the process remained in suspicion. In *United States v. Hazelwood School District,* the court noted the subjectivity of the interview. In their criticism of the interview process, the Court said:

> Principals are free to give whatever weight they desire to subjective factors in making their hiring decisions. Indeed, one principal testified that interviewing an applicant was "like dating a girl, some of them impress you, some of them don't." No evidence was presented which would indicate that any two principals apply the same criteria—objective and subjective—to evaluate applicants. (*United States v. Hazelwood School District,* 11, EPD 10854 (1976) p. 7576).

An interesting 1976 case (*Weiner v. County of Oakland*)[9] dealt specifically with the types of questions asked and their possible bias. The plaintiff had been denied employment and subsequently charged discrimination. The court reviewed the type of questions which had been asked of her and found that they were biased against women. Example questions included whether her husband approved of her working and whether her family would suffer if she were not home to prepare dinner. The court found grounds for awarding her both back pay and court costs.

Another case, *Harless v. Duck,* shows that a *well-developed interview can survive a court challenge.* A woman had brought a suit against the police department charging discrimination based on sex. The structured interview used by the department included 30 questions designed to measure the applicant's communication skills, ability to make decisions and solve problems, as well as her reaction to stress. The interview did disproportionately reject females. While 43% of female applicants failed, only 15% of male applicants did. However, in defense of their interview process, the department demonstrated that the characteristics measured by the interview were important to job performance, the questions asked related to the characteristics, and those who performed poorly in the interview also tended to perform poorly in the department's training program. Those who performed well in the interview were more likely to perform well in training. In short, the Department was able to show a link between performance in the interview and job performance.[10]

In *King v. New Hampshire Dept. of Natural Resources,* the Court found the interview indefensible because the questions asked were irrelevant to the job of state meter patrol officer. The questions included items such as could the applicant "swing a sledge hammer," and "could they run someone in?"[11]

21.4.3.   Summary

Because of its subjectivity, the trend for the future is likely to be continued surveillance of the interview. Therefore, interviewers must assure the job-relatedness of the interview process. This can be achieved by:

1. Use of a structured interview process similar to that described in the following chapter. The structured guide helps guarantee that all applicants are evaluated from the same perspective.

2. Maximizing the job-relatedness of the interview by relying only upon those topics and questions which have a logical relation to important characteristics needed to perform the job. Related to this point, avoid questions which are likely to have an unfair effect due to the race, sex, religion, age, or national origin of the applicant. The sections titled "Making the Interview Job-Related" and "Writing and Organizing Questions" (which appear in the next chapter), will aid the reader in the process of writing questions which meet the test of job-relatedness.

21.5.   THE MODEL INTERVIEWER

The following techniques characterize the effective interviewer:

Use of a Structured Interview Format.   A pattern is developed for a specific position or opening, and it is then used when interviewing all applicants for that position.

Job Relevant Questions.   Only questions which have a clear and logical relation to characteristics needed to perform the job should be included.

Supportive, Responsive Interviewer Behavior.   An effective interviewer creates an atmosphere which relaxes the applicant, gets him or her to talk, and open up. The interviewer accomplishes this through the manner in which he or she conducts the interview. The effective interviewer is an involved, active listener who is interested in the interviewee as a person and responsive during the interview.

Sharing Information.   An effective interviewer provides the applicant with an accurate, realistic picture of the organization and the job.

Follow-up and Feedback. An effective interviewer must keep track of how well the judgments made on the basis of the interview turn out. This information can then be used to improve judgments made in the future (i.e., you learn from your mistakes).

Practice and Training. Finally, the interviewing skills which are discussed in these chapters require development through training and practice. This chapter is the beginning of that training process. By no means can it stand alone. A list of training resources appears at the end of the next chapter.

Drawing on this background, the next chapter turns to those techniques and practices necessary to conduct an effective interview.

NOTES

1. R. D. Arvey and J. E. Campion, "The Employment Interview: A Summary and Review of Recent Research," *Personnel Psychology* 35 (1982): 281–322.

2. M. D. Hakel, "The Employment Interview," in K. M. Rowland, and G. R. Ferris (Eds.), *Personnel Management: New Perspectives* (Boston: Allyn and Bacon, 1983).

3. R. E. Carlson, Paul W. Thayer, Eugene E. Mayfield, and Donald A. Peterson, "Improvements in the Selection Interview," *Personnel Journal* (1971): 268–275.

4. N. Schmitt, "Social and Situational Determinants of Interview Effectiveness: Implications for the Employment Interview," *Personnel Psychology* 29 (1976): 79–101.

5. B. M. Springbett, "Factors Affecting the Final Decision in the Employment Interview," *Canadian Journal of Psychology,* 12 (1958): 13–22.

6. R. D. Arvey, "Unfair Discrimination in the Employment Interview: Legal and Psychological Aspects," *Psychological Bulletin* 86 (1979): 736–765.

7. Hester v. Southern Railway, 8, FEP 646 (1979).

8. United States v. Hazelwood School District, 11, EPD 10854 (1976).

9. Weiner v. County of Oakland, FEP 380 (1976).

10. Harless v. Duck, 14, FEP 1616 (1977).

11. King v. New Hampshire Dept. of Natural Resources, and Economic Development, 15, FEP 669 (1977).

22

PREPARING FOR
AND CONDUCTING
THE INTERVIEW

Ronald B. Morgan

22.1. INTRODUCTION

As an interviewer, your primary goal is to get the information you need to make an accurate decision about the applicant. From the discussion in Chapter 21, it should be clear that the way you plan and conduct the interview will determine how successful you are at getting this information. The following sections provide advice and examples on how to interview. These discussions are divided into four sections. The first section is focused on the actual flow of an interview. It highlights the seven key steps of an effective interview. It then takes the reader through each of these steps, providing a model which can be used to develop a structured interview guide. The second section focuses on background skills that you, as on interviewer, must use and develop to conduct an effective interview. Specifically, this section reviews methods that will help you build and maintain rapport with an applicant. The third section includes a detailed step by step description of procedures for developing and writing questions that get at relevant information. Finally, a list of training resources is included for those who wish to locate more in-depth interviewer training.

22.1.1. The Interview Guide

The interview guide is a previously developed outline for conducting the interview. It identifies the topics to be discussed and the order in which they will be reviewed. The many strengths of this approach have been described in Chapter 21. In general, the use of an interview guide forces the interviewer to develop and plan questions in advance of the interview. In addition it ensures that all applicants will experience the same interview. This is necessary to maintain the consistency and fairness of the interview process. The interview guide is particularly helpful when more than one interviewer is used to interview different applicants for the same position. Using an interview guide under these conditions increases the consistency obtained between the different interviewers. Furthermore the use of an interview guide maximizes the interviewer's control over the process by allowing him or her to establish clearly what the interview will cover, and it serves as an effective way of getting the interview back "on track" when it strays. Finally, the use of an interview guide also gives the applicant a sense of where the interview is headed.

22.2. CONDUCTING THE INTERVIEW

The following discussion describes the key steps for conducting an interview.*

22.2.1. Preliminaries

Job analysis.   Prior to interviewing any applicant, the first step is to assure that you have a detailed understanding of the job to be filled. You *must* have a clear sense of what duties and tasks will be performed in the job and what knowledge, skills, abilities, interests and other characteristics are important for performing those duties. Only then can you write job-related, EEO-safe questions. Rather than plunge into techniques for conducting a job analysis at this point, that topic is deferred to Section 22.4 ("Making the Interview Job-Related").

Resume review.   The first step in conducting the interview is to review the resume and/or application in advance. As you plan look for factual gaps

---

* The basic format for the interview guide, and many of the practical recommendations included, come from courses, industry workshops, and personal communications the author has had with Dr. Milton D. Hakel.

or ambiguities which need in-depth probing. At the same time you must be cautious and avoid developing firm "first impressions." Look primarily for areas in which you need more information.

The interview setting.   The most desirable place for the interview is a quiet office. Find a location in which both you and the applicant can relax and where interruptions will be minimized. If you are in your own office have someone take phone messages for you. Phone calls disrupt the flow of the interview, reduce your memory of important information, and are rude and distracting to the applicant.

22.2.2.   The Phases of the Interview

The actual interview consists of seven distinct phases:

1.   The opening and greeting
2.   The outlining and structuring of the interview
3.   Collecting information from the applicant
4.   Summarizing and clarifying information collected from the applicant
5.   Giving the applicant a thorough and realistic overview of the job and company
6.   Giving the applicant an opportunity to ask questions
7.   Closing the interview

This seven phase process should become the basic outline for the interview you conduct. A discussion of each of these phases follows.

Phase I: Opening the Interview.   The opening of the interview consists of the introduction or greeting. It's probably one of the most critical points in the interview. Social introductions are frequently awkward, particularly when there is a difference in status between the parties. Therefore, to get the interview off on a good start, you want to reduce this awkwardness and have this opening contact go as smoothly as possible. Remember, your goal is to create a warm and friendly atmosphere in which the applicant can relax and be candid. The opening is critical to this goal. The critical behaviors in an effective opening are:

1.   *A Smooth and Cordial Introduction.*   Pronounce the applicant's name correctly, tell the applicant what to call you, and ask him or her what he or she prefers to be called. The dialogue might go something like:

Interviewer: Good morning Mr. Jones, I'm Ron Morgan, please call me Ron. May I call you Pete? (Shake hands and offer him a seat.)

First names should be used in those instances where doing so increases the level of comfort between the applicant and interviewer. Avoid using first names if doing so is awkward or presumptuous (e.g., when there is a great difference in status between applicant and interviewer). Remember, it is your responsibility to put the applicant at ease regarding the use of names.

2. *Breaking the Ice.* After the initial introduction, an important step is to spend the initial few minutes establishing a relaxed atmosphere. At this point probably both you and the applicant are feeling a bit ill at ease. The best thing that you can do is make some relevant "small talk." To do so, pick some hobby or interest from the resume or application—preferably one that you have a shared interest in, and use it to stimulate some initial conversation. For example:

Interviewer: I noticed on your resume that you are a cross-country skier. So am I. Have you had a chance to get out lately?

The most important recommendation regarding the use of this device is to make sure that the topic is nonthreatening, and something that you can both comfortably discuss. The skill in doing this is to identify something that is simultaneously nonthreatening and relevant.

3. *Tell the Applicant About Yourself.* An important step toward building a comfortable atmosphere for conducting the interview is to spend some time informing the applicant about yourself. Information such as your position in the company, how long you have been with the company, and the kinds of things you do for the company are important at this stage of the interview. For example:

Interviewer: Let me tell you a bit about my background here at Company A. I'm the supervisor of the quality control group. I moved into this job about two years ago. Before that I was a quality inspector in our Detroit plant for six years. I also worked here part time while I was working on my engineering degree. So, I guess that I've got a pretty good idea of the kinds of things that we do here. I really enjoy working here at Company A and I especially like talking to young engineers about the kinds of things that we do here.

Remember, we're still in the opening phase of the interview and anything that you can do to reduce the apprehension of the applicant will be beneficial. In fact, studies have shown that even this little amount of self-disclosure on the part of the interviewer increases the willingness of the applicant to openly share his or her background. It makes the interviewer seem more open and less threatening.

Phase II: Outlining the Interview.   The applicant comes into the interview uncertain about how it will go—what will be discussed, the order of the topics, and how long it will last. Therefore, it is important that you give the applicant an idea of the topics to be covered and the order in which they will be discussed. Remember, you can do this because you've planned and structured the interview in advance! Specifically, you should:

First, describe the purpose of the interview. For example:

Interviewer: Joe, as you know, we've called you in because of your interest in our assistant engineer position. What I'd like to do is to discuss your qualifications and interests as well as the job requirements to see if they match.

Second, outline or highlight the major parts of the interview:

Interviewer: First, we'll talk about your education. Then, your work experience. Then other experiences you've had which will help prepare you for this job. After that, I'll tell you more about the job and the company in detail. Finally, I'll give you a chance to ask me anything that you'd like about the company or the job.

Third, tell the applicant that you will be taking notes.

Interviewer: As we go through the interview, I will be jotting down a few notes. They will help me make sure that I get all of the important facts correct.

(When you do take notes, remember to do so as unobtrusively as possible—don't allow your notetaking to interfere with the rapport you're building with the applicant.)

Outlining the course of the interview is beneficial for a number of reasons. First, it lets you assert *control* over the interview. It does so by establishing the preset pattern and order of topics. This makes it easier to guide the interview and keep it *on course.* Second, outlining the interview is a courtesy to the applicant. It gives him or her an idea of where the interview is going. Third, it builds the perception that the interview has been planned, that you take it seriously, and that it is important to you. It assures the applicant that you have a genuine interest in him or her. Finally, it gives the impression of professionalism. It's clear that you, the interviewer, know what you are doing. If you've ever suffered through a rambling interview, with no idea of what topics were to be covered by the interviewer, you can appreciate the improvement a clearly defined outline provides the applicant.

Phase III: Collecting Information from the Applicant.   The third and most involved phase is that of collecting information from the applicant. The following paragraphs discuss organization and phrasings for this portion of the interview.

Organizing questions.   It's best to organize questions by broad general categories (e.g., education, work experience, and volunteer activities). Within each category questions should move from general to specific.

Interviewer: I'd like to begin with your education. Tell me about your engineering program, the kinds of classes you took, . . .

Follow-up these general questions with *successively more specific* questions that are tailored toward getting the information *you* need.

Interviewer: Tell me more about the class that you had in applied physics.

After the applicant's response to the more specific questions, *probe* as necessary to gain all of the detail that you need. Probes are open-ended phrases which pick up on a specific point made by the applicant and encourage him or her to elaborate on the details. Some phrases which are useful when probing include:

In what way?
Tell me more . . .
How was that . . .

Interviewer: (In response to applicant's comment that the physics class was "boring") What was it about the class that was boring?

Repeat this three-stage process (general, specific, and followed by probing) for each category of questions.

Phrasing questions. How questions are phrased impacts both the EEO fairness of the interview, and the degree of rapport you are able to establish with the applicant.

1. The content of the question must have a clear relationship to performance of the job. (Section 4 describes step-by-step procedures for writing and documenting job-related questions.)

2. Avoid questions which delve into nonjob-related areas such as marital status, parenthood, and child care. When these same questions are not asked of men, they raise suspicion of discrimination.

3. Avoid questions which are likely to have a differential impact or adverse effect on certain groups. For example, arrest records are generally considered inappropriate because an arrest is not synonomous with guilt, and minorities have a higher probability of arrest. Use of this as a means of selecting employees would result in the disproportionate rejection of minorities. In the instance where the job has a "security" requirement, it would be preferable to ask for information regarding convictions rather than arrest. Other issues which should be dealt with in a similar way include spouse's salary, children, childcare arrangements, dependents, information regarding citizenship, marital status, military discharge information, information related to pregnancy, information related to whether the applicant rents or owns his or her dwelling, and inquiries that are too general such as, "Do you have any handicaps or health problems?" When questions such as these cannot be shown to have any relation to performance of the job, their use is inappropriate. This kind of background information is often needed *later* in the employment process (for enrollment in benefit programs, notifications in case of emergencies, and so on). When this is the case, collect it *after* the applicant has been hired.

4. Questions should be clearly phrased in language easily understood by the applicant. Remember, effective interviewing is based on good communication skills. This includes using words that both parties understand.

5. Avoid leading questions (those that indicate a particular answer). For example, "I guess that you don't like working with numbers," leads the applicant toward a particular answer. A more effective question would be, "Tell me how you feel about working with numbers."

6. Questions should be open-ended. Yes—no or limited-response questions fail to get and keep the applicant talking. Remember, your goal is to keep the applicant talking!

7. Avoid "emotionally toned" words in your questions. Instead use neutral words less likely to generate defensiveness in the applicant. For example, rather than noting that the applicant was "fired" from a particular job, use the phrase, "when you left that job." Similarly, rather than "gave up" or "quit," use phrases like "decided that it wasn't for you," or "decided to try something new." With this approach, you are more likely to get information which will help you determine what about the applicant led to the events you are discussing.

8. Avoid "double barreled questions" (e.g., questions that are really two in one). An ambiguous, confusing, or misleading question can lead the applicant astray.

Phase IV: Summarize and Clarify the Information Collected. It is helpful both to you and the applicant that you summarize the key points of information the applicant has provided. It is useful to do this after each segment of questions (for example, after education, after work experience). The summary demonstrates to the applicant that you have been listening and have heard accurately. This helps to improve the quality of openness during the interview and also allows the applicant to clarify or add to any information you may have omitted. For each area then, the flow will be:

1. General opening question
2. Specific questions related to the topic
3. Probing of answers as needed for more detail
4. Summarizing the major facts gathered, repeating them for the applicant, and make any corrections or clarifications needed.

Phase V: Realistic Job Preview. During this phase you describe the job and organization as they really are. A series of studies have shown that presenting a realistic view of the job helps reduce later dissatisfaction with the job, and leads to reduced turnover.[1] Anything which supports a realistic presentation (e.g., plant tours, films of actual job performance), helps prepare the applicant for making an appropriate choice about the job.

Phase VI: Applicant Questions. During the sixth phase you should give the applicant opportunity to ask questions. Remember, the applicant has to get information about the company and the job. Encourage the applicant to ask any questions needed to clarify anything on his or her mind about the job or company.

Interviewer: Now that we've talked about your background and about the job and company, let's turn the tables around and let you ask me any questions that have occurred to you.

Closing the Interview. Perhaps the most difficult phase of the interview is its closing. It is very important that you use the closing to:

1. Outline the next steps for the candidate
2. Tell the applicant when to expect a decision
3. Tell the applicant how he or she will be contacted
4. Thank the applicant for spending time with you

For example:

Interviewer: We will be talking to a few more candidates over the next few days. We expect to make a decision by next week. When we do, we will call or write you to let you know of our decision. Until then, I do want to thank you for coming by, I've really enjoyed talking with you.

(As you are finishing these comments, *stand up* and offer your handshake to the applicant. This serves as a cue to the applicant that the interview is over.

22.2.3. Evaluating the Applicant

Evaluate the candidate along those characteristics which were determined to be important for performance of the job. When making your evaluation be alert to contrast errors, halo, similarity biases, and the stereotypes you hold.

Document the feelings you have with *clear* facts or behaviors. Doing so will help you sort your biases from the person's true capabilities.

Challenge your first impressions and confront your biases. If you can't document your unfavorable evaluation of the candidate with clear facts which illustrate deficiencies in important knowledges, skills, or abilities, you may be reacting from a bias.

Remember, the most attractive applicant may not be the most qualified applicant. (Unless, of course, the job is that of model! Similarly, the applicant who reminds you of your "wayward" nephew may have little in common with him.)

22.3. SKILLS FOR BUILDING RAPPORT WITH THE APPLICANT

Remember, your goal as an interviewer is to get and keep the applicant talking. To do so you must create an atmosphere in which the applicant feels comfortable, relaxed, and open. A protective, defensive applicant will be working against your goal of obtaining accurate information. The way in which you present and conduct yourself determines, to a large degree, how self-disclosing the applicant will be. Therefore, you must be nonthreatening, genuinely interested in the applicant, protective of his or her self-esteem, responsive, and active during the interview.

22.3.1. Interviewer Behaviors

The following are examples of behaviors which affect this openness. For the most part they are behaviors which are generally recognized as effective interpersonal skills. That is, they are effective at making others feel comfortable with you.

1. Head nodding is a good listening technique. It indicates that you are following what the applicant is saying and are interested.
2. Smiling or laughing, when appropriate, indicates that you enjoy being with the applicant and are listening with interest.
3. Direct eye contact shows interest. Failure to make eye contact creates mistrust.
4. A relaxed manner, such as leaning back in your chair, reduces the level of tension present in the interview.
5. Silence can indicate a desire for the candidate to continue. You probably have noticed that when two people are talking, silence of any duration generally produces tension. Both parties feel an obligation to fill the void with conversation. You can use this strategically to keep an applicant talking. After he or she finishes a response, a pause on your part will prompt him or her to continue to elaborate on the topic. However, the use of this technique must be limited. In excess it can be both threatening and rude.

22.3.2. Behaviors to Avoid

The following are behaviors which have a negative consequence on the applicant. That is, they reduce the rapport that you are trying to build.

1. *Glancing at your watch or around the room* signals disinterest and boredom. This can be very demoralizing to the applicant. It breaks

the impression that you are genuinely interested in the conversation.

2. *Shuffling through papers* indicates a lack of interest. This may also be demotivating to the applicant.

3. *Constantly referring to a list of questions* or to the application form indicates a lack of confidence in your skill as an interviewer and that you are unprepared. Know the interview guide or outline you are using in detail. In this way occasional glances will be sufficient to keep the interview on course.

4. *Concentrating more on taking notes* than on watching the applicant shows that you are more interested in filling out a form than in getting to know the applicant.

5. *Showing expressions of surprise, shock, or annoyance* indicates that you are judging him or her. This makes the applicant regret his or her honesty and this can destroy rapport.

22.3.3.   Verbal Responsiveness

What you say and how you say it are also important to building rapport and keeping the applicant talking.

1. *Encouraging Comments.*   "Uh-huh, good, I see, I understand"—indicate that you are listening, interested, and agree with what the applicant says (as long as they are inserted at appropriate times and seem a reflection of genuine interest. An absentminded comment is far worse than none at all).

2. *Supportive Remarks and Praise.*   "It's wonderful that you were able to accomplish so much in such a short time"—indicate that you recognize the applicant's achievements. They encourage him or her to say more.

3. *Playing Down Negative Information.*   "I certainly understand how that could have happened under those circumstances"—reassures the candidate that you aren't judging him or her unfairly and will increase the possibility of revealing additional information which he or she might otherwise have tried to hide.

4. *Restatement of the Applicant's Thought.*   For example, "You were interested in that job because of the advancement potential"—indicates that you follow what has been said, and may encourage further elaboration. Use this technique cautiously. Otherwise you risk "parroting" or putting words in the applicant's mouth.

5. *Avoid Interrupting or Changing the Subject Abruptly.*   Doing so indicates a lack of interest in what has been said as well as a lack of

courtesy. This can be particularly damaging if you stop the applicant before his or her point has been made. For example:

Applicant: There was only one thing that made me stick with that job so long . . .

Interviewer: Yes, first jobs are frequently that way. Now, tell me more about your organizing the company softball team.

This type of switch makes the applicant feel a little foolish and at the same time resentful toward you for your lack of interest in his or her story.

22.4.   MAKING THE INTERVIEW JOB-RELATED

To be effective the interview must be job-related. That is, the questions *must* focus on the collection of information which is necessary and important to the performance of the job. The following three-step "job-analysis" process is an effective way to assure that the questions included in the interview are job-related.

22.4.1.   Step 1: Identify What is Done on the Job

To be effective the interviewer must have a clear idea of *exactly* what is done on the job. Otherwise, how can he or she make judgments about the applicant's capability of performing the job? If the position to be filled is one which already exists in the organization, collection of this information can be accomplished in a number of ways.

1. *Job Descriptions.*   Most organizations keep descriptions of the duties and tasks performed within existing jobs. These are often good sources for understanding the current position and how it fits into the organization.
2. *Job experts.*   In the event that the organization lacks thorough and recent job descriptions, interviews and observations of those currently holding the position can be useful. Employees are usually very willing to talk about their jobs. By asking "what they do," probing for descriptions of "how" they do these activities, and by observing the performance of important job tasks, the interviewer can acquire a fairly good sense of what the job involves.
3. *Reference materials.*   Under some conditions none of the sources cited above may be available for enriching the interviewer's understanding of the job. However, a number of reference materials can

provide information about how similar jobs in other organizations are structured. One of the most frequently used is the *Dictionary of Occupational Titles*.[2] This document consists of compilations of job descriptions and ability patterns for thousands of jobs existing in the United States. A less detailed alternative is the *Occupational Outlook Handbook*.[3] It describes jobs in more general terms than does the D.O.T. Both represent a good *starting* point for the interviewer who is relatively unfamiliar with a particular job.

The result of Step 1 should be a job description which summarizes the important activities required for successful job performance.

22.4.2. Step 2: Identify the Knowledge, Skills, and Abilities Needed to Perform the Job

After a clear understanding of the activities of the job has been acquired, the next step is to identify the characteristics needed to perform those activities. This requires identification of the knowledge, skills, abilities, interests, and personal characteristics the candidate must bring to the job. For example, analysis of a secretarial job might reveal that the following are needed:

Skill in operation of word processing equipment
Knowledge of business writing
Knowledge of grammar
Knowledge of spelling
Skill in file organization and maintenance
Skill in operation of standard office calculator
Knowledge of office accounting techniques

After the characteristics needed for successful performance of the job have been identified, a distinction must be made between those the applicant must have *prior* to employment and those which can be acquired in a short period of time. Never reject applicants on the basis of knowledges, skills, or personal characteristics which can be acquired during a short period of time on the job.

Finally, prioritize the knowledges, skills, and abilities on the basis of their importance. To do so, consider the consequences of hiring someone with insufficient development in each area. Finally, document the importance of each personal characteristic by identifying the tasks and duties which are required by it.

More detailed discussion of job-analysis techniques can be found in McCormick 4 and in the *Handbook for Analyzing Jobs* 5.

22.4.3.   Step 3: Writing Questions

Once you have a clear idea of what is done on the job, and of the knowledges, skills, and abilities needed to perform those activities, you are then prepared to write interview questions. Follow the suggestions described in Section 22.2.3 for writing and organizing questions.

22.5.   INTERVIEWER TRAINING

Interviewing is a skill, and like other skills the only way to effectively develop and improve is through guided practice. These chapters have presented the reader with an understanding of the interview process and specific suggestions for effectively dealing with the challenges presented by the interview. To further develop your skills, seek out a training program. An effective program will allow you to practice the skills which have been described here, observe (through taping) your own interviewing performance, and receive feedback and coaching from one who is knowledgeable in effective interviewing.

2.5.1.   Interviewer Training Programs and Resources

The following individuals and organizations provide either tailored in-house training programs or run public courses and seminars.

Drake Beam Morin, Inc.
277 Park Avenue
New York, NY 10172
Phone: 212 888-3134

American Management Assn.
135 West 50th Street
New York, NY 10020
Phone: 212 586-8100

Dr. Ronald B. Morgan
Burroughs Corporation
1 Burroughs Place
Detroit, MI 48232
Phone: 313 972-0118

Development Dimensions
International
1225 Washington Pike
P.O. Box 13379
Pittsburgh, PA 15243
Phone: 412 257-0600

University of Michigan
Graduate School of Business
Ann Arbor, MI
Phone: 313 763-1006

Also, many colleges and universities offer interviewer training through their continuing education and management education programs.

NOTES

1. J. P. Wanous, "Realistic Job Previews for Organizational Recruitment," *Personnel* **52** (April 1975): 50–60.

2. *Dictionary of Occupational Titles,* United States Department of Labor (Washington, D.C.: United States Government Printing Office).

3. *Occupational Outlook Handbook,* United States Department of Labor (Washington, D.C.: United States Government Printing Office).

4. E. J. McCormick, *Job Analysis: Methods and Applications* (New York: AMACOM, 1979).

5. *Handbook for Analyzing Jobs.* United States Department of Labor (Washington, D.C.: United States Government Printing Office).

23

THE ROLE OF TRAINING

Michael H. Frisch

MICHAEL H. FRISCH specializes in assisting organizations with the management of their most valuable resource: people. Dr. Frisch's expertise covers many human resource topics including staff selection and succession, performance appraisal, management development, and employee assistance. He is currently working on a wide range of organizational projects through his consulting firm, NPO Management Services, Inc., in New York City. Previously, Dr. Frisch was manager of management development for the Pepsi-Cola Company where he was responsible for human resource planning, productivity and career-development programs, and general consultation on managerial and interpersonal issues. Dr. Frisch has also served other major corporations, financial institutions, multinational companies, and nonprofit organizations.

In prior positions Dr. Frisch designed and conducted workshops on many management skills. His recent publications, "Coaching and Counseling Handbook for Managers and Supervisors" and "Hire the Best! A Practical Guide to Effective Interviewing" reflect these experiences. He has written widely in professional journals and made presentations to professional groups.

Dr. Frisch received his Ph.D. in Industrial/Organizational Psychology from Rice University, his M.S. from the Georgia Institute of Technology, and his B.A. from the State University of New York. He is a member of the American Society for Training and Development, the Metropolitan New York Association for Applied Psychology, and the American Psychological Association. Dr. Frisch is a licensed psychologist in New York State.

Training is defined as any organizationally provided learning experience intended to help employees perform more effectively in present and future jobs. This includes technical and nontechnical workshops offered by the organization's training department, structured, on-the-job experiences designed to improve skills, and open enrollment or "public" short courses sponsored by consultants or universities. Training may also include books or other reading material, self-guided instructional manuals, or even computer-assisted simulations. College and other academic programs have generally broader goals than training. These include mastering a field of study, embarking on a profession, or acquiring a basic foundation of knowledge rather than gaining skills specific to job performance enhancement.

23.1. TYPES OF TRAINING METHODS

For convenience, training methods can be divided into five categories: orientation programs, on-the-job experiences, in-house programs, external or "public" programs, and self-study materials. Each has advantages and disadvantages in terms of effectiveness, cost, flexibility, convenience, and so on. Certainly they are not mutually exclusive: Most managers will use several methods in helping an employee improve skills. Some of the trade-offs in using each are described below.

Orientation programs are targeted toward the newly placed employee; although usually applied immediately after hiring, they can also be useful after employees are transferred or promoted. The goal of orientation programs is to provide basic, organizationally specific information to help employees become productive as rapidly as possible. Such programs almost always cover the organization's history, mission, objectives, structure, departmental functions, and the names of senior managers. More ambitious programs might also provide facility tours, field trips, reading material, management presentations, or specific assignments to increase the employee's early exposure to the organization. Typically, these programs are provided to managers by the personnel department, but even in the absence of this service, individual managers should structure some type of orientation for new employees, preferably tailored to the context and responsibilities of specific jobs.

Most organizations have found that orientation programs are well worth the investment of time and money. Not only do they provide much needed information to new employees rapidly and efficiently, they also ease the transition into an unfamiliar environment. More importantly, they demonstrate to the individual that the organization is concerned about the welfare of its employees and is willing to provide whatever support is necessary to help them be productive.

On-the-job training refers to specific, structured opportunities to practice new or developing skills within the usual work setting. This method is often used to gain mastery over equipment and manufacturing processes. Highly experienced employees usually guide this type of learning, which may be very closely monitored or offered on a trial-and-error basis, depending upon cost of materials and safety. Typically, on-the-job training is not strictly regimented and can be simply described as learning by doing. Some organizations, however, carefully control on-the-job training experiences and utilize instruction manuals, specific practice assignments, and even tests, as, for example, in the development of managerial candidates by rotating them through a predetermined series of carefully monitored assignments.

The active use of supervision is also a form of on-the-job training. It may involve feedback and advice about how to handle a specific situation or it may be broader and include special project assignments designed to develop skills or offer exposure to different work environments. Although its applicability varies from organization to organization, on-the-job training is certainly an essential aspect of effective managerial practices.

In-house programs are formalized training experiences offered by the organization itself, usually through a personnel training department, and are delivered to a wide range of employees. Areas covered may include sales skills, supervisory and management skills, administrative procedures, performance appraisal, proposal writing, and many others. Sometimes these are planned so that a core sequence of generally useful programs is followed by other workshops targeted to the needs of specific individuals or departments.

In-house programs may or may not be conducted by members of the in-house training staff but attendees are always employees of the same organization, allowing material to be tailored to the language and needs of the organization. The activities in these programs are varied, ranging from lectures to unstructured group discussions, from written exercises to role playing, but the classes themselves are always conducted outside the normal work area.

External programs are similar to in-house training except that they are open to attendees from any organization. Offered by consultants, universities, and professional associations, these programs may cover basic topics, which makes them useful to smaller organizations that do not have their own training departments, or they may cover material so unusual as to be impractical for in-house handling. Topics include any imaginable training need, from assertiveness to international finance; from labor relations to managing your boss. They are expensive, at approximately $200 per person per training day, but used conscientiously, external programs can make significant contributions.

Self-study materials such as books, manuals, cassette tapes, "programmed" instruction, interactive video, and others have become very popular. They are relatively inexpensive, often reusable, and very flexible. Technological advances, particularly in microcomputers, have vastly increased the subtleties which can be built into self-study materials. For interpersonal and behavioral skill building, however, these materials lack practice and feedback capabilities which are so important to effective learning. They work best for relatively factual, objective topics.

Most organizations use some combination of all five of these general methods of training. Orientation programs and in-house or external workshops are used particularly as employees change jobs or advance. In technically demanding environments external workshops and self-study materials are used frequently to keep staff up-to-date. The bulk of development, however, is accomplished through on-the-job experiences in combination with the other methods.

23.2. TYPES OF TRAINING TOPICS

There are many different topics on which training is available—highly technical seminars on manufacturing processes, general courses on business, programs offering hints on controlling interruptions—but even with this wide diversity, training topics can be usefully divided into three general types. Although there is some overlap between them, they are distinct enough to describe both the needs of individuals and the programs offered by vendors.

The first type is technical/functional training and includes the technical skills required by a job or business, specific industrial knowledge, the use of company systems and procedures, knowledge of other functions within the organization, problem-solving and analytical skills, project administration, and other job-specific programs. Although all individuals come to their positions with technical/functional skills, additional ones are continually needed, particularly as employees change jobs. New employees moreover, often underestimate the importance of technical abilities uniquely required by an organization, believing that their academic or professional training should carry them through all work challenges. Orientation programs, self-study materials, and on-the-job training are routinely used to teach these essential technical/functional skills and are sometimes supplemented by in-house and external programs.

The second general training topic is supervisory/managerial training, which includes all aspects of leading and directing the efforts of others. Work planning, delegation, team building, giving performance feedback, motivation, group problem solving, and similar subjects fall into this category. Employees who move into formal leadership positions should always be given supervisory/managerial training, preferably on an in-house basis so as to properly reflect the unique characteristics of the organization. Supervisory/managerial training is also available externally in a wide variety of topic combinations.

The third topic is personal/interpersonal training. Complementing the downward focus of supervisory/managerial training, personal/interpersonal training covers the skills needed to interact with others across and up in the organization. Of the three types of training this one has most recently been recognized as contributing to individual and organizational effectiveness. The decline in the emphasis on formal authority, increased use of matrix and functional organizational structures, and emphasis on lean, horizontal staffing rather than many vertical layers, have all significantly increased the importance of interpersonal skills. Organizations invest large sums of money in teaching employees to communicate better, be more interpersonally sensitive, flexible, less narrow-minded, more assertive and willing to take risks, and so on. Often labeled human relations training in the past, today these types of workshops are more specific in their objectives and tailored to the realities of business interactions. Although available as in-house programs, their internal use is more limited than other topics; employees sometimes prefer the greater anonymity of external courses for these types of skills.

The paragraphs above provide a basic vocabulary with which to discuss training. The section on training methods explained the five general ways of delivering training to employees. The section on training topics described three broad types of subject material. Together they represent a grid for determining the most appropriate "How" and "What" combinations. As shown in Table 23.1, specific topics are best suited for some methods. These terms and their relationships are useful in planning and describing training. The most critical decision a manager makes about training, however, is matching training needs with specific program selections. Interestingly there is more than one process used to make these determinations. It depends on the general objectives which the training is meant to achieve.

23.3. DETERMINING TRAINING NEEDS

Choosing the most appropriate training depends on the work context and training goals. Training focused on new responsibilities should be determined differently from that designed to improve present performance which in turn should be selected differently from training used to prepare the employee for future responsibilities. The sections below examine the process for determining the training needs in each of these contexts.

TABLE 23.1.   Matching Training Topics with Methods

Method	Orientation	On-the-Job	In-House	External	Self-Study
Topic					
Technical/functional	1	1	2	2	1
Supervisory/managerial	3	2	1	2	3
Personal/interpersonal	3	2	2	1	3

Key. 1 = frequently used method; 2 = moderately used method; 3 = seldom used method.

23.3.1.   Training for New Responsibilities

Some training needs will obviously spring from work demands. Knowing that a job requires the ability to handle certain types of equipment or solve specific types of problems will automatically lead a manager to evaluate a new employee's abilities in these areas and provide necessary training. Other job components, however, may not be so clearly defined. Casual observation and/or experience with a position is often insufficient to fully understand its demands. Managers, particularly new ones, must explore all facets of the work they supervise to effectively select training for new employees.

The systematic investigation and documentation of job demands is called job analysis. It is not the purpose of this Chapter to explain job analysis methodology, but rather to emphasize the importance of gathering job-specific information before undertaking the design of training for new employees. Without such information, training may confuse more than it educates. Readers desiring a thorough explanation of job-analysis techniques should consult McCormick's excellent book on the topic (*Job Analysis,* American Management Association, 1979).

Often it is sufficient for the manager gathering job information to observe the work as it is performed and to ask questions about each step in the work process. If more detail is required, or if the job tasks are not observable, structured job-analysis interviews and job questionnaires administered to incumbents are useful. These methods identify specific tasks in the job, average time spent on each, their relative importance, and the requirements for success in terms of education, training, and experience. In some organizations job descriptions detailing this type of information are available from the personnel department. In most organizations, however, managers will need to rely on more direct exposure to jobs and the employees performing them to gain a complete picture of tasks, requirements, and training needs.

Training for new responsibilities generally utilizes orientation, on-the-job, in-house, and self-study methods directed at teaching selected technical/functional topics, such as basic job methods, procedures, and specific industrial or organizational knowledge. Training on other topics, such as managerial skills, may also be used if the new responsibilities involve directing others. Other training for new responsibilities will depend on the employee's academic background, previous experiences, and prior training compared with the abilities and knowledge required to meet the demands of the job.

23.3.2. Training for Current Responsibilities (Performance Improvement)

Determining the training needs for employees in *new* responsibilities largely relies upon the results of job analyses or other relatively objective, descriptive information. This is less true when selecting training for employees' current responsibilities, or, in other words, when helping employees to improve their present performance. As individuals acclimate themselves to their jobs, they develop work styles which include both positive and negative characteristics. Removing the deficiencies, however, is sometimes complex. Not only must objective job demands be understood but the reasons for performance limitations must also be considered if training is to be fully effective.

Training for performance improvement requires a more thorough understanding of the individual employee than training for new responsibilities. A manager may choose from a variety of approaches to helping an employee speak up at meetings or plan and organize assignments or be a better listener. The most useful interventions, however, are based not only on the ways performance needs to improve but also acknowledge *why* the individual is behaving in these ways. Unless managers understand more about their employees, efforts to improve the performance of current responsibilities will probably be superficial and ineffective.

Causes of inadequate performance can be grouped into four categories, corresponding to quite different explanations for observed performance. After the most likely cause category is determined, the most effective types of solutions and training can be identified.

23.3.3. Motivational Causes

Motivational factors are the most common causes of observed performance inadequacies. In these cases the observed performance is due to some dissatisfaction with the work itself or the work environment. This could include dissatisfaction with actual tasks, feelings of inequity concerning pay, lack of promotional opportunities, disinterest in forming relationships with co-workers, or dissatisfaction with feedback or attention from supervisors. Symptoms of motivationally caused shortfalls may be an obvious decline in quantity or quality of output, indifference to improvement, or disinterest in continued learning about the job.

23.3.4 Interpersonal Style Causes

When interpersonal style is the cause, th individual has an observable characteristic which interferes with effective interactions on the job. Examples are disruptive levels of impatience, aggressiveness, withdrawal, rigidity, or anger. Less overt tendencies, such as shyness, lack of assertiveness, or avoidance of conflict are also included in this category. These traits usually become more pronounced during times of pressure or stress and may be observed in behavior during daily interactions, at meetings, or during intergroup conflicts. Obviously in some jobs interpersonal style is more important than in other jobs. As an employee advances, however, particularly to managerial positions, interpersonal style will have an increasingly important role in effectiveness.

23.3.5. Knowledge or Skill Causes

This category includes any performance in which the employee lacks specific knowledge, skill, or experience required to perform tasks effectively. Examples include insufficient knowledge or skill in finance, law, foreign languages, computers, writing, public speaking, and so on. Knowledge-caused deficiencies are most often associated with changes in job demands or responsibilities. For example, job redesign, technological developments, company growth, and other changes may make an employee ill-equipped to meet specific demands of the job. Inadequate design or delivery of company-sponsored training may also contribute to knowledge- or skill-caused deficiencies.

23.3.6. External Influence Causes

In contrast with other categories, a performance shortfall caused by an external influence may have little to do with the employee. This occurs when the employee's productivity is interrupted by events in his or her private life or by factors within the work environment. The former may include the home situation, illness, financial deficiencies, and so on; the latter covers unusual work pressures, seasonal fluctuations in work volume, stressful co-worker interactions, pressures from other areas of the organization, and poor leadership or direction from the manager. Symptoms of externally caused problems may include unexplained latenesses or absences, requests for unusual amounts of time away from the job, sudden lethargy, or unusual irritability.

There is no fool-proof formula for determining the exact cause of a performance deficit. Several communication skills will certainly facilitate the process, however. First, descriptions of the problem performance should be specific and behavioral, rather than evaluative. If the discussion is to lead to understanding, managers must avoid value judgments, substituting instead an objective description of the difficulty. For example:

Evaluative: You can't plan well.

Descriptive: At the last two staff meetings, you failed to provide a written progress report.

Evaluative: You are harsh and inconsiderate with others.

Descriptive: During the questions segment of your presentation, you did not allow the QA manager to explain his point of view and very few questions were asked.

Objective descriptions are much more likely to lead to fruitful discussions of performance than conclusions or judgments.

Second, probing and active listening on the part of the manager is essential for exploring the cause–effect relationship. Open-ended questions and follow-up discussions will help identify the key issues and make it possible to reach agreement on the most likely cause. Specific questions might include:

How do your current duties compare with the career track you see for yourself? (Motivational probe)

How would you describe your basic interpersonal style? (Interpersonal style probe)

Tell me what subject you could learn that would improve your effectiveness? (Knowledge or skill probe)

What unusual pressures are interfering with your performance and how can I better assist you with them? (External influences probe)

Sometimes two or more causes may interact to produce the observed performance. For example, manager and employee may agree that certain responsibilities of the job are unrelated to the subordinate's career interest (motivational cause) but the end-of-year workload also causes difficulties (external influences cause). Alternatively, the manager may discover that the subordinate does not know how to perform a specific task (knowledge or skill cause) but has been too shy to ask for assistance (interpersonal style). When overlaps such as these occur, each cause should be separately targeted for improvement.

After causes are identified, solutions can be more fruitfully designed. It should be noted, however, that training is not appropriate in all cases. The likelihood that a training solution will help to improve performance depends on the cause which is operating. Following is a discussion of solutions for each cause category.

23.3.7. Linking Solutions with Causes

Managers concerned about employee motivation should focus solutions on options which "expand" or "enrich" the work along lines likely to be interesting for the employee. In some cases this may be as simple as rotating boring or routine tasks among several individuals, assigning work with employee preferences in mind, or providing assignments in new work areas. On the other hand, enrichment may require creative redesigning of the job in nonstandard, but permissible ways. New responsibilities may include managing special projects, acting in a liaison role with other work groups or facilities, teaching and guiding new employees, conducting analyses and writing reports, assisting more senior employees, or acting as spokesperson for the department. The manager should discuss these and similar suggestions for job enrichment with the employee before instituting changes.

In some cases motivation may be improved by sending an employee to an in-house or external training program. Although the main objective of such a move should be to improve knowledge or skill, selection for such a program, particularly one not offered to all employees, or one offered in a particularly desirable location, may have a beneficial effect on motivation. However, unless the work itself is also made more interesting for the individual, such attendance will have only a temporary effect. Therefore, training is not a very effective solution for motivationally caused performance deficiencies.

Although not easily changed, interpersonal style can be improved. Characteristics which may interfere with the subordinate's work effectiveness may be moderated and brought under control by suggesting specific adjustments in the subordinate's behavior. Unlike the job-focused solutions to motivationally caused problems, solutions to interpersonal deficiencies focus on behavioral changes in the individual.

Many effective training programs, usually offered externally, focus on interpersonal style. Typically these programs have a dual purpose. First, they highlight specific aspects of the employee's interpersonal style which are causing difficulties. This may be accomplished through exercises, role plays, and feedback during the program or by the inclusion of anonymous survey feedback gathered from other employees prior to the program. Second, these programs provide opportunities to practice styles which are quite different from the employee's usual tendencies. In this way, the employee's response options are expanded and hopefully improved. After attendance the individual returns to the work environment both sensitized to his or her own interpersonal issues and somewhat familiar with alternative means of interacting. Specific, behavioral on-the-job feedback from the manager will then become very important as the individual begins to apply the new understanding of interpersonal issues.

If the cause of the performance deficiency is lack of knowledge or skill, the employee should be offered appropriate materials, on-the-job training and/or internal or external program attendance. These options should be discussed with the individual, using job-description information to help select the program most targeted to the employee's need. Knowledge- or skill-caused problems are more readily solved with training than problems in any other cause category, assuming that the deficiency can be identified and effective training exists to solve it.

Once revealed, external influence causes can be handled in a variety of ways, none of which require training. The manager may decide to tolerate the external influence if it is likely to be temporary. If the influence originates in another employee, whether peer or superior, the manager may decide to intervene to improve the situation. If the influence stems from a more serious personal problem, the manager may decide to suggest that the individual seek professional help. Many large organizations have personnel professionals able to link employees with helping resources outside of the organization with complete confidentiality.

Table 23.2 illustrates the relationship between causes, solutions, and training options. Managers can use it to diagnose performance problems and determine appropriate interventions.

To summarize, training applied to existing job responsibilities almost always involves the need for performance improvements. As such, managers must understand the causes of the inadequate performance before effective solutions can be selected. Generally, performance problems may be caused by (1) motivation when a poor fit exists between the employee's interests and the job, (2) interpersonal difficulties stemming from particular behavioral characteristics, (3) lack of specific knowledge or skill, and (4) an interfering external influence. As discussed above only interpersonal style and knowledge/skill causes can be effectively addressed through training.

**TABLE 23.2. Training for Current Responsibilities
Performance Improvement Matrix**

Causes	Performance	Solutions	Training Support
Motivation	Description: The employee is dissatisfied with some aspect of the work or the work environment, such as pay, promotional opportunities, co-workers, or supervision Symptoms: Obvious decline in quantity or quality of output; disinterest in improvement or continued learning.	Solutions: Focus on modifying the job to better satisfy the individual involved. Consider ways to expand or enrich the employees' duties to include his or her specific interests. Suggestions: Rotate boring or routine tasks, revise operating methods, or consider transferring the individual. Don't attempt to change someone's basic likes and dislikes.	No
Interpersonal style	Description: The employee has a characteristic or trait (impatience, aggressiveness, withdrawal, anger, shyness, etc.) that interferes with effective interpersonal interactions. Symptoms: Poor leadership of staff; inability to direct meetings; reticence to speak before groups; interdepartmental conflicts.	Solutions: Help the individual develop specific talents he or she can use to control the expression of the ineffective trait. Suggestions: Have an impatient individual list the pros and cons of alternatives to help pace decision-making; ask the short tempered employee to list alternative responses to frustrating situations, etc.	Yes
Knowledge or skill	Description: The employee lacks the specific knowledge, skill or experience required to perform tasks effectively (e.g., poor knowledge of accounting, law, a foreign language or computers). Symptoms: Substandard performance after promotion or reassignment; aver-	Solutions: Offer training and educational opportunities, or on-the-job feedback to bolster the areas of deficiency. Suggestions: Utilize training programs, seminars, continuing education courses, technical schools, or create supervised on-the-job assignments.	Yes

TABLE 23.2.   (Continued)

Causes	Performance	Solutions	Training Support
	sion to certain tasks; resistance to any procedural changes that draw on specific skills.		
External influence	Description: The individual's normal productivity is interrupted by events in his or her private life or by factors in the work environment, such as marital or family problems, financial pressures, co-worker conflicts, or poor direction from the manager or supervisor. Symptoms: Unexplained lateness or absence; excessive personal use of the telephone; irritability.	Solutions: Encourage the employee to explain the situation to you. As supervisor, you can rearrange the workload if the situation is temporary. Suggestions: Don't criticize poor performance if its causes seem to be temporary. If the poor performance stems from a serious personal problem, recommend that the employee seek professional help.	No

23.4.   TRAINING FOR FUTURE RESPONSIBILITIES

Employees at all levels have become increasingly aware of the importance of career planning and development. Feelings of being "pigeon-holed" or constrained in the organization can lead to resentment and dissatisfaction. Training for future responsibilities helps to prepare employees for future jobs by identifying areas which will become increasingly important as advancement in the organization occurs. Some of these may overlap with aspects of current performance needing improvement. For the most part, however, training for future responsibilities focuses on skills which will be needed later but are not critical for current functioning. The shares of time managers spend on specific parts of their duties (controlling, planning, organizing, directing, staffing) change as they progress upwards in the organization. Figure 23.1 illustrates this process of changing time allocation. The most conspicuous changes are the increase in planning and staffing and the decrease in controlling and directing. Training for future responsibilities anticipates these changes and gives employees a wider skill perspective than just the current job.

Training for future responsibilities is most frequently applied to those individuals who are performing well and whose prospects for promotion are good. These individuals generally are career minded and are looking for indications of the organization's intention to promote them. However, training for future responsibilities can be applied beneficially to any employee who is an asset to the organization and whose skills are likely to become outdated for any reason. Effective training for future responsibilities encourages employees to build upon their strengths and also communicates that the organization recognizes their value.

A manager undertaking training for future duties must be sincerely interested in the employee's growth. There should be no "hidden agendas" or pressure to comply with advice or the suggested program. The manager's role should be one of "helper" by suggesting how career plans might be supported with training. If the employee's career goals and aspirations are unclear, the manager may need to provide feedback and encourage the individual to think about likely paths. Even if the ultimate career goal is unknown, skills likely to be needed for the future can be identified.

Managers often complain that training for future responsibilities is difficult without knowledge of job responsibilities and career-path options throughout the organization. This is true to some extent and in those organizations where career paths have been specified, training for future responsibilities can be more easily identified. Even in organizations where career paths have not been clearly defined however, managers can use their knowledge of how jobs in the organization change as advancement occurs. For example, technical skills typically become less important over time while managerial and interpersonal skills increase in usefulness. Preparatory

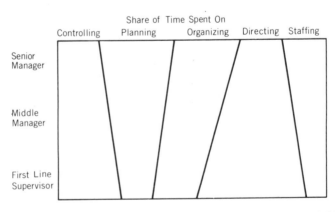

FIGURE 23.1.   Changes in functional emphasis in managerial jobs.

training in specific subjects within these skill areas is therefore likely to prove beneficial in the future for all employees who are progressing. This is also true of general business knowledge, industrial knowledge, presentation skills, familiarization with legal and social impact issues, and familiarization with the general operation of other departments and/or functions within the organization.

Training for future responsibilities can, of course, include in-house and externally offered training programs such as seminars, college courses, or other classroom-based programs. However, managers should also use their knowledge of the organization and its jobs to suggest on-the-job training experiences. In most organizations there are many possibilities for developmental exposure as a building block toward future positions. The following is a list of suggested on-the-job training possibilities which can be used to enrich current work responsibilities and expand the skill base in preparation for future jobs:

- Assign specialized reading material on technology, business, industry, or other topics.
- Conduct special projects and analyses, either internally or externally focused.
- Visit other facilities, divisions, or departments of the organization, or other organizations if possible and make a report back to the home department.
- Be the spokesperson for the department on interfunctional issues, consumer affairs, or joint departmental task forces.
- Be assigned as the analyst for the departmental Director.
- Fill in for the manager when he or she is on vacation.
- Be responsible for answering specific types of questions and inquiries.
- Be responsible for teaching and orienting new employees in the department.
- Act as the organization representative at professional meetings and conferences.
- Write position papers on topics of concern to the department.
- Make presentations whenever feasible.
- Be the designated resource for consultants who are studying the organization.

Before training for future responsibilities, however, there are two caveats to consider. First, managers should estimate how much additional time the developmental experiences will require, and whether this will interfere with the subordinate's ongoing responsibilities. If so, the manager must decide whether the individual can be released from some present duties. If no reduction or modification is possible, developmental experiences must be adapted to better suit the situation. Concern for the individual's development must be tempered with a realistic appraisal of work requirements.

Second, even with proper handling, training for future responsibilities may result in the individual feeling that he or she is being given an increased work load, regardless of the future benefit. The employee may think that, in effect, he/she is being placed at a disadvantage because of the added activities. This possibility underscores the importance of the employee's involvement throughout the process of selecting training. Plans should always be mutual, never dictated or required. In addition, it should be made clear that training for the future is always optional. The manager should exhibit flexibility about changing or adjusting these training plans.

23.5. SUMMARY

This chapter has attempted to provide basic information about training so that managers may use this important resource efficiently and effectively. Training is defined as any organizationally sponsored learning targeted at improved job performance. It comes in five broad formats including orientation, on-the-job, in-house, external, and self-study approaches, although no training is likely to be effective without opportunities for on-the-job application of skills. Training topics themselves can be divided into three broad categories: Technical/functional, supervisory/managerial, and personal/interpersonal. The first two are routinely offered by organizations, particularly as in-house and external programs, while the third category is being used increasingly, although primarily through external workshops. A more detailed discussion is given in Chapter 24.

The setting in which training is applied must also be considered. Training needs should be assessed differently for new, current, and future responsibilities. Needs for new responsibilities should be based on job requirements and the employee's previous experience. Training for current responsibilities can be thought of as performance improvement and requires an understanding of the causes of the deficient performance. Not all causes can be resolved with training. Finally, training for future responsibilities should emphasize the development of skills likely to be useful as the individual advances in the organization.

24

WHAT KIND OF TRAINING?

Robert M. Anderson, Jr.

ROBERT M. ANDERSON, JR. is Manager of Technical Education Operation for Corporate Engineering and Manufacturing, General Electric Company in Bridgeport, Connecticut. He received his Associates degree from Flint Junior College, his BSE(EE), MSE(EE), MS(Physics), and PhD(EE) from the University of Michigan. After working briefly at A.C. Spark Plug, Armour Research Foundation and Conductron Corporation, he was at Purdue University from 1967 to 1979 where he was successively assistant professor, associate professor, and professor of electrical engineering. While at Purdue, he was also Ball Brothers professor of Engineering and the director of Continuing Engineering Education. He joined General Electric in August 1979 as manager of Engineering Education and Training for Corporate Consulting Services and was appointed to his present position in September 1982. Dr. Anderson has published extensively on the subjects of electrical and optical properties of thin semiconducting films, vacuum technology education, integrated circuits education, professionalism and engineering ethics. He is also a registered professional engineer in Indiana.

24.1.   PURPOSE OF THIS CHAPTER

Engineering managers are required to make a variety of decisions concerning training for themselves and for the individuals in their organizations. These decisions have significant impact on both the short-term and long-term productivity of both the individuals and the organization. The purpose of this chapter is to help managers make training decisions rationally, effectively, and quickly.

Sections 2, 3, and 4 of this chapter should be read even in the absence of any specific training decisions to be made. Each manager should have a clear concept of why engineers should be trained and how to specify different types of training. Section 2 raises fundamental questions concerning why we do training; Section 3 outlines four specific purposes that training can accomplish for individuals and for organizations; Section 4 spells out the five key elements which must be clearly specified for any training program to be effective; and Section 4 concludes with eight key questions for faster, more effective training decisions.

The remaining sections of this chapter provide brief tutorials on the several major issues associated with the specifics of common, industrially-oriented training decisions. These sections can either be read quickly to gain some broad comprehension of the several alternatives possible or be used as references when specific training opportunities arise and specific decisions must be made.

24.2.   TWO FUNDAMENTAL QUESTIONS

Any manager who has clear answers to the following two questions will make quicker and more effective training decisions than a manager who has not invested the time and energy to consider such questions. Each manager should formulate his or her personal training policies and procedures to augment those policies and procedures which may exist in the organization.

24.2.1.   Reward or Investment?

The first question to be considered is whether training is viewed fundamentally as a reward for exemplary performance or as an investment made by an individual and the sponsoring organization. The engineering manager has obligations to his or her employees to first understand that training can be viewed as either a reward or an investment; second, to decide whether training will be primarily a reward or an investment; third, to know when training will be used in the nonprimary sense; and finally, to communicate his or her opinions and policies to his or her employees.

It is legitimate to use training as a reward for significant contributions. In such cases we typically look for short courses offered on cruise ships, at resort locations, or in the sun belt in the middle of winter. Where training is viewed as a reward, all should recognize that reward is the fundamental objective; that the specific content or quality of the short-course training experience is secondary; and that the individual will not be expected to perform differently after the training experience than before the training experience.

On the other hand if training is viewed as an investment, the manager should make the training decision with a clear understanding of what the return will be on the investment. First determine the costs of the proposed training. Consider the obvious direct expenses of fees, travel, and so on; be sure to also include the costs of salaries and benefits of individuals who train on company time. Less obvious costs include the opportunity costs; what immediate contribution could the individual make if he or she continued to work rather than participate in the proposed training? Then specify the expected return from the training. What change in behavior, in productivity, in quality of work do you expect as a result of the training? (See Section 4 of this chapter for some help with this process.) Can you quantify the expected return? With a quantified training return to compare against a quantified training investment, it is trivial to decide whether the proposed training is a good investment or not. Even in the absence of quantified training returns, thinking and analyzing proposed training in investment terms will help the manager to make better training decisions.

24.2.2.   *Ad Hoc* or Purposefully?

Training decisions can be made on an event-by-event, *ad hoc* basis, or training decisions can flow purposefully from an organizational plan and a consistent policy and procedure.

The principal problem with *ad hoc* decision making on training is that looking back over several individual training decisions we will see that significant resources have been allocated to and used by training with only a fragmented set of positive returns. The engineering manager has only limited resources to allocate to training; these resources must be purposefully concentrated and focused over several individual training decisions in order to achieve significant, positive results both for individuals and the organization.

A secondary problem with *ad hoc* decisions is that both the manager and individuals in the organization will waste much time and energy because of the confusion resulting from an *ad hoc* process. For example, individuals within the organization will not know whether any training is important, what training is important, whether specific training recommendations have a high or low probability of being approved, or even whether individuals may be rewarded or penalized for seeking additional training.

On the other hand, purposeful training flows logically from an organizational plan and consistent policy. We believe that the engineering manager has a significant fraction of the total responsibility to insure that the individuals in his or her organization have the knowledge and skills necessary to do their jobs now and in the future. Focusing on the present, the manager should know what is required today, what knowledge and skills his or her employees currently possess, what gaps (if any) exist, and what role (if any) training can play in closing the gaps. Looking to the future, the manager should go through a similar process of projecting what will be required, comparing this requirement against current knowledge and skills, determining what gaps exist, which are best filled by adding new employees and which are best filled by training. This process of analysis yields a clear framework for purposeful concentration of limited resources to achieve a maximum positive impact.

Secondary benefits from purposeful training decision making are that effective decisions can be made routinely and consistently. The principle advantage of routine decision making is that such decisions require little time and consideration from the manager. The advantages of consistency in decision making are that individual training decisions will be linked together to achieve significant results and that fewer training decisions will have to be made since individuals in the organization can learn which training recommendations have a high probability of being approved and which have a low probability of being approved.

The total responsibility for insuring that individuals in an organization have the knowledge and skills to do their jobs both today and in the future is a shared responsibility between the individual and the organization (as represented by the engineering manager). The engineering manager has the responsibility to provide a statement of the current and future needs of the organization. Both the manager and each employee have the responsibility to assess the current knowledge and skills of the employee. Where gaps exist, the manager and the individual in the organization can together devise training strategies and tactics to update knowledge and develop new skills as may be required. Both the organization and the individual should contribute something to the resulting training activities.

24.3. PURPOSES OF TRAINING

As previously discussed, training can be used to reward someone for out-standing performance or training can be viewed as an investment; with an investment view of training, training decisions can be made purposefully to achieve specifiable objectives. In this Section we discuss four general purposes of training: (1) to stay current in evolving technologies, (2) to survive revolutions in technology, (3) to earn a credential, and (4) to support organizational development. A specific training program may simultaneously address more than one of these purposes; nevertheless, we believe that each proposed training program should be evaluated in terms of which of these four purposes is the primary purpose and which are secondary purposes. With a clear understanding of the primary purpose of the proposed training, the engineering manager can more easily consider alternatives and evaluate benefits and costs.

24.3.1. To Stay Competent in Evolving Technologies

Over some period of time in the life cycle of every technology, the rate of change in the technology is slow enough that part-time continuing education and training are adequate to stay current with the evolving state-of-the-art. In such situations reading trade journals, reading professional society publications, attending trade shows, attending professional society meetings, attending one-to-five-day short courses, and so on, are adequate to acquire the new knowledge and skills. Something like 40 to 80 hours per year in purposeful learning is probably adequate. The specific activities by which individuals stay current with evolving technologies and the specific number of hours required vary from discipline to discipline and from situation to situation.

Even in this reasonably calm and steadily changing environment, a clear understanding of current and future needs of the organization is an important prerequisite for decision making. Keeping employees competent in technologies which they do not and will not use on-the-job is a waste of organizational resources—an investment with zero return.

24.3.2. To Survive Revolutions in Technology

From time to time specific technologies undergo rapid and dramatic transformation. A classic example is the series of revolutions which occurred in electronic circuits—from a technology based on vacuum tubes to one based on discrete transistors to one based on integrated circuits—all within about

20 years. Twenty years are easily within the working career of a single individual.

Frequently, in situations where the technology has undergone a revolutionary change, affected individuals require some period of intensive full-time learning activity to make the transition from competent professional in the old technology to competent professional in the new technology. Both the organization and individual must be prepared to make a substantial contribution to such a revolutionary transition of knowledge and skills. The organization must be willing to provide substantial support in terms of time-off-the-job and/or financial support, and the individual must be willing to devote substantial amounts of time which would otherwise be spent on personal activities. It may also be reasonable to expect the individual to contribute financially in terms of direct cash payment or of accepting only part-time salary while participating in an intensive training program.

24.3.3.   To Earn a Credential

One purpose for training is for the individual to earn a credential. We use the term credential in this context to include academic degrees, state licenses, certificates in specific disciplines and subdisciplines, and so on.

A certain credential may be required for practice in specific situations; hence the credential has clear organizational value and a high return on investment exists for both the organization and the individual. Having such a credential may be a precondition of employment or earning such a credential within a specified period of employment may be a condition of continued employment.

When a particular credential is not required, the organization may still benefit from the process of an individual earning the credential. As an individual learns what is required to earn the credential, his or her professional knowledge and skills are enlarged and the organization can benefit from the increased knowledge and skills. When the individual earns the credentials, his or her status in the profession is enhanced, and since the reputation of an organization is, in part, determined by the credentials accumulated by the individuals in the organization, the status of the organization is increased as more individuals accumulate more credentials.

24.3.4.   To Support Organizational Development

From time to time an engineering manager will decide (or will be told) that his or her organization needs to focus on a particular organizational issue. For example, suppose a product design manager is told that his or her organization must work more closely with an advanced manufacturing organization to speed and smooth the transition of the product design into manufacturing. Organizing a training experience on a topic relevant to both groups, structuring the training to allow for maximum attendee participation, and populating the class with individuals of both design engineering and advanced manufacturing can be a key process to achieving the desired organizational objective. In the training program, individuals from the two organizations can develop a common base of understanding and learn a common jargon. The training program can provide a neutral meeting ground on which individuals from the two organizations can begin to gain an understanding of and appreciation for each others' objectives, problems, concerns, and so on. Such improved relationships can do much to speed and smooth the transition of products from design engineering to manufacturing.

24.4. SPECIFICATIONS OF TRAINING

Engineering managers should think of training as a product and require that each proposed training program be described in terms of a set of specifications. As a minimum, the following five specifications should be described as a part of any training recommendation.

The five specifications are in a hierarchical order; that is, the engineering manager should first consider the objectives and stated benefits of training. If the manager decides the objectives and benefits are not valuable to the organization or are not consistent with the organizational plan, he or she can reject the proposal without reading the rest of it. If the objectives and benefits are desirable, next he or she should consider the costs—both direct and indirect, do a cost/benefit analysis, and again decide to either reject or continue reading. Finally, he or she should consider the proposed audience, training process, and training evaluation plans. Are they all appropriate? What changes should be made? What is vague and needs to be clarified? Finally, when the manager agrees that the benefits justify the costs and that the audience, training process, and evaluations are all appropriate, then he or she should approve the proposed training.

24.4.1. Objectives and Benefits

Requiring a clear statement of objectives and benefits for every proposed training program will substantially increase the effectiveness of training decisions. Every training program should explicitly state specifically what facts will be learned, performance skills acquired, and attitudes and beliefs developed in and through the training program. Such objectives should be stated in common-sense language and in terms which allow an evaluation to be made on whether or not the attendees achieved the objectives.

Every training program should state the benefits to the organization and these benefits should be quantified to the maximum extent possible. For example: How much will the time it takes to do a task be reduced? How much will product cost be reduced by use of the proposed techniques? How much will quality be increased? How much money will be saved by not having to hire a new person?

Assuming that the learning objectives are achieved, the engineering manager must judge whether the stated benefits would follow. For example, will learning how to analyze a project-management network diagram of tasks to determine the critical path really lead to 50% fewer projects with schedule slips? The engineering manager should rely on his or her experience and common sense in making this judgment.

The engineering manager must now decide if the stated objectives and benefits of the proposed training are desirable. Are they consistent with the current and future needs of the individual(s) and the organization? Are these "need-to-do" or just "nice-to-do"? Possible decisions after reading objectives and benefits are: (1) Not a desirable activity; don't do it even if the cost is zero. (2) Reasonable to do if costs are reasonable and other specifications are O.K. (3) High value and benefit; must find a way to do it.

24.4.2. Costs

Specifying the costs of any proposed training program is reasonably straight forward. Direct, indirect, and opportunity costs all need to be specified and quantified to the maximum extent possible.

Direct expenses include tuitions and instructor fees, books and materials, travel expenses, room and board expenses, meeting room charges, audio visual expenses, computer use charges, equipment purchases or rentals, and so on. The salaries, wages, and benefits of the attendees must be considered either as a direct expense or as an indirect expense. Indirect expense also includes the expense of those who will support or coordinate the training in some way. Opportunity cost includes a quantification of the value, beyond salary and benefits, or activities not done on the job because of participation in the training programs. For example, will the organization fail to bid on a job because a key professional is at a training program or will the probability of getting a follow-on order be reduced because the organization's readiness to service the initial order is reduced by individuals participating in the proposed training?

Now the manager must assess benefits and costs. How do the proposed benefits balance against proposed costs? Assuming the other specifications are reasonable, does this proposal make good sense from a businesses investment point of view? If you literally owned the business, would you approve this proposal? Some possible decisions at this point are: (1) If costs far exceed benefits and it seems unlikely that costs can be reduced sufficiently to align with benefits, reject proposal. (2) If costs exceed benefits by a small amount, check reasonableness of other specifications and explore opportunities to decrease costs and/or increase benefits. (3) If benefits exceed costs, continue reading the proposal to check the reasonableness of the other specifications.

24.4.3. Audience

Every proposed training activity should clearly specify the intended audience for whom the proposed training activity is designed. What are the assumptions about the incoming knowledge, skills, and attitudes? What are the job titles of the intended audience? What is the range and the average experience of the candidate attendees?

The engineering manager must decide whether or not he or she agrees that the stated intended audience is the most appropriate group. Should the group be enlarged or diminished? Do specific individuals fit the description of the intended audience?

24.4.4. Process

Every proposed training program should state explicitly the learning processes and activities in the program. Possible processes and activities include lectures, role playing, performing experiments, running simulations, working problems, writing, speaking, and so on.

The engineering manager has the responsibility to judge whether the intended participants have a reasonable chance of achieving the learning objectives by the processes and activities specified. To help with this judgment the Table 24.1 has Xs for learning activities that are particularly useful for achieving the different types of learning objectives.

24.4.5. Evaluation

Finally, every proposed training program should state explicitly how the participants will determine whether or not they have achieved the learning objectives of the training program. Depending on the nature of the training exercise, such an evaluation might include paper and pencil tests, role playing, writing proposals, solving problems, preparing designs, specifying

TABLE 24.1.   Instructional Techniques Appropriate for Different Types of Learning Objectives

Instruction Technique	Knowledge	Skills	Attitudes
Business games		X	X
Case study	X	X	X
Discussion			X
Film	X		
Lecture	X		
Problem solving		X	X
Programmed instruction	X	X	
Role playing		X	X
Simulation		X	X
Television	X		

equipment, and so on. Again, the manager has the responsibility to use common sense to judge whether the proposed evaluation scheme is sufficient to determine whether the training objectives have been achieved. The manager may also consider how the participant's assessment of achievement of the learning objectives should be reported: To the participant only, to the participant and his or her manager, to the whole class, and so on.

Eight key questions for faster more effective training decisions:

1. Does the proposed training program clearly state (1) objectives and benefits, (2) costs—direct and indirect, (3) who should participate, (4) what learning processes will be used, and (5) how learning will be evaluated.

    Yes—continue; no—reject proposal

2. Are learning objectives important to the organization?

    Yes—continue; no—reject proposal

3. Are learning objectives stated in behavioral terms?

    Yes—continue; no—reject proposal

4. Are the stated benefits of high value to the organization?

    Yes—continue; no—reject proposal

5. Do the stated benefits logically follow from the learning objectives?

    Yes—continue; no—reject proposal

6. Are all direct, indirect and opportunity costs properly described?

Yes—continue; no—reject proposal

7. Assess the benefits in comparison to the costs; is this a good business decision?

Yes—continue; no—reject proposal

8. a) Is (are) the candidate attendee(s) within the specification of the intended audience?

  b) Are the proposed learning activities appropriate to achieve the stated learning objectives?

  c) Is the proposed evaluation adequate to insure that the learning objectives have been achieved?

Yes—approve proposal; no—reject proposal

24.5. CONTENT

Hundreds—or perhaps thousands—of specific training programs exist in several different disciplines and subject categories. The purposes of this section are first, to define these broad categories of training—every specific training program falls into one of these categories—and second, to present a rationale that a typical engineering manager might use to decide whether to support specific requests for training in each of the three categories.

24.5.1. Professional Technical

Professional technical training is a term we use to describe training in technical subjects appropriate to the individual and the individual's position in the organization. Such training might be skill oriented, knowledge oriented, even just awareness oriented; but the content is clearly technical, clearly of a professional nature and relevant to the professional position of the individual(s) for whom this training is being considered.

Requests to participate in professional technical training courses should be evaluated with the process outlined in the previous sections of this chapter. Indeed, every professional should have some level of participation in programs within this category.

24.5.2.   Professional Nontechnical

This is a more difficult category to describe, but, for example, we include in this category topics such as management training for individuals now in, or soon to be in, management positions. Specific programs in this category include finance, employee relations, marketing, interpersonal skills, and so on. One might also include in this category topics such as effective communication in written or spoken formats, effective utilization of time, conducting effective meetings, and so on.

We believe that an organization can legitimately support employees to participate in training programs that fall into this category. Employees who become managers need to learn management-oriented facts, skills, and attitudes to continue to successfully contribute to the organization and to continue to advance in their careers. Technical individual contributors as well as the organization profit from improved communications, more effective use of time, more efficient meetings, and so on.

24.5.3.   Personal Development

Into this category we place all of those training subjects that the individual wants to pursue, but for which there is no value to the organization. Examples in this category include, sports training (e.g., improving your tennis game), art, history, personal financial management, and so on.

We believe that the organization has no responsibility to support training in such areas. An individual may choose to pursue such training opportunities for his or her own personal edification, but such activities should be conducted at the employee's expense and outside the bounds of the normal working time and requirements.

24.6.   GROUP OR INDIVIDUALIZED TRAINING

Occasionally engineering managers are called upon to decide between a group learning activity and an individualized activity. The discussion that follows provides some insight into the advantages and disadvantages of these two fundamentally different approaches to training.

24.6.1.   Group Training

As the term implies, group training activities are those activities which are intended to involve a group of individuals simultaneously in the same activity. The group may be large, perhaps hundreds of people at the same time;

or small, perhaps as few as two or three individuals. While the specific techniques, advantages, and disadvantages of group training are a function of the size of the group, our concern here is to differentiate group training, independent of size, from individualized training.

The advantages of group training include (1) the opportunity to share experience among participants in the same program, (2) the efficiency of presentations being made to more than one person at the same time, (3) the organizational advantages of having multiple participants from the same organization in the same training program, (4) the discipline of a scheduled group training activity which provides a strong motivation to complete required activities on schedule, and (5) the opportunity for adding an assessment element in the training activity to differentiate the participants on the basis of their skills and abilities.

The disadvantages of group training include the following: (1) Group training is a lock-step training activity generally forcing all participants to begin the training activity at the same point, progress together, and conclude at the same point. (2) Group training is less desirable when the intended audience is less homogeneous, that is, when individual participants come to the training activity with very different incoming levels of knowledge and skill, learn at very different rates, or may have very different end-of-training objectives. (3) Using a scheduled group training activity makes it difficult to accommodate divergent schedules of individual participants.

24.6.2. Individualized Training

By the term individualized training, we mean training activities that can be pursued by individual participants independent of other learning participants, but not necessarily independent of a teacher or facilitator. The most simple individualized training activity occurs when an engineer reads an article in a journal. If we stated a specific objective and benefit of reading the article, identified the costs, described the intended reader, and specified some evaluation process, then reading an article would be classified as an individualized training activity, and we could rationally judge the return on investment of the activity.

In general the primary advantage of individualized instruction programs as contrasted to group instruction programs is the increased schedule flexibility. That is, the individual can choose to pursue his or her learning activity at convenient times and will not be forced into a learning activity as dictated by a group-based schedule. Travel and peak-or-slack demand at work or home can be accommodated more easily when pursuing an individualized instruction program.

In general, the disadvantage of individualized instruction is the lack of a peer-pressure, schedule-driven group activity. Lack of peer pressure from the other individuals in the group and lack of fixed class schedules makes it easy for the individual learner to put off the training activity; hence not all those who start an individualized instruction program finish it.

The most simple individualized training activity is one in which an individual learner begins at the specified beginning of the training activity, continues along a linearly specified series of activities and concludes at a specific ending point. Every different individual learner in the individualized learning process starts at the same point, proceeds along the same path, and concludes at the same point. Such a program is called a linear individualized training program.

Branched individualized training programs are also available. In these programs different individual learners all may begin at the same point in the training program, but as each individual proceeds with activities of the training program, and responds to questions in the training activity, different individuals will be led through different instructional sequences as a function of their responses to questions. Such an individualized instruction program is referred to as a response-controlled-branched individualized-instruction program. These programs are typically designed so that every learner accomplishes the objectives of the overall training program but each student follows different paths by which these objectives are achieved. Some learners will take more time and follow longer paths while some learners will take shorter times and follow shorter paths to accomplish the same learning objectives.

The advantages of response-controlled, branched, individualized instruction programs are primarily associated with minimizing learner time to achieve training objectives. That is, those people who need less time and less practice to achieve the learning objectives are allowed a path through the instructional materials which is shorter and more efficient, while those people who need more time and practice to achieve the learning objectives follow longer paths through more material.

A third kind of individualized instruction program is that in which the materials are modularized and branched so that different individuals can choose different learning objectives. These programs are referred to as programmatically branched individualized instruction. In such programs individual learners can decide which learning objectives are important to them and which they will choose to achieve. Correspondingly, the individual learner determines the appropriate entry and exit points for the training program.

The advantage of programmatically branched individualized instruction is primarily that of allowing the learner to learn only those things which he or she deems important and desirable. Such materials can also be constructed to allow greater variety in the incoming levels of knowledge, skill, and attitude of learners. That is, those people who have lower levels of knowledge and skill are allowed to begin the programmatically branched individualized instruction at points appropriate to their level of knowledge and skill and to progress through a larger number of learning activities to achieve a particular objective which another person with more advanced incoming levels of knowledge and skill could achieve through fewer instructional modules.

24.7. MAKE OR BUY?

Frequently an engineering manager decides that a group of his professionals need to participate in some training activity and the manager must then decide whether to "make or buy" the training. Sometimes the manager can use in-house expertise to design and to deliver an internal training program to achieve the training objective for his or her organization. Sometimes the engineering manager must go outside the organization to purchase the desired training program.

24.7.1   In-House Expertise

Occasionally the organization has within itself a subject-matter expert who also has skill in organizing and communicating information. When such an expert exists and is available, a training program can be designed and delivered completely within the organization.

Note however, even in these situations the engineering manager should require the preparation of a training proposal with a clear definition of objectives and benefits, costs, intended audience, learning processes, and evaluation techniques. Frequently, when managers critically evaluate these in-house training programs, they find the benefit to cost ratio is significantly lower than they intuitively expect.

In-house training programs that use in-house expertise have several advantages: (1) Technology transfer within the organization is enhanced both in classroom-training experience and the work environment. (2) Proprietary technologies can be taught and learned in these programs. (3) Such activities frequently have very low out-of-pocket expenses. (4) Such activities can be conducted with a minimum disruption of the normal work of the organization. (5) The stature of the subject matter expert is enhanced in the organization.

The disadvantages of in-house training with in-house experts is that no new ideas and points of view are likely to be included in the training experience. This type of training is an incestuous activity.

24.7.2. Buying Training

When in-house expertise does not exist, is not available, or lacks the ability to organize and communicate the information effectively, then the engineering manager is forced into some kind of "buy" process to acquire the desired training. Generally the buy choices are (1) to buy an existing off-the-shelf training program and use it as is, (2) customize or tailor an existing course to the needs of the organization, and (3) contract for the creation of a new training program designed for the specific needs.

Off-the-Shelf.   Ideally the engineering manager would like to open a training catalog and select a course which matches the specification for the needed training. This never happens. At best, the engineering manager is able to find a preexisting, on-the-shelf training program which is "close enough" to the specification to be useful as is. The engineering manager can then choose to send people to the program or contract with the vendor to provide the program on company premises.

The advantages of sending employees to off-the-shelf training programs include (1) removing the employee from the work environment and allowing the employee to concentrate on the training, (2) allowing the employee to gain a broader perspective and expand his or her network of colleagues by having contact with employees of other organizations at the training program, and (3) using training primarily as a reward.

The advantages of contracting with a vendor to bring a program onto company premises include (1) the ability to expose a larger group of employees to the same training at the same time, (2) lower financial cost per person trained, and (3) saving on travel expense. Additionally, since all the participants are from the same organization, some company-relevant synergy between the training activity and organizational objectives will occur.

Customizing Available Programs.   When standard programs are contracted to be given on company premises (or at facilities adjacent to company premises) exclusively for individuals from the same organization, an easy and convenient opportunity exists to customize the standard program. The degree of customization may be great or small depending on the needs of the organization and flexibility of the training organization. Customizing off-the-shelf programs increases the relevance of the training program to the unique situation of the organization and its individuals. Material not relevant to the organization or its individuals can be eliminated. The material can be tailored to make it use the organization's jargon, norms, policies, procedures, and so on.

Creating Unique Programs. In the extreme the engineering manager may contract for the creation of a unique training program. Unique training programs require more expense and longer lead time than do off-the-shelf programs. Even with unique programs some risk exists that the resulting training activity may not turn out to be what was originally envisioned by the engineering manager.

The advantages of creating unique programs include (1) the opportunity to focus sharply on organizational objectives, (2) tailoring to the specific individuals attending the program, and (3) the opportunity to develop a competitive advantage for the organization.

24.8. SOURCES OF TRAINING

Assuming that the engineering manager is in a "buy" mode for acquiring the desired training for a single individual or group of individuals within the organization, where can he or she buy such training?

24.8.1. Within the Company

The engineering manager should contact appropriate company resources to determine whether or not some component of the company can provide the training. Typical places to look include training organizations at corporate, group, division or local levels; employee relation functions; customer training organizations, and so on. Research and development organizations or engineering organizations in other product businesses of the same company may be willing and able to provide the training. Frequently managers are surprised at the breadth and depth of training resources available within their own company.

24.8.2. Suppliers

Organizations from which you buy hardware or software will frequently provide training in the use of that hardware or software. Even when such training programs are not included in the usual customer services of the supplier organization, the engineering manager can creatively explore opportunities to use the expertise of the supplier organization. Such exploration frequently results in highly effective yet low-cost training programs.

24.8.3. Schools

A common source of training is the local university, college, community college, vocational school, or even high school. Traditional educational organizations are increasingly anxious to develop continuing education programs and training programs for industry. By working with an educational organization in the local community, training needs can be met and simultaneously, community relationships can be improved. Academic professionals are frequently the best source for training programs whose content is more basic, fundamental, or theoretical. Great care must be exercised when schools and their faculties are used for training programs where content is state-of-the-art or more engineering-practice oriented.

24.8.4. Professional Societies

Every technical professional discipline has its own technical professional society. Increasingly professional societies recognize an obligation to provide continuing education programs as part of their professional services. Such programs might include short courses, correspondence courses, or even satellite-delivered teleconferenced courses. To determine what services might be available, just call the professional society headquarters. Even without a structured continuing education program, the professional society can still be used to identify subject matter experts who might be willing to develop and deliver a training program.

24.8.5. Training Companies and Entrepreneurs

Adult education and industrial training are becoming big business. Many large and small business enterprises are establishing technical training as one of their product areas. Off-the-shelf short courses and correspondence courses are typically marketed through direct mail solicitations. If you are a member of a professional society or if you request information on any training program, your name will appear on mailing lists used by those enterprises trying to sell training. Hence your daily mail will include flyers, brochures, catalogs, and other "junk mail"—aimed at getting you to buy training materials.

To procure customized training you may contract with a subject-matter expert to design and deliver the desired training program. In this case the program is to identify candidate subject-matter experts. A few phone calls to the appropriate professional societies, nationally recognized universities, or nationally recognized companies will usually lead to the identification of several candidate subject-matter experts. After describing the desired training, the candidate subject-matter experts can be invited to submit proposals to do the training.

24.9. MEDIA CONSIDERATIONS

In this section we discuss some of the advantages and disadvantages of the several different instructional media available for the delivery of training.

24.9.1. Live Instruction

The most common media for instruction is live instruction, that is, a subject-matter expert presenting in-person to a group of learners.

The most significant advantages of live instruction are the personal contact with the subject-matter expert and the opportunity to interrupt the presentation to ask questions. However, this advantage is reduced as the size of the learning group increases; the more people in the room at the same time, the more difficult it is for an individual student to interject a question. We all tend to be most comfortable with live instruction because we have the most experience with it.

Live instruction also has some disadvantages: (1) Using live instruction, the subject-matter expert and learners must come together at the same time. (2) Few individuals are both subject-matter experts and good teachers. Further, since only a few such persons exist in the world, the probability is low that such a person will be available at the local learning site.

24.9.2. Televised Instruction

Three different kinds of televised instruction are discussed in this Section. While television is universally accepted as an entertainment medium, it has not achieved similar, widespread acceptance as an instructional medium. Nevertheless we see an increasing utilization of televised instruction and encourage the engineering manager to carefully consider its advantages and applicability.

Two-way Video and Two-way Audio.   An alternative description for two-way video and two-way audio is full two-way video teleconferencing. In this media two or more geographically separated sites interchange video and audio signals. For example one subject-matter expert can be at one location, another subject-matter expert can be at another location, and groups of students can be at several other locations. Speaking one at a time, each person at every site can see and hear whoever is speaking at the moment. In short, through the wonder of two-way video and two-way audio, a super classroom can be created electronically. The advantages of full two-way video teleconferencing are obvious. Subject matter experts who also possess extraordinary teaching skills can be used from central locations to reach out to students at different geographical locations. Students are able to question the instructor. Intimacy between subject-matter expert and students is maintained because students are visible to the instructor when they ask questions.

The primary disadvantage of full two-way video conferencing is cost. The expense of fully switched two-way audio and video is extraordinarily large and virtually never can be justified for use in training. A secondary disadvantage is the rigid time schedule that must be maintained; instruction must start and stop at times dictated by the transmitting network schedule; students cannot "stay after class" to continue the discussion with the teacher.

One-way Video With Two-way Audio.   A simpler and more cost-effective arrangement of teleconferencing is one in which students see the instructor on video but their own link back to the teacher is only by audio. The video and audio from the instructor is frequently broadcast on ITFS (Instructional Television Fixed Service) within the local community (line of sight between transmitting and receiving antennas is required) or is beamed to an earth satellite and transponded back to earth to reach larger geographical regions. The audio links from the students back to the subject-matter expert are frequently along common carrier phone lines; however, sometimes within ITFS systems, an FM radio transmission technique is used.

The principle advantage of this technique is that a subject-matter expert who is a master teacher can be linked to a large number of geographically dispersed students. The audio link from the students to the originating studio allows the students to question the instructor.

The disadvantages of one-way video and two-way audio teleconferencing are similar to those for full two-way video conferencing in that subject-matter expert and students must all be present and be participating in the learning experience at the same time. While cost of this instructional delivery system is reasonably high, and while the logistics associated with operating such an instructional delivery system are reasonably complex, still the expense of the system is significantly less than full two-way video conferencing and can frequently be justified with a rational comparison of nonteleconferencing alternatives.

Video Tape or Disk.   It is possible to capture a subject-matter expert's presentation by using videotape or videodisk. Adding some printed text and perhaps some computer-aided exercises or simulations can result in an effective multimedia instructional package.

The advantages of video-recorded instruction include the following: (1) The learner is able to schedule the learning activity at a convenient time. (2) Learning can take place individually or in small groups which may be self-led or which may be facilitated by someone of only moderate subject-matter expertise. (3) The recorded instruction allows the opportunity to stop and discuss a point or replay a portion of the presentation. (4) This instructional delivery media is typically less expensive than either of the two teleconferencing techniques previously described.

Disadvantages associated with this technique also exist. The learner cannot interject questions. At worst students cannot contact the instructor at all—not by telephone, not even by mail—at best they can telephone their questions to the instructor. Such instruction can be obsolete if the state-of-the-art changes significantly.

24.9.3.   Audio Instruction

Audio instruction means instruction delivered to the student by radio broadcast or by audiotape. Descriptions of three types of audio instruction follow.

Two-way Audio.   Two-way audio instruction is instruction in the radio talk-show format. The lecture is broadcast on radio frequencies and students telephone their questions or comments to the instructor in the studio.

While this technique has limited applicability, it can be used creatively. The instructor can prepare key visual material ahead of the broadcast and mail it to the students before the broadcast. Thus visual material can be included in an audio delivery medium. With this technique students have the opportunity to interact directly with the instructor. The technique is significantly less expensive than any kind of video-broadcast- or videotape-approach and is particularly suitable for local geographical regions using standard AM or FM broadcasts.

Disadvantages of this technique includes: (1) Exchanging visual information, while possible, is difficult. (2) The subject-matter expert and students participate simultaneously in this learning activity; hence no schedule flexibility exists.

One-way Audio.   A one-way audio instructional mode is the same as a standard radio broadcast with no call-in capability. The advantage of this system is that the instruction can be radiated across the local radio broadcast region at very low cost. But since no student interaction is permissible in this delivery mode and visual information is exchanged only with the greatest difficulty, this technique is not recommended for industrial training programs.

Audio Tapes.   An audiotape delivery format typically combines audio cassettes and printed text. All instructional materials are bound into a notebook or some other suitable package for delivery to a single learner or a small group of learners.

The advantages of this technique are several. (1) A low cost instructional package can be prepared which is both educationally effective and interesting. (2) The program has great schedule flexibility. The student can listen to the audio cassettes at the office, at home, or riding in his or her car between the two. Similarly, the student can read the printed materials at any place and time of convenience.

The obvious disadvantages of this media include: (1) It is not possible to show video motion. (2) The student is not able to interject questions or comments.

24.9.4.   Printed Instruction

We should not forget the educational value of printed materials without any accompanying audio- or videotapes. Printed text is the oldest form of instructional technology available to civilized man.

Books and Articles.   Individuals can and do learn by reading textbooks, journal articles, advertising literature, instruction books, and so on. Such learning activities are typically unstructured and pursued on an individual basis. However, small groups of individuals can be organized to read journal articles or sections of books and then participate in a review of the reading.

An engineering manager can provide the leadership to establish such learning activities. For example, the manager may assign particular individuals in his or her organization to read each issue of certain journals as they are published and extract those articles relevant to the work of the group. The manager may then conduct semi-formal review and discussion of these articles among members of his or her organization. When these semistructured learning activities are managed in a purposeful way, large organizational benefits result at small cost.

Correspondence Courses.   Classical correspondence courses are entirely in printed format. The remote learner receives a series of lessons either all at once or sequentially, he or she completes the reading and assigned exercises, and mails the responses back to some central point where they are reviewed (graded) and then returned. Such correspondence courses are usually available at extremely low cost and are available at virtually any point in the world.

The principal disadvantage of correspondence courses is that they are frequently boring and uninteresting. They usually require an extraordinarily high level of motivation to complete; typically only about 30% of those who start a correspondence course will complete it.

24.9.5. Computer-aided Instruction

Much jargon exists in this instructional arena and no set of definitions of the terms is universally accepted. Computer-based instruction typically means that the instructional materials are presented using a computer in some way. Computer-managed instruction typically means that a computer is used not only to present some of the instructional material, but also to do some of the administrative functions such as testing or record keeping. Computer-aided instruction may mean the same as computer-based or computer-managed instruction.

Overview. Computer-aided instruction (CAI) can be either individualized or group instruction. When CAI is individualized, each learner interacts directly with the computer programs and progresses at his or her own rate. Most importantly, each learner, regardless of geographical location, has access to the same learning program. When computer-aided instruction is used in a group-instruction mode, frequently the computer is used only to provide a simulation or do some of the administrative functions for a group of learners.

Computer-aided instruction may present text or graphical information directly from the computer or indirectly by controlling some audiovisual device such as a videodisk or videotape player. The CAI program may be linear or branched and may include deterministic or probabilistic simulations of some process. The program may include practice exercises together with testing and scoring activities. The computer program may be on a personal computer, minicomputer or mainframe computer.

Off-the-shelf computer-based instruction is becoming increasingly available. Almost every packaged software program now includes a tutorial for beginners. Traditional publishing organizations are now selling an increasing array of computer-based learning programs. Control Data Corporation is continuing to increase the instructional software available on their PLATO system, and other typically smaller entrepreneurial organizations are attempting to identify and secure niche positions in computer-based instruction.

Instructional authoring systems allow reasonably easy creation of instruction to meet unique needs. PLATO is the oldest and most well known system, but now well over a dozen other systems are available. One or more authoring systems are usually available for each different computer.

Simulations.   One big advantage of computer-based instruction in a training environment is the ability to use a simulation as an integral part of the training. Having a computer model of a process, that is, a simulation, allows instruction to be designed such that learners can be taught a segment and then asked to use what they have just learned to make a set of decisions which become inputs to the simulation model. The simulation is then run and the output illustrates the results of the input decisions. If the simulation is realistic, then the learners can gain the equivalent of real-world experience within the training program. Using the simulation learners are able to get quick feedback on the consequences of their decisions so that good learning is positively reinforced as quickly as possible; poor learning is detected and corrected just as quickly.

Creating realistic simulations is a most difficult task and should not be undertaken without a careful and detailed assessment of the expected expense and the expected benefit. Many general business simulations are available for use in training programs; however good simulations of technical processes are not generally available.

Sometimes simulations are used in a game format wherein learners play the game, some kind of score is kept to measure performance and ultimately "winners and losers" are identified. The introduction of a competitive element within a training activity almost always increases motivation to learn. Generally it is better to structure the competition against reality, as represented by the simulation, rather than directly against other learners in the same group. Assessing the learning of individuals or groups can still be done by having each compete against the simulation and then comparing results.

Interactive Videodisk.   Linking a computer to a random-access videodisk player yields the most flexible instructional-media system imaginable. The instruction can be linear or branched; it can use deterministic or probabilistic simulations (audio, still frames, normal motion, slow motion, time-lapse motion, full color, text, photographs, charts, graphs, animations, and so on) as appropriate for maximum instructional effectiveness.

An increasing array of interactive videodisk instruction programs will become available. Initially these will be in broad personal development and business areas, but in time technical content will also become available.

25

THE COST EFFECTIVENESS OF TRAINING

Michael H. Frisch

25.1. INTRODUCTION

In most organizations, particularly larger ones, attendance at company-sponsored training programs is almost taken for granted, with little thought given to the ultimate payback for the time and money spent. With direct costs of training conservatively estimated at $100 per person per training day, and much more for individualized or hands-on technical training, organizations would do well to consider how best to increase their return on the investment of training dollars. In addition, employees have become much more demanding of the training offered to them. No more are they likely to hungrily attend dry lectures about management theory. Stimulating, relevant, creatively presented content is what employees require, and they will be vocal about not getting it.

This chapter focuses on ways individual managers can evaluate the effectiveness of training and obtain maximum benefit from programs which are used. These recommendations emphasize practical suggestions rather than methodology of evaluation research. The underlying assumption is that individual managers can, by their own observations and actions, greatly improve the value of training both to the attendee and organization.

25.2. Why Train?

Organizations have three major motives for training their employees. First, in technical environments particularly, relying exclusively on often incomplete and potentially dangerous on-the-job learning is contraindicated. More formalized training helps new employees or employees in new jobs become more productive in less time. In addition the work methods taught are typically built upon the collective experience of many skilled employees and therefore represent far greater productivity potential than any individual employee might acquire on the job.

This rationale applies equally to organizations faced with new or changing production technologies. It has been said that five years is the average half life for technical information and methods. Obviously, training is the primary vehicle for making new methods and new equipment operational.

Studies of occupational obsolescence in engineering knowledge show that as expected, significant "erosion" occurs as the time since graduation increases. However, they also provide evidence that the loss has been much more rapid in recent years than in the past. Figure 25.1 shows this increasing rate of decline in the applicable knowledge of graduating classes of engineers from 1935 to 1960. Not only do some elements of education go unused and are forgotten, but the explosion of new information means that recent graduates lose ground faster relative to earlier graduates. There is every ex-

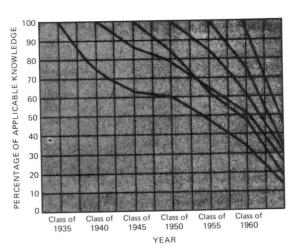

FIGURE 25.1. Erosion curves showing occupational obsolescence in the field of engineering. From S.B. Selikoff, "On the Obsolescence and Retraining of Engineering Personnel," *Training and Development Journal* (May 1969): 3

pectation that this trend will accelerate. These types of analyses clearly demonstrate the ever-present need for technical training.

Second, even if technology was not changing rapidly, employees would still require training as they increased their own level of mastery. Individual growth and development within the same job, and promotions which may occur, require that training be applied at moments when an employee is ready to assimilate and use more advanced methods. This applies not only to technical training, but also to managerial and interpersonal skills as well as professional growth. Career enhancement and skills development for each employee has become an essential element of retaining a motivated work force.

Finally, there is a purely economic justification for training. In most cases it is less expensive to develop skills in current employees than to hire skills from the outside. Locating candidates, interviewer time, travel and relocation costs, orientation time, and other factors make recruiting individuals with needed skills a very expensive proposition. Costs of hiring vary depending upon the organization and job, but can easily range from $5000 to $10,000 per hire even for entry level jobs. External search fees can double this. When viewed in this light, training current employees is an unbeatable bargain.

25.3. EVALUATING TRAINING

Regardless of the logical arguments supporting the use of training, the relative effectiveness of a particular program should never be ignored. Ineffective training not only ultimately wastes time and money; it also may confuse, frustrate, or even demoralize employees. Even a program with a strong following and good payback may have room for significant improvements, if they can be identified. Training evaluation helps target program redesign efforts and is essential to maintaining the quality of existing training offerings.

In most organizations it is the responsibility of the personnel or training department to assure that company-sponsored training programs are up-to-date, skillfully delivered, and adequate to the specific needs of managers in the organization. Although there are several formal research paradigms for training evaluation, it is not the purpose of this Chapter to explain these methods. Instead, more pragmatic approaches are emphasized.

Individual managers have an important role in assuring high-quality training. Often it is only through the informal evaluation of training, coming from the reactions and complaints of attendees and their managers, that company-sponsored programs are revised and upgraded. Also, when utilizing external individual-by-individual programs, it is solely the manager's responsibility to evaluate their effectiveness and determine their payback. Therefore, managers must be aware of practical methods for quality control of the training which they utilize within their departments.

A useful and concise approach to training evaluation is based on four levels of impact that training can have. As shown in Figure 25.2, the four levels are: satisfaction with the program, knowledge gained, skills developed, and organizational payback. Even without rigorous evaluation of each of these levels, managers will find the scheme useful.

Level I is the most simple and often the only level of evaluation that is applied. Level I asks how satisfied the participants felt about the program. Sometimes labeled the "applause meter," Level I evaluation is typically made following attendance at a program, using a questionnaire. A sample form for this is shown in Figure 25.3. These questionnaires are particularly useful if they ask participants for reactions to each segment of the program. This allows program developers to make selective improvements even in a program that was satisfactory.

It is valuable to know how much participants enjoyed a program, and it is unlikely, although not impossible, that a disliked program will have a positive effect on job performance. However, even if participants are satisfied with the course, important questions about its quality still remain. Level II asks what participants learned from the training experience. A well-constructed program will have clear and specific instructional objectives which can be compared with the actual learning that occurred. Sometimes this is evaluated by means of pre- and post-program testing. Score increases on the second test would largely be attributed to the training. For individual managers, however, Level II analysis becomes a matter of discussing course

FIGURE 25.2.   Levels of training evaluation. Based on D.L. Kirkpatrick, "Evaluating Training Programs: Evidence vs. Proof," *Training and Development Journal,* 31 (1977): 9–12.

Training Program Evaluation   Date: _____

Evaluated by: _____   Program title: _____
Title: _____   Offered by: _____
Location: _____   Instructor(s): _____

Instructions: Please mark "X" on that part of the scale that best fits your opinion on each item. Complete every item.

Part I

1. Subject matter

A.	Meaningful	Meaningless
B.	Uninteresting	Interesting
C.	Relevant to my job	Unrelated to my job
D.	Difficult to understand	Easy to understand

2. Instructor(s)

A.	Boring	Stimulating
B.	Challenging	Not challenging
C.	Considerate	Inconsiderate
D.	Ineffective	Effective
E.	Impatient with us	Patient with us
F.	Responsive	Unresponsive

3. Course material

A.	Useless	Useful
B.	Appropriate	Inappropriate
C.	Readable	Unreadable
D.	Out-of-date	Up-to-date

4. Applicability

A.	Fit my mgr. role	Didn't fit my mgr. role
B.	Relevant to my job	Not relevant to my job
C.	Can't apply it on job	Can apply it on job

5. Clarity of presentation

A.	Clearly presented	Confusing in presentation
B.	Not logically presented	Logically presented
C.	Systematic	Unsystematic
D.	Saw course direction	Saw no course direction

6. Involvement

A.	Felt fully involved	Never felt involved
B.	Did not participate	Actively took part
C.	Remained quiet throughout course	Openly questioned and offered ideas
D.	Felt connected to others during course	felt unconnected to others during course

FIGURE 25.3.   **Sample evaluation form for Level 1.**

7. Self-esteem

A.   Came away with | | | | | | | | Came away with
      higher self-confidence   lowered self-confidence

B.   Course didn't help me | | | | | | | | Felt helped by course

C.   Better understanding | | | | | | | | Understand my job
      of my job   no better

D.   More hopeful about | | | | | | | | Less hopeful about
      future in my job   future in my job

8. Overall seminar evaluation

                    Low | | | | | | | | High

Part II

Incomplete sentences: Please finish the incomplete sentences below with the first phrase that comes to your mind. Complete every sentence.

1. The useful part(s) of this course for me

_____

2. The least valuable part(s)

_____

3. When I get back I plan to implement

_____

4. If I could change anything in this course, it

_____

5. With respect to this course, the training department should

_____

6. Other comments

_____
_____
_____
_____

FIGURE 25.3   (*Continued*)

content with employees after they return from the training to determine whether effective learning has taken place.

Probably the most important measure of training success is improvement in performance. Level III evaluation asks how well participants are applying newly learned skills to their jobs. In other words, what observable changes in work behavior have occurred? Again, formalized pre- and post-training measurements can help answer this question. More realistically, existing performance appraisal systems can be used to investigate the impact of training. At the minimum, prior to training managers should have identified aspects of performance for improvement, and after training they should look for those changes. Any recognition of improvement should be discussed with employees within 30 days of the training.

Level IV evaluation, the broadest assessment, inquires how the organization's bottom line has been affected by training. This level is not particularly applicable to individualized training. Where groups of employees are being trained, however, both departmental and larger unit performance may be analyzed for Level IV benefits. Level IV analysis is most frequently the responsibility of a centralized training department or headquarters group which can track results at many points in the organization. For example, a newly implemented sales training program might be evaluated by a task force of senior sales managers using data supplied by accounting sources guided by a study designed by the human resources department. For all of the difficulty in conducting Level IV evaluation, if training does not in some way positively influence an organization's bottom line its basic value is open to question. The difficulty lies in identifying and quantifying the specific contribution made by training as opposed to all other factors which influence organizational performance.

Although managers may never design and conduct research studies on the success of a specific training program, the levels of evaluation previously discussed have significant practical usefulness. Level I and II evaluation should be part of every manager's discussion with an employee who has returned from a training experience. Discussing subjective reactions to the program and reviewing knowledge gained is useful to the manager in understanding that particular program and to the employee assimulating the experience and applying it to the job. Level III evaluation asks managers to be astute observers of improvements in the actual performance of employees who have attended training. It is the sole level of evaluation for which managers can be held accountable, using results documented in employee performance evaluations required at least annually by the organization. Level IV evaluation represents a practical overall view and asks, "What is the actual return on the investment in training and development?" When managers lose sight of this focus, the impact of training always suffers.

25.4. IN-HOUSE VERSUS EXTERNAL TRAINING

Managers should be conscientious consumers of the training they use. A few facts about the training industry will help. Training programs and workshops are offered by a wide variety of training and consulting firms, universities, and professional associations. These vendors usually offer "open enrollment" or "public" courses which employees from many organizations may attend. Although these programs are useful for organizations without internal support for training and for more unusual training needs, they are much more costly per person than training conducted in-house usually by the personnel or affiliated department. External training costs approximately $200 per person per training day, not including travel and lodging expenses; in-house programs can be estimated at half that amount per day and usually involve fewer additional expenses.

Any vendor who offers training publicly will normally conduct programs internally if a group of 8 to 12 participants can be formed. Not only does this result in substantial per person saving, but it also allows the training material to be tailored to the particular needs of the sponsoring group. Frequently managers will use occasional attendance by employees at external programs to help identify the best programs for in-house use, and subsequently they will form a contract with the vendor for wider application within the organization. All of these benefits depend upon the ability of a manager to identify individuals with the same training needs at approximately the same time of the year and to coordinate the in-house logistics.

Managers interested in external training should obtain information from relevant professional associations and the continuing education departments of local colleges and universities. The names of other local external training vendors and consultants are best identified by asking other managers about their use of training. Vendors operating on a national basis should also be contacted for information. Several are listed below:

American Management Associations
135 West 50th Street
New York, NY 10020
(212) 586-8100

BNA Communications, Inc.
9439 Key West Avenue
Rockville, MD 20850
(301) 948-0540

Drake Beam Morin, Inc.
277 Park Avenue
New York, NY 10172
(212) 888-3134

Fred Pryor Seminars
2000 Johnson Drive
P.O. Box 2951
Shawnee Mission, KS 66201
(913) 384-6400

Forum Schrello
555 East Ocean Boulevard #222
Long Beach, CA 90802
(213) 437-2234

NTL Institute
P.O. Box 9155
Roslyn Station, VA 22209
(703) 527-1500

Quality Circle Institute
1425 Vista Way, Airport Industrial
Park
P.O. Box Q,
Red Bluff, CA 96080
(916) 527-6970

Scientific Methods, Inc.
P.O. Box 195
Austin, TX 78767
(512) 477-5781

Wilson Learning Corporation
6950 Washington Avenue, South
Eden Prairie, MN 55344
(612) 944-315)

In addition, directories of training vendors and consultants can be obtained from American Society for Training and Development (202-484-2390) and Training Magazine (612-333-0471). Finally, a valuable source of both general and evaluative information about training vendors is Mantread, Inc. (612-293-1044), but an annual membership fee is charged.

Before selecting a training program, the following should be explored carefully:

- Read brochures or other descriptive materials specifically for program content. If a comprehensive outline of the program is not included, request one. As a general rule, the more detailed the description the more complete the training.
- Look at the range and scope of the material to be covered in the time allotted for the program. If it seems very ambitious, it probably is meant as an overview course or assumes prior experience with the content. Remember, training in basic skills takes time; even the most efficient design cannot teach microprocessor circuitry in a three-hour seminar.
- Check references. Talk to people quoted in the brochure and colleagues who have taken the course, or ask the vendor for a list of recent program participants. Ask past participants open-ended questions about the program in order to get a broad picture of their experiences. Questions such as "Why did you select this particular seminar?" "What did you expect from it?" "How did you benefit from having attended?" should clarify what to expect from the program and whether it will realize its promises.
- Find out how frequently the program has been offered and how many years the vendor has been in business. The longer the program and/or the company have been in existence, the more valuable the experience is likely to be.
- Ask the vendor if hand-out materials will be provided to reinforce the training. If possible, review these documents before making your final decision about using the program.
- Check to see if the vendor provides an opportunity for follow-up consultation and/or questions to the consultant after the training pro-

gram. If they do provide such follow-up support, be sure to find out if there is a charge for it.

- Don't be convinced or dissuaded by the program's price alone. The cost may not be an accurate indicator of quality. Decision to buy should be determined by all of the factors listed above.

25.5. GETTING MAXIMUM BENEFIT FROM TRAINING

Once a program is selected, there are many things managers can do to assure that maximum benefit is derived from attendance.

First, it is critical that managers prepare employees properly prior to the training. Learning, and training in particular, work best if the learner:

1. Clearly understands his or her need for training
2. Wants to correct or improve the deficiency
3. Views training positively and wants to attend

Each of these points should be discussed with the potential trainee. If the individual disagrees or seems hesitant about any of them, a more probing discussion to compare the individual's view of his or her development needs with the manager's perspective is indicated.

Second, check the logistical details of the selected program. There is a saying that, "It is the little things that often make the biggest difference." Small errors in dates, locations, or registration requirements can cause major confusion and disappointment. Managers must be sure that all necessary information about the selected program has been communicated to the attendee:

- Specific name of the program and the vendor (if it is not an in-house program) including address and phone number.
- Location (city and meeting site)
- Dates and scheduled times of the program
- Registration procedures
- "Prework" materials, if necessary
- Travel and lodging arrangements, if needed

Managers should make registration, travel and lodging arrangements, or direct the employee to do so, as far in advance as possible.

Managers should also consider the dates that the employee will be attending the program in terms of workload. If there are any doubts about the department's ability to operate without the individual for the required time, reschedule attendance for another session. Pulling someone out of a training program or cancelling attendance at the last moment wastes time and money and is very poor for morale.

Third, set learning objectives with the employee. Several days or a week before the training program, the manager should meet with the attendee to clarify expectations and establish clear objectives for the training. What makes an effective objective? As with work accountabilities, objectives should be:

☐ Specific, for example, "be able to conduct *campus* interviews"
☐ Behavioral, for example, "be able to make presentations to upper management"

Managers should suggest that the attendee make a list of specific job-relevant questions he or she would like answered during the workshop.

This goal-setting effort is very useful because it highlights specific aims for the attendee. It also coordinates the manager's expectations of the training with those of the employee. "But I thought you wanted me to learn . . ." is an unfortunate and frustrating discovery after one has invested $1200 in a workshop.

Fourth, follow-up with the attendee immediately after the program. Once the employee is back on the job, training has a strong tendency to fade very quickly. Within a day of two following the workshop the manager should review the general content of the program with the attendee, as well as the specific points that had been targeted for learning. Additionally, manager and employee should discuss ways in which the newly learned material can be applied directly to day-to-day work. Identifying specific opportunities for attempting to use new skills and discussing how the employee should apply them are very helpful follow-up procedures. As part of this follow-up process, the attendee should complete a training evaluation form, and the manager should add the form to others in a central training file. As previously discussed Level I evaluation is a valuable aspect of assuring quality control. In this way future attendees will benefit by each employee's participation in a training program.

Finally, continue the follow-up process. Periodic brief discussions with the employee about the application of new skills remain critical to learning, even long after the training. As a manager observes progress, development can be further accelerated by giving the employee greater opportunities to apply the new skills. These might include:

- Special projects that would use the new skills
- Enlarged or enriched job responsibilities (even if temporary) that depend more heavily on the skills.
- Exposure to other parts of the organizations in which the skills may be emphasized.
- Skill practice opportunities
- Assignments with experienced "pros" who will model and further reinforce application of improved methods

Similarly, an occasional "pat on the back" or word of encouragement goes a long way toward making new skills operational. Managers should remember that using new and unfamiliar skills can be very frustrating. Support from

1. Preparation of subordinate   *Check*
   Does the subordinate:
      Clearly understand the need for training?   ☐
      Want to correct/improve the deficiency?   ☐
      View training positively and want to attend?   ☐
2. Details of Program
   Does the subordinate have information regarding: ☐
      Name of program/program vendor?   ☐
      Location (city and meeting site)?   ☐
      Dates to times?   ☐
      "Pre-work" materials, if necessary?   ☐
      Travel and accommodation arrangements?   ☐
3. Objectives of Training
   Is the subordinate clear regarding:
      Objectives of the training?   ☐
      What skills he/she can expect to learn?   ☐
      How the skills relate to the job?   ☐
4. Follow-Up After Program
   As a manager have I:
      Reviewed content of the program with subordinate?   ☐
      Discussed specific application of skills to the job?   ☐
      Identified opportunities for subordinate to try new skills?   ☐
      Agreed target dates for application?   ☐
      Shown support and enthusiasm?   ☐
5. Ongoing Maintenance of New Skills
   As a manager, have I considered:
      Special project for subordinate to apply new skills?   ☐
      Exposure to other parts of the organization?   ☐
      Practice opportunities?   ☐
      Assignments with "pros" who will model effective performance? ☐
      Complimenting subordinate on progress made?   ☐

FIGURE 25.4.   Checklist for attending training programs.

the "boss" can make the difference between expanding capabilities and complacent resignation to existing skill levels.

Several of these suggestions do require limited amounts of time from the manager. Unfortunately, there is no substitution for conscientious pre- and post-training support. The checklist provided in Figure 25.4 should help managers keep track of the steps needed to optimize the benefits of a training program.

25.6. SUMMARY

Measuring the cost-effectiveness of training programs is a requirement that few would dispute, particularly in the face of changing technology and increased worker interest in self-development. However, demonstrating the effectiveness of training programs is more complex than it may seem. A four-level schema was discussed for training program evaluation, covering satisfaction with the program, knowledge gained, new skills applied to the work, and actual organizational payback. These levels were discussed in terms of managers' practical use of them rather than presented as research designs. In addition advice was given about using external training, including names of vendor information sources. Finally, methods were offered for maximizing the benefits derived from training by structuring the communication between managers and employees before and after the program.

26

MOTIVATION

Roy Udolf

ROY UDOLF received his B.E.E. degree from New York University, J.D. from Brooklyn Law School, and Ph.D. in Psychology from Adelphi University. He is a diplomate of the American Board of Forensic Psychology and is a teaching fellow and professor of Psychology at the New College of Hofstra University. He is the author of five books and numerous professional journal articles on forensic psychology, hypnosis, human engineering, and logic design. Dr. Udolf has been at Hofstra University full time since 1967 and is in private practice as a clinical psychologist and industrial consultant.

26.1. BASIC CONCEPTS OF MOTIVATION

To a psychologist all human behavior is theoretically as predictable as the behavior of an electronic circuit. If we knew all of the details of a person's past history and genetic makeup, the environmental influences presently acting on him, and if we understood perfectly of the principles of human behavior we would be able to predict his or her actions in a given situation without error. Freud labeled this notion that all human behavior is lawful and therefore predictable (and thus by implication controllable), "psychic determinism." Lewin alluded to the same situation in his oft-quoted equation $B = f(E, P)$ which indicates that all behavior is a function of factors in the environment and within the person. It is these latter internal determinants of behavior that psychologists refer to by the term personality.

Of course in a real situation we never even approach having this much information about a person whose behavior we are interested in predicting or controlling and we are a long way from fully understanding all of the principles governing human behavior. However, there are enough similarities in the genetic makeup and life experiences of most people so that despite substantial individual differences, it is possible, even with our present day limited psychological knowledge, to make fairly good predictions of human behavior. This permits us to conduct our daily business and personal, human interactions in a practical and effective manner. If all of us, even people untrained in psychology or management, were not able to predict human behavior, it would be impossible to function in daily life.

Is there any employee who is unable to predict how his or her boss would react to being greeted in the morning with a cheerful, "Good morning, Stupid," or any husband who would be unable to gauge his wife's reaction if he came home at 3 a.m. with a strange lipstick on his collar?

The most fundamental factors that affect the interaction of a person with his or her environment are called primary drives. This is basically a biological concept and relates to the fact that all living things are complex, biological systems that require constant interchange of chemical substances with the environment in order to survive. For example, we all need to take in food to repair tissue and provide energy for our daily activities. Since about 95% of our bodies is composed of water which is constantly being lost through perspiration, evaporation, urination, and other means, we constantly need to replenish this loss. Oxygen is needed to liberate the chemical energy in food and support the chemical processes involved in metabolism. In addition we must periodically eliminate the waste products of metabolism before they accumulate and poison us.

If an organism were born into the world without some innate mechanism that caused it to seek out these environmental products, or react appropriately when its storage capacity for them was in danger of depletion, this organism would be unable to survive. It is this inborn mechanism that is referred to as a primary drive by psychologists. It is called a drive because if you need something you have to do something to get it. Hence a drive activates or drives a person into behaving. It is called primary because it is innate. Examples of primary drives would include such things as hunger, thirst, air hunger, pain avoidance, elimination, sex, and so on.

For theoretical reasons psychologists often make a technical distinction between the term "need" which refers to an environmental requirement and "drive" which refers to a subjective condition of the organism designed to facilitate the attaining of this need, but often, particularly among management personnel, the terms are used interchangeably.

It should be noted that all primary drives, with the exception of sex, are survival drives in that if they are not gratified the organism will die. Sex is, of course, a survival drive for the species, if not for the individual, for if individuals were not driven to reproduce themselves a species would die out in one generation.

As a result of an unfortunate choice, the German word "Trieb," used by Freud, was translated into English as "instinct" instead of "drive." Thus to Freudians the term id instinct means exactly the same as the more modern term, primary drive. It has no connection at all with the biological meaning of the term instinct which refers to a complicated series of behaviors often controlled by environmental cues which are both unlearned and occur species wide, for example, nest building by robins.

People often erroneously refer to human instincts in the biological sense of the term, for example, the so called "instinct of self-preservation." There is no evidence that there are any true biological instincts in people and what is sometimes referred to as an instinct is usually a need. If we had a real instinct for self preservation we would behave in a stereotyped manner anytime we were in danger. All one has to do is to consider the Kamikaze pilots of World War II to recognize that they had a much higher need for self-esteem than for self-preservation.

Sexual behavior in rats is instinctual. All rats copulate in the same stereotyped manner whether sexually experienced or not. Human sexual behavior on the other hand is extremely varied and hence learned. In fact authors write best selling books designed to teach the readers varying techniques of sexual behavior.

Probably the reason that people do not have instinctual behavior patterns is that our highly developed cognitive abilities have taken over the role of generating adaptive behavior sequences in response to problems presented by a changing and often unpredictable environment.

An interesting insight into the nature of primary motivation or drives can be had by an examination of Freud's notion that all neurotic symptoms are caused by frustrated sexual motivation. This he believed produced unconscious sexual conflicts that expressed themselves indirectly through maladaptive and symbolic behavior called symptoms.

The body has varying storage capacities for the various environmental products that it requires. For example, depending on environmental conditions and a person's activities, it may be possible to survive for about two months without food, two days without water, and for an athlete in prime condition, one minute without air. On the other hand, since sex is not a survival requirement for an individual, a person could be deprived of a sexual outlet for a lifetime and suffer no physical damage. Thus biologically speaking, sex is the weakest of all primary drives while air hunger is one of the strongest. Why should sexual deprivation play such a large role in the generation of psychological problems while the much stronger air hunger drive is given no consideration at all by clinicians?

The answer of course is that we live in a society that has many strict rules concerning the expression of a sexual drive with respect to when, where, with whom, and under what circumstances such a drive may be expressed while it has little or no concern about the gratification of most other primary drives which are ordinarily immediately gratified. Hence, because of social rules, a sex drive is often deprived for long periods, which permits it to become very intense despite its biological weakness. Indeed, the fact that sex is not an individual survival drive is what permits this situation to occur at all. If we lived in a society that considered breathing or eating to be immoral and hence restricted the free expression of these drives we would not live long enough to develop neurotic symptoms.

Freud considered primary drives or id instincts to be the basic source of what he called "psychic energy" or the energy necessary for all human mental activities. He assumed that this energy resulted from the metabolism of the body and modern physiological psychology confirms this view.

There is evidence that the seat of the primary drives is a region of the brainstem called the hypothalamus which contains chemoreceptors that are sensitive to the chemical composition of the bloodstream. For example, these receptors produce feelings of hunger in response to a low blood sugar level or thirst in response to a high blood salt level. Appropriate motivated behavior in rats has been produced by electrically stimulating hunger, thirst, sex, or elimination centers in the hypothalamus.

While the details of this research are fascinating, they are not as important to executives concerned about manipulating and modifying the behavior of others as are the functions of drives.

There are five major functions served by all drives:

26.1.1. Drives Activate an Organism

As previously noted, this is the most basic function of a drive—if a person or an animal needs something, he must do something in order to attain it. In general, animals cannot be trained to run a maze or do anything else unless they are activated by being deprived of food or water. In the absence of some drive state, they will be so inactive that they will never make the desired response that must be followed with a reward or reinforcement to be learned. Similarly, people who are unmotivated are inactive. This is the reason that managers are themselves motivated to learn about the principals of motivation.

26.1.2. Drives Determine Reinforcement

The basic method of learning, which involves the voluntary or skeletal nervous system as opposed to the involuntary or autonomic nervous system, is called instrumental conditioning and is described by the following paradigm:

$$S \rightarrow R \rightarrow \text{Rein.}$$

This paradigm states that any time a stimulus is followed by a response and that response is followed by one of a class of events called a reinforcement, the stimulus becomes more likely to be followed by the response on subsequent presentations. Put another way, the stimulus gains an increment of evocation control over the response.

The question of what events constitute a reward or reinforcement is an empirical one. Some events have reinforcing properties, others do not. It turns out that whether a particular object or event is a reinforcement is a function of what drive the organism was acting under at the time of making the response. Food is a reinforcement to a hungry animal and it will learn to negotiate a maze to attain it. Water is not a reward to such an animal and it will never learn a task to secure it. On the other hand if the animal was activated by making it thirsty instead of hungry, water not food is needed for reinforcement. Thus the drive currently producing behavior determines what is or is not a reinforcement at any particular time.

The relationship between a drive and a goal is very much like the relationship between the head and tail of a coin. A drive like hunger is something internal that is present at the beginning of a task. The goal is some external object that is sought. Often the goal of a primary drive is some object which gratifies or reduces the drive, and there are some theorists who believe that all primary reinforcements must be drive reductions.

If an animal had no operating drives it would be impossible for it to learn anything for two reasons. Firstly, it would be inactive and hence would make no responses that could be reinforced. Second, even if a response were made it could not be rewarded and thus learned. Fortunately for managers, who deal with human as opposed to animal behavior, people are never without the presence of some active motivation due to the existence of learned or secondary drives or motives.

26.1.3. Drives Function as Stimuli

Drives not only produce or activate behavior; they may also direct behavior. People generally behave differently when they are hungry from when they are sleepy. This is because they can differentiate the inner feelings produced by the various drive states and have learned that different sequences of behavior are necessary to gratify different drives. Thus drive states may function as the stimulus term in the foregoing conditioning paradigm.

26.1.4. Drives Interact

The interaction of drives is more properly described as a characteristic rather than a function of drives.

Often more than one drive may be active at the same time. If both drives can be gratified by the same course of action then this behavior becomes more likely to occur than if either of the drives were acting singly. Their effects may be said to be additive, although they are not necessarily additive linearly.

On the other hand, if two or more drives coexist and each requires an action to gratify it that is incompatible with the gratification of the other, a state of conflict is said to exist. In order to have a true conflict it is necessary for the opposing drives not only to require incompatible behavior but to be approximately equal in intensity. If one drive is much stronger than the other, then the result is likely to be the gratification of the stronger at the expense of the weaker.

The strength of the opposing drives is basically a function of the biological strength of each drive and length of time that each has been deprived. For example, since a thirst drive is biologically stronger than a hunger drive (there is a longer storage capacity for food than for water), if both drives were deprived equally the organism would be more thirsty than hungry and would simply give up food to get water.

In a real conflict situation, in which the opposing drive states are equal as well as opposite, the person is in a state of paralysis in the sense that no action can be decided on and neither drive is gratified. In effect such conflicts produce frustration (i.e., the blocking of goal-directed activity) until resolved.

Lewin classified human conflicts into three broad categories:[1]

1. Approach–approach conflicts
2. Approach–avoidance conflicts
3. Avoidance–avoidance conflicts

In an approach–approach conflict there are two goals and both of them are equally positive, but to attain one goal it is necessary to abandon the other. For example, a person would like to become a college professor but also desires to provide a high standard of living for her family. In animal research on Lewinian conflicts, Dollard and Miller found that this type of situation was not a true conflict as it was intrinsically unstable.[2] The slightest amount of movement toward either goal, promptly resolved the conflict and the animal proceeded directly to the chosen goal while ignoring the other. This is the reason that we do not generally agonize for prolonged periods when forced to choose between several positive decisions such as which model car to purchase or which of two potentially good employees to hire. The remaining two types of conflicts were experimentally demonstrated to be quite stable and disruptive of effective action.

In an approach–avoidance conflict there is only one goal but it has both positive and negative attributes. That is, it has both reinforcing and punishing aspects. An example of such a situation in an industrial setting would be where the only qualified applicant for a critical job makes salary demands that may strain the firm's financial limitations. A more common example might involve the desire to consult a physician about some troublesome physical symptom, which is opposed by the fear of the doctor discovering some serious illness.

An avoidance–avoidance conflict involves two goals but here they are both aversive, and avoiding one requires accepting the other, for example, having to vote for one of two candidates for public office when each is perceived as being unfit for the job. This is the situation usually referred to as the selection of the lesser of two evils. As is often the case in a stable conflict situation, the person may solve the conflict by "leaving the field" or abandonment. For example, the disgruntled voter may simply not vote.

26.1.5.   Primary Drives Form the Basis for Learning Secondary Drives

The most important function of a primary drive is that it is the basis for the learning of all secondary drives or needs. Secondary drives are of enormous importance in human behavior because the vast majority of human activities are secondarily, not primarily, motivated. The goal underlying secondarily motivated behavior is referred to as an incentive.

The reason that so little human behavior is primarily motivated (e.g., most people do not even eat lunch because they are hungry) is that in our culture, with the exception of sex, most primary drives are readily gratified and hence they are maintained at too low a steady-state level to have much influence on behavior.

Secondary drives, on the other hand, are much more difficult to satiate and there are some who believe that they may actually feed on themselves and become even stronger when gratified. For example, the need for money is clearly a learned need. An infant given the choice between a shiny trinket and a wrinkled old $100 bill will generally chose the former. As an adult this same person may destroy his life by spending 40 hours a week for 40 years working at a job he detests simply to acquire money periodically.

Without going into the details of the technical psychological theories concerning the mechanism by which secondary needs and drives derive from primary drives, it is obvious that since money is used as a medium of exchange in our culture (primarily because it is more convenient than the bartering of goods and services), it has been associated with primary reinforcements such as food and other necessities and comforts it can purchase. Hence it has derived its value from these primary needs and can be seen to be symbolically representative of them.

But how can this account for the common observation that financially secure people, even billionaires who have financial resources that could not be spent in several lifetimes, whatever their level of personal extravagance, continue to strive in business to acquire even more wealth. After all, regardless of how wealthy a person may be, there are limits to how much he or she can eat and how many other physical resources he or she can consume.

Obviously money is not the only motive that people work for and, as will be shown presently, when a person makes enough money to comfortably gratify most of his or her physical needs, it may become a relatively minor motivating factor in his or her job performance. This is a fact of which most executives are cognizant. Neglecting the role of motivators other than money, however, the need for it still appears insatiable. Wealthy individuals who are retired will still risk their assets in investments designed to make more money or at least to slow its rate of depletion.

What seems to happen is that while money is originally perceived as a means to the end of obtaining the physical necessities of life, when these ends are attained it becomes a symbol of other, learned needs such as power or prestige which it is also instrumental in attaining. This gives the illusion that money has become an end in itself.

Perhaps it is a mistake to overemphasize the use of money as an example of a secondary drive. While it is one of the most common incentives it is not as important in motivating the job performance of most highly paid professional employees as many people believe it to be.

Among the most important secondary motives that play a role in industrial performance are such learned needs as the need for achievement and mastery, recognition and group acceptance, self-esteem, and personal development and growth.

The significant point about learned needs is that they make both learning and the manipulation of behavior possible at all times regardless of whether a primary drive is currently operative. If a child did not learn to love or need its mother so that a kind word or some attention from her could function as a reward for a new behavior, he or she would be virtually untrainable unless periodically starved. He or she would be unable to learn to speak his or her native language and hence would never acquire the ability to think in abstract terms and thus fully develop his or her human potential.

Some readers will take the view that controlling the behavior of others by granting or withholding goals and incentives has connotations of thought control or totalitarian tactics. However, all people are concerned with modifying the behavior of others as part of their daily lives. Parents try to control the behavior of their children, which is what is meant by child rearing, and children try to manipulate the behavior of their parents. The same type of interaction is seen in every kind of human interpersonal relationship be it between teacher and student, employer and employee, or two lovers. This is the essential nature of an interpersonal relationship.

In industry, engineering supervisors are generally appointed from the ranks of successful engineers as a reward for their superior job performances. This practice has negative as well as positive aspects, for often the qualities that make a good executive are different from those that produce a good engineer. It is common to find that an excellent engineer may turn out to be a poor supervisor. The basic reason for this regrettable state of affairs is that these two jobs are so different. An engineer works with machines. He or she is an expert in the physical laws of the universe and is creative and innovative enough to apply these laws to the solution of practical problems in the design of equipment.

A supervisor, on the other hand, while needing as much technical knowledge as the engineers who work for him in order to communicate with them and understand their professional problems, is basically concerned with people rather than machines. What the supervisor needs is an understanding of the laws of how people behave and the creativity to be able to apply these laws to solve people-type problems. Thus, rather than being concerned about the issue of the ethics of manipulating people, the supervisor must realize that this is the sole reason for his or her job. A supervisor might better be termed a psychological engineer for he or she is in fact an applied scientist or engineer just as much as is an electrical engineer. The only real difference is that the machine the supervisor must work with is a human machine or personality rather than a relatively simpler electronic circuit.

26.2. THEORETICAL VIEWS OF WORK MOTIVATION

Theoretical views concerning motivation, in general, and the motivation of workers, in particular, generally deal with either the nature of the major motivational factors and their relationship to each other or with the dynamics of how these factors affect job performance. While it may be useful to sample a few of the more influential of these theoretical positions for exposure to the concepts they involve, they are probably more important to psychologists interested in learning principles of human behavior than to managers who are primarily interested in solving practical day-to-day problems in human relationships.

The reason for this is that theories are based on normative data and statistical analysis. Statistics enable us to make excellent predictions about groups of people but are invalid when it comes to making predictions about individuals. An engineering supervisor does not work with large groups of people but with a small collection of individuals. Group surveys showing the major motivational factors of workers may have little relationship to the special personalities of these individuals. An executive is in a sense a clinician and there is no substitute for his or her learning all about the individual characteristics of each member of his or her staff. To function effectively the group leader must know not only the professional strengths and weaknesses of each group member but also each group member's personal idiosyncrasies and individual motive structure.

With this limitation in mind, the first theoretical position to be examined is Maslow's notion of a hierarchy of needs.[3] This extremely popular view holds that all motivational factors can be classified into one of the following five hierarchically arranged categories:

1. Physiological needs (primary drives)
2. Safety and security needs
3. Love and belonging needs (interpersonal needs)
4. Status and self-esteem needs
5. Self-actualization needs

These needs are listed in the order of their priority. Thus the individual is motivated to gratify the more basic primary drives in Category 1 before he is interested in gratifying the less urgent secondary needs in the lower categories. The theory implies that the level of motivation that a person is influenced by is a function of which of the higher priority levels of needs are currently being gratified and at which hierarchical level frustration occurs. Thus self-actualization needs will not be gratified until after status needs are met.

Since self-actualization and personnel growth are continuing processes it is unlikely that these needs could ever be sufficiently gratified to leave the individual in a satiated or motivationless state.

The theory implies that workers at different levels in an industrial organization will work for different incentives. At the lower levels, money and material benefits will be more important than at managerial or professional levels where status and a sense of achievement become prominent factors.

Herzberg in 1959 proposed a two category theory of work motivation based on worker responses to questions concerning job satisfaction and dissatisfaction.[4] The first category he called maintenance factors. This included such items as salary, working conditions, security, interpersonal relationships, and so on. These are lower level or more basic needs and their gratification does not lead to feelings of satisfaction as much as their frustration leads to feelings of dissatisfaction.

The second category of needs, whose attainment does produce feelings of satisfaction, Herzberg calls motivators. They include such things as achievement, recognition, advancement, responsibility, growth, and job satisfaction.

While the number of categories postulated and specifics of the motivational factors involved in this view vary from Maslow's theory, it is basically the same type of approach. The implication is that if the worker's basic maintenance needs are not being supplied by the job, then the worker will be so dissatisfied that he or she will probably leave, but if these needs are gratified, this will not assure that the worker will strive to enhance his or her performance. To achieve this end it is necessary to supply the motivational incentives.

An effective way of achieving this is suggested by the nature of the motivational factors that the theory proposes. Most of these are intrinsic factors that relate to the work itself as opposed to extrinsic motivation such as money or fringe benefits. Hence the approach advocated by adherents of this viewpoint is job enrichment. That is, manipulate the design of a job so that the work itself provides reinforcement in the form of interest, challenge, and the feeling of accomplishment.

A good illustration of this can be seen in the contrasts between the working environments provided for a young engineer in small and large manufacturing companies. In a small company a beginning engineer generally gets to work on all phases of several projects. As a result the engineer learns a great deal of the various phases of engineering work, sees the big picture of what the work will lead to, and develops feelings of accomplishment and job satisfaction. By contrast, in a large company (which may stockpile engineers with a view to improving its competitive position on proposals for government contracts) a new engineer is often given segments of isolated projects such as the layout of printed circuit boards that have been designed by another specialist. Such an engineer has little understanding of how his or her work fits into an overall major project, learns little, and typically gets no satisfaction from work. Such a situation is correctable by a project manager who makes work assignments that take into account both the technical and also human requirements of the organizational structure.

Vroom's approach deals not so much with the classification of motivational factors but with how they affect job performance.[5] In brief, he views job performance as a function of ability and effort, and motivational factors increase effort. Porter and Lawler would modify this view to include the effect of how the employee perceives his or her job, although it would seem that this really relates to job training and might be subsumed under the heading of ability.[6]

The notion that increased motivation improves task performance is, like most generalities, subject to certain limitations. There is evidence that in some cases the relationship between performance and motivation may be curvilinear and that while moderate increases in motivation will enhance task performance, an excessive amount may impair it. Cognitive or skilled tasks are probably more vulnerable to this effect than more routine ones. A common example of such a situation is where test anxiety prevents a well-prepared student from doing well on a critical exam.

Vroom's expectancy theory holds that motivation is determined by two factors: (1) the valence or value that a worker attaches to the incentives that he or she is likely to get for his or her work and (2) his or her view of the probability of attaining them. Thus if a worker believes it is very likely that a highly valued reward will be received for extra effort, such extra effort is likely to be made. If either the probability of attaining the reward or its value appears low, the extra effort will be less likely to occur. From a practical management point of view, to motivate workers, management must offer rewards that the individual worker values (which, in terms of the foregoing theories, depends on which of his or her needs are currently being gratified or frustrated and his or her individual personality) and convince him or her that the probability of attaining these incentives by enhanced job performance is high.

It is for the latter reason that it is critically important for all supervisors to be aware of what the members of their staff are doing, what problems they are having, and what their accomplishments are. By making constructive suggestions to help an engineer solve a problem or improve substandard job performance, and more importantly, by recognizing an engineer's accomplishments and contributions to a project, a supervisor conveys to the engineer the idea that he or she is aware of the engineer's day-to-day efforts. If an employee believes (as many employees in large companies probably correctly believe), that his or her boss does not have any idea of what is being done, what could possibly induce such a person to believe that he or she will be rewarded for extra efforts or improved job performance?

The major problem with this theoretical approach, particularly as it relates to engineering personnel, is that it places too much reliance on the extrinsic type of incentives that are more appropriate to workers who are lower in the industrial hierarchy and deemphasizes the intrinsic job satisfactions that are vital to most professional people and engineers.

The situation is analogous to the conflict between what has been called the X theory of management, as described by McGregor, and theory Y or the worker participation viewpoint.[7]

Theory X, which has been called the stick-and-carrot approach to management takes the position that workers are basically lazy and unreliable and must be forced to work by the imposition of a set of rewards and punishments administered through close supervisory scrutiny. This is essentially the Captain Bligh approach to management and it is doubtful that it is likely to produce good results even in procuring forced labor from convicts. It might conceivably have some utility in animal training but if applied to the supervision of engineers it is likely to prove disastrous.

Theory X managers would advocate compensation in the form of piecework payment for work actually performed as opposed to an hourly wage paid without regard to productivity. The theory would also advocate production bonuses paid on the basis of performance beyond an established quota for hourly workers. This is often ineffective because many workers will experience peer pressure not to exceed production quotas for fear that this will result in their being raised. It is also a much simpler matter to set quantitative production standards than it is to set qualitative standards for use with such a reinforcement scheme.

Theory Y management emphasizes intrinsic job satisfactions and seeks to manage by designing work assignments, often with the participation of workers, in a manner that provides staff with a sense of participation, interest, and accomplishment. This is clearly the method of choice in dealing with engineers and other professional workers.

26.3. PRACTICAL MANAGEMENT CONSIDERATIONS

While it is vital for every supervisor to learn as much as possible about each employee as an individual, there are certain generalities that may apply to many if not all engineers, and at this point it might be useful to pass along a few clinical observations that the author has made in the course of 20 years of experience as an engineer and engineering manager.

Generally engineers are intrinsically motivated. Most of them take pride in their work and have a strong need for recognition as the creative artists that they believe themselves to be.

Young engineers go through a period of several years before they fully realize what an engineer really is, namely a person who can develop and guide through to completion, on schedule, a technical product manufactured for a predetermined price. They were trained in engineering school in only the technical phases of their work, which in a typical development project is often only a minor aspect of the job. Hence many young engineers consider that being given administrative work, such as the preparation of a parts list or the writing of a report to military specifications, is a "put down" and an indication of management's low opinion of their technical abilities. They typically find such work boring and unsatisfying and believe it appropriate for technicians.

The remedy for these attitudes, which are very common, particularly among the more gifted technical people, is an educational process by which the supervisor eventually gets the engineer to realize both the true nature of engineering and the importance of these seemingly routine jobs to the overall project. With a good supervisor to guide him or her, a young engineer will eventually come to understand that he or she is fundamentally a business person, for the company is in business to make a profit and his or her job was created to help it achieve that goal. With time a young engineer will come to recognize that it is not possible to manufacture or maintain equipment without the supporting paperwork, and that the customer is just as interested in this aspect of the total project as he is in the hardware. On the other hand, until the importance of this type of work is appreciated, the supervisor should try to provide the engineer with enough diversionary technical assignments to provide the necessary intrinsic job satisfactions.

Engineers are often highly sensitive with regard to their perceptions of their professional status. While most texts on industrial psychology tend to ignore it there is a very pronounced caste system in most industrial organizations which is analogous to the military distinction made between officers and enlisted people. For example, shop workers generally punch time cards, have defined coffee break periods negotiated by their union, and are paid on an hourly basis for work actually performed. Engineers generally do not punch time clocks, take breaks more or less at will, and are paid a weekly salary even when late or absent from work. Indeed most engineers will refer to their salary in annual rather than weekly terms and never as an hourly rate unless job shopping.

The requirements of one government contract—that engineers sign in when late—created a morale problem in one large engineering staff that probably cost the government and the company many thousands of times the money sought to be saved by such an ill-advised procedure.

There are a variety of psychological tests available, such as the Thematic Apperception Test (TAT), to measure an individual's motivational structure. This instrument requires the subject to make up stories about people shown in ambiguous social situations on a series of 20 pictures. The tester then looks for recurrent themes in the various stories. For example, if several of the stories have an achievement theme, this is taken to indicate that the subject's need for achievement is not being gratified in his or her daily life and hence it is active as a strong element in his or her personal motivation.

Such a test is not particularly useful in an industrial setting as it must be administered by a psychologist. A more practical way of discovering what is important to an individual employee is by becoming familiar with him or her as a person through daily informal contact. After a reasonable interval on the job, a good supervisor should know enough about each employee to write a book about each one. A supervisor should be familiar not only with the professional background, strengths, and weaknesses, but also with the personal and family problems, future goals, and aspirations of each employee.

Knowing the relative importance of a variety of incentives to each employee is only half of the battle. Often a supervisor will be extremely limited in the ability to dispense rewards for superior job performance. For example, the supervisor's total budget for salary increases may force him or her into some difficult decisions concerning how much if any should go to any given person.

While on occasion, and for a variety of reasons, inequities will occur, such as an engineer being grossly underpaid (and a good supervisor must constantly fight to remedy this situation), in most cases engineers, being in demand, are well paid and are not primarily motivated by physical needs. To them the amount of a raise may be more important as a statement of the esteem in which they are held by management than as an increment in purchasing power. With such personnel a sincere statement by a supervisor that he or she would give twice the amount if it were possible, may be more important than the raise itself.

Another problem in monetary rewards is that due to the current income tax structure a large expenditure in salary increases often results in a much smaller benefit to the employee. This is the reason for the popularity of fringe benefits and such things as profit-sharing plans and stock options among more highly compensated executives.

Fortunately, because of the complex and varied nature of acquired human needs, management has a large number of alternative incentives—some of which are considerably more effective than money, provided that some minimal level of financial security has been attained.

The most important and effective incentive is the intrinsic satisfaction that comes from an interesting and challenging job well done. Careful consideration of the needs and personalities of the individual engineer and the matching of the job to the man or woman can optimize this type of job satisfaction.

In the area of extrinsic incentives, perhaps the most readily available but most difficult to effectively use, is the recognition of superior job performance by management. To be effective, praise must be both sincere and appropriate to the level of accomplishment. It ought to be public when practical. The difficulty is that it must be perceived by the employee as a reflection of the supervisor's genuine appreciation and not as a manipulative technique. To be of any value, praise must come from a supervisor whom the employee respects and perceives as being cognizant of his or her daily problems and efforts.

Just as it is important for an engineer to respect the supervisor both professionally and personally, if the supervisor's recognition is to be of any incentive value, so it is necessary for the supervisor to both experience and convey to the employee feelings of respect for him. This attitude of mutual respect is vital to the proper functioning of any professional relationship and without it few employees will have much enthusiasm for the most important aspect of any job. This is of course making the boss look good so that he or she will be promoted and create a vacuum for the employee to move up and fill.

Punishment for a poor job performance tends to be ineffective and destructive for a variety of reasons. It is basically a sign of ineffective management in that it ignores the cause of unsatisfactory job performance. It is stupid to punish an employee who performs poorly because he or she does not understand the requirements of a job. This will result in resentment, not improved performance. What such an employee needs is job training. Similarly, poor performance caused by lack of motivation requires motivation rather than alienation.

In general an employee who lacks the intelligence to perform well on the job should be given some job within his or her capabilities, but this will rarely be a problem when working with an engineering staff because most engineers are bright. Less gifted people simply do not graduate from an engineering curriculum.

A very important incentive that management has at its disposal, at a comparatively low cost, is a variety of status symbols that appeal to very powerful human motives especially among engineers. One such device is a graded series of job titles. Going from being an engineer to a senior engineer or from the latter to a principal or project engineer may be a very important event to the person involved. People in a university setting often go to a great deal of trouble to get "promoted" to the rank of full professor when no increase in compensation is involved at all!

Another important status incentive is provided by the work area assigned to an individual. Engineers in a large company who work in a "bull pen" environment where dozens of desks are aligned in columns and rows generally have lower morale than those who work in areas divided up into four-desk partitions. In such an office more senior people may be assigned two to a partition and those with the most seniority may have a private area.

A private office is indicative of even higher attainment and it is a rare company indeed in which the relative standing of the executives can not be gauged by a visitor in terms of the size, furnishings, and number of windows in their offices.

Similarly, the assignment of staff can determine status. An engineering supervisor who has a private secretary has more status than an engineer who must wait to get reports typed by a departmental secretary. In turn the engineer who has a departmental secretary has more status than one who must bring work to a typing pool and hope it is assigned to someone competent.

In the final analysis, the art of motivating a staff is learning what each person wants most, being able to negotiate with higher levels of management to maximize what you have available to give, and then giving it in the most effective manner possible.

Material rewards, such as raises, can generally be given only periodically, often long after the behavior that supervision seeks to reinforce, but recognition can and should always be given immediately when it is most effective. Just as an engineer must be an artist in dealing with the problems posed by machines, an engineering supervisor needs to be an artist in dealing with the human problems posed by engineers.

NOTES

1. K. Lewin, *A Dynamic Theory of Personality* (New York: McGraw-Hill, 1935).
2. J. Dollard and N. E. Miller, *Personality and Psychotherapy* (New York: McGraw-Hill, 1950).
3. A. H. Maslow, "A Theory of Motivation," *Psychological Review* **50** (1943), 370–396; see also his *Motivation and Personality* (New York: Harper & Row, 1970).
4. F. Herzberg, *The Motivation to Work* (New York: Wiley, 1959).
5. V. H. Vroom, *Work and Motivation* (New York: Wiley, 1964).
6. L. W. Porter and E. E. Lawler, *Managerial Attitudes and Performance* (Homewood, IL: Irwin-Dorsey, 1968).
7. D. McGregor, *The Human Side of Enterprise* (New York: McGraw-Hill, 1960).

27

MANAGING CREATIVITY

Bert Holtje

BERT HOLTJE is president of James Peter Associates, Inc., a Tenafly, New Jersey, firm that creates books and periodicals for publishers, business firms, and professional associations. He has a B.S. and an M.A. in psychology. He is the author of 27 published books. One of his books, *Theory and Problems of Marketing,* is in print in four languages and is widely used in universities in the United States as well as in Europe and Latin America.

27.1. INTRODUCTION

When creative people are asked to describe how they produce ideas, they frequently give considerable credit to the influence of social and environmental factors. Unfortunately, however, traditional research on creativity has centered heavily on trait theory. Trait theory has been conceptually productive, but its restrictive view has not resulted in much practical advice for the person faced with the day-to-day task of managing a creative effort. Focused narrowly on the personality characteristics of creative people, this work implies that only people with certain traits are capable of being creative. It further implies that creative personality traits are difficult to modify.

Fortunately for the engineering manager, this is not the case. Creativity is not the province of only a few. It can be guided, modified, and enhanced, as so many heuristic systems have shown. Although an understanding of the characteristics that creative people have in common is helpful for the engineering manager, this chapter will concentrate on productive ways to structure a creativity-enhancing social environment.

27.2.   CREATIVITY IS A PROCESS

Because current research has identified a number of internal and external elements that enhance creative effort, it's entirely practical to view creativity not as a cluster of traits, but as a dynamic process that can be cognitively and behaviorally modified. Trait theory will continue to enrich the theoretical underpinnings of the process. But for the practical purposes of the engineering manager, the results of research on environmental and social influences will be more fruitful.

Social and environmental stress, especially that produced by the awareness of performance evaluation, can hobble even the most creative person.[1] Einstein, for example, found it virtually impossible to consider any scientific problems for nearly a year after cramming for his exams. In even the most low-key engineering environment, members are expected to produce. The expectations of employers and the self-imposed expectations of the engineers themselves, result in stress that some people can handle and others cannot. This is difficult for individuals doing creative work and it's even more difficult for managers who must be the interface between engineers and managements that are frequently impatient for new ideas.

27.3.   AN OPERATIONAL DEFINITION OF CREATIVITY

Since we are less concerned with the spontaneous generation of a creative product than with the careful nurturance of creativity, we will work with an operational definition of creativity. By general consensus, a creative product is one which is based on a novel solution to a problem. The solution must be appropriate to the situation, and it must have been generated heuristically, not by the use of known algorithms.

In a typical engineering project which doesn't require a novel solution, the desired result is understood before the task is undertaken and the achievement of the goal is usually based on the application of known algorithms. On the other hand, when something new must be created, and no known processes seem appropriate, the task becomes heuristic. For example, designing a circuit to do a specific job by using commonly available active and passive components is an algorithmic process. It may require considerable knowledge and cognitive activity, but the process is not creative. However, when the same result is achieved without using any conventional components, the task is heuristic. The algorithms must be created to achieve the results. It is the heuristic approach that separates creative engineering from the routine application of technology.

27.4. EVALUATING CREATIVITY

In most cases a product or an idea will be judged as being creative when appropriate and independent observers agree on its merit. However, the subjective element of the operational definition is frequently in conflict with traditional engineering that is firmly rooted in objectivity. Since a creative idea is unique, there will probably be no objective standards by which it can be judged. Therefore, an operational definition of creativity must be based on subjective experience. It must include agreement by appropriate and independent observers who are thoroughly familiar with the problem. For our purposes in this chapter creativity will be considered to be the collective judgment by qualified people that an idea or product is indeed unique. Keep in mind that the ultimate commercial usefulness of the product is not in question at this point. For now we are concerned only with the process by which creative solutions are generated. The commercial usefulness of an idea will follow the well-worn paths bounded by human, capital, and physical resources.

Of course when a person who is unfamiliar with the appropriate algorithms produces a novel solution heuristically, his or her effort must be considered as creative. For this person, developing the algorithms needed to achieve the result was a truly creative effort.

27.5. THE CREATIVE CONTINUUM

Subjective evaluations frequently contain conditions which those trained in objective sciences find difficult to deal with. For one thing, responses are themselves evidence of creativity. For another, although it's impossible to explain the process used, appropriate observers can recognize and agree on the nature of a creative idea. Finally, even though objective standards may not exist, different responses can be judged to be more or less creative by independent observers.

The last point—that ideas may be judged to be more or less creative—

leads to an important assumption that is critical for the effective management of creative work. It leads to the notion that there is a creative continuum that ranges from ordinary solutions to problems to extraordinary solutions. For the engineering manager the task is one of striving for the extraordinary—managing people to produce on the high end of the curve. The solutions don't have to be complex. The elegance of a simple solution frequently outshines the kluge-type ideas that seem to entrance less creative and more technology-oriented individuals.

27.6. THE ROLE OF EDUCATION

Based on the concept of creativity as a process that exists on a continuum, it follows that most people can produce new and original ideas. The extent of individual creativity is generally determined by cognitive abilities and the restraints and enhancements that are imposed by intrinsic and extrinsic conditions.

Although education is considered to be essential to creative engineering output, not all well-educated people are capable of the same level of creativity. Other internal and external conditions contribute to success. For example, perseverance and a deep involvement with a project generally correlate with a high degree of creativity. Freedom from external constraints, a condition which engineering managers can provide for their people, is also important for success. The freedom to engage in intellectual playfulness is also very important.

It's worth nothing that one researcher, Gough, found virtually no correlation between on-the-job creativity of research scientists and their performance on intelligence tests, or on tests for engineering ability.

Perhaps the most striking of the omissions from the criterion correlated variables are the ability measures. The correlation for the Terman Concept Mastery Test was −.07, for the Minnesota Engineering Analogies Test +.13, and for the General Information Survey +.07. None of these values is significant, and one is in the negative direction.

The relationship between intelligence and creativity is by no means a simple one. Where subject matter itself requires high intelligence for the mastery of its fundamentals, as in mathematics or physics, the correlation of measured intelligence with originality in problem-solving within the discipline tends to be positive, but quite low. Again, however, it must be remembered that commitment to such endeavors is already selective for intelligence, so that the average I.Q. is already a superior one.[2]

27.7. THE ROLE OF MOTIVATION

Unfortunately motivation is a theme that has been overplayed in business. Virtually every manager has his or her pet theories of how to get the most out of people by motivation. As practical as behavioral psychology has been, its legacy to business has been to imply that externally applied systems are the answer to virtually every problem of improving human performance. Externally applied motivation is valuable in the management of creativity, but not in the usual ways.

The externally imposed motivation that is most effective is that which provides an environment that nurtures creative activity.[3] The promise of raises and bonuses can be effective ways of getting short-term projects through to completion, but they really don't stimulate creativity.

There is, in fact, considerable evidence to show that external attempts to motivate people can hinder creativity. The more intense the external pressure is to achieve a specific goal, the less attention will be paid to seemingly irrelevant elements of the problem. When intellectual focus is narrowed by such intense external motivation, the latent, incidental learning that is so important for cognitive exploration is hindered.

During periods of creative investigation, it's important for individuals to back away from their goal-directed behavior to occasionally look at the seemingly incidental elements of the project. Strong extrinsic motivation seldom leaves room for this creative rumination, which means that traditional paths are traveled more frequently than those which could lead to creative solutions. The stronger the external pressure, the more likely an individual is to rely on existing algorithms rather than to explore problems heuristically.

This is a situation that can cause dissonance in the person charged with managing a creative project. Corporate management is frequently unaware of the dynamics of the creative process and will push hard for results. The harder management pushes, the less likely that the solution will be unique. However, left to their own devices, creative people can wander intellectually and lose sight of the goals. It's a difficult situation for a manager, but there are some practical answers. However, the suggestions given later in this chapter are based on the reader accepting the premise that there is not a sharp distinction between ordinary behavior and creative behavior. The assumption is that engineering behavior can be viewed as ranging from work bounded by the reliance of familiar algorithms to heuristic attempts to solve problems.

As important as it is for engineering managers to understand the dynamics of creative work, it's equally important for them to understand and appreciate the view that those who do creative work have of authority. Getzels and Jackson explained it this way:

> Highly creative people are more likely than others to view authority as conventional rather than absolute; to make fewer black-and-white distinctions; to have a less dogmatic and more relativistic view of life; to show more independence of judgment and less conventionality and conformity, both intellectual and social; to be more willing to entertain and sometimes express their own 'irrational' impulses; to place greater value on humor and in fact to have a better sense of humor; in short to be somewhat freer and less rigidly controlled.[4]

The implications of the view of authority held by most creative people have important implications for engineering managers. The suggestions given later will be made with this understanding in mind.

27.8. GROUP CREATIVE EFFORT

Explicit in the title of this chapter is the fact that some people will manage the creative efforts of other people. However, for some creative people the idea of "groupthink" conjures up images of thinking-by-the-numbers. In the hands of a management with little or no understanding of creativity, this is more than a possibility. However, a small group of people working on the same project can bring to bear individual perceptions and intellectual specialties that would result in making output difficult or impossible for most individuals.

Properly guided, group members support each other, motivate each other, and contribute to a higher level of success than an individual might obtain in the same amount of time. Brainstorming techniques, for example, were created to harness group effort, and to reduce the friction that can often occur within a creative group.

To be most effective, a group working on an engineering problem must frequently function as an individual. The group must be able to identify problems, develop possible solutions, and make many decisions. While trying to function creatively, it must also maintain a level of stability. Yet the group itself, as well as the individual members, will produce complex interactive effects. These interactions can elicit the entire range of responses and emotions that are so often difficult to manage in others. The goal of the creative group is to get the best creative effort from each member, yet the group must function as though it was of one mind.

In any field most of the creative work is usually done by a small minority of individuals. Dennis explains it this way:

In one study of diverse fields, the top 10% of contributors produced about 50% of the work. If the men in the bottom 50% of each group are combined, their total output is, in every instance, less than the contributions of the highest decile.[5]

For the engineering manager with no exposure to the subject of creativity, the task would appear to be one of trying to identify the most productive people and not to work with the others. As stated earlier, the search for creative individuals by trying to identify traits observed in other creative individuals has been about as productive as the attempts to correlate personality types with morphology. The nurturing of everyone by providing a supporting environment is far more practical and productive.

Being part of a group, especially one that has been given creative responsibility, can evoke strong ego and affiliative responses in some individuals. Some members may seek to be liked and accepted by others they perceive as being important to them. Members with a strong need for affiliation often require special attention if they are to be as productive as they are capable of being.

In a situation where new ideas are being sought, each individual will want to defend his or her contributions. But some individuals whose affiliative needs are stronger than others will, on occasion, give in to satisfy their own psychological needs. The fear of failure in the eyes of a respected colleague can be quite distressing to productive creative effort. Research done by Paul Hare indicates that:

Individuals are more likely to conform to the opinions of other group members if they are uncertain about the facts, are pressured by a majority or some person of power, are concerned that someone in the group may not like them, or hold values which support conformity. In the various group techniques designed to increase creativity, such as "brainstorming," an attempt is made to reduce pressures to conform and to substitute pressure to be a non-conformist.[6]

To illustrate the pressures, in one experiment, a small group is asked to estimate comparative lengths of groups of printed lines. Although a group is involved, only one person is actually the subject of the experiment. The others are shills who have been instructed to select comparisons that are obviously incorrect. The individual subjects with a need for group acceptance usually change their judgments to conform with the group, even though the visual differences are obvious.

As a manager of creative effort it is important for you to identify these situations as early as possible, and to make sure that every person in the group retains his or her individuality, regardless of how different their ideas are from others in the group. The individual need for affiliation can be channelled productively when the goal of exceptional group performance is stressed by a manager to all of the members.

27.9. A PRODUCTIVE MANAGEMENT STYLE

Traditional western management style is a top-down approach. In many ways this can be productive. Furthermore, in virtually all situations it provides definite ego gratification for the person at the top who is doing the managing. However, when people are frequently told what to do, their interest in doing anything other than that which is specified can be reduced. This of course reduces the possibility that a person will produce creative responses. In addition to reducing the desire to do anything other than that which has been specified, the top-down management style can foster more than a small amount of hostility. These feelings are usually expressed by doing nothing more than that which was requested—if that.

As unwieldly as a collaborative style of group management can be, it does stimulate creative effort. Members of a collaborative group feel a responsibility for group success, which in turn, enhances individual ego strength and leaves members with a sense of positive group identification. But the management of a collaborative group is often more difficult than other forms of management, especially for those who are less sensitive to people's needs. Group success, however, is reinforcing, and managers who operate collaboratively frequently state that they get more personal satisfaction from their work than those who manage authoritatively.

27.10. CREATING A PRODUCTIVE ENVIRONMENT

The management of a creative engineering group is really the nurturing of individuals working together to achieve group goals. As stated earlier, the stress created in a situation which includes periodic evaluation can be a major block to creativity. When an individual is placed in a situation where he or she is expected to be creative and the work is reviewed directly or indirectly, the personal involvement in the situation can produce stress. It has been shown that excessive stress can produce performance decrements when evaluation is involved. The manager of a creative group can fall victim to this circular trap unless specific steps are taken to control for it.

People who are anxious over their performances are frequently concerned that they will fail to meet the expectations of the group leader or other significant individuals in the group. They tend to become preoccupied with themselves and to have considerable self-doubt. Worrying about the evaluation of the leader and the opinion of others in the group can produce task-irrelevant cognitions that interfere with the mental effort needed to be creative. These intrusive, interfering thoughts can reduce the attention paid to the creative task of the group.

Leaders of creative groups can help members retain their task focus simply by stressing the importance of maintaining a goal orientation. This may seem like a blinding glimpse into the obvious, but it isn't. More often than not, group leaders who want to encourage members who are experiencing difficulty will resort to supporting personal reassurance that everything will be all right. This approach is not only ineffective with the more anxious members of the group, it has the effect of decreasing performance in individuals with lower levels of anxiety.

27.11. HOW TO ENCOURAGE CREATIVITY

Keep in mind that you can expect creative effort from just about anyone—not just a group that may be designated to explore novel ways to solve a problem. And remember that although people appreciate the traditional rewards usually given for achievement, the rewards are seldom sufficient in themselves to foster the desire to achieve.

Most creativity-enhancement systems have concentrated on heuristic devices, exercises, and tricks that individuals and groups can use to approach problems. The principals of brainstorming and synectics, for example, contain elements which have proven to be quite effective in a wide range of situations. The present emphasis on environmental management isn't meant to be a substitute for these and other systems. It's meant to enhance whichever system will work best for your specific situation. It's beyond the scope of this chapter on management of creativity to review and evaluate each of these systems.

James Adams, an engineer and researcher on the subject of creativity has provided some very interesting insights into the role of the manager.

> Bosses with answers are a particular problem in the engineering profession. Many productive problem-solvers are strong-headed. They can carry a concept through to completion in spite of apathy or hostility from others, and difficulty of finding support for a new idea. If they happen to have good judgment, they are able to accomplish noticeable achievements in a company environment and are often promoted to management. One therefore often finds that engineering managers are successful idea-havers who are stubborn enough to push their ideas through to completion. They tend to continue in

this mode when managing others. Although a manager such as this can be an effective problem-solver, he is essentially operating with his own conceptual ability and in-house service organization—he is probably not going to make much use of the conceptual ability of his subordinates. In order to maximize the creative output of a group, a manager must be willing and able to encourage his subordinates to think conceptually and reward them when they succeed. He should, of course, conceptualize on his own. But he should do it somewhat in tandem with the other members of the group, if he is attempting to use them to the fullest. This is an obvious piece of advice that is surprisingly often ignored. Time and again I have seen design groups operating mainly on the concepts of the group leader. Such a group admittedly can be successful if the leader is an outstanding conceptualizer and the members of the group are content to develop his ideas. However, our concern is with environmental blocks, and such a working situation is hardly an environment conducive to conceptualization on the part of group members.[7]

These are some techniques that have been proven to help increase creative output:

1. Encourage your people to keep records of all of their ideas. Inspiration comes at odd times, especially when people seem to be occupied with thoughts other than those involving the project. Many creative people feel that their best ideas come to them during periods of relaxation, especially after concentrating heavily on other projects. Those managing creative groups used to encourage the members to keep paper and pencil handy to record thoughts. But with the new microtape recorders it's a lot easier to supply equipment for member use.

2. Present new and interesting questions and problems to your group regularly. Even though your creative group may be working hard at solving one problem, it's helpful to provide other projects that encourage intellectual playfulness.

3. Make sure that every individual in the group maintains his or her professional competence. In some fields information becomes obsolete almost as fast as it is created. Supply your people with journals; send them to seminars; and provide the opportunity for them to maintain university connections.

4. Encourage your people to expand their knowledge in other areas. Today's highly specialized people tend to narrow their focus and restrict their activities far too much. This limits the intellectual excursions that are so important for creative thinking.

5. Plan the group activity so that rigid patterns are avoided. For example, if you feel as though weekly staff meetings are necessary, don't hold them at the same time on the same day of every week. Encourage your staff

to look for other ways to do things, other than the projects on which they may be working. Many competent managers have organized informal competitions among their staff to solve interesting, but not job-relevant problems. One manager started a kite-design competition with a team of aeronautical engineers that was not only effective in providing intellectual stimulation; it brought them together socially as well.

6. Be very open to all ideas that are presented to you, regardless of your first impressions. One of the main tenets of brainstorming is the nonjudgmental acceptance of ideas from others. Since many new ideas, especially those conceived in an engineering environment, are seldom ready to use initially, try to encourage shaping and refinement—but without being judgmental.

7. Encourage your staff to be keenly perceptive at all levels. Many ideas occur when new relationships can be seen in existing systems. The more of these relationships you can encourage, the better the chances that group members will produce innovative solutions.

8. Encourage outside activities. Although it's impossible to strengthen the brain by use, as one would a muscle, the participation in creative activities can result in positive reinforcement and transference on the job. More than a few very creative scientists have found relaxation and stimulation in activities such as painting, writing fiction, and cooking.

9. Try to help your staff develop and enjoy a sense of humor. The stereotype of the humorless engineer is not without precedent. However, a sense of humor and creative output have been shown to correlate positively. Humor can relieve tension, and it's easier to do creative work in a relaxed state.

10. Encourage self-confidence. Even though you may be leading a group effort, each individual should be personally secure enough to think boldly and to express his or her wildest ideas.

11. Encourage risk-taking. When people know that you prefer that they risk and fail rather than not take risks and produce mediocre ideas, you will have taken a big step toward encouraging creativity.

12. Understand your own needs. To be a successful manager of a creative engineering effort, you simply can't be a traditional top-down manager. You must adapt a collaborative style. Your personal satisfaction should come from group success. When your staff understands that you are not manipulating them for ego gratification and that you genuinely want them to succeed, you will have taken a big step to motivate them.

24.12.   BUILD STRONG RELATIONSHIPS

The most productive managers of group creative effort usually emphasize that their success is based strongly on working closely and personally with each member of the group. Leon Wortman stressed this:

> The creative person has an ego, just as all people do. His need is not for idle flattery, for he will immediately recognize an artificial compliment or social ritual. If his progress appears to be slow—and you know it because he will tell you with a sense of frustration—give encouragement that offsets the letdown feelings. Redefine the objectives with him and make certain the original goals are still being pursued and that energies are not being diverted onto a scientifically fascinating route, but one that is apart from the real purpose of the effort. Tell him how others feel about his work, especially the thoughts of those for whom there is mutual respect. Even though he doesn't complain, be alert to signs of depression or resentment; these might reveal some of his innermost feelings, those that are being protected by ego defense mechanisms. And, of course, attend to the hygiene factors, the environment, the physical facilities and other systems that effect morale and attitude. He may be a brilliant, ingenious, creative person. At the same time, he is very human.[8]

Perhaps the most important personal characteristic a manager of creative people can have is patience. As I discussed earlier, the job of managing creative people is mainly that of providing an encouraging environment. A major part of that environment must include the patience of those who must manage and evaluate creative work.

24.13.   SUMMARY

Although trait research has contributed much to the theoretical understanding of creativity, it has been of little help to engineering managers. In fact, armed only with a list of traits exhibited by creative people, many managers have failed simply because they spent time looking for creative people when they should have been encouraging everyone.

Since independent judges can agree not only on whether or not something is creative, but whether it is more or less creative, it can be assumed that creativity exists on a continuum. And this continuum should include everyone—not just those singled out for creative work because they seem to have traits in common with people who have produced creative work.

Creativity in individuals can be modified and enhanced, but the traditional methods of motivation have little or no effect on the creative process. In fact strong attempts to motivate creativity have been shown to have negative effects. However, when the environment in which people are expected to do creative work is made supportive and nonjudgmental, creative output can be enhanced and when the stress of evaluation is controlled, the results on creativity are positive.

NOTES

1. I. G. Sarason, "Stress, Anxiety and Cognitive Interference: Reactions to Tests," *Journal of Personality and Social Psychology* **46** (4) (American Psychological Association): 929–939.

2. H. G. Gough, "Techniques for Identifying the Creative Research Scientist," in *The Creative Person* (University of California, Liberal Arts Department, 1961).

3. T. M. Amabile, "The Social Psychology of Creativity: A Componential Conceptualization," *Journal of Personality and Social Psychology* **42** (2) 357–376.

4. J. W. Getzels and P. W. Jackson, *Creativity and Intelligence: Explorations With Gifted Students* (New York: Wiley, 1962).

5. W. Dennis, "Variations in Productivity Among Creative Workers," *Scientific Monthly* **80** (1955): 277–278.

6. A. P. Hare, *Creativity in Small Groups* (Beverly Hills, CA: Sage, 1982), pp. 158.

7. J. L. Adams, *Conceptual Blockbusting: A Guide to Better Ideas,* 2nd ed. (New York: Norton, 1979), pp. 68–69.

8. L. A. Wortman, *Effective Management for Engineers and Scientists,* (New York: Wiley, 1981), p. 71.

28

COACHING AND COUNSELING

Roy W. Walters

ROY W. WALTERS is president of Roy W. Walters and Associates, a consulting firm he founded in 1967 which specializes in redesigning work and work systems and improvement of productivity and the quality of products or services. The firm serves a diverse array of industrial firms, both large and small, as well as government agencies. Previously, Mr. Walters had been with Bell Telephone Company of Pennsylvania and AT&T in a variety of line, staff, and executive positions; at AT&T he was director of Employment and Development, responsible for the first empirical research into redesigning work systems.

He is a frequent speaker at professional meetings and has published widely in professional journals. He was a contributor to the *Encyclopedia of Management* and *The Handbook of Modern Office Management and Administrative Services* and is the author of the 1965 book *Job Enrichment for Results.* He is also the publisher of *Behavioral Sciences Newsletter,* a monthly compendium of social science applications in business and government.

28.1. THE COMMUNICATIONS IMPERATIVE

The process of communication is common to all organizations regardless of whether they manufacture a product or perform a service. When structure is provided, that is, people are arranged for the purpose of achieving a common goal, they must communicate with each other in order to achieve any results. In this sense communication between people is the central process of an organization. No products can be produced, services rendered, objectives set, or decisions made without communication between individuals who are members of the organization. Furthermore, managers cannot fulfill their coaching and counseling responsibilities without good communication.

Employee communications is the reciprocal "giving and receiving" of information by employees when necessary to carry on a job function. But information received must be translated by the receivers perception of the information. Perception is not logic, it is experience. Things received are always part of a total picture. The body language, physical and facial gestures, tone of voice, and the total environment, not to mention the cultural and social aspects, cannot be dissociated from the spoken language. Realistically, without them the spoken word has no meaning and hence does not communicate.

It is necessary for managers to understand what communication should be in order to integrate that process into the total managing function.

Communications must provide for a two-way exchange of ideas and information between workers and management at all levels of the organization hierarchy. At no time should it be a vehicle for imposing management views or directions on subordinates. On occasion directions do have to be given in order to comply with certain laws and requirements. These however, are directives—not communications.

Communications must not be singled out as a special function. It should be part of the normal, day-to-day routine of the organization.

Communications must be an integral part of each manager's and each employee's job. It should not be the responsibility of a staff specialist. If such a specialist's position exists, that individual's responsibility is to assist managers and employees in their communications responsibility.

Starting at the top, management must constantly be aware that the success of the organization depends on the answers to two questions:

1.   What do my employees need to know to be as effective as possible in their jobs?
2.   What do I need to know from my employees that will help me do my job better as well as help my employees do their jobs better?

28.2.   FEEDBACK

Feedback, or knowledge of results, is the cement that holds an effective organization together. It is critical to the learning and development of all employees. Without it workers and managers alike would be making errors, wasting time and energy, and doing unnecessary tasks—all items that will lead to the eventual collapse of the organization.

Feedback has two sources. The first and most potent is feedback that comes directly from the work itself. It requires that the mechanisms necessary to communicate the results of someone's effort are built right into the work itself. The person doing the job will be able to answer the question "How did you do yesterday?" He or she will have easily accessible data, both quantitative and qualitative, that immediately tells how he or she is doing in the work. This feedback is referred to as being intrinsic.

The second source is feedback that comes from outside the work itself. It could come from a client, peer, or boss. If you are a highly structured person, the concept of a series of discussions with your subordinates about every 3 to 6 months probably has considerable appeal. If on the other hand you are more opportunistic and venturesome in your approach, you may prefer that more responsive feedback be given on the spot when you see something being done successfully or unsuccessfully.

Managers must be aware that subordinates need feedback in order to meet standards. This assumes that the manager's standards are consonant with those set by even higher management and that operating at that level will meet the goals and objectives of the organization.

Research over the last few years has led to the development of factors that if practiced provide good feedback. These are:

1.   The feedback is descriptive rather than evaluative. The sender details activities, either positive or negative, in a descriptive, detailed fashion without adding his or her personal evaluation.

2.   The feedback is specific rather than general. In order to improve upon performance, feedback must be specific enough that the recipient knows what detailed changes are needed.

3.   The feedback takes into account the needs of both the receiver and giver of feedback. The receiver needs information that will lead to improved performance and the giver needs assurance that he or she has lived up to their responsibility of assisting subordinates in their growth and development and that the performance will improve.

4.   The feedback must be directed toward behavior that the receiver can control. Giving feedback about situations beyond the receiver's control is of no use. This leads to frustration and even to dysfunctional behavior.

5.   The feedback is solicited rather than imposed. The giver must adopt the attitude that the feedback being offered is constructive and in the best interests of the subordinate and the organization. It should not be imposed in an "either-or" fashion.

6.   The feedback is well-timed. A guiding principle is that it be as immediate as possible. Stale feedback is of little value because it becomes distorted and less relevant.

7.   The feedback should be checked to insure clear communication. The best way to insure this happening is to ask the recipient to repeat the communication as he or she understood it. If there is a disparity it should be immediately cleared.

28.3.   SELF-ASSESSMENT—A CRITICAL STEP

If you are sincere about developing a solid coaching and counseling relationship—regardless of whether you are on the giving or receiving end—there should be no surprises. For there to be no surprises, there must be complete honesty. If you are a supervisor, this means honesty with your subordinates and if you are a subordinate, this means honesty with both your supervisor and yourself. The latter is most difficult.

Self-assessment should not be thought of as an "extra" something you really shouldn't do on the job. This is work. This is an effort to make yourself a better engineer or a better supervisor. It is an investment of time intended to help you get better results for the organization you serve and to prepare yourself for more responsibility should you decide this is what you want.

A good starting point is to make an objective appraisal of your abilities, interests, and personal values, or what you wish you could put first and actually what you do put first in your daily life (these may be different). While doing this you should be able to recognize that this is not a one-shot affair in which you arrive at all the answers. You change constantly, circumstances change constantly, new jobs and functions are created, and new technologies come into the picture. So your self-appraisal should be current and relative to your situation today.

A format for this might be:

1.   Review your daily planned work
   a.   Significant contributions you have made to the organization's goals
   b.   Unusually difficult problems you have solved
   c.   Work areas in need of change or improvement
   d.   Your major knowledge, skills, experience strengths, or experience gaps
2.   Review your total position responsibilities
   a.   Significant responsibilities not yet undertaken
   b.   Working relationships that have been strengthened

   c. Working relationships that need strengthening
3. Review any special projects or emergencies you handled with distraction and those you could have handled better.
4. Did you ask for help from your manager? Was it received? Could you have asked for more?
5. Look toward next year
   a. Improve specific performance areas if necessary
   b. Add to knowledge, skill, and experience
   c. Improve relationships with your manager, peers, and subordinates if necessary
   d. Make better use of your talents if necessary
   e. Change working methods or behavior if necessary
6. What assistance from your manager or others will you need in order to carry out your plans for the next year?

If you have made the effort to go through the work connected with such an analysis, you should have a realistic view of yourself, your work, and plans for continuing growth and development. There should be no surprises when you and your boss (or you and your subordinate) sit down to review your situation.

Of course, the perceptive subordinate should constantly be picking up signals of how he or she is doing. Daily there are numerous clues regarding others' evaluations of your performance. If these clues register effectively, the effort required to make a self-evaluation should be greatly reduced. A few examples follow:

Do you meet deadlines?

Do you miss deadlines?

Is your work a finished product?

Is your work less than a finished product?

Is your judgment accepted?

Is your judgment never accepted?

Do you work on assignments alone?

Do you receive help from others in order to complete assignments?

Answers to these few sample questions—and there are hundreds more— should give you much data on how you are doing.

The mature individual searches for all information that can be of assistance in gaining insight into his or her abilities. Such an individual does not find such information threatening. He or she effectively uses it in a constructive manner.

After you have done all this there may exist large gaps between your view of your performance and your superior's view of your performance. This is the function of coaching. Until these two sets of perceptions are made consonant, neither boss nor subordinate will be able to communicate to the other's satisfaction. Dysfunctional behavior often occurs when this happens.

To coach another means to "train intensively by instruction, demonstration, and practice." Since it is the subordinate's requirement to meet the standards of the superior, those standards must be made clear and be completely understood by the subordinate. In order to be completely understood the superior may have to give detailed instructions about methods and procedures, may have to demonstrate proper methodology in specific instances, and may have to restructure assignments as well as organizations in order to permit practice. The superior is not doing his or her managerial ("coaching") job if this role is not performed.

28.4.   UNDERSTANDING YOUR PEOPLES' NEEDS

An old axiom in development circles is that "all development is self development." What this really means is that no one can develop another person. The person must want to develop and have the internal drives (motivation) to accomplish this. The boss or coach can create new situations, devise new assignments, place the person in new ventures or undertakings, outline new learning, and so on. But it is the individual who must develop.

So many persons use Vince Lombardi, the former coach of the Green Bay Packers, as the model of "developer of men." Lombardi never developed any of his people. He did give them a complete understanding of the high standards he expected to be met in order to play for him. Then he gave potential players the opportunity to demonstrate their capabilities. But they developed themselves. He didn't do it. If their internal motivation was sufficient and they had enough natural ability they could play on his teams. If they were lacking in either of these, then they were not good enough to play for him. But it was their decision and their efforts that enhanced their development. Not his!

The starting point for any manager-development effort is the performance appraisal. It should be made by the coach prior to any verbal discussion and should focus on what the subordinate does well and what limitations must be overcome. Such an appraisal review should always be a joint effort. It requires much work on the part of the subordinate (see previous section), it has to be a valid self-appraisal. Remember, if this is done well there should be no surprises. But it also requires active and caring leadership by a person's supervisor.

In appraising themselves people tend to be either too critical or not critical enough. Most likely they will see their strengths in different places than those seen by the supervisor. People often also have a tendency to pride themselves on nonabilities rather than on abilities.

Engineers often judge themselves to be good managers because they are "objective," "analytical," and "good at solving problems." But, to be a good manager other traits are necessary such as high degrees of empathy, ability to understand how others do their work (it may differ from the way you do it), understanding of "nonrational" factors such as personality, and emotional courage to make difficult decisions.

Any review of performance should be based on performance objectives which a subordinate has set for himself or herself in conjunction with a superior. It should start with actual performance against these objectives. It should never start out with any aspect of "potential." Reviews should focus on what the person has consistently done well. This should lead to an awareness of the strengths of the person and of any factors which prevent the person from making these strengths fully effective.

The other side of this review is that of the subordinate who is constantly asking, "What do I want for my life? What directions should I go? What are my aspirations? What are my values? What must I do to continue to learn, change, and equip myself to live up to my demands on myself and my expectations out of life?" These questions are better asked by an outsider, by someone who cares about the person's development but at the same time is sufficiently perceptive to have the insight that most persons do not have about themselves.

Any self-development usually requires the learning of new knowledge, skills, and styles of operation. These are only acquired by new experiences. Anything less than actual experience is game playing. There are two main factors in self-development. One is job experience. The other is the example or model of the superior (teacher-coach).

Any appraisal of one's self should always lead to conclusions regarding needs and opportunities with respect to what another person can currently contribute, as well as with respect to additional experience needed. Superiors should constantly be asking themselves "What are the best job experiences for this person so that the person's strengths can develop the quickest and go the greatest distance?"

As a coach or teacher, working on your own development sets a splendid and incontrovertible example. Such an example encourages subordinates to develop their own strengths, overcome deficiencies, and gain the experience they need. The opposite is true. The superior who constantly reminds subordinates of what they cannot do, who constantly discourages, and constructs assignments that never allow subordinates to learn or experiment will stunt self-development.

It is a well-known fact that no superior can develop himself or herself unless he or she works on the development of others. It is by doing this that managers raise their demands on themselves. In any profession the best managers or performers always look upon the persons they have helped develop through their coaching as the brightest and largest monuments they can leave behind.

28.5. VERBAL TECHNIQUES FOR EFFECTIVE COACHING

Most of our coaching requires extensive communications with our subordinates. We must pass information as well as understanding to others. Effective communicators reach out beyond themselves to the people with whom they communicate. They avoid communicating with a self-centered focus. They know that good communication is a search for meaning, a process through which the communication affects others and is in turn affected by others. All communication must center on others—not on words. The good communicators know that communication is an interaction between them and the people with whom they communicate and that the real message is the sum and total of all the ways they affect their receivers and of all the ways the receivers affect them.

The main communication medium is our voice. We have another medium—our body. While there is an interrelationship between the two—simply because we cannot separate them when communicating in person—let's examine them separately.

Our verbal expressions tell much more than just words. Tone of voice, modulations, pace of speech, articulation, cohesion of thought—all communicate. So managers and supervisors must give much thought to their voice and speech patterns.

While we do talk to our subordinates under a variety of conditions, and therefore communicate a variety of things—probably some bad along with the good—we must be certain that we pick a proper time and place when we wish to be optimally helpful in our coaching role. When subordinates are busy, or managers are busy, that is not the proper time. When there is noise and confusion in the workplace, that is not the proper time.

Time should be set aside for a coaching discussion. Both parties need to reflect on what it is they wish to talk about. Both parties need time to think over what it is they want to say and how they want to say it. There is a great danger in talking off the top of one's head. Wrong words and phrases can tumble out of ones mouth unintentionally. Therefore, preparation is necessary.

A quiet, secluded spot should be selected. It would be better if it is not in the manager's or supervisor's office. Offices represent power bases, and while the manager whose office is the scene of the coaching session will feel comfortable and secure, it's still his or her turf. The manager and the subordinate know this and the latter will be uncomfortable.

A neutral spot would be better. Then both parties will be equally comfortable. There will be nothing in the surrounding area that denotes a power base. Never forget that both parties are aware of who has the power. This need not be exacerbated.

We previously said that in a face-to-face discussion the words cannot be separated from the body. Body language is as strong a communicator as the spoken word. Facial expressions, posture, and hand and eye movements all communicate. Looking directly in the eye when speaking to someone conveys full attention, directness, and personal concern. The crossing of arms or speaking from behind a desk can be viewed as barriers. Dress and posture both convey messages. All of these nonverbal communication manners have meaning. They are usually sent simultaneously with verbal communication, so its most difficult to separate the dichotomy of messages that are transmitted.

The coach must not only be able to understand and translate the body language of the subordinate, but must be careful to control his or her own behavior. The most effective counseling is in most instances listening, and the most effective listening is nonevaluative. Effective listeners neither agree or disagree with people. They communicate both verbally and nonverbally, "I'm trying to understand you and what you say. I'm listening hard. I'm interested."

When coaches express agreement or approval they build dependency. When they express disagreement or disapproval they stimulate defensiveness or aggression. When they communicate understanding and acceptance, they establish the base for meaningful communication.

One very disturbing nonverbal phenomenon is silence. Silence is a natural part of any conversation. In our daily casual conversations we expect and readily accept pauses without attaching any special significance to them. But in a more formal situation, such as a coaching session, we fear silence. There is a great tendency to fill in silence gaps with further comment, a question, or restatement.

Allowing a pause to occur naturally is not something to fear. It has value. It provides both the subordinate and coach with an opportunity to recapitulate what has gone before the pause. Both parties can pause to think, reflect on what has been said, organize their thoughts and feelings, and search for the right words with which to express themselves.

The total meaning of words must come from an integration of both verbal and nonverbal behavior of the subordinate; and the coach's search for meaning must take into account all available information.

A desire to understand, accept, and to be of help is of greater value than any technique.

28.6.   BEHAVIOR MODIFICATION

One of the oldest maxims in education holds that learning by observation and doing (practice) is the most effective way to learn. This principle is now being widely used in management education. The label given to this type of education is "behavior modification." The name is not an accurate one, since label indicates that some behavior needs to be modified. That may be the case, but in using it to teach proper or preferred behavior in situations never previously encountered, the designation is less than adequate. "Behavior modeling" is a better term.

Most of us have experienced this type of learning experience. This is the way we were taught to drive a car or type. It is after all nothing more than showing someone how to do something, getting them to try the new skill, giving them some coaching, and then having them apply what they have learned in a real situation.

The main thrust of "behavior modeling" training has been in the teaching of managerial or supervisory skills and mainly in the development of human-interaction skills.

Many kinds of situations can be dealt with in using "behavior-modeling" programs in supervisory training. The orientation of new employees, handling of communications problems and employee grievances and complaints, and the giving of recognition for accomplishments are some of the situations that lend themselves to this kind of training. It has been demonstrated that by training managers to effectively handle a series of such interactions in specified ways, trainees not only learn to perform specific tasks in an acceptable fashion, do actually transfer this style of interaction with subordinates to other situations.

The format used in training managers via "behavior modeling" is typically a TV tape (or similar medium) showing a model manager handling some specified interaction in an effective way. Workbooks associated with the demonstration tape contain a number of learning points which are the specific behavioral steps the manager should employ in handling the situation. The model shown clearly demonstrates the utilization of these points in a sample situation.

It is recommended that following this phase much practice opportunity be given. This is done by role playing. Each trainee is given the opportunity to display his or her ability to follow the designated learning points, using a situation similar to that that was just viewed on the tape. This practice should continue until the trainee demonstrates the preferred behavior. There is some difference between this practice method and typical role playing. Here the trainee has very specific instructions (given via the learning points) as to what is to be accomplished. The trainee then practices the specified skill until "it is right." Skill rehearsal is a better description than role playing. Leading the trainee through the proper set of preferred behaviors makes the on-the-job application of these skills relatively well-assured.

As in any learning situation, reinforcement should be used at the appropriate place and in the appropriate manner. When the trainee has difficulty in handling the new skills, coaching should be given by the instructor. Reviews, practice, and reinforcement should continue until the trainee feels thoroughly adequate.

The new learning must be transferred via application to the job situation. In subsequent trainee meetings, the trainee should be required to report back to other trainees what transpired when the new behavior was tried out in the job situation. These experimental sessions with appropriate critique and reinforcement are very valuable to the modeling process.

It has been clearly demonstrated that "behavior modeling" can be effective as a means of training managers to change their behavior in their on-the-job dealings with their subordinates. But if the program undertaken contains taped programs, lectures, readings, and so on, it is something other than behavior modeling. To fit under that definition it must contain four critical elements: they are model viewing, practice, reinforcement, and transfer. Full attention must be given to these four elements. Each of them represents unique challenges to the developer as well as the trainee.

29

PERFORMANCE APPRAISAL AND COMPENSATION

Patrick J. Montana and William V. Bellando

PATRICK J. MONTANA has been associated with IMA since March 1980, after founding his own human resources consulting firm a few months earlier. For the past six years he served as president of the Professional Institute, and of the National Center for Career Life Planning for the American Management Associations. Previously he was director of Planning and Manpower Development for the Sperry and Hutchinson Company, and he continues to serve as professor of Management at Hofstra University's School of Business. During 1973, he served as a presidential interchange executive with the United States Department of Labor in Washington, D.C. He received the Secretary of Labor's award for "Outstanding Work in Executive Development and Pioneering Work in Executive Counseling." He received an M.B.A. from the University of Cincinnati and earned his Ph.D. from the Graduate School of Business at New York University. He is the author of 10 books and numerous articles, and is listed in *Who's Who in America*.

WILLIAM V. BELLANDO received his B.S. degree in Civil Engineering from Manhattan College in 1977 and has more than seven years of experience with an engineer consultant. There he handled many assignments, both in the field and office, including that of project engineer. He was also involved in project management and was an active participant in the company's MBO program. Mr. Bellando also has an M.B.A. in Finance from Hofstra University and is a professional engineer in the State of New York. He is currently employed by the accounting firm of Deloitte, Haskins & Sells.

29.1. INTRODUCTION: THE BASIC SETTING

Since engineers are generally more inclined by nature to speak and think in the language of production and efficiency rather than in the language of human behavior and motivation, this chapter on the performance appraisal and its relationship to compensation and other areas of management will be in the form of a dialogue among an engineering department supervisor, engineer, and member of the personnel department. The authors believe that in this way the reader will obtain a better appreciation of the behavioral and motivational aspects of performance appraisal, which the engineering manager can apply in any situation. In a sense the case method is being applied here in order to show the recommended procedure.

The setting is the systems engineering department of a fictitious company, Nortran Company. Jim Collins, the department supervisor, has been discussing the role of the manager (Jim was a former engineer) and other members of the managerial team with an associate engineer, Steve Wright. George Phillips, the personnel manager is trying to obtain a better understanding of the process of performance appraisal and has decided to talk to Jim and Steve about it since they work very well together as a team and their department serves as an excellent model of management in the Company.

To begin with Jim and Steve understand that in order to get the group to work together effectively, the job of managers, consist to an important extent of getting results through others. Therefore, the task of performance appraisal requires an examination of job descriptions, performance contracts, statements of responsibility, and standards of performance.

The scene is Jim Collins' office. George Phillips, the personnel manager, begins the dialogue:

George: Shall we start with job descriptions? Useful ones, it seems to me, are surprisingly rare. The typical example may include title, reporting relationships, and a somewhat haphazard list of duties, but it too often fails to spell out the job's objectives, much less detailed responsibilities.

Jim: I'm not surprised. Writing a sound job description is a chore I used to despise, whether it meant writing a completely new description or just updating an existing one. That was before I learned from experience how important an adequate description is to satisfactory performance on any job.

George: But writing it is still hard work, yes? Unfortunately a lot of managers think of it as a chore—a needless chore. You're unique only in that you can take it in stride and actually benefit from it. Steve, you and Jim have told me rather colorfully how the two of you struggled—I don't think that's too strong a word—to reach an understanding about what was expected of you. However, you also worked up a formal job description, didn't you?

Steve: We certainly did. We used some of our regular meetings to draft it, discuss it, and then make the final version just as clear and concise as we could.

George: That's the job description for a senior engineer that I see in the manual?

Jim: Figure 29.1, that's it. Of course not all Senior Engineers' jobs are alike and not all managers are alike in the way they want their offices to be run, but the figure follows a common job-description

TITLE: Senior Engineer
DEPT./DIV.: Systems Engineering
DATE PREPARED: 5/81 DATE OF MOST RECENT REVIEW: 4/82
STATUS: Exempt __x__ Nonexempt _____

Objective: By performing diversified engineering duties, assist department supervisor in carrying out assigned responsibilities effectively and thereby helping to achieve departmental and company goals, both short- and long-term.

Reporting relationships: Reports directly to department supervisor. May have technical aide assistance but does no supervising.

Experience: Requires at least a four-year degree in Mechanical Engineering supplemented by three or more years of technical engineering.

Responsibilities:

1. Have the ability to read piping, mechanical, electrical drawings and flow diagrams.
2. Have a working knowledge of applicable codes, regulations and standards as necessary for a specific assignment.
3. Review and evaluate technical reports and specifications.
4. Prepare, for the department supervisor's review and approval, technical specifications, purchase orders and technical reports utilized in the purchase of equipment.
5. Provide technical assistance to other departments as required.
6. To participate in meetings with vendors and suppliers to assist in resolving engineering questions.
7. As authorized and required, write and sign routine correspondence such as letters and memorandums.
8. For each assignment maintain proper documentation and correspondence in accordance with quality assurance standards.

Figure 29.1. Position description. From S. B. Selikoff, "On the Obsolescence and Retraining of Engineering Personnel," *Training and Development Journal,* **May 1969, p. 3.**

format and should give the beginning job-description writer an idea of the kind of language used.

Steve: Incidentally, I think the language may be the big stumbling block for people who—quote—"hate" to write job descriptions. They assume that their writing has to be very dry, technical, and pompous-sounding; so the minute they pick up a pencil and try to get something on paper, they go completely blank.

Jim: Well, boiling your ideas down till you can put them in just a few, simple, easily understood words is precisely what makes the process so difficult, as Steve and I know.

George: You're right I guess, but it's too bad that people feel that way. Because the language isn't all that important in comparison with the job duties and responsibilities you're describing. It doesn't have to be fancy. The important thing is to make the words mean to the person holding the job exactly what they mean to the boss.

Steve: I agree. Jim and I stewed for too long over that first job description of mine. Unnecessarily.

George: You say you've done others?

Steve: Of course. Revisions of the first one, that is. They weren't nearly so painful, I'm thankful to report.

Jim: A job description isn't anything you can write once and for all and then proceed to forget. It has to be taken out and looked at from time to time and probably changed in spots. All jobs change in response to changes in the organization and the work itself, the person in the job and the person to whom he or she reports, the environment in which you're operating—all kinds of factors. But I can't imagine why I'm telling you this, George! You're in personnel; you know better than I do that job descriptions can't be set in concrete.

George: (laughs) Maybe I just can't forget those interviewing techniques it's taken me so long to get used to. I'm supposed to make the other person do the talking! But look here, both of you. Would you agree with this summary statement? A good job description should include clearly delineated responsibilities. It should state the extent of the initiative and authority that the incumbent may exercise and should cover any expectations regarding interaction between the incumbent on the one hand and superiors or subordinates on the other hand.

Steve: Yes. Except that *I* don't have any subordinates. In my case, I

	guess maybe you mean *Jim's* subordinates. I have no authority over them, but I do have to pass along orders from Jim occasionally—and I try to be just as tactful about it as I can.
Jim:	There've been no complaints from that direction. None whatever, although some engineers have been known to try to assume responsibilities before they were ready and ended up getting themselves thoroughly resented, disliked, and distrusted. But everybody likes and respects Steve—both inside and outside the company.
Steve:	Thanks a lot, Jim.
Jim:	I'm sorry George—I think we may have interrupted you there. You did mean to add outsiders to those people with whom Jim is expected to have a certain amount of interaction, didn't you?
George:	Oh absolutely! Clients, for example—although since you're not in sales, I don't suppose you get too many of them.
Jim:	More than you might expect. We do special work for quite a few clients.
Steve:	And there are always vendors and suppliers. A lot of them I know by now, but some just stop in cold, hoping for a little business. Jim isn't the type that refuses to see visitors or keeps them waiting for hours; but there are times when I have to sit down with them and discuss their products, if Jim needs a breather or is not available.
Jim:	(a little embarrassed) Well the goodwill generally—
George:	—pays off when you least expect it, I agree, and often when you can use it most. But shall we move on? Besides an up-to-date job description, are there any other written documents that serve as guides for Steve?
Steve:	Jim, what about the performance contract?

29.2.   WRITING AND COMMUNICATING PERFORMANCE OBJECTIVES

George:	I'm not sure I quite understand what you mean by that one, Jim, but it sounds like something I might recommend to other managers and subordinates. Just what is a performance contract?
Jim:	A performance contract defines the expectations of both the manager and subordinate, so far as the subordinate's performance is concerned during any specified period. A logical time span might

be from one performance appraisal to the next. For Steve that used to be every 6 months; now, as I'm sure you know, it's once a year.

George: That's policy, yes—fair or not, depending on how you're affected. New employees are brought along relatively fast till they reach the midpoint; then appraisals and compensation reviews are less frequent. How are your and Steve's joint expectations defined, Jim?

Jim: Through (1) statements of responsibility and (2) standards of performance.

Steve: Upon which both of us agree.

George: I see.

Jim: I hate to sound like a broken record, but it *is* true—just as it is whenever you're talking about a person's performance—that the key to a successfully fulfilled performance contract is the fact that it was mutually agreed to by manager and subordinate.

George: Let's stop here for a moment. Do I understand correctly that a job description is a permanent document—although subject to updating when necessary—whereas a performance contract is intended to cover a particular segment of time and, often probably some particular aspect of the work that needs attention?

Jim: I think that's a fair way of expressing the difference. Yes.

George: And responsibilities are explicitly stated in both the job description and the performance contract.

Jim: That's right and in both cases each responsibility should be accompanied by an equally explicit standard of performance. For instance: Steve, do you remember any of the responsibilities we spelled out for you in the performance contract we've used in the manual?

Steve: How could I ever forget a single one? They were positively terrifying! The worst had to do with maintaining proper documentation. I was supposed to go through a quality assurance audit with a small number of findings, and I was sure I'd exceed that the very first time I was audited.

Jim: O.K. Here's the way we worded that responsibility, as nearly as I can remember: *For each assignment maintain proper documentation and correspondence in accordance with quality assurance standards.* A job description will list any number of such responsibilities one after the other in just this way. You can see though, how unsatisfactory the statement is. The holder of the job—you, that is—will be forced to ask, "But how will I *know* when I comply with QA standards?" What does "maintain proper documentation" *mean?* So we add a standard of performance, which is a kind of target that you aim at: *Performance shall be considered satisfactory when the outcome of a quality assurance audit shall result in no more than three "outstanding items."*

Steve: (groans) At first I thought that was going to be impossible. I realized that you did need to have proper documentation to insure that our vendors were fulfilling their contracts and we were getting deliveries on time. But three outstanding items? To me that was unreasonable. Still, I appreciated your letting me decide for myself how I would go about cutting down on the amount of deficiencies I was being cited for. I enjoyed reorganizing my system and once I'd finally succeeded it really made it easier to keep track of costs and delivery dates. It resulted in my spending less time on the phone and allowed me to undertake new responsibilities in new areas.

George: What other responsibilities did you have to pay special attention to?

Steve: Writing reports. That one read something like this: *Prepare, for the department supervisor's review and approval technical specifications, purchase orders, and technical reports utilized in the purchase of equipment.* Well it was a while before I got to drafting even the most simple report in my own words for Jim to sign. Meanwhile there was that terrible standard. My technical and analytical skills were pretty good, you understand, but when it came to putting it down on paper I would end up staring at that blank page. It took a while before I was able to overcome this stumbling block.

Jim: Oh, come now—that standard wasn't so tough. *Performance shall be considered satisfactory when only one draft is required prior to final approval.*

Steve: Only one draft! At a time I may have been averaging three drafts per report—and sometimes having to redraft the same section more than that!

Jim: But that took care of just the "review" in the statement of responsibility. There was also the matter of promptness. Because our business is so integrated, everything—specifications, purchase orders, reports—has to be finalized very quickly to avoid scheduling delays. So we added a rider to the standard: *Performance shall be considered satisfactory when—further—all drafts are approved within five working days after submittal.*

Steve: In the manual you will find these special-attention responsibilities as Figure 29.2. Then, in Figure 29.3 we've added the corresponding standards just as Jim spelled them out and I—very dubiously, as I've admitted—agreed to them. Figure 29.4 shows the complete performance contract. And finally, Figure 29.5 is the original job description of Figure 29.1—here supplemented with performance standards since many companies insist on this procedure.

Jim: They're right. A list of responsibilities, no matter how detailed, has little impact on job performance if the required level of performance isn't specified by means of valid standards.

George: Jim, those standards you suggested to Steve strike me as being fairly tough to meet: but I guess for a senior engineer they are justified. If Steve had been a beginner and the job less demanding, your expectations supposedly wouldn't have been so high.

Steve: Even so, Jim's standards have always been tough. Those standards were meant to make me *stretch,* and that's just what they did.

George: Aren't we seeing an old truism here—that a person doesn't know what he or she can do till he or she really tries?

Steve: You're moralizing a little, but yes, it certainly was true for me.

Jim: If there's one thing I've learned as a manager, it's that performance standards, to be valid, *have* to stretch a person. Of course they also have to be attainable, and I do try to keep them that way. They must never be constraining. They must tie in directly with the responsibility concerned and be as specific as possible in stating the results you desire and time period with which you are concerned. For example, you don't just say that "correspondence and documentation comply with QA." What does that mean? Nothing. Instead you say that *no more than three* outstanding items are acceptable per audit. Only then can you hope to measure your results.

Steve: Jim, you've just mentioned some very important points for managers and engineers to keep in mind when they discuss standards. Do you think you could sum them up in an orderly way?

George: Yes, do that.

For each assignment maintain proper documentation and correspondence in accordance with quality assurance standards.

Prepare, for the department supervisor's review and approval technical specifications, purchase orders and technical reports utilized in the purchase of equipment.

Figure 29.2.   Typical statements of responsibility.

For each assignment maintain proper documentation and correspondence in accordance with quality assurance standards.

Performance shall be considered satisfactory when the outcome of a quality assurance audit shall result in no more than three "outstanding items."

Prepare, for the department supervisor's review and approval technical specifications, purchase orders, and technical reports utilized in the purchase of equipment.

Performance shall be considered satisfactory when only one draft is required prior to final approval and is approved within five working days after submittal.

Figure 29.3.   Statements of responsibility with added standards.

From 4/1/82 to 7/1/82

Special attention will be paid, during the period here specified, to the following job responsibilities:

1. For each assignment maintain proper documentation and correspondence in accordance with quality assurance standards.

    Performance shall be considered satisfactory when the outcome of a quality assurance audit shall result in no more than three "outstanding items."

2. Prepare, for the department supervisor's review and approval, technical specifications, purchase orders, and technical reports utilized in the purchase of equipment.

    Performance shall be considered satisfactory when only one draft is required prior to final approval and is approved within five working days after submittal.

3. Review and evaluate technical reports and specifications.

    Performance shall be considered satisfactory when all "Technical Document Review" forms are completed within three working days after receipt of the documents.

In addition, all other regular duties, as well as any special tasks assigned, will be carried out in accordance with the agreed-on standards and expectations.

Date: March 28, 1982   Signed: Steve Wright

                                                        Jim Allen

Figure 29.4.   Performance contract.

TITLE: Senior Engineer
DEPT./DIV.: Systems Engineering
DATE PREPARED: 5/81 Date of Most Recent Review: 4/82
STATUS: Exempt __x__ Nonexempt _____

Objective: By performing diversified engineering duties, to assist the department supervisor in carrying out assigned responsibilities effectively and thereby helping to achieve departmental and company goals, both short- and long-term.

Reporting Relationships: Reports directly to department supervisor. May have technical aide assistance but does no supervising.

Experience: Requires at least a four-year degree in mechanical engineering supplemented by three or more years of technical engineering.

Responsibilities:

1. For each assignment maintain proper documentation and correspondence in accordance with quality assurance standards.

   Performance shall be considered satisfactory when the outcome of a quality assurance audit shall result in no more than three "outstanding items."

2. Prepare, for the department supervisor's review and approval, technical specifications, purchase orders and technical reports utilized in the purchase of equipment.

   Performance shall be considered satisfactory when only one draft is required prior to final approval and is approved within five working days after submittal.

3. Review and evaluate technical reports and specifications.

   Performance shall be considered satisfactory when all "Technical Document Review" forms are completed within three working days after receipt of the documents.

Figure 29.5.   Position description.

Jim: *First,* decide as precisely as you can the result you're after, so that
            the employee's initiative and judgment are not hampered. *Second,*
            make sure that the performance required by each standard has
            one or more aspects that can be measured accurately; time, for
            example, or units of work. And *third,* be absolutely certain that
            both parties agree on the standards that are being established so
            as to ensure real commitment to implementing them.

George: Then we have all the elements of an effective results-oriented per-
            formance system. Figure 29.6 shows the sequence from objective
            setting, through the negotiation of responsibilities and standards,
            to those two important documents—job description and perform-
            ance contract. Maybe for emphasis we should read off the nine
            steps listed?

Jim: O.K. *Step 1:* Manager and engineer discuss and agree on the ob-
            jectives of the engineer's job.

STEP 1:   Manager and engineer discuss, and agree on, the objectives of the engineer's job.

STEP 2:   The engineer writes down, in draft form, his or her understanding of the job's key responsibilities, using as the basis for this draft the agreed-on objectives.

STEP 3:   The manager reviews the draft to make sure that the engineer's perception of the responsibilities agrees with the manager's own.

STEP 4:   Manager and engineer meet to close any gap between their respective perceptions.

STEP 5:   The engineer drafts standards of performance for the key job responsibilities.

STEP 6:   The manager reviews these standards to determine how closely the engineer's standards meet the needs of the job as the manager sees them.

STEP 7:   Manager and engineer negotiate final standards that will reflect both parties' expectations.

STEP 8:   Manager and engineer complete the formal description of the engineer's job, which will be supplemented periodically by performance contracts and will document their understanding of the engineer's responsibilities and performance standards.

STEP 9:   At stated intervals, manager and engineer sit down together to discuss the engineer's progress in meeting standards and thus objectives, a process that leads naturally to considerations of skill development, planning for the future, and, at the proper time, compensation review.

Figure 29.6.   Establishing a results-oriented performance system for manager and senior engineer.

Steve: *Step 2:* The engineer writes down, in draft form, his or her understanding of the job's key responsibilities, using as the basis for this draft the agreed-on objectives.

George: *Step 3:* The manager reviews the draft to make sure that the engineer's perception of the responsibilities agrees with the manager's own perception.

Jim: *Step 4:* Manager engineers meet to close any gaps between their respective perceptions.

Steve: *Step 5:* The engineer drafts standards of performance for the key job responsibilities.

George: *Step 6:* The manager reviews these standards to determine how closely the engineer's standards meet the needs of the job as the manager sees them.

Jim: *Step 7:* Manager and engineer negotiate final standards that will reflect both parties' expectations.

Steve: *Step 8:* Manager and engineer complete the formal description of the engineer's job, which will be supplemented periodically by performance contracts and will document their understanding of the engineer's responsibilities and performance standards.

George: *Step 9:* At stated intervals manager and engineer sit down together to discuss the engineer's progress in meeting standards and thus objectives, a process that leads naturally to considerations of skill development, planning for the future, and—at the proper time—compensation review.

Jim: If a manager-engineer team can achieve all this, I can practically guarantee a marked improvement in their results.

George: Right. Yet it isn't going to be easy—they should understand that this result-oriented performance system has to be worked at.

29.3.   PRINCIPLES OF PERFORMANCE APPRAISAL

George: Throughout this meeting, as we've considered the manager-engineer team from numerous angles, we've had occasion to mention the performance appraisal. I think it's time now to look at it in detail—particularly since a great many people on both sides of the typical appraisal would be happy to forget it. Bosses ignore it when they can and, if driven to it by the personnel department, fumble their way through the most perfunctory kind of review. Whereas the subordinate endures it all stoically: "Well if this is what it takes to get a raise around here . . . ."

Steve: But that's so silly, George! I definitely want feedback on how I'm doing. Not what I think about my performance but what Jim thinks. I need that type of summing-up. Oh, I squirm sometimes when Jim is critical, but I know it'll help prevent problems later if he tells me frankly what I'm doing badly; and I feel great when he tells me what I'm doing well.

Jim: I too need my annual appraisal session with Steve so I can prepare better for the coming year. I need his help in evaluating not just his performance but the overall effectiveness of our teamwork. Then I can plan how to improve that teamwork in the future. Also, I'm frequently inspired to set new objectives for which to strive. That's something I always enjoy.

George: Jim, can you briefly describe the appraisal process, since managers so often try to avoid it?

Jim: Sure thing! The first step of course is getting ready for the session. I remind Steve that it's appraisal time again, and we set a convenient day and hour for our meeting. Once or twice we've arranged to have the small conference room down the hall, but we find that my office is just as satisfactory if we shut the door and warn the secretary that we're not to be interrupted. Those are the important things, you know, in deciding where to hold the session—quiet, privacy, and no interruptions.

Steve: Once the date is set, George, I know I have to make a list of what I think were my major accomplishments during the past 12 months.

Jim: I believe that Steve should have this chance to state his achievements formally. Besides it gives me a chance to reinforce his successes.

George: And during the appraisal session you compare his list with your own ideas about his performance and sometimes adjust your judgments accordingly?

Steve: That's right. We review both the tasks that have gone well and those that haven't and discuss anything that needs to be improved or changed.

George: What about this face-to-face process, Jim? I realize, of course, that many managers feel most uncomfortable about it—which is why they avoid it when they can. How do you feel about it?

Jim: Well pretty comfortable on the whole, although I wasn't at first. I think there'll always be situations where even the most experienced manager will feel uneasy about leveling with a particular individual. In general I believe that the key to making the appraisal session an amicable one is not so much what you tell your subordinate as how you go about it. I try to choose only a few areas of performance so as not to overwhelm the person with too much information. I try to be absolutely objective and I give recent examples, making these as specific and clear as I can.

29.4.   USING THE APPRAISAL PROCESS TO MOTIVATE

Steve: What Jim's saying is true, George. The way he handles my appraisal, I really *want* to improve. He's honest with me, but he's tactful and diplomatic too—and it's a lot easier to take criticism, I can tell you, when it's given nicely. Also, I like the way Jim involves me in the appraisal session. If I don't agree with anything he says, I know I can always explain my point of view and I know he will really listen.

George: It seems to me that you two are very considerate of each other. And you respect each other. That may be the reason this appraisal procedure works for you. But tell me—what types of things do you discuss together?

Steve: It depends of course on what has happened during the year. Sometimes we talk about work methods, scheduling, or priorities. Or we may get into the behavioral or attitudinal aspects of the job—or, yes, interpersonal relationships.

Jim: Whatever the topic, its essence is improved job effectiveness. So we do some pretty careful analysis: What is the present situation with regard to, say, putting together the technical specification I give my superior? What can be done to make it less time-consuming in the future?

George: And the end-result of your discussion is—what?

Steve: Jim and I always develop what we call our plan for improvement. It's not complicated. This will give you the basic idea. We set up objectives for the year ahead, including target dates and checkpoints along the way where we're supposed to review our progress.

Jim: Let me just add that in order to be successful, the improvement plan must be put in writing. It must be realistic, with specific, measurable objectives, and both the manager and employee must agree upon and commit themselves to it.

George: As a representative of the personnel department, I have to ask about the actual appraisal form we send you. As you know it has two parts. The first, which goes to the personnel department for filing, merely declares that an appraisal has taken place and states what training and/or development steps are recommended. The second part is simply a worksheet for the use of manager and employee; it never goes to the personnel department for inclusion in the employee's file; in fact it is supposed to be destroyed after the appraisal session. Jim, I suppose you do use this form—not only with Steve but with the other engineers reporting to you?

Jim: Oh, yes, I use it. I like its emphasis on development, but for my taste a boss should go a little further. I mean you could fill out that first page and sign it, but never once think about that employee's development needs again. So Steve and I have a regular improvement plan with areas where we hope to do better and specific target dates and checkpoints.

George: You're saying you do more than the form requires.

Jim: Yes, we do. And we believe it pays off.

George: What about the organization's attempt to separate performance appraisal from compensation review? Do you buy that?

Jim: In theory I guess I buy it. Yes. If you're going to be realistic though, you have to admit that a salary increase depends—or should depend—on performance; so both sides still have money in the back of their minds all through the appraisal session. They're bound to. The employee may be concentrating so hard on the probable amount of the hoped-for increase that it's literally all he or she hears. In other words all the boss's criticizing, planning, and counseling will go to waste.

Steve: I agree. It's awfully hard to sit there trying to pay attention when all you can think about is "How much is my increase?"

Jim: Exactly. Oh, separating the two processes is a step in the right direction, if not the whole solution. At least the employee knows that money won't be mentioned during the performance appraisal; so even though the salary recommendation may follow only a few days later, there's a chance that *part* of the appraisal, anyway, may sink in during those few days of waiting. Probably the effectiveness of the system, whatever it calls for, depends on the skills of the individual manager in using it.

George: Increasingly appraisal systems are being expanded to include career counseling—on the theory that an organization is only as effective as its members and for best results must help those members to develop and grow. One way to motivate them to develop and grow is of course to mesh the organization's goals and objectives with those of the individual. Do you, Jim and Steve, ever discuss Nortran's goals—in terms of your own jobs, that is?

Jim: I won't say we talk about them every day, but I think Steve is pretty aware of the company's goals as they filter down through my department. In fact I believe that's part of the orientation that every employee is entitled to. Knowing, I mean, why Nortran is in business and just how you fit in.

Steve: It would be awfully dreary if you just were going through the motions of your job, doing what you were told without knowing why.

29.5. USING THE APPRAISAL PROCESS FOR CAREER LIFE PLANNING

George: But what about your own personal life and career goals? Do you ever discuss your aspirations with one another?

Jim: Sometimes. It isn't always easy. I've learned from experience with other subordinates that the subject has all kinds of emotional overtones. You shouldn't attempt to talk about it, I think, until the two of you have established a solid foundation of trust.

Steve: I suppose when it comes to talking about career aspirations I'm a fairly good example. Jim knows I'm concerned about going up and up in the organization.

George: True, but you're still looking for more from your job, aren't you? Self-fulfillment, if that isn't too grand a word? The satisfaction of proving to yourself that you can handle a tough piece of work, learning a variety of marketable skills, getting to know new people, and what not?

Steve: Yes, naturally, but who knows, circumstances might change. Or I might change.

Jim: All the more reason for my not assuming that you're completely lacking in ambition, that you ask only to be a senior engineer forever and ever.

Steve: (laughs) That could be decidedly risky. For instance, what would you do if Mr. Hartmann suddenly picked me to be in the production department? Can you see me working on the machine floor, either of you?

George: Why not?

Jim: (simultaneously) Oh, no!

Steve: Joking aside though, Jim goes so far as to urge me every now and then to envision some goals for the future, both short- and long-range. He's asked me where I think I'd like to be 5 and even 10 years from now, not just here at Nortran but at home with the family. That gave me some interesting things to ponder because I'd never deliberately sat down and thought so far ahead.

George: Did you come up with any new ideas?

Steve: Some. I went home and tried them out on Phyllis, my wife. We lay awake most of the night, I remember, planning how I would take early retirement and start a business of my own with jobs for her and the boys and even their wives once they get the children safely raised. Anything seemed possible. But, George, I happen to know that Jim does an excellent job of career counseling with his other engineers. Several of them are very ambitious, also.

29.6.   MAKING MANAGEMENT AND THE APPRAISAL PROCESS WORK

Jim: I feel it's to Nortran's advantage for me to be interested in their goals and hopes. Furthermore, there are two very specific steps I can help them to take in realizing those goals and hopes. First, I can encourage them to set reasonable near-term objectives within the organization; that is, identify the next higher position to which they can logically aspire. And second, I can suggest that they develop plans and skills that will improve their chances for advancement.

George: (thoughtfully) It would be a shame, for instance, if Paul Moriarty, one of our outstanding managers, were to quit Nortran out of sheer discouragement in a year or two.

Jim: Much less a few months? I hate even to think about it.

Steve: Jim's not one to hold a person back. Not at all.

George: Well then, Jim, wouldn't it be an equal shame to lose Steve out of sheer discouragement?

Jim: It would indeed, and I do see what you're getting at. I should make it my responsibility to keep up to date at all times on his thinking about me, the job, Nortran, himself, and the future. I should never be complacent about him.

George: O.K. Now let's turn this proposition around, shall we? Steve, do you know what Jim's own goals are for the future? Has he ever confided in you?

Steve: Not in so many words, I'd have to say. But I think that almost any good engineer gets to know a whole lot about the boss; you simply absorb a great many little bits of information, and eventually they add up to quite a fair picture. So I do have some idea of Jim's aspirations.

Jim: (lightly) It would surprise me more if you didn't.

Steve: Of course I do know the logical steps for Jim at Nortran. I'd be a fool if I didn't. And as I see it, a good engineer should be helping the boss up the managerial ladder—not in any underhand, conniving sort of way, but simply by doing a bang-up job, so the boss knows that part of the work will be taken care of and he or she can concentrate on the really important parts. And ... well ... a lot of little things. Like being especially helpful when the job gets tedious, never abusing a confidence, showing where you stand if there's real trouble . . . you know. I've already said I expect Jim to be a vice-president eventually, and I guess I'm working toward that goal in my small way as hard as he is. Or isn't that the sort of thing one talks about? Jim, am I embarrassing you too much?

Jim: (clears his throat) Steve, if and when I achieve that goal, wherever I am, I hope you'll be around to see the day.

George: Then you think it should work two ways. Manager and engineer should both be alert to help to help each other reach their goals.

Steve: That's exactly what I think!

Jim: I agree with you, Steve. We owe it to ourselves to help each of us grow—so we'll have a chance at realizing those goals and ambitions. Which means first of all, creating a climate in which this growth can occur.

George: Once again I can see that your success lies in your ability to communicate effectively with one another. Maybe even without words at times, each of you seems to know by instinct just what the other thinks and feels. And beyond that I sense a strong commitment toward working together as a team to realize both organizational and personal goals. In essence that is what makes the management process as well as performance appraisal work effectively.

30

EQUAL EMPLOYMENT OPPORTUNITY AND AFFIRMATIVE ACTION

Bernard E. DeLury and Francine M. Keneborus

BERNARD E. DELURY is staff vice president of Labor Relations, Sea-Land Corporation based in Menlo Park, New Jersey. Mr. DeLury served as United States Assistant Secretary of Labor for both employment standards and labor-management relations from 1973 and 1977. He was also with the New York State Department of Labor as the Deputy Industrial Commissioner for the New York City Metropolitan Area. Mr. DeLury received his B.A. from St. John's University and M.A. in Sociology from C. W. Post College.

FRANCINE M. KENEBORUS is currently an equal opportunity affairs specialist with Sea-Land Corporation. She was assistant manager of Corporate Equal Opportunity for Johnson & Johnson. Her corporate experience was preceded by six years with the United States Department of Labor, first as a compliance officer with the Labor Management Services Administration, and then as an equal opportunity specialist with the Office of Federal Contract Compliance. She is a graduate of Douglass College, Rutgers University.

The views expressed in this article are strictly those of the authors.

30.1.   INTRODUCTION

Equal Employment Opportunity and Affirmative Action have their impact on the engineering business enterprise. Throughout the entire employment process, one or both of these concepts affects day-to-day business decisions. While we generally tend to think of minorities and women when considering these issues, almost all employees are covered by one or more of the legislative provisions. It is critical that managers understand what their responsibilities are in implementing the regulations and safeguarding their organization from charges of discrimination.

What is Equal Employment Opportunity? How is it different from Affirmative Action? Who is covered? Equal Employment Opportunity (EEO) guarantees to all individuals equal access to employment based on his or her qualifications, disregarding race, color, sex, religion, age, national origin, or handicap. In "thou shalt not" language, discrimination in all employment phases, including recruitment, hiring, training, promotion, placement, benefits, and compensation, is prohibited by federal, state and local statutes. All organizations with 15 or more employees are covered by EEO laws. Affirmative Action however, is applicable only to those firms with government contracts or subcontracts. Affirmative Action goes beyond the concept of equal access; it means that not only will men and women, minorities and nonminorities be treated equally and equitably, but that companies will actively seek out and employ women and minorities. Extra measures are taken to ensure that target group members (minorities, women) are employed in the workforce at a rate commensurate with their availability in the labor area, and once they are employed, that they participate in training and promotions at representational levels.

This chapter will deal primarily with minorities in the workforce (Chapter 31 considers the special problems of women in engineering firms). With nearly every kind of engineering firm reporting a shortage of engineers,[1] it is short-sighted if not foolhardy to ignore minorities and women in the engineering talent pool. Demographic trends make it imperative that we examine the implications of a multicultural workplace within the framework of Equal Opportunity and Affirmative Action, for it is becoming more and more apparent that minorities will increasingly hold degrees in quantitatively-based fields.

30.1.1   Demographic trends

The 1980 United States Census reported that 2.6% of all engineers were black; 2.3% Hispanic; and 4.4% Asian and Pacific Islanders. However, these minority underrepresentations will be altered by projected representational trends in the makeup of engineering degree recipients (Table 30.1). Thus educational data show an overrepresentation of Asian-Americans; they choose quantitative majors at double the selection rate of Whites. This minority group has at least double their representational share of engineering degrees at all levels. Significantly, the percentage of Asian-Americans earning Ph.D.s in 1979–1980 (data not included on Table 30.1) rose to 17.1%.[2] For other minority groups the degree information reflects no great shift projected in representation of the 1980s, either relative to nonminorities or to their share of the cohort age population. However, studies have demonstrated a strong correlation between parental college experience and children's choice of postsecondary education in quantitative fields which could lead to increased Black and Hispanic engineering representation in the next decades.

TABLE 30.1. Representation of Racial and Ethnic Groups by Engineering Degree and Year (1975/76 to 1978/79)[2]

Year	Racial and Ethnic Group	Percentage Earning Degrees			
		Associate	Bachelor	Masters	Ph.D.
1975/76	Whites	86.2	91.3	91.9	91.3
	Blacks	7.9	3.2	1.9	1.1
	Hispanics	4.2	2.9	1.9	0.9
	American Indians	0.6	0.4	0.3	0.2
	Asian-Americans	1.1	2.3	4.0	6.6
1976/77	Whites	84.8	92.0	90.0	89.9
	Blacks	8.1	3.0	1.9	1.3
	Hispanics	4.7	2.0	2.0	1.5
	American Indians	0.7	0.3	0.2	0.1
	Asian-Americans	1.7	2.7	6.0	7.2
1978/79	Whites	84.0	90.8	88.3	85.8
	Blacks	8.0	3.1	2.1	1.5
	Hispanics	4.7	2.7	1.9	1.4
	American Indians	0.7	0.3	0.2	0.1
	Asian-Americans	2.6	3.2	7.5	11.2

The Asian-American representational leaps are the result of educational experience: 40% attend universities and 60% of that group are enrolled in the "most selective" universities; an additional 13% are in the "most selective" four-year colleges where they overwhelmingly select quantitative majors. A sociological reason has been suggested that because of difficulties they experience with Western languages, Asian-Americans have utilized a skill advantage in mathematical rather than verbal areas.[3] Whether or not this hypothesis is valid, the fact remains that the pool of minority engineers is increasing, and engineering companies must recognize that the complexion of their workforces is changing and will continue to change through the 1990s: It is anticipated that by 1995 occupations in engineering are expected to experience a 48–52% job growth. Students, non-minority and minority, will be attracted to an area which presents excellent job prospects as well as high reward—in 1984, average starting-salary offers ranged from $22,764–$29,568.[4]

30.2. WORKING WITH THE AFFIRMATIVE ACTION PLAN

A firm with government contracts or subcontracts is required by executive order to prepare an Affirmative Action Plan (AAP). The AAP contains analyses that determine percentages of minority and female representation within internal organizations by department and occupational category. "Minorities" include Blacks, Hispanics, Asians, and American Indians; separate representations may be calculated for each minority subgroup. Estimations are then made of the percentages of minorities (and minority subgroups) and women in the external labor force who possess the requisite skills utilized in the company. These resulting "availabilities" are compared to the internal representations. For those occupational categories whose representation levels are below the availability rate, "underutilization" is declared. The AAP addresses companies' efforts to identify and correct underutilization through the use of goals and action plans that outline recruitment efforts, training programs, career pathing, and so on.

Goals are not quotas. Goals represent targeted opportunities for which action plans are employed in order to effect representation commensurate with availability in an occupational category or "job group," as opposed to quotas which are hard and fast numbers with no thought given to actual opportunities. Tokenism, the hiring or promoting of an unqualified target group member in order to achieve a goal, is not a solution to Affirmative Action efforts. It is counterproductive to the organization's work, demoralizing to the work group, and results in a disgruntled and unhappy employee. Goals for affirmative action are really no different from project or production goals: They are attainable, reasonable, and clearly measurable. If they are not attained, explanations can be made of what efforts were made toward goal achievement and what circumstances prevented it.

At the beginning of each year (or each AAP year), managers should meet with the person responsible for preparing the AAP (generally someone in the personnel department) to review the potential opportunities for hires, promotions, retirements, transfers, and any other activity which might result in personnel movement. The managers will have their work group's activity divided up into the various occupational categories. Discussion should include a summary of personnel activity of the previous year, number of opportunities used to work toward/achieve goals, outreach and developmental efforts, conditions which stymied goal attainment.

It is critical that each manager understand what his or her goals are, outline a plan of action to effect those goals, and know that he or she will be held accountable for Affirmative Action responsibilities in performance evaluation. The AAP exists not only to identify the organization's underutilizations, but to present the steps to be undertaken to correct representation. The AAP contains recruitment sources which specialize in target group members and organizations for professional minorities and women— sources for qualified individuals which can assist the manager in meeting goals.

The AAP contains contacts for minority outreach efforts. Recruitment of minorities, especially Blacks and Hispanics, can be difficult when availability figures are low. There are several organizations which are dedicated to increasing the number of Black and Hispanic engineering students in the educational pipeline. Generally these organizations provide guidance, tutoring, internships, scholarships, and encourage students to pursue an engineering career. They include (but are not limited to): National Action Council for Minorities in Engineering (NACME), Graduate Engineering for Minorities (GEM), Inroads, Inc., Minorities in Engineering (MEP), and Consortium for Minorities in Engineering.

The Office of Scientific and Engineering Personnel of the National Academy of Sciences reports that new job openings and replacement demands create approximately 110,000 engineering jobs per year. This figure is projected to rise during the 1980s. Newly graduated engineers account for an estimated 47% of the positions filled.[5] The remainder must come from alternative sources. The Affirmative Action Plan also helps to pinpoint those sources. In the engineering firm's workforce are workers with related skills, technicians who can be trained and upgraded; the AAP will have typical lines or progression and promotion paths for these positions. Individuals with engineering technology degrees and workers from foreign countries may also be recruited. The wide ranges of sources are reflected in the AAP.

30.3. DISCRIMINATION AS A SUB-SURFACE ATTITUDE

Discrimination in the workplace is seldom overt, rather it is a subsurface attitude which may manifest itself in ethnic jokes, radical-stereotyped comments, assumptions of limitations of an individual's abilities or promotability. Individuals are not automatically rid of their racial prejudices when they join your organization, but the company can demand that whenever employees are performing in the workplace or managing a work group, that their behavior reflect respect, not prejudice, and that all personnel decisions be based on the merits and talents of peers and subordinates.

Discrimination may be manifested in a number of ways. The use of language that could be interpreted as being racist or ridiculing of someone's religion or national origin may be evidence of discrimination, i.e., a decision made on the basis of an individual's race, sex, religion, age, national origin, disability, or status as a veteran. It may affect hiring, placement, promotion, discipline, termination, training, compensation, transfer, or any other condition of employment.

Discriminatory behavior is not limited to treatment from managers. Subtle discrimination may also come from the minority engineer's peers in the form of racial slurs or jokes, ostracism, lack of cooperation, patronism. While it is virtually impossible to prevent incidents or behaviors occurring among a manager's subordinates, it is the manager's responsibility to ensure that actions do not persist or to permit an employee to infer that discriminatory behaviors will be tolerated. The manager must be firm and clear with people of his or her race, sex, or ethnic background that any language, jokes, or attitudes which are discriminatory will not be permitted or supported. Managers should be alert to the interactions of their staff, keep a careful watch on the work-place atmosphere, morale, and activity. Sensitivity to employees' perceptions and assumptions is extremely important to maintaining a productive work atmosphere.

Often it is difficult to separate a legitimate complaint of actual discrimination from poor management practices or oversensitivity on the complainant's part. Generally, discrimination falls into the following categories:[6]

Disparate impact. An apparently neutral employment practice which disproportionately screens out or negatively affects a target group's representation. Qualifications for a position which stipulate minimum height and weight restrictions might have disparate impact; for example, qualifications requiring that all candidates for an opening in a particular department be six feet tall and weigh 180 pounds or more would effectively eliminate most Hispanics, Asians, and women from consideration. Instead, construct qualification requirements must be job-related and task-oriented; for example, incumbent is responsible for stocking supply room, so he or she must be able to lift 50 pounds. It is also necessary to ensure that job prerequisites are job-related and necessary to perform the job. Unreasonable qualifications might include advanced degrees, demonstration of language and writing proficiency, disqualification for criminal records, and dress and grooming requirements. Ability to perform the job is the "bottom line" when making hiring, transfer, developmental, or promotional assignments.

Disparate Treatment. Employment practices vary in implementation based on an individual's race, sex, religion, national origin, disability or status as a Vietnam era or disabled veteran. For example, when filling a position, educational requirements are waived for a nonminority applicant, but a minority candidate must have earned a specific degree.

Present Effect of Past Discrimination. Ostensibly neutral employment practice whose current application is nondiscriminatory, but when examined in the context of past application, is found to be discriminatory. The individuals who were discriminated against in the past continue the effects of the discriminatory treatment into the present; for example, all minorities who were hired before 1964 were placed into two departments with lower-paying job classifications. Their seniority was tied to the department; if they bid into another department, they would lose their seniority. The seniority system effectively froze the target group into inferior positions.

Reasonable Accommodation. Employers are required to make reasonable accommodations for the religious beliefs or handicaps of their applicants and employees, eliminating the impediments to their employment such as work scheduled on the Sabbath or architectural barriers. Exceptions to the accommodation requirement are based on cost, nature of the enterprise, and undue hardship on the conduct of the employer's business.

While it would be helpful for managers if discriminatory situations manifested themselves only in one of these four categories, there are others more difficult to discern. In addition to the subsurface attitudes and the four categories, managers must guard against acting upon notions of "where one belongs" because of the employee's race, sex, age, disability, or veteran status. Individuals must be differentiated only by their abilities, talents, and performances. In addition managers need to be aware of the implications of a multicultural workforce. Differences in culture can create differences in the communication process—it is crucial that your message be clearly conveyed. Check to verify that the information was interpreted in the way it was intended. Sensitivity to subordinates' perceptions and conceptions demonstrates a manager's respect for the staff. A manager's willingness to discuss decisions and problems with subordinates creates an atmosphere where potential problems and conflicts may be resolved at a primary level.

30.3.1. Special Problems for Minority Engineers

In discussions with minority engineers on their perceptions of difficulties encountered by minorities, common themes emerge, the primary one being the conflict between the objective and subjective. J. Russell Elliott, director of Industrial Engineering at Ortho Pharmaceutical Corporation, explained that engineering is unique in its orientation to the objective—if something cannot be measured, then it is lost. An engineer's background has not prepared the engineering manager to deal with "people problems." It becomes awkward and difficult to deal with subjective areas, especially in performance appraisals. Evaluations of projects completed in terms of deadlines, budget, and end product, present discrete, measurable objectives which lend themselves to more bias-free ratings. The difficulty arises with those subjective areas that require an evaluation of an employee's "potential." The basic principles of performance evaluations are treated in Chapter 29.

There also exists a tendency for engineering managers to insist that they will hire (promote, etc.) the person who can "do the job" regardless of their race or sex; job requirements are so objective that they do not permit discrimination. Minorities themselves have perceived that this belief may be operationally valid during the initial 5–8 years of engineering employment when one is given in-place promotions from "Junior Engineer" to "Senior Engineer." Bias and a loss of objectivity arise when the professional engineer is a candidate to become a managerial engineer because this requires an assessment of leadership skills and career potential which are areas prone to subjective judgments.

At this stage, the promotability of minorities depends on their acceptance in the organization. If the firm's operating philosophy, compensation standards, and performance rating system were designed to benefit each employee's growth, they will determine the degree to which affirmative action and equal opportunity will be implemented. One must be sensitive to the fact that more than one minority engineer has expressed frustration and anger over being mistaken for the mailroom clerk when on the elevator or otherwise away from his or her own department.

While communications and sensitivity to employee-relations issues are problems particularly encountered by engineering managers, it should be noted that the other employment difficulties that exist for minorities in all other industries and occupations also exist in engineering. Minority employees are constantly being tested and find that they must prove themselves each time they are given a new assignment or position. There exists for them a sense that management is giving them the opportunity (that they deserve) with the attitude of "we'll see if they can do it" and make judgments on this subject on a group basis, rather than by evaluating individual performances.

Whatever the justification or justice of such a view, minority professionals and managers know that in order to gain recognition and reward, they must perform at least 50% better than their peers. They are employed in an environment where the system will not be working for their benefit, where they will watch their co-workers promoted in half the time.

Minorities recognize that finding a mentor, someone who recognizes potential and acknowledges in the establishment of the mentor relationship the individual's professional potential, is important to their careers. Within a team or department structure, it is very easy for the majority manager to take credit for the minority member's project without sharing it.

The solution to such problems is to manage the team so that each member is successful in his or her job. Remember that success travels upwards: one's subordinates' progress reflects very well on the boss who manages it. Convey to all that they are important to the team and that you as manager want to help them succeed. This individualized plan for each person does not mean that minorities (or females, or any other target group) receive any special treatment over white males, nor does it allow for different, more lenient performance standards or disciplinary treatment. It does however, mean that managers recognize and treat each team member as an individual with differing talents, abilities, strengths, weaknesses, perceptions, and sensitivities. It is management's task to manage each person so that they optimally and productively perform.

30.4.   WHAT YOU NEED TO KNOW ABOUT THE LAW

There are a number of laws and executive orders which comprise the legal foundation of equal opportunity and affirmative action. While the thrust of this chapter has been on minorities in the workplace, it will become evident in our discussion of the statute that nearly all employees are covered and protected by at least one of these laws.

Most EEO complaints arise from day-to-day decisions made by managers. The potential liability for both the company and the manager is great, so it is very important that each person who makes decisions affecting employees, from hiring to compensation to discipline/termination, understands the components of the EEO laws and their impact on managerial responsibilities; each law's basic content will be listed along with an example of its application.

Title VII of the Civil Rights Act of 1964 prohibits discrimination on the basis of race, sex, color, religion, or national origin in any aspect of employment by companies with 15 or more employees. Individuals are guaranteed equal treatment and are not to be considered as a member of a group. One application of Title VII is the use of qualifications for a position as cited in the example for Disparate Impact where the qualifications must be job-related and must not effectively screen out groups of people. Caution should be exercised to verify that all requirements are demanded by the job tasks. In another situation it would be unlawful to deny interviews to all Hispanic secretarial candidates because the last Hispanic secretary you had "did not work out" (treating an individual as a member of a group).

Executive Order 11246 of 1965 (as Amended by Executive Order 11375) requires that all federal contractors holding contracts and subcontracts of $10,000 or more must take affirmative action to include women and minorities in the workforce at a rate comparable to their representation in the relevant labor area. Written affirmative action plans (AAPs) are updated annually for all firms holding federal contracts and subcontracts of $50,000 or more and who employ 50 or more people. The purpose of E.O. 11246 was to utilize the influence of federal contract awards as an impetus to contractors to increase minority and female employment opportunities in areas where they had historically been underutilized. As practiced under the regulation requirements when preparing your company's AAP, availability of minority electrical engineers is determined by the company to be 7% (for example) and an analysis is made of minority electrical engineer representation in the firm. If utilization of minority electrical engineers is found to be less than 7%, a goal of 7% representation would be set along with an action plan to achieve that goal. Success at goal achievement is measured annually by degree of goal attainment and the efforts aimed at recruitment, training, promotion, and so on.

The Age Discrimination in Employment Act of 1967 (as amended in 1978 and 1984) bars employers with 20 or more employees from discriminating against individuals 40–70 years old on the basis of age. This law offers protection to applicants and employees; whenever any employment decision is being made, age should never be the determining factor. Currently, this is the most prevalent basis for legal action brought to the courts by white males. Action tends to arise from forced retirement, especially when the company is attempting to reduce the size of its workforce.[7] Whenever there are attempts made to provide incentives for older workers to retire, provisions of the offer should be in writing and emphasize that election of participation in the retirement offer is totally voluntary.

In another situation if there are two completely qualified candidates for the position, one of whom is 32 and the other 46, do not select the 32 year old on the assumption that you anticipate that he or she will be available to work for you for more years than the person 46 years old.

The Equal Pay Act of 1963 covers employers with two or more employees and guarantees that men and women performing work requiring equal skill, effort, and responsibility under similar conditions at the same location, be paid the same salary or hourly wage. Exceptions are permitted for shift differentials, seniority, level of sophistication, and increments granted for performance. When hiring entry-level designers, men and women should be paid the same salary. The salary offered should not be based on what each designer earned with his or her previous employer.

The Rehabilitation Act of 1973 requires companies with federal contracts or subcontracts of $2,500 or more to take affirmative action for the employment of handicapped individuals. Although one or more of their "major life activities" may be impaired, the handicapped person must be qualified to perform the duties of the position. Employment of the disabled might require an accommodation on the part of the employer: A hearing-impaired employee is provided a phone with a device that amplifies the sound transmission; the person is able to complete job tasks, and the low-cost accommodation assists the otherwise qualified person to better perform their job. Remember that "handicap" does not apply only to mobility-impairment; diabetes, medically controlled high blood pressure, recovery from a heart attack, and epilepsy (to name a few of the "hidden" handicaps) are also included.

The Vietnam Era Veterans Readjustment Assistance Act of 1972 provides that companies holding $10,000 or more in federal contracts and subcontracts take affirmative action for employment of Vietnam-era veterans and disabled veterans of all wars. Implementation centers on the use of the Major Job Listing Program of the State Employment Service. Employers utilize the program to recruit externally for position openings with salaries of $25,000 or less.

The Pregnancy Act of 1978, as an amendment to Title VII, provides that a woman experiencing pregnancy, childbirth, and all related medical conditions be treated as an individual experiencing any other medically disabling condition. In addition to providing for equal medical coverage benefits, all other aspects of employment must be equal: terms of leave, provisions for job reinstatement, accrual of seniority, and so on. This law does not require a company to provide benefits or conditions of employment to pregnant women that it does not offer to all its other employees, but it does require that whatever treatment is provided to an employee who becomes medically unable to work, it offers to the woman disabled by childbearing.

In addition to the federal statutes just mentioned, there are state laws and municipal ordinances that prohibit discrimination on the basis of race, sex, religion, age, national origin, handicap, or veteran status. Many locations have also extended the ages covered under age discrimination to 18–70 and the scope of antidiscrimination laws to include sexual orientation or sexual preference. Facilities located in the areas with expanded coverage are subject to the terms of both federal and local legislation, so it is worthwhile to keep abreast of the possibly broader local ordinances and statutes.

The Equal Employment Opportunity Commission has issued interpretive guidelines on implementation of protections offered under Title VII for national origin and religious discriminations. While no statistics are maintained by a company on employees' ethnic background or religious affiliation, the firm ensures that employees function in a work environment that is free from harassment or discrimination.

The National Origin Guidelines[8] focus on discrimination in employment practices where opportunities are denied based on an employee's national origin or citizenship status. According to them, national origin discrimination may be manifested in (1) administering tests in English to applicants whose native language is not English when English language skills are not a job requirement, (2) discrimination against persons married to or who associate with a person of a particular national origin, (3) adverse action taken against a person who belongs to an organization which promotes the interests of a certain nationality, (4) unequal treatment based on an individual's church or school attendance which is frequented by people of a specific national origin, (5) discrimination solely because the person's name or name of their spouse is indicative of a particular national origin, or (6) imposition of height and weight requirements which effectively bar participation of a class of people whose height and weight norm fall outside the requirements when these specifications are not job performance related.

The imposition of rules requiring workers to speak only English while on the job have been reviewed by various courts. Findings of nondiscrimination have resulted when the employer demonstrated that the usage of English was necessary for workplace safety and communication with customers, supervisors, and co-workers. In situations where it is possible to perform all job duties without speaking English or using it very little, "English-only" rules have been disallowed; conversely, when proved difficult or impossible to execute job tasks without use of English, "English-only" rules have been permitted.

A white, Anglo-Saxon, Protestant female may not be passed over for an interview for a job because she is married to an Asian and has adopted his Asian surname when the interview is not granted because an assumption is made that she is Asian. It does not matter what her true nationality is—the discrimination rests with what she is perceived to be. Nor may she be denied the interview merely because her husband is Asian.

If persons are legally able to work in the United States, they may not be denied an employment opportunity based on their non-United States citizenship. Likewise no employment decisions should be made on the basis of birthplace, that is, not born in the United States. Exceptions to discrimination based on citizenship exist only when it is demonstrated that national security is involved.

The Guidelines on Discrimination Because of Religion[9] provide that employers make reasonable accommodations to permit their employees to fulfill their religious obligations. The degree to which accommodations must be made is regulated by the amount of hardship it would impose on the conduct of business. Examples of accommodation include permitting employees to use their vacation or personal time to observe religious holidays or allowing employees to swap shift duties or days off to eliminate conflict of work and Sabbath. Undue hardship on the company's business may exist when a trouble-shooting technician is unable, because of religious beliefs, to work from sundown on Friday to the end of the Sabbath if the systems which he or she repairs are operated 7 days a week and no other technician is available locally. In addition, sometimes business operations and customer safety require that no accommodation can be made. It is important to keep in mind whenever terminating or refusing to hire a person because of his or her religion, that the burden of proof of undue hardship rests with the employer.

In some instances religious observance entails adherence to "alternative" styles of dress or appearance, for example, the wearing of long clothing, head coverings, or beards. When the company objects, management must likewise show that its objections are grounded in the safe or efficient conduct of business or job performance.

Equal opportunity laws extend protections to virtually all employees. As a manager it is important to convey to peers and subordinates that the legal requirements imposed on the employment process are not an unnecessary burden, but that they have resulted in fair treatment for all employees, consideration for advancement based on abilities and talents, the recognition of individuality, and the incorporation of all people into the workplace.

Managers must be aware of what the laws require because they are responsible for implementation on a daily basis in their departments and project teams. Ignorance and insensitivity will almost certainly lead to blunder and an EEOC charge.

30.5.   PREVENTING POSSIBLE LITIGATION SITUATIONS

There is no magic formula that can guarantee that not a single applicant or employee will ever file a complaint with the EEOC. Court awards and monetary settlements publicized in newspaper and television accounts have made people aware of their rights under employment laws and of the potential liability companies bear when they violate them. Remember that terminated employees have nothing to lose by lodging an EEOC charge if they are female, minority, over 40 years old, nonminority (reverse discrimination), handicapped, disabled veteran, veteran of the Vietnam era, male (reverse discrimination), and so on. Disheartening? Perhaps, but there are defenses available to managers and their companies, and the best defense is to practice sound, fair, good management techniques. This Section will not tell you how to circumvent the laws, but will list practices which could lead to an EEOC investigation or courtroom case.

30.5.1. The Old-Boy Network

In many organizations there was an "old-boy-network" through which personnel decisions were made outside of formal channels and depended on who knew whom, or who was the vice-president's latest protege. This informal network appeared to work in the gym locker room, on the golf course and at the country club. It is no surprise that people know and associate with people similar to them—the "totem effect"—so it is also no surprise that the old-boy network is comprised of and promotes white males. Historically most of companies' decision-making powers rested in the hands of white males and was passed on to more of the same.

Minorities and women are at a disadvantage within this system because white males feel they have more in common with other white males—shared schools, churches, clubs, sports, and so on. With bonds of commonality, they associate more with one another both on and off the job. Because of the greater exposure offered, the white male's abilities and accomplishments gain more visibility. When opportunities arise, their names become discussed as potential candidates. Even if minorities or females are added to the list for consideration, the white male has an advantage because he is the known quantity. There are other advantages which this person has: Through his association, he has an awareness of concerns on a broader organizational scope, understands the concerns of his management, and in effect has picked up the "company lingo." At many firms in which this system has been ensconced, minorities and women are strangers in a strange land.

While the old-boy network will not simply disappear, adherence to more formal channels for promotions will help to ameliorate its effects. Developmental assignments are an excellent means of increasing visibility and skills: Delegate to all your employees, including minorities and women, duties for special task forces. Management should also make the effort to incorporate minorities and women into these informal network systems and become a mentor. Mentoring is preferable for some minority member or female who is not your immediate subordinate. Your focus will not be on their job duties but on imparting awareness of the organization's philosophy, structure, and politics. See to it that they become assimilated into the company culture and fully comprehend and absorb the feedback they are given on their performance.

30.5.2. Elimination of Floating Criteria

Nothing is as frustrating (and potentially illegal) as floating criteria for a particular position. John Jones, a white male, has been a project manager for 4 years. He is promoted and his position becomes vacant. Sam Smith, a black male, who has worked for Jones and Jones' predecessor, is certain he is the most qualified to succeed Jones. When interviewed he is told that the successful candidate should have a masters degree. Neither of the two past incumbents (nor Smith) holds an advanced degree. Smith believes he is being discriminated against. If upgrading the educational requirement of the job is a pretext of awarding the job to someone else, Smith may be correct and the company held liable. As a result Smith may be awarded back pay in the amount of the promotional increment from the day he would have become project manager and be guaranteed the next open project manager position.

This situation can be avoided by ensuring that each position's requisite criteria are job-related and performance-demanded. Consider the applicability of substituting experience for education (and vice-versa). Eliminate subjective assessments and ensure that all requirements are not unnecessarily screening out candidates. Apply the criteria for requisite skills and educational attainment uniformly; once an exception is made, it is the exception which sets the standard. The best defense in the example above is a solid, demonstrated reason for changing the criteria. Once Jones was promoted, his old work group was reorganized and his former position was given additional responsibilities, including a long-term project necessitating knowledge gained in the masters program or equivalent experience (exceeding Smith's tenure). The job description and job analysis should be revised to reflect the increased duties. To avoid a charge or litigation, one more thing must be done: Explain to Smith why he did not get the promotion. If that exchange does not occur, he will probably go ahead and act on his perception that he has been discriminated against.

30.5.3. Sensitivity to Perceptions

People carry emotional baggage which helps to form and color their perceptions of the world. Those perceptions are not necessarily valid, but that does not prevent people from acting on those perceptions or misconceptions. Managers should be open and sensitive to the perceptions of their subordinates. As a manager you should exhibit a willingness to discuss subordinates' viewpoints, and objectively attempt to counter and correct their observations if situations have been misconstrued. Should their perceptions be on target, you as manager will have to take corrective actions. Empathy with minority or opposite sex perceptions may be difficult initially, but do not discount observations merely because you have not perceived situations in the same way. Keep in mind that cultural differences will affect a person's perceptions; in your rational discussion with an employee, listen for differences relating to his or her culture, and respect that perception.

30.5.4.   Provide Feedback

In the context of this chapter, feedback is important in conveying to subordinates the reality of their employment situation—their promotion potential, problem areas, and career path options. Complaints often arise when individuals believe they are not moving as quickly as they should. Feedback sessions allow risk-free reality testing of their professional aspirations. Their unrealistic expectations can lead to disgruntlement, discontent, and low morale. Be honest about informing subordinates when you feel they are ready for promotion. Minorities often feel that they do not receive the critical feedback they need to become next-job ready because majority managers are uncomfortable in presenting them with negative information. An individualized approach to managing each subordinate will counter any discomfort in dealing with minorities and females; remember they all have the same objectives of performing to the best of their abilities and maximizing their career growth within the company. Employees must also understand that there are often limited numbers of promotional opportunities, and that even once one is considered "promotable," the next position may not be available. Limited numbers of opportunities can also mean strong competition for jobs. When one of your subordinates does not receive the job for which he or she was a candidate, find out why they were not selected. Let your employee know, and if it is a performance or experience deficiency, try to incorporate the corrective action into his or her developmental plan. Honest feedback also means a candid discussion of career growth potential. When the employee is unrealistic about his or her strengths, be candid about your assessment. Again, institute efforts directed at correction or assist in a readjustment of career aspirations.

When employees have confidence in you as their manager; view you as a source of reliable, available, accurate information; count on your feedback for an honest, objective assessment of their performance, then they will also come to you first with their complaints and perceptions of unfair/unequal treatment. When someone does complain, treat the allegation seriously and institute an investigation quickly, approaching the issue without any preconceived notions. It is much better for the company and interactions within your department or team to settle the issues within the parameters of the organization. Your investigation should be more attractive to the aggrieved because the allegation can be resolved speedily, and to the organization because the settlement options are generally more favorable, less costly, and the strength of the resolution system is enhanced.

One last word on retaliation or "winning the battle and losing the war": Whenever an employee has complained, whether internally, to a federal or state EEO agency, or in an investigation, of a practice he or she believes to be discriminatory, you must make sure that no retaliatory action is taken against him or her. Retaliation can take the form of any negatively perceived activity: termination, suspension, demotion, transfer, threats, exceedingly heavy workload, shunning, or any other disciplinary action. The charge or complaint based on the entire gamut of fair employment practice legislation may have been found to have no merit or been settled, but if retaliation related to the complaint is proved, the liability can be just as onerous: back pay, reinstatement or reassignment, damages, and attorney's fees.

To avoid the retaliation trap, do not take any adverse actions towards any complainant until allegations have been resolved. If a performance problem has existed, make sure that you continue to follow your company's documented performance improvement plan and procedures or progressive discipline procedures. The individual's personnel file should contain written records that are objective and consist of examples of specific behaviors. One word of caution: Do not attempt to construct a whole file at once as a response to a discrimination complaint, finally putting in writing discussions or incidents which occurred weeks earlier. Maintain identical kinds of files on all your subordinates; do not set up files only when someone becomes a "problem." Do not subject one person to a degree of scrutiny or performance that is not imposed on all co-workers.

Practicing good management principles administered fairly, objectively, and sensitively in an environment of established policies and performance standards with criteria uniformly applied will result in a work team operating productively and discrimination-free.

NOTES

1. Division of Science Resource Studies/STIA, Proceedings of the 1984 Meeting of the Scientific Manpower Commission and the Engineering Manpower Commission, May 15, 1984.
2. S. E. Berryman, *Who Will Do Science?* The Rockefeller Foundation, November 1983.
3. *Ibid.*

4. "Who Makes What Where," *Working Woman,* January 1985.

5. A. Fechter, Proceedings of the 1984 Meeting of the Scientific Manpower Commission and the Engineering Manpower Commission, May 15, 1984.

6. B. Schlei and P. Grossman, *Employment Discrimination Law,* Bureau of National Affairs, 1976.

7. S. E. Parnes, *What Every Personnel Assistant Should Know About EEO Law,* Bureau of Law and Business, Inc., 1984.

8. *EEOC Guidelines on Discrimination Because of National Origin,* 29 CFR Part 1606, March 18, 1974.

9. *EEOC Guidelines on Discrimination Because of Religion,* 29 CRF Part 1605, July 13, 1967.

31

THE WOMAN EMPLOYEE IN ENGINEERING

Ada I. Pressman

ADA I. PRESSMAN is Engineering Manager for the Western Power Division of Bechtel Power Corporation in Norwalk, California. Ms. Pressman received a Bachelor of Mechanical Engineering from The Ohio State University and a Master of Business Administration from Golden Gate University. She served as a national vice-president of The Instrument Society of America and national president of the Society of Women Engineers. Honors include Distinguished Alumna from Ohio State University, Achievement Award from the Society of Women Engineers, and the TWIN Award from the YWCA. She is a registered professional engineer in California and Arizona.

31.1. BARRIERS AND PROGRESS

31.1.1. The Past

In the past, two reasons for a scarcity of women professionals and managers in the technical work force were (1) the limited supply of those qualified for selection and (2) the so-called "conditioned bias and antiquated views of male management."[1]

The limited supply was the result of uninformative counseling by parents and school professionals. Male children are inculcated from the beginning

TABLE 31.1.   Women Engineering Graduates

Year Ending in June	Bachelors	Masters	Ph.D.s	Percentage of Total Engineering B.S.
1962	125	40	4	0.36
1963	130	32	11	0.39
1964	164	34	7	0.41
1965	139	44	10	0.38
1966	146	76	9	0.41
1967	184	78	11	0.51
1968	177	58	5	0.47
1969	328	107	23	0.82
1970	358	170	16	0.83
1971	353	158	25	0.82
1972	525	299	35	1.19
1973	624	226	48	1.44
1974	744	393	36	1.80
1975	878	380	56	2.30
1976	1376	557	56	3.62
1977	1961	646	67	4.89
1978	3280	814	51	7.11
1979	4716	890	61	9.00
1980	5631	1086	88	9.70
1981	6557	1232	90	10.42
1982	8140	1549	126	12.15

Source. Engineering Manpower Bulletin, No. 18, September 1983, published by Engineering Manpower Commission of the American Association of Engineering Societies, Inc.

that they will work all their lives or at least pursue an active vocation. Female children in the past and even today are not generally given such definite goals. With little or no encouragement, lack of confidence in their ability to do something "nontraditional" prevented women from fully exploring their options. Society emphasized independence and career orientation for men. Society dictated that women were not allowed to work. Their joys in life were intended to be derived from the achievements of their husbands and children, and too often they were taught that there would always be a man to turn to for advice and decisions. Only in very recent years has it been considered "somewhat acceptable" for women to be independent and have careers in the professions that have traditionally been held principally by men. Today's young women have much more of a positive attitude toward their own potential, and in general such awareness is beneficial to everyone.

Up through the 1970s the challenge for women engineers was breaking down the barrier of acceptance. In the 1950s there were insufficient women in engineering to qualify as a statistic and female enrollment in engineering in the universities remained at less than 1% throughout the 1960s. In 1970, (Table 31.1) there were 358 women who graduated with a bachelor's degree in engineering or 0.83% of the total bachelors degrees in engineering.

As the cultural attitude toward women and by women changed, society was challenged to remove constraints that had previously been placed on women with regard to choosing a career outside of the home. Females in high school were preparing themselves for a technical career by studying mathematics, physics, and chemistry. Women were accepted by the universities and encouraged to enroll in engineering.

In the mid-1970s, Purdue University began a four-year effort to boost the number of women to about 25% of total engineering undergraduates.[2] The campaign included seminars, brochures, and posters directed at high school females.

Many colleges increased their liaison with high school teachers and guidance counselors so that these people would direct more females in the field. The Lawrence Hall of Science at Berkeley, which had to work hard to get several hundred students to attend a technical-career symposium in the early 1970s, attracted 2000 and turned away over 600 in a similar event in the late 1970s. The University of Iowa held an annual conference for high school and college students entitled "Women in Engineering: Why Not You?"[3] The American Society of Engineering Education reported that over 100 schools had special programs to attract women into engineering during this period.

The engineering schools extended beyond attracting youngsters. The University of California at Davis, California State University at Northridge, and the University of Dayton[4] all had programs to encourage women with degrees in mathematics and science to continue their education and switch to engineering careers. Most of the cost of the continuing education programs was covered under grants by the National Science Foundation to provide these graduates with the necessary skills for entry-level engineering jobs.

The rapid growth of the Society of Women Engineers provides another indicator. There were less than 50 student sections in 1974 and over 125 four years later. In 1984 there were over 9000 students in 220 student sections of the Society.

The demand for women engineers increased in the 1970s, thanks in part to legislation. The Equal Pay Amendment of 1963 to the Fair Labor and Standards Act requires the same pay for men and women doing substantially the same work requiring equal skill, equal effort, and equal responsibility under similar working conditions in the same establishment. Title VII of the Civil Rights Act of 1964 prohibits discrimination based on race, color, religion, sex, or national origin. This Act also established the Equal Employment Opportunity Commission (EEOC) to monitor employers in their employment practices.

Industry and government was required to seek minorities and women as employees in all jobs if they qualified. For those organizations that did not comply the government took appropriate action. In 1973 only 2.8% of American Telephone & Telegraph Company's highly paid craft jobs were held by women when the corporate giant was forced by the government to begin an aggressive affirmative action program.[5] After the company's 5-year consent decree had expired in 1979, the number had reached 23,567 or 10.1%. The International Union of Electrical, Radio, and Machine Workers has won several major sex discrimination lawsuits against the electrical industry in recent years, forcing companies such as Westinghouse and General Electric to upgrade many jobs held primarily by women.

On June 23, 1972, a section of the Education Amendments of 1972 expanded coverage of the Equal Pay Act to include an estimated 15 million executive, administrative, professional employers, and outside sales people, an estimated 4 to 5 million of whom were women. These amendments mean that the Equal Pay Act now covers all employees of all private and public institutions at all levels.

The Equal Employment Opportunity Act of 1972 provided that prohibition of discrimination against any Federal government worker on the basis of race, color, religion, national origin, or sex is enforceable by statute. The Civil Service Commission has statutory authority to see that all personnel actions in the Federal Government are not only free from discrimination, but also actively oriented toward equality of opportunity.

Also in the 1970s women who were aware that engineers earn higher salaries than most other entry level corporate employees decided to enroll in engineering, engineering technology, or associated fields such as computer science.[6]

By the mid-1980s the picture of entry-level acceptance was brighter than ever. As each new wave of young women takes its place in the engineering community, it is that much easier for the next graduating class. Earlier, women students complained of isolation or of difficulty in coping with what was seen as a masculine environment. However, such problems seem to have faded as more women took to the campus. Most upperclasswomen feel they are fully accepted by their classmates and most professors confirm this fact.[7] As young male engineers learn to work with women in the classroom, they are more accepting of women professionals on the job. It is still up to the individual woman to prove that she is capable once hired, but now she is sought after and given the opportunity to succeed. As in other professions today, entry-level positions and the initial promotion are about even for men and women.[8] Clearly, atoms, mathematical symbols, or chemical elements have no preference for either masculine or feminine treatment.

Academia on the other hand is not growing. Despite a desire to increase the participation of women engineers on academic faculties and in administration, little progress is apparent. A 1983 study made by the University of California System[9] shows that the total number of women in all fields who received doctorate degrees was 32.4% of the total from 1982–1983. Yet women made up only 12.5% of the nearly 7000 professors who comprised the regular University of California system in that year. Women earned over 9% of the computer science doctorates and less than 5% of the engineering degrees. At Purdue University a 1983 update of a 1971 report on the status of female faculty shows that over the decade the percentage of full-time, tenured and tenure-track university-wide female faculty has stayed at 12.9% plus or minus 0.2%.[10] In the school of science, 9.8% of total faculty were women in 1971 but the proportion dropped to 6.4% in 1982. At the University of Maryland 18% of the campus's 1324 full-time faculty members are women, up from 13.2% in 1973. However, by the fall of 1983, 41% of men hired in 1973 and employed full time had been promoted to associate professor compared with 33% of the women. Similarly, 6.5% of the full time teaching faculty at Georgia Tech have been female during the past few years.[11]

Assuming this data is consistent at other academic institutions, and unless there is some additional incentive, there is no reason to believe that the numbers will improve. High starting salaries in industry are luring men and women engineers into industry and away from graduate study, teaching, and research. From 1983–84 the average faculty salaries for men were about $9000 or 35% higher than women in engineering and computer science.[12] This is little change from 1972 when women in academia earned 77% of the salary of the males.[13] This makes industry salaries even more attractive for women and without a shift in this pattern, we can expect to see little improvement in the ratio of female to male engineering faculty members in the immediate future.

31.1.2. The Future

The challenge for women engineers is changing from acceptance in the 1970s to pay and promotion in the 1980s.

Pay. Historically, statistics indicate that women have earned about 60% of men's pay for decades. Although faculty salaries are lagging as noted, the salary gap in industry is beginning to narrow. A Bureau of Labor Statistics survey in mid 1983 showed women earning 66% of men's weekly salaries across the board, with women engineers earning 80.9% as much as their male counterparts.[14] The starting salary differential for new graduates declined from 3.8% in 1977 to 1.3% in 1982.[15]

It is doubtful that equality in pay will become a reality between the female and male population that graduated before the equal pay laws went into effect. Although there are some lawsuits, any rights or wrongs are difficult to prove when it comes to the subjective evaluation of a particular position and subjective evaluation of an employee's performance.

As noted previously, in 1973 American Telephone and Telegraph, then the nation's largest private employer, representing its 22 subsidiary companies entered into a landmark settlement with the Department of Labor and the Equal Employment Opportunity Commission.[16] AT&T agreed to pay $15 million in restitution to an estimated 15,000 women and minorities for alleged discrimination in job assignments, pay, and promotions. AT&T also agreed to establish goals and timetables for the hiring and promotion of women and minorities into its management ranks.

Other companies are operating under legal settlement to meet goals and timetables; however, the differentials in salary indicate the existence of pay discrimination. Women and scholars express surprise at the lack of progress in wage equalization despite the women's liberation movement.

Quite often women are not always informed about their own situations, even when they are aware of the broad issue of sex discrimination.[17] Additionally, some of the mature women who were forced to accept the salary differential in their earlier working years do not choose to make an issue even though they may feel that the condition still exists. It has been noted that the injustice exists, in part, from women letting things slide and refusing to acknowledge the extent that sex discrimination hurts them as individuals. The current disadvantages women suffer result mostly from people whose actions are suggested by a traditional society.

The recently graduated female engineers are more assertive and will be more vocal about less pay for equal work. Also, as will be discussed, women will move into supervisory and management positions where they can and will have some input to the promotion and salary administration of younger females and males.

Forces favoring affirmative action and equal opportunity have raised the consciousness of most employers who are attempting to adhere to the requirements of the Federal Law to provide equal pay for equal work.

Promotion.   Promotions up the career ladder from entry-level positions are still slow in coming. Women managers have become a significant part of the corporate pyramid; however, a recent research report indicates that 50%

are in entry management, 25% in middle management and almost none in upper management.[18] Table 31.2 shows supervisory responsibility for the years 1972, 1974, 1977, 1979, and 1982 as reported by the Society of Women Engineers. The results are the tabulation of answers to a questionnaire sent to each member of the Society. The surveys indicate a decrease in the percentage of women in the higher levels of supervision/management in the period from 1972 to 1982. This is to be expected with the large number of recent graduates entering the work force. As the new engineers develop and mature, the supply of qualified candidates for promotion will also increase. Analogous to the subject of pay, the recently graduated women engineers are more assertive and vocal about their goals. Additionally, similar to the male population, some women prefer to remain in the purely technical arena and not move into management.

The female and male college graduates enter the work force with equivalent technical credentials; therefore, they will need as a minimum the same tools to advance in their chosen profession. They need the same training, covering the day-to-day on-the-job responsibilities, in-house classes sponsored by their employers, and advanced formal education available outside of the company.

Today employers recognize the need to provide training for women employees to become managers; unfortunately many of them do not have the proper perspective on what or how it should be provided. Male managers often fail to give their female subordinates the kind of supervisory feedback that all employees need to improve their performance and rise in the ranks. When a male engineer does something wrong, he is usually reprimanded. But when a female engineer makes a mistake, it may be ignored, because some male bosses still feel they have to be careful about what they say to females. Women need constructive criticism and feedback in order to pre-

TABLE 31.2. Supervisory Responsibility

Description	1972	1974	1977	1979	1982
No regular supervision	39%	37%	39.5%	40%	41.4%
Direct or staff supervision	21	23	22.5	23	22.6
Supervision of team or unit	14	14	11.5	13	11.7
Supervision of project or section	11	13	15.5	14	16.4
Management of major department, division or program	10	8	5	6	5.2
General management of organization	5	5	2	2	2.5
No response			4	2	

Source. "A Profile of the Woman Engineer." Society of Women Engineers.

pare for advancement. Both private business and government agencies have made some progress in identifying key women and preparing them for executive positions. Slowly, the combination of affirmative action pressures, availability of qualified women engineers, and action by women engineers will lead to career advancement.

Many women engineers enjoy an advantage in today's engineering office in that they are more likely than men to have acquired keyboard skills. The modern engineering office has computer terminals proliferating for many functions, and most have a typewriter-like keyboard. Whether performing engineering analysis, design, drafting, or specifications, it is highly probable that good keyboard skills enable the engineer to outperform those without these skills. Theories have been circulated claiming various superiorities or inferiorities between men and women in engineering; for example, capabilities to analyze functional relationships and visualize spatial relationships.

The "conventional wisdom" is that women cannot conceptualize physical design. In the computer-aided design and drafting units, male and female employees may be found randomly dispersed among the various functional groups. There are men and women programmers, operators, trainers and physical plant designers all working together with three dimensional electronic models, data bases, and analysis programs. Some men can handle the keyboard better than some women and some women are more proficient with the three dimensional graphics than their male counterparts. Thus, there is daily opportunity to observe that the differences in engineering ability are individual, not based on gender.

Women are making known their aspirations and developing the skills to achieve their goals. They are making sacrifices when necessary to manage career and family. Women have often dropped out of the work force at critical points in their careers to have children.[19] Some researchers suggest that employers expect the women to interrupt their careers and are thus reluctant to promote or invest in training them. However, more than half of all married women whose children are younger than 6 years old and husbands are present in the home—are now working full or part time. This is a substantial increase from 1970 when the figure was 30.3% and from 1960 when it was only 18.6%. But career advancement still depends on such things as subjective performance appraisals and being the chosen candidate when an opportunity arises. Men are still making most of these subjective decisions; they are just beginning to get accustomed to helping women advance.

Women who are interested in advancement to supervisor or a managerial position should cultivate an understanding of the various circumstances required in managerial styles and work toward a situation which could be enjoyed without excessive stress. Women must develop their skills in the area of decision making, which automatically involves taking risks. Women should actively seek opportunities for increasing their abilities and status. They need to assess the alternatives and decide on courses of action which may involve further studies, new skills, different environments, or more time on projects. They must recognize the need and assume the responsibility for their own technical and psychological readiness for career advancement. Women who are competent, motivated and achievement-oriented can convince management of their right to be supported.

In technical organizations there are basically four styles of management[20] and women do not fit all four equally well. A number of factors help to determine what management style best suits the leader, workers and circumstances. For example, in the middle of a fire, no meeting is called to discuss how it should be put out. Instead the leader tells people what to do. On the other hand a new manager fresh from graduate school will not be too successful in telling an engineer with years of experience how to do simple tasks. What works well in one environment is completely wrong in another environment.

Worker maturity is the basic factor that determines management style. In terms of worker maturity, it is not a question of being mature or immature in professional development but a question of degree. Additional factors that have some influence on the various styles include senior managers' style, employees' attitudes, time, and extenuating circumstances such as confidentiality of information.

With a mature, well-motivated group of engineers, management can adopt a posture of organizing and delegating functions that require brain power and where physical size and strength are of little importance. Women can perform well in the "delegating" environment. A manager defines the objective; however, the authority to complete the task is transferred to the subordinates even though the responsibility remains with the manager. This is very effective where the subordinates have the experience and expertise in technical details in areas where the manager has limited, recent, state-of-the-art experience.

Managers use the "participating" style of involving employees opinions on how to best approach certain decisions. In a developing organization with moderately mature, motivated engineers, management can adopt the posture of participative, supportive leadership. Women often have much talent in subtly eliciting cooperation, from experience gained by providing indirect guidance to their male friends, brothers, husbands or associates. They do well in "give-and-take" development of policies and plans.

In the same developing organization with less mature employees, the manager will make the decisions either singularly or by direction from the superiors. The manager is then responsible for "selling" the objective to the subordinates, explaining why the task is needed, what the ultimate benefit is, what the implementation details are, and convincing the workforce that this was the best choice. Here too, women are good at eliciting the cooperation of the subordinates.

Women are at a disadvantage in management only in those immature, poorly prepared, or badly motivated groups that require highly directive, dictatorial, or tightly controlled management. "Telling" subordinates what to do, when to do it, and how to do it with little or no explanation is not a style that comes easy to women since it has not been a skill developed through their adolescent and maturing years. It is not easy for women engineers to manage field manual labor forces in construction of large projects. Even though the women may know the what, when, and how of the task, manual employees have not yet accepted the idea of total direction from a female supervisor. This too will change as women develop their management skills and become more acceptable in this environment.

To date most engineering organizations do not have women in upper management as role models for those aspiring to upward mobility. Since the managerial qualities a company looks for are in general the same for men and women, females would do well to study the managerial traits of the male managers in their own organizations. Another approach is for women to prepare themselves by studying management techniques and following the precedent used by women in the past. Not knowing the ropes nor having a model, network, or mentor, it was necessary for some of us to learn the theory and profit by mistakes in applying the theory to practice. Some individuals are creative and develop new routes to management levels and this is a possibility that women as well as men should explore.

Since there are no instant supervisors or managers, either women or men, it is the responsibility of all companies to identify those persons with potential and provide the appropriate training (technical, political, and social) for advancement within the company. Management would profit by eliminating the barriers (identification, training, and opportunity) in order to achieve a better fit between women and the managerial environment.

Another area where women engineers are visibly absent is the executive board room. As noted earlier women engineers have just reached the status of acceptance and now are moving into management. However, except for those few who own their own small business or have a family connection, there are few if any women engineers participating in the decisions of the United States industrial corporations.[21] There are some female directors in the industry; however, their background is, in general, business. It is estimated that about 1000 women and 49,000 men hold the top-management jobs in the country's major corporations. Only 367 women and 15,500 men sit on the boards of the country's top 1300 public companies in the mid 1980s.[22]In a recent study of women vice-presidents of Fortune 1000 companies, it was noted that fewer women make it to the top of the corporate ladder in industrial corporations than in other arenas such as retailing or banking. The study also pointed out that a significantly large number of women executives of the industrial firms rose to the top-management posts through legal positions. Very few made it through engineering paths.

Again the Society of Women Engineers can be cited as an indicator of the changing times.[23] "Women in Engineering" was the theme of their Conference held in 1971 and "Career Guidance for Women Entering Engineering" was the theme in 1973. In 1976 the change in direction started with the Conference on "Upward Mobility and Professional Development for Women Engineers" followed by the 1980 "Leadership Conference."

31.2. WORK WITH PEERS, SUPERVISORS, AND SUBORDINATES

31.2.1. Peers

Women engineers entering the work force today have little change from the university atmosphere with regard to the proportion of females and males. A large percentage of the professors are male and so is a correspondingly large percentage of the supervisors and managers in industry. Additionally the percentage of female to male peers is very similar in the classroom and industry. Women engineers have been accepted by their peers in the university and at starting levels in industry. The status is such that at the entry level and the first promotion, the competition is even in most large companies.[24] However, higher in the organization, women still have to be better to be promoted.

Women who have risen to the management level do not have the same acceptance by their peers. Many of the women managers today do not feel that they are part of the management team even though they are part of the management organization. Many male managers have difficulty relating to woman peers and women managers are usually not included in the informal communication channels that are necessary for further development and advancement. It is unnecessarily difficult and very isolating for a woman to be the only one in the managerial ranks of an organization. As more women move into management in the future it will probably become more comfortable for the women and men. In the meantime it behooves existing male management to eliminate this barrier for legal reasons in addition to the possibility that it might help their own careers. As the channels of communication open up, talented and experienced women will be able to offer constructive help to male managers in addressing tasks that may be less difficult for the female population—the same as the male managers presently do.

31.2.2. Supervisors

In general the supervisors of engineers in most organizations are not concerned with the gender of the personnel under their direction. Each supervisor is evaluated on the performance, development, and technical output of the subordinates in addition to his or her own performance. This is the starting point where women have an equal opportunity for self-development, professional development, and learning to work with other engineers with varying talents.

Misconceptions about women in engineering and in management are not easily corrected, and the views held by many male managers in the past have not changed. However, the "old school" of male managers are slowly being replaced by younger managers, many of whom are less biased and more willing to accept women professionals on an equal basis.

31.2.3. Subordinates

Women as well as men advance to supervisor or management positions because they have been recognized for their past technical performance and potential for upward mobility.

As previously noted women need to prepare for managerial positions by developing their leadership styles and being aware of the types of organizational situations they will encounter in their industries. They must continue in the area of self-development in order to supervise and develop the new subordinates. This position may be a difficult assignment for a female especially if one or more of the subordinates is an older male engineer. It means thinking through situations, assessing needs and wants, anticipating possible outcomes, and developing a style that is appropriate for the situation.[25] Cultural conditioning has made it difficult for older men to accept the idea that women can function in all types of positions, particularly management positions.[26] It is a mistake for the male subordinate to see his problem as one of "adjusting to a female boss" when the real problem is "how to work productively with a particular individual."[27] It is better for a man to be aware of his feelings of hostility than to deny them and have his dealings colored by below-surface emotions. Quite often the lack of acceptance of a young female manager is little different from the acceptance of a young male man-

ager; however, this point is often ignored or dismissed. A good working relationship depends ultimately on the individuals—not on their gender. The
female supervisor must establish and maintain a professional atmosphere to
confirm that her assignment was appropriate and that she has the ability to
meet the challenge.

Breaking into the old networks, changing established patterns, and altering male attitudes are the biggest remaining challenges for all women at
work.

31.3. HARASSMENT—SEXUAL AND OTHERWISE

Harassment in the office, shop, and construction site has been with us for a
long time. In spite of forceful legislation and heightened awareness, it per-
sists in diminished form.

The root causes of harassment can be the subject of psychological trea-
tises.[28] Insecurity and the need to exert power over women or humiliate
them, or an unwillingness to control sexual impulses are possible explanations. Confusion of roles may also enter into such behavior. The inability or
unwillingness to differentiate social versus work settings can contribute to
social/sexual behaviors. A manager who treats a woman at work as a potential date rather than a coworker, may make little distinction between interacting with a woman in a bar or in a committee meeting.

Insecure male peers have been known to use profane or obscene language
to make their female co-workers feel uncomfortable. The behavior is observed most frequently among the subprofessional ranks or men with less
formal education than others of their rank. Some male supervisors have
been known to resent female subordinates forced into their groups by higher
management and have harassed the females into leaving. Sexual harassment
has been discovered to take many forms in engineering offices. It may be as
mild as a remark with a double meaning or as brutal as a threat of dismissal
for refusal of favors. Some male engineers get away with rude ogling or remarks because the typical targets are young and inexperienced females who
are reluctant to complain and "stir up trouble." If the harassment continues
intolerably, the young women may seek an older woman or trusted older
man to be their spokesperson. In addition to their legal rights of freedom
from harassment, women engineers should realize that upper management
is their ally in opposing such activity as counter-productive.

Federal Law has defined a number of identifiable forms of sexual harassment but no list can be all encompassing. The term "sexual harassment" is
used to denote a variety of unwanted behaviors that call unwarranted attention to another's sexuality.[29] It may include unwanted comments, looks,

gestures, touching, propositions, and sharing of intimate information. The definition of sexual harassment has been expanded rather[30] dramatically, as exemplified by EEOC Guidelines. EEOC Guidelines set out three criteria for determining whether an action constitutes unlawful behavior. These criteria are (1) submission to the conduct is either an explicit or implicit term or condition of employment, (2) submission to or rejection of the conduct is used as a basis for employment decisions affecting the person who did the submitting or rejecting, or (3) the conduct has the purpose or effect of substantially interfering with an individual's work performance or creating an intimidating, hostile, or offensive work environment.

Employees who must work in an environment which they feel is sexually intimidating can charge sexual harassment, even if they have never been asked to engage in sexual activity or been threatened with loss of job for refusing sexual favors. Workers who are subjected to foul language, gestures, or unwanted lewd or suggestive material are also being sexually harassed, according to the more liberal definition of sexual harassment.

In 1978 Heelen alleged she was fired from Johns Manville Corp. for refusing her supervisor's demands for sexual favors.[31] Denver District Court Judge Finesilver found Johns Manville guilty of sexual discrimination on the basis that an employer is liable when refusal of a supervisor's unsolicited sexual advances is the basis of the employee's termination.

In the case of *Kyriazi v. Western Electric Co.* (1978), Western Electric was found guilty based on "top-to-bottom" discriminatory sex policies. The court held five of the plaintiff's male coworkers and supervisors liable for tortuous interference with her employment contract. The coworkers had tormented the plaintiff with loud, derogatory, sexually explicit remarks about her, and with an obscene and humiliating cartoon left on her desk. The supervisors had acquiesced in this course of conduct despite her complaints.

Enough cases of sexual harassment have occurred, and enough consciousness has been raised that companies now feel responsible for investigating and resolving cases of sexual harassment, especially because they know that failure to do so can result in costly law suits. EEOC guidelines state that employers are responsible for the behavior of their supervisory employees. Pleading ignorance of the existence of sexually harassing behavior is not an adequate explanation for failure to act to prevent it. EEOC assumes that a company should know about any sexually harassing behavior committed by supervisory personnel. Companies are also responsible, albeit on a more limited basis, for the behavior of subordinates toward other subordinates within the organization. The law must be upheld with firm, fair discipline and enforcement in the engineering office as well as in other workplaces.

In some engineering companies upper management has sponsored awareness seminars to combat sexual harassment. These seminars are intended to alert managers and supervisors on how to observe infractions and inform employees of the company commitment to eliminate harassment, sexual or otherwise, in the workplace.

NOTES

1. Alfred J. Gardner, *How It Is—How It Has Been—And How It Will Be: Women in Engineering and Management,* Proceedings of Engineering Foundation Conference, Henniker, New Hampshire, 1972

2. Mark D. Rosenzweig, "Tomorrow's Chemical Engineer," *Chemical Engineering,* May 8, 1978

3. Newsletter, Society of Women Engineers, May 1978

4. Rosenzweig, *op. cit.* Newsletter, Society of Woman Engineers, March/April 1977. Newsletter, Society of Women Engineers, October 1977. Newsletter, Society of Women Engineers, September 1978.

5. Karen Tumulty, "Wage Gap: Women Still the 2nd Sex," *Los Angeles Times,* September 13, 1984

6. Rosenzweig, *op. cit.*

7. *Ibid.*

8. "Corporate Woman," *Business Week*, October 1, 1984

9. Anne C. Roark, "Women and Blacks Trail in UC Hiring," *Los Angeles Times,* January 20, 1984

10. *Manpower Comments,* Scientific Manpower Commission, 21 (5) June 1984

11. *Fact Book,* Georgia Tech, A Unit of the University System of Georgia, 1982–83, 1983–84

12. *Manpower Comments,* Scientific Manpower Commission, 21 (3) April 1984

13. *Woman Engineer,* 4 (4) Summer 1984

14. Debra Kaplan Rubin, "Fifth Annual Salary Survey," *Working Woman,* January 1984.

15. *Engineering Manpower Bulletin,* Engineering Manpower Commission of the American Association of Engineering Societies, Inc., New York, New York, September, 1983.

16. Margaret Hennig, and Anne Jardim, *The Managerial Woman* (New York: Simon and Schuster, 1977).

17. Faye Crosby, "Selective Vision" *Working Woman,* July 1984

18. "Corporate Woman," *op. cit.*; J. S. Diamond, "Women on the Job: Surge Widely Felt," *Los Angeles Times,* September 9, 1984

19. Tumulty, *op. cit.*

20. Paul Hersey, and Kenneth H. Blanchard, *Management of Organizational Behavior: Utilizing Human Resources,* (Englewood Cliffs: Prentice-Hall, N.J. 1982).

21. "Women Executives In Industrial Companies Most Scarce," *The Engineer of California,* December 1983.

22. Corporate Woman, *op. cit.*

23. Proceedings, Society of Women Engineers.

24. Corporate Woman, *op. cit.*

25. Hennig, *op. cit.*

26. Edith M. Lynch, *The Executive Suite—Feminine Style,* (New York: AMACOM, 1973).

27. Lauren Uris, *The Executive Deskbook* (Princeton, NJ: Van Nostrand Reinhold, 1976).

28. Mary Coeli Meyer, Frederick J. Collins, Jeanne Oestrich, and Inge Berchtold, *Sexual Harassment,* (New York: Petrocelli, 1981).

29. *Ibid.*

30. Barbara Gutek, *Sex Role Stereotyping and Affirmative Action Policy,* Institute of Industrial Relations, (Los Angeles: University of California, 1982).

31. Meyer, *op. cit.*

32

QUALITY STANDARDS AND ENGINEERING MANAGEMENT

Mahesh Chandra

MAHESH CHANDRA is an associate professor in Business Computer Information Systems and Quantitative Methods at Hofstra University and chairman of the department. Prior to coming to Hofstra, he taught at the University of Iowa in the department of Industrial and Management Engineering and the University of Delhi in the department of Operations Research. His research interests include reliability and maintainability, failure data analysis, quality control, time series analysis and forecasting, and operations research. Dr. Chandra has published papers in *Mathematics of Operations Research,* the *Journal* of the *American Statistical Association,* and in other journals. He has a D.Sc. degree in Operations Research from George Washington University, an M.Sc. in Operations Research from the University of Delhi, and a B.S. from Agra University. Dr. Chandra is an ASQC certified reliability engineer and certified quality engineer.

32.1. INTRODUCTION

It should need no emphasis that quality is at the center of all engineering work. It enters the domain of engineering management at two levels. First, there is the quality of engineering work itself; how many mistakes are made? What is the quality of the design process itself? Unfortunately such questions are not easy to answer in a way common to other aspects of quality standards, control, and enforcement which enter into broader engineering and manufacturing management. The main reason is that engineering tasks tend to lack the kinds of large batches or regularities which make regularized procedures feasible elsewhere. Projects may not be all completely different, but they may lack common features which would make a more formal quality control structure feasible. If such regularities do exist, then some control procedures may be possible; an example drawn from routine clerical work is presented in Section 32.7.1. Such work in engineering is, however, exactly the sort of job for which computer-assisted design (CAD) is most suitable (see Chapter 14) or if it is not at present, it very likely soon will be. Thus in assuring the quality of engineering work itself, traditional forms of checking and control still have an important part to play.

Second, engineering management must be concerned with the quality system devised for the product being turned out. Defining quality levels, devising control procedures, and changing the design in response to quality problems are all essential jobs for engineering departments and their managers.

From a time when a sort of dichotomy between the two was sometimes encountered, the close link between quality and productivity is no longer in dispute. Two key elements in the survival and success of any enterprise are flexibility and quality assurance. Flexibility means the ease and capability of the system to face the changing environment and quality assurance means to provide goods and services that meet the specifications of performance and the customer perception of these in use. Quality is not only a function of what raw materials, tools, production processes are used, and what type of work is performed by operators and supervisors, but is also a function of product design and managerial decision process. Feigenbaum[1] defines quality as "the total composite product and service characteristics of marketing, engineering, manufacturing and maintenance through which the product and service in use will meet the expectations of the customers." The total commitment to quality involves everyone. It involves top management, middle management, operation management, engineers, supervisors, and operators.

Can the quality of a product be assured as a result of its being properly inspected and tested? Probably yes, but at what cost? One hundred items are tested and 25 of these are observed to be defective. The defective items are either destroyed or reworked so that the customer gets 100% defect-free items. But what is lacking is the attitude of doing things right the first time, the cost of which is far less than the cost of reworking or replacing the defective items. Crosby[2] emphasizes that "Quality is free. It's not a gift, but it is free. What costs money are unquality things—all the actions that involve not doing jobs right the first time." Emphasis should not be laid on correcting faults and failures but on preventing these from happening. Although faults and failures cannot be completely eliminated, with proper planning, training, and rational decision-making process, they certainly can be minimized. A quality assurance approach should not be inspection oriented but production and design oriented.

32.2. TOTAL QUALITY CONTROL (TQC) SYSTEM APPROACH

In many firms quality assurance means control of production processes and inspection and testing schemes to identify "bad" and "good" items. However, a total quality-control system-approach emphasizes an integrated approach to marketing research, product design; manufacturing; service standards and provision of spare parts; training of managers, engineers and operators; and a quality cost program to assure quality of goods and services. As shown in Figure 32.1, quality assurance is considered an integral part of the whole system providing support to various functions of the enterprise to produce quality items and provide quality services in the most economic manner.

32.2.1. Quality and Marketing

A company must assess the market for the product and the cost to produce it at desired quality level before the product design and manufacturing stages. Quality techniques should be used in (1) determining warranty and service policies, (2) training of service personnel, (3) customer satisfaction surveys, and (4) postwarranty audits. Many companies in the United States have

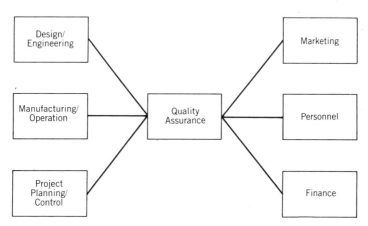

FIGURE 32.1.   Total quality control system.

started emphasizing the word quality in their product promotional programs. Ford, for instance, advertises that "quality is job one." Chrysler's offer of *five-year, 50,000 mile* warranty is an example of assuring quality of products to customers through extended service plans. Caterpillar Tractor Company and L. L. Bean, Inc. are among the companies that emphasize customer satisfaction surveys to measure customer perception of the quality of their products. The surveys help the companies to identify design problems and offer the opportunity of correcting these and informing the users of the problems associated with the use of the product, if proper care is not taken.

32.2.2.   Quality in Product Design Engineering

Since design is the first step in the product, effective control of quality control must begin with the design of the product. The use of quality techniques in product design can save a company from heavy losses occurring as a result of product recall and product liability. Quality in product design involves (1) establishing drawings and specifications, (2) developing an engineering prototype, and (3) testing the new product and analyzing potential problems in reliability, maintainability, and how well the product would meet customer needs.

32.2.3.   Quality in Manufacturing

Quality control techniques have been successfully used in various stages of manufacturing. Quality in manufacturing involves (1) vendor control, (2) receiving inspection, (3) metrology, (4) establishment of product specification and process control, (5) in-process inspection and testing, (6) final inspection and testing, and (7) personnel qualifications.

32.3.   QUALITY AND MANAGEMENT

It is absolutely necessary that a company has a clearly stated quality philosophy translated into an explicit quality policy. Management must demonstrate total commitment to quality through consistent leadership. At the National Conference for Quality held on March 31 through April 2, 1982 in Washington, D.C., Philip B. Crosby said that "Senior management is the primary cause of the company's quality status. Senior management must find out what it is costing them in money and reputation to produce nonconforming products and services." At the same conference, Armand V.

Feigenbaum emphasized the role of top management by stating that "Certain very successful firms that have recognized that their most immediate available cost improvement potential is by means of strong quality programs. This is in sharp contrast to other firms in which there still remains the very old managerial belief that quality levels must cost more and somehow make production more difficult." Management must consider quality assurance as a management process as well as technological process. Crosby, Deming, and Juran emphasize that most quality problems are "management" problems.[2] Crosby states that effective quality control requires rising of an organization from the stage of "uncertainty"—no comprehension of quality as a management tool, to the stage of "certainty—considering quality management as an essential part of company system.

32.3.1. Quality Planning

Quality assurance plan is developed to identify the tasks that a company must perform to meet its quality objectives. Quality planning deals with development, implementation, and management of the quality system. It involves the engineering function and the management function in order to produce quality products and provide quality services. The quality assurance planning includes the preproduction quality tasks (prevention), and the quality tasks during production (conformance).

Preproduction quality assurance planning includes:

Design reviews and evaluation of engineering prototype to detect quality problems and seek corrective action before production.

Process control development to prevent production of defective products.

Design of test equipment and methods to assure that the equipment will perform its required function.

Establishing criteria for selection and evaluation of vendors to assure compliance with requirements.

Training requirements to assure that persons performing their jobs have the required skills.

Designing a quality cost system to monitor costs associated with prevention, appraisal, and failure, and to identify areas which need action.

Development of a quality reporting system which emphasizes reporting of relevant results on a current basis.

Development of corrective action program which should emphasize procedures for identifying reasons for nonconformance and solution techniques to prevent nonconformance in the future.

Quality tasks during production include such things as implementing process control, testing and inspection, vendor control, product evaluation, material review, field performance, corrective action, and quality audit.

For extensive discussion of quality planning and related topics, the author has referred to an excellent book by Caplan[3] and Chapters 3, 5, and 6 of *Quality Control Handbook* by Juran[4].

32.3.2.   Corrective Action Program

Corrective action is necessitated as a result of production of items that fail to meet the required specifications. Correcting the immediate problem by way of rework, repair, and so on is only a temporary solution that fails to remove causes of nonconformance. An effective corrective action should deal with correcting the cause of a defect and prevention of the same defect from that cause in the future. Hagan[5] states "The establishment of better controls or disciplines—which is the substance of improved management action—should be the ultimate objective of a company's corrective action program."

The establishment of a corrective action system should be based on:

Timely reporting of quality problems
Avoiding quick fix solutions
Diagnosis of all possible causes
Prevention of causes that lead to nonconformance

The nonconformance observed may be due to operator or managerial action. Figure 32.2 is an example of a typical corrective action report used to

To: _____ Date: _____

From: _____ SDR no. _____ QAR no. _____

Was the error/Nonconformance:

Operator controllable ☐
Management controllable ☐

Operator controllable (Answer all of the following)

Did the operator:	Yes	No	N/A*
O1. Have written instructions?	☐	☐	☐
O2. Use the correct written instructions?	☐	☐	☐
O3. Follow the correct written instructions?	☐	☐	☐
O4. Understand the written instructions?	☐	☐	☐

FIGURE 32.2.   Corrective action report.

O5. Use the correct equipment?
(Machine/Fixture/Gages/Instrument)?

O6. Use the equipment correctly?

O7. Understand how to use the equipment?

O8. Check his or her own work?

O9. Submit the first piece to inspection?

O10. Wait for inspection approval before producing product?

Management-controllable (answer all of the questions)

M1. Does the applicable documentation serve its purpose?

M2. Are there written instructions which apply to the operation?

List document number(s) and revision levels_____

M3. Are they to the latest applicable revision?

M4. Were they given to the operator at the start of his or her assignment?

M5. If they had them in their file, were they checked by the supervisor for correctness & currency?

M6. Does the operator understand the instructions?

M7. Did the supervisor review the instructions with the operator?

M8. Was the necessary equipment provided to the operator?

M9. Was the equipment calibrated; is it stickered?

M10. Is the equipment within its calibration period?

M11. Was the operator told to submit a first piece to inspection?

M12. Was the operator told to receive inspection approval before producing product?

M13. Was the operator to check his or her own work?

M14. Was the operator told how often to check his or her own work?

M15. Is the operator required to record his or her checks?

M16. Was the operator informed of possible pitfalls and/or previous problem to be avoided?

M17. Was the operator notified of the errors or discrepancies uncovered by inspection?

M18. Was the operator notified of the quality audit findings?

* All N/A answers require reason for not being applicable.

FIGURE 32.2. (*Continued*)

identify whether the nonconformance was operator controllable or management controllable. Preventive measures may require changes in design, process change, tool adjustments, operator training, and so on.

Meaningful preventive action would require:

Awareness of the problem to be solved

Location and definition of the problem

Examination of the problem

Discovery of possible solutions

Establishment of criteria for testing solutions

Choosing the best solutions

32.3.3.   Quality Cost Program

Any cost incurred as a result of failure to produce quality products and provide quality services, contributes to the cost of quality. The establishment and implementation of a quality cost program enables the management in the production of quality goods at a minimum quality cost. ASQC publication, "Quality Costs — What and How"[6] states, "The basic concept of quality costs is recognition and organization of certain quality-related costs to gain knowledge of their major contributing segments and of the direction of their trends." The four categories of quality costs are:

1.   *Prevention.* Costs associated with prevention of discrepancies such as the cost of quality planning, quality training, quality measurement and control equipment, data analysis, and quality information system.

2.   *Appraisal.* Costs incurred to determine the degree of conformance to quality requirements, such as the cost of receiving or incoming test and inspection, acceptance testing, product quality audits, maintenance and calibration of test and inspection equipment, and field performance testing.

3.   *Internal Failure.* Costs arising when products fail to meet quality requirements prior to delivery of products to customers, such as cost of scrap, rework and repair, trouble shooting or failure analysis.

4.   *External Failure.* Costs incurred when products fail to meet quality requirements after delivery of these to customers, such as cost of servicing rejected material, warranty replacements.

Detailed discussion of these cost categories is given in the ASQC publication mentioned earlier in this Chapter. Other ASQC publications on this topic are "Guide to Reducing Quality Costs"[7] and the "Guide for Managing Vendor Quality Costs."[8] The April 1983 issue of *Quality Progress* was devoted to quality costs to highlight the necessity of quality cost program in the development of total quality cost system.

32.3.4. Analysis of Quality Costs

The previously discussed four categories of quality costs tell us how a company is spending on prevention, appraisal, and internal and external failure. However, for comparing these costs over a given period of time, it is necessary to have comparison bases which relate quality costs to business changes. Three measurement bases that have been used successfully are— net sales billed, direct labor costs, and the quality cost per unit. See Figure 32.3 for a typical quality costs reporting form. Figure 32.4 provides a way to observe quality-costs trends over different periods.

32.4. QUALITY STANDARDS

The standards are set to establish criteria against which actual performance is compared. National and international standards of performance have been established by government and professional associations for various industry groups, (for example, nuclear, chemical, and electronics) to reflect public acceptance, technical excellence, and business viability. Additionally each firm sets its own quality standards—usually referred to as workmanship standards, cost standards, and product performance standards. It is important that precise and realistic standards be set to measure performance of workers and products. Quality analysis using statistical techniques should be performed to determine realistic quality standards for various quality characteristics of the product. Quality standards should be constantly revised to reflect:

government regulations
progress in technology
customer's perception of the product in use
changes in socio-economic conditions
quality levels in relation to production costs and customer needs

Generic and MultiLevel Quality Standards. These are quality management standards which list and summarize elements of a quality assurance system. The 'generic' quality standards such as ANSI Z1.15 1979 are

Year_____   Quarter_____

1. *Prevention Costs*   _____
   Quality control engineering $ _____
   Process control engineering $ _____
   Quality measurement equipment $ _____
   Quality training $ _____
   Other $ _____
   Total prevention costs $ _____

2. *Appraisal Costs*
   Incoming inspection and test $ _____
   In-process inspection and test $ _____
   Final inspection and test $ _____
   Calibration $ _____
   Quality audits $ _____
   Other $ _____
   Total appraisal costs $ _____

3. *Internal Failure Costs*
   Scrap $ _____
   Rework and repair $ _____
   Material review & corrective action $ _____
   Other $ _____
   Total internal failure cost $ _____

4. *External Failure Costs*
   Complaint expenditures $ _____
   Service of returned goods $ _____
   Warranty replacements $ _____
   Other $ _____
   Total external failure costs $ _____
   Total quality costs $ _____

   Measurement Bases
   Net sales $ _____
   Direct labor $ _____
   Quality cost per unit $ _____
   Total quality cost target $ _____

FIGURE 32.3.   Quality cost report.

market-oriented standards and are prepared from the point of view of the manufacturer who wants to install a quality assurance system. The 'multi-level' quality systems such as NATO AQAP-1 1981 are prepared from the perspective of the purchaser and list elements required in the establishment of a quality program by the vendor to assure compliance with the requirements of the contract. The major headings of ANSI Z1.15 are:

Policy, planning, and administration

Design assurance and design change control

Control of purchased material
Production quality control
User contract and field performance
Corrective Action
Employee selection, training, and motivation

National and international quality associations and organizations such as the American Society for Quality Control (ASQC), the American National Standards Institute (ANSI), the British Standards Institution (BSI), the Union of Japanese Scientists and Engineers (JUSE), the European Organization for Quality Control (EOQC), the International Organization for Standardization (ISO), and the International Electrotechnical Commission (IEC) have played a very significant role in the development of quality assurance systems. Tables 32.1 and 32.2 list the generic and multilevel quality standards respectively. These standards are being reviewed by the Technical Committee 176 of ISO in preparation of international generic quality system standards. ISO committees' tasks deal with all technologies except electrical and electronic, which fall under the purview of IEC.

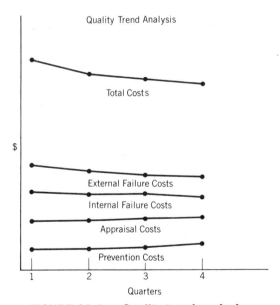

FIGURE 32.4.   Quality trend analysis.

TABLE 32.1. Generic Standards

Country	Document Identification		Available From
	Number	Title	
United States	ANSI/ASQC Z-1.15–1979	Generic Guidelines for Quality Systems	American Society for Quality Control 161 West Wisconsin Avenue Milwaukee, WI 53202, U.S.A.
United Kingdom	BS4892:1972	A Guide to Quality Assurance	British Standards Institution 2 Park Street London, W1A 2BS, England
France	AFNOR NF X50–110	Recommendations for a System of Quality Management for the Use of Companies	AFNOR Tour Europe Cedex 7 Paris La Defense 92080, France
Germany	DIN 55–355	Basic Elements of Quality Assurance Systems	Deutsches Institut für Normung e.V. Burggrafenstrasse 4–10 1000 Berlin 30, West Germany
Netherlands	NEN 2646	Quality Assurance–General Conditions for Quality Systems for Designing, Manufacturing, and so on	Nederlands Normalisatie—Instituut Postbus 5810, 2280 HV Rijswijk (ZH)
United States	ANSI–ASME NQA–1–1979	Quality Assurance Program Requirements for Nuclear Power Plants	American Society of Mechanical Engineers 345 East 47th Street New York, NY 10017, U.S.A

TABLE 32.2. Multilevel Standards

Country	Document Identification		Available From
	Number	Title	
United Kingdom	BSS750: 1979	Part 1. Specification for Design Manufacture and Installation	British Standards Institution 2 Park Street London W1A 2BS, England
Canada	CSA 2299.1–1978	Quality Assurance Program Requirements	Canadian Standards Association Standards Division 178 Rexdale Boulevard Rexdale, Ontario H9W 1R3, Canada
Norway	NVS–5–1594	Recommendations of the Contractor's Quality Assurance Program—Part 1 Quality Assurance System	Norges Standardiseringsforbund Haakon VII's Gate 2 N–Oslo 1, Norway
South Africa	SABS 0157 1979	Code of Practice for Quality Management Systems	South African Bureau of Standards Private Bag X191 Pretoria, Republic of South Africa 0001
NATO	AQAP–1–1972	NATO Quality Control System Requirements for Industry	U.S. Naval Publications & Forms Center (NPFC–43) 5801 Tabor Avenue Philadelphia, PA 19120, U.S.A
Australia	AS 1821–1975	Suppliers Quality Control System (Level 2 is AS 1811. Level 3 is 1823).	Standards Association of Australia Standards House 60 Arthur Street North Sydney, N.S.W. 2060, Australia

32.5. QC CIRCLES—A MOTIVATION FOR QUALITY THROUGH PARTICIPATION

The QC circle movement started in 1962 in Japan; since then, American industry has sought to make use of the idea as a key element in providing quality goods and services. The QC circle concept is based on the problem-solving idea which in turn leads to job enrichment and increased productivity. A QC circle is a small group of people doing similar work who meet regularly to identify, analyze, and solve product quality problems. The features that distinguish the QC circle program from other motivational approaches like the Stakhanovism system, the Scanlon system of joint committees, and the zero defect (ZD) program are:

Group participation

Participation on a voluntary basis

Emphasis on analyzing the causes of nonconformance and finding solutions

Emphasis on training

Emphasis on nonfinancial rewards

Support of top management, initiative of middle management, and training of supervisors and operators in group dynamics, and use of statistical and other problem-solving techniques are key elements to the successful implementation of the QC program in a company. Juran[9] states that "The QC circle starts with a different set of beliefs: (1) We really don't know the cause of our quality troubles; we don't know which are the main troubles. (2) Hence, we must teach people how to analyze the trouble pattern in order to identify the main troubles. (3) We must teach people how to list the suspected causes of the main troubles and how to discover which are the real causes. (4) We must help to secure remedies for these real causes. (5) Finally, we must teach people how to sustain these gains through modern quality control methods."

32.5.1. Development of QC Circle Program

The steps involved in the implementation of a QC circle program include:

1. *Selling the program to top management.* This is done by arranging a day or so seminar for the management to highlight the results of QC circle programs at other companies using actual and survey results and testimonials.

2. *Training of facilitators, leaders, and managers.* The facilitator has the overall responsibility for the development, implementation, and administration of QC circles. The facilitator introduces the concept to management and personnel in the company and should be a people-oriented person. The leader is usually the department supervisor and, among others, has the responsibility of keeping records and assisting in the preparation of management presentations. Facilitators and leaders are given extensive training in QC circle techniques and in turn train members in these techniques.

3. *Measuring Results.* Once the QC circle program is started, it is important to assess its effectiveness. Lower absenteeism rate, higher production rate, reduction in nonconformance rate, and savings in terms of time and costs are some of the measures which can be used to reflect success of the program. Attitude questionnaires are used to know the feelings of the members about the program.

32.5.2.   QC Circle Techniques

Various techniques which are used in problem identification and finding solutions include:

1. *Brainstorming.* The QC circle concept emphasizes that people will take interest and pride if they are permitted to participate in the decision made about their work. Brainstorming is an idea-generating technique which encourages the members of the circle to suggest ideas to improve the quality of the product or service and solutions to quality problems being encountered. No one is criticized or discouraged. All suggestions are given consideration.

2. *Pareto Analysis.* Pareto diagrams are used to identify the elements which contribute most to quality problems. These elements can be tools, machines, parts, people, decisions, and procedures. It has been observed that most problems are caused by a small percentage of one or more of these elements. It pays more to concentrate on the 'vital few' than to waste resources in analyzing all the causes, most of which may be insignificant. Pareto analysis identifies "vital few" items from the "trivial many" in the process of solving a problem. A typical pareto diagram is shown in Figure 32.5.

3. *Cause and Effect Diagrams.* These diagrams represent the problem (effect) and factors (causes and subcauses) which contribute to the problems. This technique is a systematic procedure for examining the problem, identifying the causes, finding possible solutions, and

choosing the best ones. A typical cause and effect diagram is shown in Figure 32.6.

Other QC circle techniques include check sheets, graphs, histograms, multivari charts, and so on.

32.6. COMPUTERS AND QUALITY ASSURANCE

Rapid advances have taken place in computer-aided design and manufacturing (CAD/CAM), automatic test equipment (ATE), computer coordinated measuring machines (CMM), computer graphics, image analysis via computer, and computerized management information system (MIS), and so on. Computer-aided quality assurance based on these technological developments would help to improve efficiency, quality, and productivity of goods and services. The use of CAD/CAM can provide a paperless system for receiving inspection by elimination of folders containing information on revision of drawings, inspection history, and other documents. Through the use of CAD/CAM, the true effect of the out-of-specification parts can be studied by overlaying images of the parts, and a decision made with regard to acceptance or rejection of these. For the electronics industry, quality control requires the use of automated testing. Examples of ATE used in the electronics industry are semiconductor testers, linear semiconductor testers, and circuit board and continuity testers.

Automated MRB corrective analysis would result in quick and correct

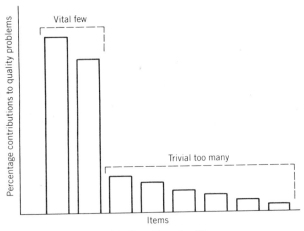

FIGURE 32.5.   Pareto diagram.

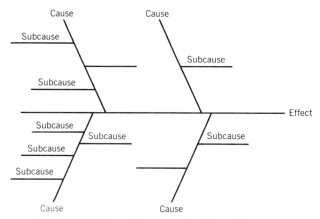

FIGURE 32.6.   Cause and effect diagram.

examination of the problems. Computerized quality information system helps the top management in strategic planning and the operation management in traceability and accountability by providing timely and accurate information. Thus the computer provides us with the means to have quality assurance built into various functions of the manufacturing system.

32.7.   STATISTICAL QUALITY CONTROL TECHNIQUES

"When you can measure what you are speaking about and express it in numbers, you know something about it; and when you cannot measure it, when you cannot express it in numbers, your knowledge is of a meager and unsatisfactory kind" (Lord Kelvin).

The science of statistics deals with analyzing data which are subject to variation with the objective of drawing inferences and making decisions. Statistical quality control deals with the use of statistical methods in the control and improvement of quality in manufacturing and service. The works of Walter A. Shewhart, H.F. Dodge, and H.G. Romig of Bell Telephone Laboratory laid the foundation of statistical quality control. A brief discussion of some of the statistical quality control techniques follows. Extensive coverage of these can be found in Juran (See note 2); Burr; Duncan; Box, Hunter, and Hunter, and Grant and Leavenworth.[10]

32.7.1   Process Control Charts

In any production process a certain degree of variation is inherent. The basis for using process control charts, also known as Shewhart Charts, is that variation observed is due to either random causes or assignable causes. Central, upper, and lower control limits on the charts are established on the basis of the distribution of quality characteristic. Samples of given size are taken at frequent intervals of time and measures of a quality characteristic, such as the mean, plotted on the chart. Points outside the control limits may be indicative of the presence of assignable causes and attempts are made to identify and correct these. Assignable causes include use of faulty raw material, operator error, malfunctioning of machine, incorrect setting of tools, and the reasons which lead to such conditions.

Control charts may be based on measures such as the mean, standard deviation, or range of a sample; these are variables charts. Others control attributes, that is, the number or proportion of errors or defective items. Control charts for attributes can be successfully used in nonproduct cases, for example, engineering work, clerical errors, absentees, industrial accidents, and highway accidents. We illustrate the use of control charts for defects (c-Charts) for evaluating the quality of the work performed by considering the following example adapted from Ullmann.[11]

An engineering project requires the preparation of a large number of relatively similar but complex detail drawings. The number of errors in 35 recently prepared drawings were: 3, 2, 5, 1, 0, 2, 5, 6, 1, 9, 6, 2, 7, 9, 3, 2, 1, 9, 5, 3, 8, 9, 6, 5, 4, (5, 7, 4, 9, 11, 3, 7, 4, 6, 4). Of these only the first 25 observations were available at the time of introducing the control chart. The control limits are given by $\bar{c} \pm 3\sqrt{\bar{c}}$ with the lower one taken as 0, whenever a negative value results. The numbers defective in the first 25 drawings sum to 113. Therefore,

central line $\bar{c} = 113/25 = 4.52$
upper control limit $UCL_c = \bar{c} + 3\sqrt{\bar{c}} = 4.52 + \sqrt{4.52} = 10.9$
lower control limit $LCL_c = \bar{c} - 3\sqrt{\bar{c}} = 0$

The control chart thus prepared is shown in Figure 32.7. Since none of the 25 observations (not plotted on the chart) were outside the control limits, the standard number of errors c' was taken as $\bar{c} = 4.52$ and a control chart with central line of 4.52 and upper control limit of 10.9 was used to monitor errors in later prepared drawings.

An examination of the plot of the number of errors in drawings 26 to 30 on the control chart revealed deterioration in the quality of work performed and called for corrective action. As a result of the corrective action, the number of errors in the last five drawings 31 to 35 indicates an improvement in the quality of the work. At this stage it is desirable to recompute the control limits to maintain better standards.

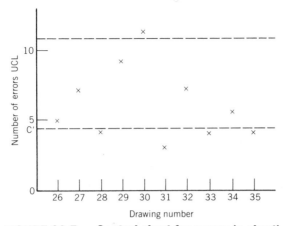

FIGURE 32.7. Control chart for errors (c-chart).

32.7.2. Acceptance Sampling Schemes

These schemes are used to make a decision to accept or reject the lots based on sampling inspection or testing. Various types of sampling plans used are:

1. *Single Sampling Plan.* A sample of size n is taken from the lot and the lot is accepted if the number of defectives found in the sample is less n or equal to the acceptance number c.

2. *Double Sampling Plan.* A sample of size n_1 is taken and a decision made to accept the lot, reject it, or else take another sample of size n_2. If the decision is made to take another sample, the criterion to accept or reject the lot is based on the combined results. A multiple sampling plan is extension of double sampling plan involving more than two stages in the decision making.

3. *Sequential Sampling Plan.* In this scheme, units are drawn from the lot one by one and after each unit a decision is made to accept or reject the lot or continue sampling.

4. *Continuous Sampling Plans.* These plans have been developed for use in continuous production inspection and provide criteria for changing between 100% inspection and sampling inspection.

The two most widely used standard sampling inspection tables in industry are MIL-STD-105D (for attribute data) and MIL-STD-414 (for variable data). These tables index the plans by "Acceptance Quality Level"

(AQL) and provide guidelines for changing between tightened and reduced sampling inspection. Though a good amount of theoretical work has been done on sampling schemes based on economic aspects, these schemes have not found wide acceptance in industry.

32.7.3.   Techniques for Industrial Experimentation

Engineers are often interested in studying the effects of using different types of raw material, machines, procedures, and production processes on the quality characteristics of products. Proper experimental design is as important as the analysis of data using sound statistical techniques. Statistical techniques used in industrial experiments include:

Tests of hypotheses
Regression and correlation analysis
Analysis of variance
Analysis of covariance
Analysis of means
Response surface methodology
Design of experiments
Evolutionary operation (EVOP)

NOTES

1. A. V. Feigenbaum, *Total Quality Control,* 3rd ed. (New York: McGraw-Hill, 1983), p. 7.
2. P. B. Crosby, *Quality Is Free* (New York: McGraw-Hill, 1979), p. 1. W. E. Deming, "What Happened in Japan?" Industrial Quality Control, *24,* 89 (1967). W. E. Deming, Report to Management, *Quality Progress 5, 2, 3, 42* (1972). J. M. Juran, and F. M. Gryna, Jr., *Quality Planning and Analysis,* 2nd ed., (New York: McGraw-Hill, 1980).
3. F. Caplan, *The Quality System—A Source Book for Managers and Engineers,* (Radnor, PA: Chilton, 1980).
4. J. M. Juran, *Quality Control Handbook,* 3rd ed., (New York: McGraw-Hill, 1974).
5. J. T. Hagan, *A Management Role for Quality Control* (New York: American Management Association 1968) p. 104.
6. Quality Cost Effectiveness Technical Committee, ASQC, *Quality Costs—What and How,* 2nd ed., (Milwaukee: American Society for Quality Control 1971), p. 5.
7. Quality Cost Technical Committee, ASQC, *Guide for Reducing Quality Costs* (Milwaukee: American Society for Quality Control, 1977).

8. Quality Cost Technical Committee, ASQC, *Guide for Managing Vendor Quality Costs,* (Milwaukee: American Society for Quality Control, 1980).

9. J. M. Juran, QC Circle Phenomenon, *Industrial Quality Control 23,* 329 (1967).

10. I. R. Burr, *Statistical Quality Control Methods,* (New York: Marcel Dekker, 1976). A. J. Duncan, *Quality Control and Industrial Statistics,* (Homewood, IL: Richard D. Irwin, 1965). G. E. P. Box, W. H. Hunter, and J. S. Hunter, *Statistics for Experimenters* (New York: Wiley, 1978). E. L. Grant and R. S. Leavenworth, *Statistical Quality Control,* (New York: McGraw-Hill, 1980).

11. J. E. Ullmann, *Quantitative Methods in Management,* (New York: McGraw-Hill, 1976).

33

SAFETY AND COMPLIANCE
WITH OSHA

Paul V. Bonfiglio

PAUL V. BONFIGLIO has been a Compliance Officer with the Occupational Safety and Health Administration (OSHA) of the U.S. Department of Labor since 1975. From 1964 to 1975, he was a construction safety inspector for the Industrial Safety Service of New York State. He has served as an instructor in the Continuing Engineering Education Programs of Hofstra University, and at the Training Institute of the U.S. Department of Labor in Chicago. He has a B.S. degree in Labor Studies from Empire State College. Mr. Bonfiglio is presently the executive vice-president of Local 2513, American Federation of Government Employees as well as a representative to the Central Labor Council, AFL-CIO, in New York City.

33.1. INTRODUCTION

Occupational safety and health has been addressed from many directions and by many interests including government agencies. Gains and improvements have been made for the protection of the factory worker, maritime worker, migrant farm worker, construction worker, and to some degree even for the office worker. However, little if anything has been said or done for the engineer, as an employee who is required at times to leave the comfort of his or her office to visit unfamiliar environments.

This chapter will examine the unsafe conditions that an engineer may be exposed to while working in his office, while visiting a client's office and industrial plant for tests or sampling, or on a visit to a construction project. Finally, we will discuss the employer's responsibility and obligation to employees safety.

To understand the mandate of the Occupational Safety and Health Administration (OSHA), its history has to be reviewed to bring it into perspective. In the 1960s concern with occupational safety and health reached the point of legislative action. Among the 90 million Americans then working, it was estimated that job-related accidents accounted for more than 14,000 deaths each year, with nearly 2.5 million workers disabled and approximately 300,000 new cases of occupational diseases reported yearly. In response to such statistics, the Ninety First Congress of the United States enacted the "Occupational Safety and Health Act of 1970." Its full description reads:

AN ACT

To Assure safe and healthful working conditions for working men and women; by authorizing enforcement of the standards developed under the Act; by assisting and encouraging the States in their efforts to assure safe and healthful working conditions; by providing for research, information, education, and training in the field of occupational safety and health; and for other purposes.

(Public Law 91-596, 91st Congress, S. 2193, December 29, 1970.)

In general, coverage of the Act extends to all employers and their employees in the 50 states, District of Columbia, Puerto Rico, and all other territories under Federal Government jurisdiction.

OSHA is responsible for promulgating legally enforceable standards, based on conditions, or the adoption or use of one or more practices, means, methods, or processes reasonably necessary or appropriate to protect workers on the job. It is the employers' responsibility to become familiar with standards applicable to their establishments and to ensure that employees have and use personal protective gear and equipment required for their safety. Employers are responsible for following the intent of the Act's "General Duty Clause" even where specific standards have not been promulgated. It states that each employer "Shall furnish a place of employment which is free from recognized hazards that are causing or are likely to cause death or serious physical harm to his employees."

Employers are responsible for keeping employees informed about OSHA and about the various safety and health matters with which they are involved. OSHA requires that each employer post certain materials at prominent locations in the workplace.

OSHA is authorized under the Act to conduct workplace inspections of all establishments covered by the provisions. Compliance Safety and Health Officers are chosen for their knowledge and experience in the occupational safety and health field and are vigorously trained in OSHA Standards and in recognition of safety and health hazards.

With few exceptions inspections are conducted without advance notice. To give advance notice of an OSHA inspection can bring a fine of up to $1000 and/or a six month jail term. Under special circumstances, OSHA may give notice to an employer, but this will be less than 24 hours.

33.2. OFFICE SAFETY

The hazards of office work are obviously less dramatic than those in heavy industry and therefore have been largely ignored or regarded as not serious in the past. This however, is badly in error. Recent evidence has established that one out of six office workers will experience an occupational injury in any one year. Through education this statistic can be reduced and conceding that much of the responsibility for compliance lies with the employer, all workers collectively can contribute to making the office a safe workplace. The following are some of the principal items requiring serious attention, together with the appropriate citations of the full regulations in *Code of Federal Regulations,* Vol. 29 (29 CFR).

33.2.1. Walking and Working Surfaces

Most of the falling accidents are results of slipping or tripping. Slips can be caused by wet or highly polished floors. Tripping is often caused by obstructed aisles, poorly placed electric cords, and neglected objects, such as pencils, tacks, or paper clips. (29 CFR-Subpart D 1910.)

33.2.2. Electrical Equipment

Electrical equipment, refrigerators, fans, and water coolers shall be grounded and electrical tools shall be grounded or double insulated. All phone lines, wires, and extension cords should be fastened under desks and covered if they cross aisles. Equipment shall be inspected regularly and power shall be disconnected during cleaning and maintenance. (29 CFR-Subpart S 1910.)

33.2.3. Tools

Paper cutters, scissors, letter openers, staplers, and hole punchers can cause injury if not handled correctly. (29 CFR-Subpart P 1910.)

33.2.4. Ventilation

Poor ventilation systems may in addition to maintaining the office too hot or too cold, may circulate the effects of chemicals and fumes. Problems of asbestos, dust, smoke, and even germs can be compounded by a faulty ventilation system. Filters must be changed at least twice a year and ventilation systems need to be cleaned and maintained regularly. (29 CFR-Subpart G 1910.)

33.2.5. Chemicals

Chemical fumes emanating from office machines, such as photocopy and duplicators, can cause throat irritation, coughing, head-ache, chest pains, eye irritation, and vision problems. Other office products containing chemicals that can cause unsafe environments for workers includes: Paper; both carbon and carbonless, duplicating fluids, ink, ink remover, and rubber solvents. (29 CFR-Subpart Z 1910.)

33.2.6. Lighting

Poor lighting can result in tired and sore eyes. Headache, neck and back pain can also result from straining the eyes. Visual fatigue can be caused from fluorescent lighting glare as can reading small figures or unclear print in poor light. (29 CFR-Subpart S 1910.)

33.2.7. Noise

Office noises seldom reach the decibel levels of a hazard however, the combination of telephones, human voices, office machines, and typewriters can create unpleasant sounds which could cause distractions and possible stress. Noise levels can be reduced considerably with acoustically treated walls and ceilings, carpeting, mats under typewriters, and other noise reducers, including plants. (29 CFR-Subpart G 1910.)

33.2.8.   Ladders

Well-constructed ladders and stepstools are the best devices to use to reach high objects. Chairs, desks, desk drawers, or file cabinets cannot be considered safe methods of reaching high items. (29 CFR-Subpart D 1910.)

33.2.9.   Exits

Exits shall be clearly marked and identified. Exit doors shall be maintained unobstructed and unlocked. The walkways and stairways to exits shall be adequate in width and marked with direction signs. (29 CFR-Subpart E 1910.)

33.2.10.   Fire Fighting

Fire extinguishers shall be clearly marked and readily accessible. All extinguishers shall be inspected yearly and charged when necessary. The class of extinguishers shall meet requirements for fire types. Fire drills, emergency instructions, and fire brigades should be established. (29 CFR-Subpart L 1910.)

33.3.   INDUSTRIAL PLANTS

As an employee of an engineering firm, the engineer may be called upon to visit a plant. The visit will probably start in the plant's office. In the plant itself, the normal hazards may be exacerbated because the engineer may be concerned with new or experimental machinery or processes or installations under construction (see Section 33.4). If the engineer is concerned with plant design or operation, his or her professional responsibility must extend beyond his or her own safety. He or she may spend time near spray booths and in storage areas for flammable materials. An engineer will be close to machinery and on or near scaffolds and ladders. He or she will observe the fire protection, means of access, electrical power system, walking surfaces, and last but not least, he or she will check his own personal protective equipment. This section will focus on these and other items and on the employer's responsibility in relation to them.

33.3.1.   Office Areas

Little difference exists between the conditions in an engineer's office and in the office of a plant. The same types of tripping and slipping hazards will be present; refrigerators, water coolers, and fans will have to be grounded; the exits identified, marked, and maintained unlocked. The stairways and passageways maintained clear and free of obstructions. The ventilation system may present a problem depending on the materials used in the plant. The fire fighting equipment, ladders, and electric wiring shall be checked and kept in good repair. It should be pointed out that items usually found in an office need not be considered unsafe and hazardous. Much depends on the attitude of the management and relationship between the management and employees. Through education and a positive program of cooperation, offices can be a safe place to work. (29 CFR-Subpart D 1910.)

33.3.2.   Personal Protective Equipment

The engineer should be aware of the hazards in the plant and take the necessary precautions. The employer shall provide the safety equipment for his or her employees. If welding and cutting work is in progress, eye and face protection has to be provided. Noise, above allowable levels can be reduced with ear-protective devices. (Plain cotton is not acceptable.) The ventilation system if operating, should be clearing the environment of harmful dusts, fog, fumes, mists, gases, smoke spray, or vapors. The engineer shall be provided with an approved respirator, including proper filters. Proper shoes are required when walkways and workareas may be wet and slippery due to the operations. Protective clothing such as coats, boots, gloves, and safety helmets should be chosen for their effectiveness and the employer shall assure the equipment's adequacy. He or she shall also be responsible for the employee-owned equipment. (29 CFR-Subpart I 1910.)

33.3.3.   Spray Booths

In the plant the engineer may spend time in the spray booth area. This area must be visited with caution, particularly if dangerous quantities of flammables are present. To remain cognizant of the potential hazards, the engineer should have an understanding of spray booths. Spray booths, designed and operated within limitations provided by law and regulations, will be as safe as any other area. A point to ponder might be the unsafe act as compared to the unsafe condition. The unsafe condition is easier to correct, while the unsafe act is caused by a worker and requires more attention. This will be addressed at length later on in this section. The engineer should be familiar with the many types of spray booths. The most common, the dry spray booth, uses baffles or filters which collects the overspray and allows an even flow of air through the booth. The waterwash spray booth uses a water-washing system designed to minimize dusts or residues from entering the exhaust ducts. Electrostatic fluidized type aerates the material from below to form an air-supported cloud which is charged electrically, opposite to the charge of the object to be coated. The engineer should be aware of the construction requirements for spray booths. Booths shall be smooth and free of edges to prevent build-up of residues. The floor of the spray booth, if combustible, shall be covered with noncombustible materials. The electrical wiring within the spray area shall be explosion-proof type, approved for Class 1, Group D, locations. Exhaust ducts and piping systems conveying flammable or combustible liquids shall be effectively and permanently grounded. Each spray booth shall have an independent exhaust system discharging to the exterior of the building. Spray booths shall be designed with the air velocity at least one hundred feet a minute and with approved automatic sprinklers on the upstream and downstream sides of the filters. (29 CFR-Subpart H 1910.)

33.3.4. Ventilation

Whenever hazardous substances, such as dusts, fumes, mists, vapors, or gases are present or are produced in a plant, the concentrations shall not exceed specific limits. The engineer visiting a plant where high concentrations of contaminants may exist can be protected with an air-line respirator or by an air-purifying respirator which removes most of the dust or fumes.

To better understand the hazards that may exist, the engineer should have some knowledge of gaseous contaminants. Some substances are gas in their normal form while some substances are solids at very low temperatures and liquids at high pressures. An example, carbon dioxide, is a gas at room temperature, a solid at low temperatures, as in dry ice, and a liquid when pressurized. Vapors are formed by evaporation of substances which ordinarily exists as a liquid. Dusts are solid particles produced by grinding or crushing. Mists are tiny liquid droplets given off when a liquid is sprayed or mixed vigorously. Fumes are tiny metallic particles given off when metals are heated. Fumes would be found near soldering, welding, or brazing operations. The same can be found by casting and galvanizing.

Oxygen deficient atmospheres are a condition found in areas with poor ventilation. The oxygen can be pushed out of an area by another gas, it can be used up as the result of a fire or can be displaced by rusting iron. It can be created using dry ice at low temperatures. By identifying what hazards may exist and in what concentrations, the engineer can then choose the proper respirator. Respiratory protective devices shall be approved by the United States Bureau of Mines or by the National Institute for Occupational Safety and Health (NIOSH).

Local exhaust ventilation can be used in isolated areas. The system shall have freely moveable hoods located as close as practicable to the work and it shall be of sufficient capacity to remove contaminants from the breathing zone.

Ventilation is the preferred method used to control airborne contaminants which adversely affect employees. An effective, well-designed ventilation system will lower the employees exposure to toxic contaminants and will clear the general environment of flammables and explosive atmospheres.

Ducts of the ventilation system shall be made of sheet metal or black iron. The interiors of ducts shall be smooth with no obstructions and the joints and seams welded or soldered. Fans of ventilation systems in flammable locations shall be explosion proof and the blade shall be of nonferrous material. The system shall discharge to the outer air and in such a manner that it cannot re-enter the building. Maintenance shall be such that the airflow shall be maintained at all times in the ventilation ducts. (29 CFR-Subpart G 1910.)

33.3.5. Hazardous Materials

While in the plant the engineer will be in areas that may have compressed gas cylinders, flammables, flammable storage areas, LP gas, and possibly explosives. None of these materials have to be unsafe if handled, stored, or used correctly. If the ventilations system is operating as designed and the spray booth is functioning correctly, most of the hazards have been removed from the areas containing hazardous materials.

The compressed gas cylinders shall be maintained in a safe condition. Each cylinder shall have a pressure relief device. Cylinders shall be stored in an upright position, secured in place with valve protection caps covering valves. Defective or damaged cylinders shall be removed from service.

Inside storage rooms for flammables shall be constructed to meet the fire-resistant rating. Openings to other rooms or buildings shall be provided with noncombustible liquid-tight raised sills at least four inches in height. Openings shall be provided with approved self-closing fire doors. Electric wiring and equipment in inside storage rooms used for Class 1 liquids shall be approved for Class 1, Division 2 hazardous locations. Each inside storage room shall be provided with an exhaust ventilation system.

Flammable or combustible liquids shall be stored in tanks or closed containers. They shall be drawn from or transferred into vessels, containers, or portable tanks within a building only through a closed piping system, from a safety can, by means of a device drawing through the top, or from a container or portable tank by gravity through an approved self-closing valve.

When pouring from one container to another, containers shall be bonded together. Precautions shall be taken to prevent ignition of flammables. Care should be practiced to eliminate the problems of; open flames, lightning, smoking, cutting and welding, heat, and static, electrical and mechanical sparks. Heat producing chemical reactions have to be guarded against as well as radiant heat. (29 CFR-Subpart H 1910.)

33.3.6. Means of Egress—Fire Protection

When considering the means of access, the engineer, should also bear in mind the three distinct parts: First, the way of exit access, second, the exit, and third, the way of exit discharge. All should terminate out of the building to a public way. Every building shall be provided with exits of kinds, numbers, location, and capacity appropriate to individual building with due regard to the occupancy. Every exit shall be arranged and maintained to provide free and unobstructed egress from all parts of the building. Each exit shall be clearly visible and each path of escape shall be marked. No lock or fastening shall be installed on any exit door that can prevent free escape from the inside of building. Mental, penal, or corrective institutions are exceptions.

Fire brigades, although not mandatory, can be established if firm prepares a statement or written policy listing the basic structure, the type, amount, and frequency of training. The members shall be physically capable for the duties required. Training and education shall be provided for brigade members before they perform emergency activities and frequently enough thereafter to assure quality similar to that of major state fire training schools. Fire fighting equipment shall be maintained and inspected by the employer at least annually. Each member shall be equipped with an approved self-contained breathing apparatus.

Portable fire extinguishers shall be mounted, located, and identified so that they are readily accessible to all employees. The extinguishers shall be selected based on the class of workplace fires anticipated. Portable fire extinguishers shall have an annual maintenance check.

Automatic sprinkler system shall have a main drain flow test each year and system piping shall be protected against freezing and exterior surface corrosion.

An alarm system shall provide warning for emergency action. It shall be perceived above ambient noise and light levels. The system shall be explained to each employee. The system shall be maintained in operating order and a manually operated actuation device shall be readily accessible for use in conjunction with the employee alarms. (29 CFR-Subpart E/L 1910.)

33.3.7.   Machinery and Machine Guarding

The engineer when in the machine area of the plant has to be aware of other hazards in addition to the noise levels. Point of operations of machines, ingoing nip points, rotating parts, flying chips, and sparks all require protection. One or more methods of machine guarding shall be provided to protect the operator and other employees in the machine area. Machines can be guarded at the point of operation by barrier guard, electronic safety devices, two-hand tripping devices or hand restraints. Each type designed for the sole purpose of keeping the operator's hands from the point of operations.

The blades of fans less than seven feet above the floor shall also be guarded.

Machines designed for a fixed location shall be anchored to prevent walking or moving.

The moving parts of woodworking machinery, including blades, belts, pulleys, gears, and shafts shall be adequately guarded. Each machine shall have a mechanical or electrical power control shut-off switch within reach of the operator. Motors shall not restart after a power failure if injury to the operator might result. Swing cutoff saws shall return to the back of the table when released. The saw shall be restrained from swinging beyond the front edge of the table. Grinders, both bench type and pedestal type have the same potential problems. The spindle end, nut, and flange projections shall be covered. Off-hand grinding machines shall have work rests to support the work. Bench grinders shall be secured to the bench to prevent walking or moving and pedestal grinders designed for fixed locations shall be anchored.

Having referred to just a few machines, it is well to point out that every machine has moving and rotating parts that can cause problems to the operator and to others in the area. Guards and devices can protect against accidental contact with the moving parts or point of operation, but its the knowledge of the hazards that will prevent the operator from taking chances. (29 CFR-Subpart O 1910.)

33.4.   CONSTRUCTION SITES

When the engineer is instructed to visit a construction site for whatever reasons, he or she most probably will become apprehensive. This strange world contains mysteries for the neophyte who doesn't know what hazards to expect. The truth of the matter is that construction sites, with their constantly changing conditions, are no more unsafe than any other workplace. With diligence and a firm commitment to safety, the construction work site can be maintained free of most hazards.

In this section, we will be offering the engineer the knowledge necessary to visit these areas in comfort. It is well to state at this point that some types of construction are by their very nature hazardous. However, being aware of what to expect and being prepared for any eventuality will reduce the surprises.

Too much cannot be said of the conditions to be found in tunnel operations, but the engineer will do well to remember that the tunnel project was designed by engineers, and that their prestige and integrity rests with the structural stability of the tunnel. Tunnels can in many ways be compared to caverns, and few people hesitate to pay money to visit these underground wonderlands.

Excavations and trenching will also be addressed in this section. These areas are not viewed with great trepidation by the author, as very little in protection is required to convert the unsafe trench or excavation to a safe one.

A demolition site on the other hand has inherent hazards for the life of the project. The constantly changing conditions at demolition sites give rise to the unsafe conditions and the unsafe acts. The best that can be said about a visit to a demolition site is that most of the inspection work on major demolition projects is completed prior to the start of demolition. In the preparation stages actual presence in the structure may be required, but once the demolition work has started, the inspections can be continued from a distance.

Hoisting equipment is another area in which the engineer should have working knowledge and understanding. The devices used for hoisting on a construction site are an obvious necessity, but they have also created a number of hazards.

Personal protective equipment used by the engineer while on a construction site will be examined as well as the electrical requirements along with any potential electrical problems. Powder-activated tools are another source of concern, as are housekeeping and floor and wall-opening requirements. Finally, ladders and scaffolds will be reviewed in detail.

Together, these items should not only help an engineer develop the confidence to visit construction sites safely, but also to learn to notice hazards that might jeopardize the project or operations.

33.4.1. Personal Protective Equipment

Most people realize that an approved-type safety helmet is a necessity whenever visiting a construction project. The possibility of being struck by falling or flying objects, being struck by something being carried, or of bumping ones head against something is always present on a construction site. The engineer will do well by remaining aware of these potentials. Adequate shoes for the engineer are an obvious need because of widespread tripping and cutting hazards, such as scrap lumber with protruding nails. Gloves should be included because splinters and splits can be expected on ladders and on safety railings. Safety glasses and eye protection gear are required near welding or in proximity to masonry cutting operations. Foul weather gear, coats, and rubber boots are likewise often indicated. With the previously mentioned equipment and adherence to the regulations, we can assure that the engineer will feel safe on a visit to a construction work site. (29 CFR-Subpart E 1926.)

33.4.2. Housekeeping

Housekeeping on a construction site must be viewed as a continuing and ever-present problem. The engineer should view this item as a job-to-job condition and feel privileged if he or she finds the passageways and stairways clear and unobstructed. Construction progress is coordinated by the construction superintendent and each phase is allotted the necessary time and effort. The clean-up in most cases is given a low priority. Debris on a constantly progressing construction site will always be a problem, but with care and caution the engineer need not feel threatened. (29 CFR-Subpart C 1926.)

33.4.3. Floor and Wall Openings, and Stairways

The engineer must consider the protection or the lack of protection of floor and wall openings on the construction site. Railings in and around stairways have to be considered as well as the guarding of open-sided floors. The need for protection in these areas should be obvious, but thought must also be given to the metal stairways with hollow pan-type treads. Open-sided floors six feet or more above adjacent floor or ground shall be guarded by a standard safety railing. A single half-inch wire rope will meet the requirement of periphery protection, while floors are temporary-planked or temporary-metal decked during structural steel assembly. (29 CFR-Subpart M 1926.)

33.4.4. Ladders and Scaffolds

The engineer visiting a construction site may spend little or no time on ladders or scaffolds, but it is to his or her advantage to know and understand the requirements. Ladders shall be provided as access between levels, and they shall be secured against being displaced. Ladders shall not be placed in passageways, the side rails of ladders shall extend at least 36 inches above the landing and ladders with broken members shall be removed from service. Scaffolds shall be erected, moved, or dismantled under the supervision of a competent person. Guard rails shall be installed on open sides of work platforms of scaffolds over 10 feet in height. Scaffolds shall have an access ladder or equivalent safe access. Planking on scaffolds shall be of approved grades and overlapped at least 12 inches or secured from movement. Work platforms on mobile scaffolds shall be fully planked and planks shall be secured in place. (29 CFR-Subpart L 1926.)

33.4.5. Powder-Actuated Tools

Powder-actuated tools such as are used for driving anchor bolts into concrete, present special hazards. OSHA regulations provide that only people who have been trained in the operation of the particular tool shall be allowed to operate a powder-actuated tool. The tools shall be tested each day and any defective tools shall be removed immediately. Powder-actuated tools shall never be left unattended or used to drive pins or fasteners into easily penetrable materials. (29 CFR-Subpart I 1926.)

33.4.6. Electrical

All temporary wiring on a construction site shall be installed on proper insulators and wiring shall be effectively grounded. Temporary lighting shall have guards to prevent accidental contact with the bulbs. Electric wiring shall be kept clear of working spaces, walkways, and similar locations. Portable electric tools protected by an approved system of double insulation need not be grounded but shall be distinctively marked. Fixed equipment, including motors, generators, and electrically driven machinery shall be grounded. All 120-volt, single phase, 15 and 20 ampere receptacle outlets on construction sites, not part of the permanent wiring of the building shall have approved ground-fault circuit interrupters or instead, an assured equipment-grounding conductor program shall be established by the employer. (29 CFR-Subpart K 1926.)

33.4.7 Excavations and Trenching

Visits to trenching and excavation sites should present no great problems for the engineer. Seldom will he or she actually enter a trench, and being aware of the potential hazards, he or she can provide for his or her own safety. He or she should know that trenches more than 5 feet in depth shall be shored, sheet piled, or sloped to an angle of repose. A ladder shall be provided as access and all loose boulders, rocks, and trees shall be removed from the banks of trenches and excavations. The excavated material shall be stored at least 24 inches back from the edge of the trench. Utility companies shall have been contacted prior to the opening of the excavation. The location of all gas, electric, and water lines shall be exposed carefully and supported if necessary. Excavation shall be inspected after every rain storm and additional protection provided if necessary. (Subpart P.)

33.4.8. Steel Erection

During the erection of structural steel on a construction project, the engineer would be ahead of the operations with the following information: Iron workers are conceded falls up to 25 feet as compared to 10 feet for all other workers. This becomes understandable when we consider that steel columns are fabricated in two-story modules, which makes providing protection at lesser heights impracticable. Perimeter guarding is required when the planking or metal decking has been installed and the engineer should be aware that two bolts shall be installed at each connection of beams and that the bolts should be drawn up, wrench tight before the structural member is released from the hoisting line. Tag lines shall be used to control unwieldy loads. Containers shall be used for storing bolts, drift pins, and rivets on structural elevations. Decks shall be planked within 2 stories or 30 feet below the section where steel erection is being done. (Subpart R.)

33.4.9. Tunnels

Tunnel work may be considered the most hazardous of all construction work but while this may have been true in the past, it need not be true now. By using better and safer construction methods, with proper knowledge and understanding of the hazards, and with the compliance assistance being offered to the tunnel operator and the tunnel worker, tunnels can be made a safe place to work.

There are five major tunnel-building methods. Digging by hand and pushing shields is a method that has been around for centuries with several variations in use. Freezing the ground is a relatively new method that has not been widely adopted so far. Compressed air is a tested and proven method. Its high cost limits its use, but at times it could be the best and only method. Drilling and blasting and tunnel boring machines are now two of the most common types of tunneling.

The drilling and blasting method consists of drilling holes in the rock, loading and tamping the holes with explosives, firing the explosives, and then excavating the loose rock. This procedure is repeated until the tunnel is completed. This method also requires support of any loose rock in the roof or walls by some type of roof control, the lining of the tunnel, the ventilation system, the emergency systems, and hauling operations. The use and safeguarding of explosives is not considered here; it is obviously a large subject, but normally, we do not expect the engineer to visit a tunnel operation during the blasting stage. Tunnel boring machines are being increasingly used; they may save time and cost and are suitable for deep tunnels in hard, but stable rock.

The following is a list of requirements which the contractor or operator must meet in order to have a safe project. First, someone must be assigned to oversee safety. This person or a subordinate will be responsible for taking air-quality samples as well as tests for flammables, toxic gases, dusts, mists, and fumes that occur in tunnels. Tests shall be conducted as often as necessary and with the records maintained. Rescue crews and evacuation plans shall be developed. A safe means of access and egress from all tunnel work areas shall be readily available. The operator shall maintain a check-in and check-out system that will provide positive identification of every employee underground. Each tunnel shaft more than 50 feet in depth shall have an emergency hoisting facility operated on independent power. A self-rescuer shall be available for each worker near the advancing face of the tunnel. An independently powered telephone or other signal communication shall be provided between the work face and the tunnel portal. A ventilation system shall be installed and moved forward with the advancing face and the fans of the ventilation system shall be reversible. There are limitations on smoking or carrying open flames in areas where fire or explosion hazards may exist. No gasoline or liquefied petroleum gases shall be stored or used underground. The roof and walls of the tunnel shall be examined at the start of each shift, and any loose rock shall be supported or removed. The shafts of tunnels shall be lined to at least 5 feet below the start of solid rock. All workers on skips and platforms used in shafts shall wear safety belts, unless guard rails or cages are provided. Hauling equipment in tunnels shall be equipped with warning devices, and lights provided front and rear. No workers shall be permitted to ride on jumbos (the movable scaffold at the drilling face) and guard rails shall be provided on jumbos over 10 feet in height. (Subpart S.)

33.4.10. Cranes

To assure the safety of all workers on a construction site, each crane at the site shall be inspected by a competent person prior to each use. Cranes shall have a thorough annual inspection by a competent person or certified government or private agency. Cranes shall never be operated within 10 feet of high-voltage power lines. The accessible areas within the swing radius of the rear of the rotating superstructure of the crane shall be barricaded to prevent employees from being struck or crushed by the crane. The crane shall have a fire extinguisher at all operator stations.

33.4.11.   Material Hoists

No person shall be allowed to ride the material hoist. Entrances to the hoistways shall be protected by substantial gates and the roof of material hoist cage shall be protected with two inch planking. The operator of the material hoist shall have overhead protection. Material hoist cars shall be equipped with arresting devices in case of rope failure. A licensed professional engineer shall design all material hoist towers.

33.4.12.   Personnel Hoists

Hoist towers designed for personnel shall be enclosed for the full height on the sides used for entrance and exit to the structure. The towers shall be anchored to the structure at 25 foot intervals. Hoistway doors shall be provided with mechanical locks accessible only to the person on the car. Overhead protection of two inch planking shall be provided for the top of the personnel car. Electric contacts shall be installed to prevent movement of hoist when door or gate is open. Car of personnel hoist shall be provided with an emergency stop switch and a capacity limit chart posted in the car. Personnel hoists shall be inspected prior to being put into service, and at least every three months thereafter. Records of the inspections shall be maintained for the duration of the job. (29 CFR-Subpart N 1926.)

33.4.13.   Demolition

At a demolition work site, an engineering survey to detect hazards shall be maintained by the employer. All electric, gas, water, steam, sewer, and other service lines shall be shut off, capped, or otherwise controlled outside the building. All floor openings not used as material drops shall be covered over. Only stairways used for access shall be used; all other stairways and passageways shall be entirely closed at all times. Areas outside the building where material may be dropped shall be effectively protected. Material chutes shall be provided with a substantial gate at the discharge end. A competent employee shall be assigned to control the gate and the loading of trucks. Interior floor openings used for the removal of debris shall not be opened more than 25% of the total floor area. Employees shall not be permitted to work on the top of a wall when weather conditions constitute a hazard. During mechanical demolition only those workers necessary for the operation shall be permitted in the area during the demolition. Continuing inspections by a competent person shall be made to detect hazards resulting from weakened walls or floors. No employees shall be permitted to work where such hazards exist. (29 CFR-Subpart T 1926.)

33.5. COMPLIANCE AND THE RIGHTS OF EMPLOYEES

The following is a brief discussion of how compliance is assured and some of the problems that may arise.

33.5.1. Who is Responsible for Correcting Hazards?

In an office or plant the answer is simple: The employer is responsible for providing a safe work place for his or her employees and for correcting all hazards. The problem becomes complicated when we consider a visitor to the plant who is not an employee. If the problem is imminent danger, the engineer can refuse to enter the plant and should request an immediate OSHA inspection. The visiting engineer's employer, under most situations cannot be cited for unsafe conditions where he or she has no control. The question becomes more complicated on a construction site where there may be many employers. Ideally it would be simple if the general contractor or owner/builder were charged with the responsibility of providing the protection and correcting the hazards. However, the OSHA Act provides that the employer of exposed workers will be cited under most circumstances. Although it may be appropriate to issue citations to each employer with employees exposed to hazards, employers meeting the following criteria need not to be cited:

1. The employer did not create the hazard
2. The employer has neither the ability nor the authority to correct the hazard
3. The employer discussed the condition with the controlling employer
4. The employer has trained his employees to avoid hazards and has taken measures to protect his or her employees

The engineering firm should have no problem meeting all of these criteria when their employees visit a construction site.

33.5.2. Can Employees Refuse to Enter Unsafe Workplaces?

The engineer may be faced with the decision of whether or not to enter a plant or site that he or she knows to be unsafe. If there is imminent danger, then there is no question that the employee has the right to refuse to work in such circumstances. OSHA rules protect employees from discrimination if the employer failed to correct the danger after being informed, if the danger is so imminent that there is no time to eliminate it through normal procedures, if the danger is so grave that death or serious physical harm will result, and if the employer has no reasonable alternative. The Act does not grant the right to refuse to work under circumstances which may constitute a threat of injury or death. The question remains however, whether such a right will be implied. In a court case in the United States District Court, Northern District of Georgia in 1975, the Court ruled that the employee did not have a right to refuse to work under conditions which he or she felt were hazardous.

The original Occupational Safety and Health Bill reported out of the House Education and Labor Committee in early 1970, had a subsection which allowed workers to walk off the job with pay if they were exposed to toxic substances that could have harmful effects. This was attacked by opponents of the Bill as a guarantee for workers to be paid while striking. The Bill as passed resolved the immediate danger problem by giving the employees the right to request in writing an immediate OSHA inspection.

In a test case on this provision, The Court of Appeals for the Sixth District ruled that a worker should not have to choose between his job and his life without the reasonable safeguard provided by the OSHA regulations. The United States Supreme Court affirmed the United States Court of Appeals' decision and ruled that if an employee with no reasonable alternative refuses to expose himself to dangerous conditions he should be protected against subsequent discrimination. However, it has to be stressed that the reversal was based on the fact that the employer did not provide the employee the opportunity to work at other safe assignments. It was the employer's own retaliatory and discriminatory conduct which prevented the employee from rendering any further service.

33.5.3.   Can an Employer be Cited?

The owner of an engineering firm can be cited if violations of the OSHA standards are found on an inspection of his office. The same would not be the rule if the engineer was visiting a plant or factory. If the engineer was exposed to hazards during these visits the main concern would be getting the conditions corrected and bringing them to management's attention. On the construction site the employer will not be cited if he meets the four criteria of Section 33.5.1. However, if the employer fails to inform the controlling employer of the hazard, his firm can be cited. This could happen even if the other three conditions are met. The same applies if the engineering firm does not offer its employees alternative means of protection from hazards short of walking off the job. It is important to note, therefore, that responsibility for safety in engineering projects may be professional in nature as well as part of the usual conditions of employment.

REFERENCES

1. *Occupational Safety and Health Cases* Published in conjunction with Occupational Safety and Health Reporter, by The Bureau of National Affairs, Inc. Washington, DC, Volumes 7, 8, and 9.

2. Public Law 91-596 91st Cong., December 29, 1970. Washington, DC, 31 pages.

3. State of New York Department of Labor, Board of Standards and Appeals *Tunneling Operations, Industrial Code Rule 30,* December 1, 1957, 29 pages.

4. U.S. Department of Labor, Occupational Safety and Health Administration *All about OSHA,* 1982 (revised) OSHA 2056, 46 pages.

5. U.S. Department of Labor, Occupational Safety and Health Administration, *OSHA Safety and Health Standards* (29 CFR 1926/1910) OSHA 2207, 1983. Washington, DC, 230 pages.

6. U.S. Department of Labor, Occupational Safety and Health Administration, *OSHA Safety and Health Standards* (29 CFR 1910) OSHA 2206, 1983. Washington, DC, 838 pages.

7. U.S. Department of Labor, Occupational Safety and Health Administration, *Industrial Hygiene Field Operations Manual;* Chapter V, April 10, 1979.

8. U.S. Department of Labor, National Institute for Occupational Safety and Health, Division of Technical Services. DHEW (NIOSH) Publication No. 78-193B, Cincinnati, October 1978.

9. *Office Safety,* American Federation of Government Employees, AFL-CIO. Washington, DC, 17 pages.

34

THE EXIT INTERVIEW AND TERMINATION ASSISTANCE

John J. McCooe and Arthur Kellner

JOHN J. McCOOE is president of McCooe & Associates, Inc. A graduate of Fordham University with a degree in Industrial Relations, he has been involved in manufacturing management and employee relations since his career's beginning with GTE, followed by ITT, and Bristol-Meyers. He has completed extensive postgraduate studies at New York University, Cornell University, and Syracuse University.

ARTHUR D. KELLNER is an industrial psychologist with comprehensive experience in managerial assessment and development. He has worked in Europe, Latin America, and other parts of the world in addition to the United States. Dr. Kellner's professional publications include articles on psychological assessment, employee attitudes, and selection interviewing techniques. For a number of years he was in the corporate management of a major international corporation where he was responsible for the corporate psychological assessment program and executive development activities. He is now a management consultant on human resource matters. He has a Ph.D. in Industrial Psychology from Case Western Reserve, is a member of the American Psychological Association, and is a Licensed Psychologist in New Jersey and New York.

34.1. THE CRUCIAL FINAL TOUCH

The exit interview is a special kind of meeting between two persons to exchange information. In that sense it is an interview just as much as a hiring interview. There are some essential differences in their interview approaches, however. For one thing, the exit interview deals with a situation that is inherently negative—a participant is leaving the company. The situation can lead to emotionality, misunderstandings, and, in some cases, outright hostility.

Usually the exit interview is conducted after the termination action has been initiated. If it is a voluntary termination, the person leaving has made the departure known to his or her superior. If it is an involuntary separation, the superior will have talked to the person about the termination. The exit interview is then conducted for these purposes: (1) clarification and reinforcement of the status of the terminee; (2) explanation and clarification of the terms of separation; (3) reduction of emotionality; (4) obtaining of information about the company, organizational environment of the terminee, and reasons for leaving, and (5) counseling employee on next course of action.

From these weighty objectives it can be seen that the exit interview is not easy to conduct. Here are some examples of termination situations where the exit interviewer played a key role in the process.

Example A

A unit general manager was being terminated from a large corporation because of poor performance. The president of the corporation, who was to inform the general manager of his termination, met with him for two hours. They talked about a wide range of topics—horses, football, and real estate. Only in the final stage of the discussion did the president obliquely refer to the general manager's departure. It was all very pleasant. In the exit interview with a personnel staff member later, the general manager said, "I *think* I've been fired." The personnel man said in a benevolent manner, "Joe, let there be no doubt; you have been fired."

Example B

A consultant was asked to exit interview an executive who was being terminated. In the course of the discussion the consultant learned that the man had just taken on extensive financial obligations—a new house and a boat—not anticipating such a sudden threat to his income level. The executive was 25 pounds overweight. He said he planned to go on a diet and begin jogging five miles per day. The exit interviewer, recognizing a possible stress situation, advised the man to see his doctor before undertaking this new regimen.

Example C

A personnel staff member sat in on a meeting in which an executive was informed of her dismissal. Subsequently the personnel man was asked to exit interview the executive and assist in her outplacement. The exit interview did not go well because the executive associated the personnel man with the "bad guys" who were terminating her. The static in the relationship made it impossible for the personnel man to render any real assistance. The executive left with an attitude of bitterness. She later sued the company and won. There was speculation that if the personnel man had not been in the dismissal meeting, he would have been more effective in the exit interview role. The lawsuit might not have occurred.

To achieve the objective of the exit interview, the person carrying it out must do it in a well-planned and skillful manner. Preparation is especially important. The interviewer should be thoroughly familiar with the background of the person's termination based on discussion with the supervisor and other persons who are involved. He or she should have information on the terms of the separation, including benefits and other personnel programs which affect the person.

In conducting the interview it is crucial to have privacy. Practical considerations are important and privacy is a fundamental requirement for a good interview. The interviewer should close his or her door and hold off on phone calls and other interruptions. The situation requires that the exit interviewer give complete attention to the terminee. Interviewing requires a concentration of effort with freedom from distractions. The idea is to provide a physical and psychological atmosphere in which a full exchange of information can take place.

To start out the exit interview, the interviewer should describe his or her role in the situation. He or she should explain why the conversation with the interviewee is taking place and what they will try to accomplish in the interview. In the early stages of the interview, the interviewee may want to do much of the talking, and the interviewer should encourage him or her to do so. Since one of the purposes of the exit interview is to obtain information on the interviewee's view of the organization, his or her talkativeness will provide such information which, hopefully, can be used constructively. Also, if the situation involves an involuntary termination, considerable emotion may be involved. In that case the interviewee's talking about his or her termination will help bleed off some of the negative feelings.

The loss of one's job can be a very traumatic occurrence. The job is a vital part of the person's life space. What does one tell the neighbors? How does one spend one's time? What does one do when getting up in the morning?

The loss of job is an embarrassment, and a person in this situation needs to preserve his or her ego. A number of defense mechanisms go into play. First, denial, "They're not really going to do that to me. They couldn't really mean it." Then projection, "It's not my fault. It's the boss's fault. It's the Marketing Department," and so on. Finally, anger and hostility can come into play and the exit interviewer needs to keep cool in dealing with such reactions. How? By allowing an interviewee to express and get his or her feelings out on the table. By so doing, the interviewee may reduce the emotionality and also gain better insight into the realities of the situation.

All temptations toward arguing or countering the assertions of the interviewee must be resisted. The exit interview is no time for debate or persuasion. A cool, supportive, nondirective approach will work better in helping the interviewee to gain a better understanding of himself or herself and the problem he or she is faced with. Here is an opportunity for constructive guidance, particularly in helping the interviewee to deal with the stress experienced upon losing a job and experienced during the transition into a new activity.

It helps the exit interviewer to be aware of these potent forces in the termination process. In one company, an exit interviewer was talking with a woman of some sixty years of age who was being offered early retirement. The offer, which was quite attractive, presented a financial conflict to her because, being a widow and living alone, she had no idea of what to do with her time after retirement. There was concern in the company that she might revert to alcohol abuse, an earlier problem, as a means of coping with this stress situation. The exit interviewer counseled her on this kind of coping mechanism and took some other steps, helping her to avoid slipping into such a behavior pattern without realizing it.

In handling these situations in the exit interview, it is essential for the interviewer to be a good listener. Listening is actually an active process. It takes effort and does not come naturally. Good listening means *understanding* what is being said; it is more than just hearing the words. That is where the effort comes in. Listening in the exit interview means putting oneself into the mind of the interviewee, feeling what he or she is feeling and seeing the world through his or her eyes, at least for the moment.

A supportive, encouraging posture always works best for a good exit interviewer.[1] He or she resists the urge to react or expound on his or her point of view, but rather tries to understand the other person and help that person to understand the present difficult situation. Listening carefully in the exit interview also helps the interviewee to be aware of what is going on—and there is a lot going on. Important feelings and attitudes are being expressed. They should be recognized and dealt with. The listener should recognize the full significance of what the interviewee is saying. The role is one of understanding and of helping through this understanding.

On the brighter side, the exit interviewer should review in detail the terms of the separation with the person leaving the organization. It is important that they have a good mutual understanding of the terms in order to avoid later problems. Some companies use check lists and termination agreements and ask the terminee to sign an agreement showing his or her acceptance of the terms. Frequently, when separatees fully understand the details of the separation package, they can realize that there might be positive aspects to the arrangement. Benefit programs can also be somewhat complicated, for example, retirement, medical coverage during severance period, and so forth. These programs should be explained carefully. The exit interviewer may wish to arrange another meeting with the appropriate staff specialist to cover such items. These specialists should restrict their discussion to their specialty, referring questions back to the exit interviewer.

34.2. OUTPLACEMENT ASSISTANCE

With persons who are leaving the organization involuntarily, the exit interview may be the last step before a review of outplacement actions. More and more firms are undertaking programs using their own staffs or consultants, to assist terminees in making the transition to a new career situation. In many cases the forced termination provides an occasion for the affected person to review his or her entire life situation, reconsider what he or she really wants to do in life, and find a new career which is more rewarding than the previous activity. Such changes are not easy to make, but are happening more and more often. It is helpful for the exit interviewer to be aware of these new attitudes and trends so as to deal with the terminee in a constructive manner.[2] Involuntary retirement may call for special counseling in this sense. Several major self-help books have tried to deal with these problems. They may be particularly difficult in cases where large numbers of employees are let go at once, especially in small labor markets.

34.3. GROUP TERMINATION

Closing out a major engineering project, branch office, or company where many are involved is significantly more complicated than the individual separation. The plan must take into consideration the company organization before and after the reduction in forces and ways for minimizing the broad impact on employees. One aspect that may, in a sense, be helpful is that such changes are often at least hinted at in advance, so that the often troublesome element of surprise may be lessened.

Beginning with the current organization, the manager reestablishes what the current departmental or company structure is, with all its functions. Manpower data concerning all employees and their positions should be assembled for company review. A new organization plan is made for each operation at its new or reduced levels, estimating staffing levels for the upcoming year. Managers of each function must clarify their operation to assure that "on the Monday after" business will continue to function normally.

Personnel policy and practice should be reviewed to ascertain how employees will be affected. Considerations such as normal or early retirement, bonus incentive to retire early, severance pay, vacation pay are evaluated. A determination is made after conferring with benefits experts, labor relations counsel, and communications staff as to which approaches are appropriate.

Communications with all employees—those remaining and those departing—are equally important for success. The community, family, individual, and remaining company are all affected. A decision must be made as to when and how the plan will be implemented. Drafts of communiques and announcements should be discussed and approved in advance.

A clear strategy must be set as to the manpower reduction and possible rehire program. Some considerations are: planned resignations, leaves of absence, individual termination, entire department terminations, promotions, and transfers to other divisions. Employees may consider the purchase of operations or companies.

The Group Termination program is implemented with trained personnel. Operating executives identify the approach, using internal human resources personnel, operating managers trained for this purpose, or consultants in outplacement.

A coordinator clarifies who actually will be terminated. This person is a focal point for action. The list is verified and approved by the senior executives. The goal is to eliminate any confusion, changes at the last minute, or worse, after the fact. Such events create a feeling of "They don't know what they are doing!"

The coordinator is prepared. All are then advised—employees, managers, customers, suppliers, government officials, and the press. The letter to employees and news releases are sent; the manager is prepared to work with human resources personnel and consultants. On the day the announcement is made, conference and meeting rooms, as well as private offices, have been readied for the meetings. The outplacement organization—internal staff and consultants—have discussed the program with management. So all are prepared, hand-out material is available. Benefit executives are briefed to meet with potential retirees. The outplacement program is underway.

Managers meet individually with each member of their group to advise them of the situation. They attend an orientation meeting, group training sessions, and then individual counseling sessions. The terminee has interview training sessions and is now prepared for the job search.

The coordinator advises other companies about the availability of the employees. Ads may be placed in local or national newspapers. This coordinator will assure continuous implementation of the program through to conclusion. The goals and achievements are continuously reviewed to assure communications are good for the individual, as well as the remaining organization. The Plan and its concerned expert execution are vital.

34.4. VOLUNTARY SEPARATIONS

This chapter has focussed mostly on involuntary termination. Not every exit interview involves such an action. In any organization there are voluntary resignations as well, and these terminations should also prompt an exit interview. The counseling aspects, while of special importance in involuntary terminations, are of less significance in exit interviews with persons who are resigning voluntarily. In this case more emphasis is placed upon using the opportunity to examine the reasons for the person's leaving the organization. Is his or her departure part of a pattern of employee unrest? What is the problem: Compensation levels? Supervisory practices? Company policies?

Many companies undertake active turnover control programs. In these programs they systematically obtain data on reasons for termination, look for patterns, and seek to make prompt correction of situations which are out of balance.

To conduct this kind of exit interview the interviewer should control the meeting. He should gently but firmly guide the conversation in the direction he or she wants so as to obtain useful information. This may require some moderately aggressive probing in some cases. The interviewer should look for specifics and not settle for broad generalities in the terminee's comments. He or she should ask follow-up questions to get at real issues: "Would you elaborate on that point, please? What makes you feel that way?"

The terminee may state that he or she is leaving only because of a better job offer. That response does not seem to leave much to talk about. However, there may be more information beneath the surface of such a statement. What is it that makes the new job more attractive? What is there about the current situation that made the employee susceptible to another offer or caused the employee to look outside the company? Chances are the interviewee will choose to say as little as possible in this situation; when one is moving to new pastures, why talk about the past? The best approach for the interviewer is to convince the interviewee that his or her comments will be helpful and that they will be kept in confidence. The interviewee should want to leave on a positive basis. This aspect should encourage cooperation. Once again a polite, probing approach is needed.

In doing such exit interviews, the interviewer should take notes. This step is important because one cannot recall later everything that has transpired in the interview. There is some concern that the note taking may inhibit the interviewee's self-expression. Experience shows this is not a problem. The interviewee will generally become accustomed to the situation rather quickly and become unconcerned with the fact that notes are being taken. Of course this process requires discretion on the part of the interviewer. If the interviewee says something of a sensitive nature, it may be wise to allow some time to go by before writing down that comment, so as to make the note-taking less obvious and threatening.

Upon completion of such an interview the interviewer should record the observations. This record could be on a standard form, if available, or perhaps in a memo. The important point is that the information provided by the terminee be evaluated along with other data as a basis for appropriate organizational decisions.

34.5.　GENERAL GUIDELINES

Clearly, there are different exit interview situations. With some differences in emphasis, the basic approaches are the same in all termination interviews. Let us summarize these approaches as follows:

1. Prepare for the exit interview. Become informed on the reasons and context for the termination. Review the terminee's personnel file. Do your homework on the separation terms. Be fully informed so you can explain the terms and reach an agreement with the terminee.

2. Conduct the interview in a proper setting. Assure privacy and freedom from interruptions. Direct your full attention to the interviewee. Establish a good basis for communication.

3. Let the interviewee do the talking when he or she feels inclined to do so. If appropriate, bleed off emotion and hostility. Be a good listener.

4. Be supportive, helpful, and positive. Avoid arguing or persuading. Do not counter negative statements by the interviewee. Rather, let the terminee gently come around to see more positive views on his or her own volition. Reinforce these positive views.

5. At the appropriate point in the interview explain the termination arrangements. Obtain the interviewee's understanding and agreement.

6. Be alert for signs of stress reactions. Help the interviewee to recognize

inadequate coping behavior. Guide him toward positive readjustment efforts.

7. In cases of voluntary separation, probe aggressively, if necessary, for information on reasons for leaving. Do not accept generalities; dig for specifics.

8. Keep notes on your interview. Save them for later reference. They could be important, especially if, as seems to occur more frequently nowadays, litigation results from the termination action.

9. Close the interview on a positive note. Try to avoid leaving unresolved issues hanging. Close the loop by obtaining additional information. If necessary, but only if absolutely necessary, arrange another meeting, or close out the discussions through a phone call.

Management is learning that employee termination of any kind, and especially involuntary separation, must be handled very carefully. It is to the company's advantage to have terminated personnel leave in a positive frame of mind. Also, information derived from exit interviews can be of value to the organization, an opportunity for reexamination, perhaps, of controversial issues. Employee turnover seems bound to occur for one reason or another, even in the most well-managed companies. When company representatives are called upon to interview out-going employees, they can render a valuable service by giving careful attention to the process. The exit interview should be more than a casual kiss-off. There is a lot going on in the exit interview situation and the effective interviewer is on top of it all the way.

NOTES

1. L. Zunni, *Contact—The First Four Minutes* (New York: Ballantine, 1975).
2. R. N. Bolles, *What Color Is Your Parachute?* (Berkeley, CA: Ten-Speed Press, 1984).

35

LABOR RELATIONS
IN ENGINEERING

George S. Roukis

GEORGE S. ROUKIS received his Ph.D. in History and Russian area studies from
New York University. He is a professor of management at the Hofstra University
School of Business and a professional labor arbitrator. He formerly served as United
States Deputy Assistant Secretary of Labor (1973–1975) and presently serves as a
neutral referee on the National Railroad Adjustment Board and numerous public
law boards in the railroad industry. He is a contract arbitrator in the United States
Postal industry. He has recently coedited a book entitled *Managing Terrorism: Strat-
egies For The Corporate Executive* and has published articles in *Labor Law Journal,
Journal of Collective Negotiations in the Public Sector,* and *The Arbitration Journal.*
He has settled over 1500 labor disputes and is a member of the National Academy of
Arbitrators.

35.1. INTRODUCTION

Labor relations for engineers have been characterized by two major concerns. The first is that of collective bargaining. Many engineers work in
manufacturing industries which are the center of industrial unionism.
Others are associated with the construction industry which is the traditional
focus of the craft unions. The actual and potential role of collective bargaining in engineering employment is therefore an important subject.

Second, the working conditions of engineers, in particular job security, are also involved in labor relations. By its nature, much of the large part of engineering that has to do with capital projects or equipment involves finite jobs with likelihood of job loss, unless some continuing activity can be found when the work is done. This problem has lent urgency to current changes and interpretations of the traditional employment at will, in which essentially no assurance of continued employment exists. These two problem areas form the subject matter of this chapter.

35.2.   LEGAL FOUNDATIONS OF COLLECTIVE BARGAINING

In the United States today, collective bargaining directly affects approximately 22 million union members and indirectly influences the compensation, terms, and conditions of employment of tens of millions of nonunionized employees. By comparison with the industrial relations of other countries, the American system is distinguished by its singular emphasis on business unionism. Simply stated, this means that American unions have, for the most part, sought specific organizational and economic objectives within the context of a capitalist economy rather than the broader ideological political goals of the socialist unions in Western Europe. This is not to assert that collective bargaining has been without conflicting viewpoints and emotions, which would certainly be at odds with our industrial relations history, but merely to point out that collective bargaining in the United States is pragmatic and business-oriented. In fact, the laws and regulations governing the process further define and reinforce this emphasis.

Employees are concerned with securing higher incomes, better working conditions, and importantly, in this day of rapid technological change, some form of job security. The doctrine of employment at will is not readily acceptable to most employees. Moreover, in a free society collective bargaining provides an opportunity for employees to satisfy their psychological and sociological needs. If the workplace can be structured to offer the employees greater control over their work, productivity can be increased. Collective bargaining is an instrumental process to this end. If employees perceive their condition to be poor or visibly less than comparable with their peers employed elsewhere, they will explore or seek some form of collective representation to improve their condition. It is usually these perceived inequities that precipitate union organization drives. Collective bargaining is the vehicle to right these inequities. The law regulates the conflict process.

During the past five decades a plethora of labor laws were enacted to address particular economic problems. The Railway Labor Act, 1926 as amended, which regulates collective bargaining in the railroad and airline industries, provided the legislative backdrop for the later passage of the National Labor Relations Act (NLRA) in 1935. The NLRA extended collective bargaining rights and administrative enforcement protection to all other covered workers engaged in interstate commerce. The National Labor Relations Board (NLRB) created by the NLRA is the administering federal agency. Engineers and scientists who seek collective bargaining status would petition this Agency to enforce their statutory rights. Section 7 of the amended act provides:

Employees shall have the right to self-organization, to form, join or assist labor organizations, to bargain collectively through representatives of their own choosing, and to engage in other concerted activities for the purpose of collective bargaining or other mutual aid or protection, and shall also have the right to refrain from any or all of such activities except to the extent that such right may be affected by an agreement requiring membership in a labor organization as a condition of employment as authorized in Section 8 (a).

This provision establishes the basic parameters of the collective bargaining process. The 1935 law was further amended by the Taft-Hartley Act in 1947 and the Landrum-Griffin Act in 1959. In 1974 the National Labor Relations Act was again amended to extend collective bargaining rights to health care workers. Postal workers are covered by the NLRA as well as professional football and baseball players.

The NLRA does not extend coverage to public sector workers, but many states have enacted comprehensive laws providing collective negotiating rights to public employees. In New York State, for example, the Public Employment Relations Act was passed in 1967 which extended collective negotiation rights to all public employees at the water district, fire district, school district, village, town, city, and state level.

In addition to the laws regulating the collective bargaining process, superminimum wage laws such as the Davis-Bacon Act (1931), Walsh-Healy Act (1935), and Service Contract Act (1965) were enacted to insure prevailing wage schedules for private-sector employees engaged in work for the federal government. These laws complement the collective bargaining process. In 1965, when President Lyndon B. Johnson issued Executive Order 11246, the federal contract procurement process was used to effectuate social goals. Private-sector contractors, as a condition precedent for federal contracts, were required to develop detailed affirmative action plans to insure more minority hiring. Executive Order 11375, issued in 1967, extended this mandated hiring requirement to women. Other major laws passed during the 1960s and 1970s were the Equal Pay Act (1963), Civil Rights Act (1964), Age Discrimination in Employment Act (1967), Occupational Safety and Health Act (1970), and the Employee Retirement Income Security Act

(1974). In total, there are about 137 federal labor laws and executive orders which regulate the employment relationship.

35.3. ENGINEERING UNIONS AND PROFESSIONAL SOCIETIES

While some mass-employed professionals like teachers have strongly espoused unionism, engineers have been much less receptive to the union message. For the most part, engineers have been indifferent or hostile to unions. Approximately 5.4% of the engineering profession are represented by unions and this figure is variable depending upon whom we include in the aggregate calculation.

The largest known independent union is the Professional Engineering Employees Association (SPEEA). The largest union affiliated with the AFL-CIO is the International Federation of Professional and Technical Engineers (IFPTE). Other unions would include the United Automobile Workers (UAW), Marine Engineers Benevolent Association (MEBA), Professional Engineers and Scientists Association (PESA), National Engineers and Professional Association (NEPA), and the International Association of Machinists and Aerospace Workers (IAMAW).[1]

Among the professional organizations, the National Society of Professional Engineers (NSPE), founded in 1934, has a prime lobbying function limited to the enhancement of the profession, but it is not a union. It neither engages in collective bargaining nor any of the other activities normally associated with NLRB-certified employee organizations. Initially, it began as a society of licensed or registered engineers, but it expanded its membership to include all degreed engineers, and now has about 95,000 members. Some of the other major professional engineering societies would include: the American Society of Civil Engineers (ASCE), the American Society of Mechanical Engineers (ASME), the American Institute of Industrial Engineers (AIIE), the Institute of Electrial and Electronic Engineers (IEEE), and the American Institute of Chemical Engineers, (AIChE). The IEEE is the largest of them, with approximately 160,000 members. Altogether there are about 28 member societies, 7 associate societies, and 8 regional societies.[2]

In recent years, however, the engineering profession has been faced with problems that cannot be settled by the traditional pronouncements of the Societies. Layoffs, erosion of income standards, lost pensions, and diminution of status have caused some engineers to look closely at their predicament. In large organizations where work may become less professional and a loss of autonomy is perceived, engineers might begin to see value in organized collective effort. If job insecurity is also viewed as a realistic contingency and older, more senior engineers feel slighted because newer, younger ones are offered more money, the impetus for collective action increases. It is further heightened when engineers believe their earlier training is less relevant to changing technology.

In response to this situation some professional societies have developed employment guidelines which set forth in labor contract language, terms and conditions of employment, salary and benefit scales, performance review standards, and grievance resolution procedures. These guidelines are not binding upon the employer and cannot be enforced by any process of arbitration or public judicial determination. The rationale underlying their advocacy was to reduce union sentiments and regularize the employment relationships of engineers. If engineers clearly understood the level of their entitlements and were provided at least in appearance, some modicum of job security, they would be less inclined to seek union representation. Employers, on the other hand, do not view the employment guidelines concept with the same degree of equanimity. Instead, they believe that guidelines would lead to unnecessary litigation and cause divisiveness within the organization. They argue that other nonunionized employees would view the guidelines as discriminatory and contrary to their interests and the professional societies might, by definition, take on the role of *de facto* employee bargaining representatives. By and large employers consider engineers as managerial personnel and beyond the pale of organized workers.

The role of professional societies in determining working conditions for engineers is further complicated by the fact that they are open both to rank and file engineers and to their managers. The latter have, in many societies, been prominent among the officers. Membership societies do, of course elect their officers by some process in which members have direct or indirect input, but usually there is a slate of candidates nominated by a committee of the current society board and members have a substantial communication problem in mounting a challenge. Thus, efforts at electing officers more sympathetic to concern over working conditions have run not only into electoral problems but into inevitable role conflicts among rank and file and managerial members.

35.4. ENVIRONMENTAL FACTORS AFFECTING UNIONIZATION

From the viewpoint of prospective employment for engineers, the 1980s were expected to be very good in the area of defense work. The Reagan Administration projected defense expenditures of about $1.8 trillion, with roughly half of that sum for weapons procurement, military construction and research and development. Such conditions may not be conducive to unionization, because times of expansion and prosperity tend to push the grievances that encourage it, into the background. However, many other technologically oriented industries in the United States are in a state of severe decline where the very opposite conditions prevail. In any case, even the defense projections are not likely to be realized, as Congress is unlikely to sanction military spending levels beyond inflation and may not even do that; that, at least, was the clear prospect that emerged from the 1985 Federal budget debates.

While defense expenditures will undoubtedly benefit the engineering profession, it is by no means certain that salaries, fringe benefits, and prerequisites will increase at a concomitant rate. A significant part of the defense budget will be spent in the Sunbelt area where salaries and benefits have been traditionally lower. Unionization is weak or nonexistent in the Sunbelt states, but this will not preclude discontent if professional engineers see that their peers in the Northeast quadrant are earning more money. If technicians who are more likely to be union members are also enjoying better compensation and benefit packages, the discontent will be more pronounced.

By 1990 the proportion of the labor force between 25 and 45 years old will have risen to more than half (65 out of 118 million), compared with one third in 1975. This will magnify the concerns over career blockage and related problems which is traditionally strongest in this group. Moreover, professional and technical employees currently make up about one seventh of the total labor force. By 1990 they will comprise one sixth.

As more of these individuals who work in hierarchical organizations find their career paths blocked, and their work less professionally independent, they might be more inclined to experiment with unionization. This will be evident if unionized workers in their firms or in contiguous areas earn more money, and are perceived as having more control over their work. The problem will further be compounded by the growing number of minorities and women in the workforce. Role conflict will increase as employees adjust to new work relationships. A diminished sense of image could increase prounion feeling. Such tactics as recasting of the workplace to comport with Japanese management techniques might not be enough to allay such fears and sentiments.

Employers will vigorously contest unionization attempts since the historical record shows that management usually prevails in union certification elections. By arguing that professional engineers are managers and associating unions with bossism and other negative connotations, employers have been successful in defusing enthusiasm for unions. But there are areas of vulnerability that might induce engineers to consider the union alternative.

Under the Fair Labor Standards Act engineers are considered managers, but this status becomes questionable under actual job conditions. For instance, under the Act, engineers may not be entitled to overtime pay, and may thus receive less than the drafters who work for them and who do get extra pay. They may be unionized either independently or as part of the principal union in the establishment. Engineers may also view negatively inequities in performance evaluations and work assignments. The company may have disparate *ad hoc* grievance procedures that produce contradictory and unfair results. These employment problems, whether real or extensively perceived, may dispose employees and engineers, in particular, to attempt unionization.

In the absence of collective bargaining, resort to the courts is an expensive method to enforce individually asserted statutory or constitutional rights. The protective machinery provided by a collective bargaining agreement, or the enforcement bureaucracy of the Equal Employment Opportunity Commission and the United States Department of Labor, is not available to the individual engineer. In many respects, principled and competent engineers have problems in protecting their positions. If they are justifiably dissatisfied with the work product or the technological environment, they could be summarily fired or disciplined for their actions. This is by no means the normative pattern, but engineers have been dismissed for well-intentioned whistle blowing. It should be noted, however, that legislation and court decisions have considerably expanded the latitude allowed whistle-blowers when they report legal violations by their employers (including government agencies) or refuse to carry our orders that violate public policy. By 1985, laws specifically protecting them were in force in 21 states, with other states considering them seriously.[3] This topic will be further addressed in Section 35.8.

35.5. RIGHTS UNDER UNIONIZATION AND APPLICABLE PROTECTIVE LEGISLATION

The small percentage of engineers represented by a collective bargaining agent does not provide the data base needed to accurately determine the efficacy of unionization. If engineers are satisfied with their employment conditions they will obviously be less disposed to seek a bargaining representative. Such contentment is widely found, but added protection is offered by such provisions as those in the Age Discrimination in Employment Act and by those governing labor organizing itself.

Under the National Labor Relations Act, for example, an employer is prohibited from interfering with the rights guaranteed in Section 7 (*supra*). If an employer attempts to dominate or interfere with the formation or administration of a labor organization, it is considered an unfair labor practice.

If the employer discriminates with respect to hire and tenure, encourages or discourages membership in any labor organization, or discharges or discriminates against an employee who has filed charges or given testimony under the Act, such behavior is also considered improper and illegal. It is also an unfair labor practice for the employer to refuse to bargain collectively with the representative agency designated or selected by the majority of employees in the bargaining unit. These rights are spelled out in Section 8 (a) (1)-(5) of the NLRA.

In 1947, when the Act was modified by the Taft-Hartley amendments, employee organizations were enjoined from committing specifically defined union unfair labor practices. These prohibited practices were delineated in Section 8 (a) (1)-(7) of the Act. A union was barred from coercing an employee from exercising his or her rights granted by Section 7. A union was further barred from discriminating or forcing an employer to discriminate against an employee under a union shop contract for any reason other than nonpayment of uniform dues or membership fees. Thus if an employee was required to join a union after a specified period of employment and did not pay these fees and dues, he or she could be lawfully discharged. The labor agreement would provide a special adjudicative procedure to insure that the employee was properly removed from the payroll, but the employee could not refuse to join the union where a union shop was in place. A union was prohibited from exacting excessive or discriminatory fees and dues under a union shop agreement and from causing or attempting to cause an employer to pay or deliver any money or other thing of value for services not performed. In 1959 the internal affairs of unions were more fully regulated when the Landrum-Griffin Act was enacted. Collective bargaining has become part of the nation's institutional fabric, but it is regulated to comport with the needs of the public interest. The right of employees to organize unions and bargain collectively has not been denied or abridged, but it is not an unfettered right either. The law imposes rights and obligations upon employers and employee organizations alike. These rights are further defined when the parties agree on the terms and conditions of employment governing their collective relationship.

In the case of *Local 368, United Federation of Engineers, International Union of Electrical, Radio and Machine Workers, AFL-CIO v. Western Electric,* the United States District Court in New Jersey granted the employer's motion to dismiss the union's claim. The union argued that the employer had violated the controlling labor agreement when it laid off eleven engineers and replaced them with other personnel. The company argued that the separations were not subject to the grievance arbitration procedures of the agreement. The questions posed were procedural and substantive: (1) Procedural—Did the union fully abide by the contract's grievance procedures? and (2) Substantive—Were the layoffs covered by the labor agreement?

In deciding this issue, the federal court held that the employee organization did not exhaust fully the grievance arbitration procedures of the applicable collective agreement, and consequently was barred from initiating a court claim. The exhaustion of remedies doctrine was applied. If the union had completely utilized the agreement's grievance process, the Court would have considered the dispute on its merits. This, of course, does not mean that the Court would have ruled in favor of the petitioning organization, merely that the claim's substance would have been carefully considered. If the type of layoff implemented in this instance was unprotected by the agreement, the court would have ruled for the employer unless the layoff violated the provisions of the Age Discrimination in Employment Act or Title VII of the Civil Rights Act of 1964. In any event, whatever the disposition would have been, it is important to note that a properly written agreement can protect employees from unreasonable employer actions. Job security provisions are permissible subjects of bargaining.[4]

In *Engineers v. General Dynamics Corporation,* the National Labor Relations Board was asked to determine the composition of the prospective bargaining unit. By law, a labor organization may only act as the legitimate bargaining agent if it represents a majority of the employees in a bargaining unit which the NLRB considers appropriate. In order to secure a certification election, it must demonstrate by verifiable evidence that it represents a "substantial" number of the affected employees. The NLRB has construed "substantial" to mean at least 30%. In considering the question of "appropriateness" the Board must consider such factors as any past bargaining history, bargaining custom in the industry, interests and desires of employees, interchangeability of employees, and organization and representation of employees. Thus, if a particular grouping of employees constitutes an appropriate unit, such group may properly constitute an appropriate unit, even though they might be more appropriately represented in a more inclusive unit. Managerial and supervisory personnel would be excluded from the unit. Professional employees, under Section 9(b)(1) of the Act, have the right to vote separately on their desires for representation.

In *Engineers v. General Dynamics Corporation,* the parties were unable to agree on what constituted an appropriate bargaining unit. The issue posed was whether the employer's permanent line or functional organization was the sole determinant regarding employee inclusions or exclusions or whether the employer's ever-changing project-oriented organization governed the structure of the unit. The NLRB considered the more valid type of organizational structure as the pertinent measure which recognized implicitly the possibility of overlapping roles. In an engineering environment where job security is a paramount concern, engineers who perform periodically managerial functions and then exclusively professional functions may find organizing a union a trying task. An organic, highly flexible organization would represent this type of environment. In organizations where engineering roles are more staid and routine, organizing should not be as difficult. Defining roles and functions is easier.[5]

Similarly, engineers who feel discriminated against because of age considerations are protected by the Age Discrimination in Employment Act. The Act could be invoked when an older engineer is bypassed for promotion on the basis of asserted skills obsolescence, or the belief that the employer prefers a younger worker for a reason unrelated to competency and merit. Like sex discrimination, age discrimination is not readily admitted by employers, but must be proven by a close examination of personnel policies and decisions which favor one group over another. The burden of proving that a personnel policy is neutral in effect is on the employer. In *A.P. Wilson v. Sealtest Foods Divisions of Kraftco Corporation* the federal circuit appeals court held that an age discrimination claim was improperly dismissed when the petitioner had shown that as a member of a protected class, he was entitled to the protection of the statute. The person was able to demonstrate that he was doing satisfactory work when his employer asked him to retire early while replacing him with a younger employee.[6] In the *Friendly Ice Cream Corporation* case affecting 380 restaurants in six New England states, the employer agreed it would affirmatively recruit older workers and expand its efforts to recruit older applicants for management training programs. The employer also agreed to pay $40,000 in lost wages to the persons who had suffered from the initial discriminatory act.[7] In two cases involving involuntary retirement and layoffs, which older engineers are prone to experience, the employers agreed to reinstatement and back pay remedies when the evidence palpably showed that protected older workers were disproportionately laid off, retired, or assigned to inactive status because of age. Pan American Airways agreed to pay 29 former employees $250,000 in damages, while Standard Oil Company of California paid approximately $2 million to 160 adversely affected persons.[8]

The Act provides an exception from its provisions when there are bona fide retirement plans that are not a subterfuge to evade the manifest purposes of the statute. An employee may be required to take early retirement, that is, before age 70, if he or she is a participant in a qualified, retirement plan that provides for it.

The Act places rigorous administrative duties upon the employer. The employer must keep for three years payroll or other records containing data on each individual employed. This information includes the name, address, date of birth, occupation, compensatory rate, and amount of pay earned each week by the employee. Additionally, the employer must keep for a period of one year records relating to hiring, promotion, demotion, transfer, selection of training, layoff, recall or discharge, job orders submitted to an employment agency or union, advertisements or notices to the public or to employees relating to job openings, training programs, and promotions or overtime opportunities. Copies of seniority or merit-system plans must be kept for the full time the plan or system is operational and for at least one year after its termination.

In 1978 enforcement of the Age Discrimination in Employment Act was transferred from the Wage-Hour Division, Employment Standards Administration, United States Department of Labor to the Federal Equal Employment Opportunity Commission.

35.6.   EMPLOYER RESPONSES TO ENGINEERING UNIONS

In a manner reminiscent of teachers in the 1950s and 1960s, the emphasis on professionalism is pronounced among engineers. Many engineers have opposed unionism because they believe that the creative qualities of the profession cannot be transferred to a union milieu. In the 1960s when blue-collar workers were earning more money than teachers, the education profession could not see how the factory method of employee relations could be transferred to the classroom. According to educators, the individual teacher was not a production worker inextricably tied to an integrated manufacturing system; the teacher was an independent professional solely in control of his or her work environment and accountable to the profession. Notwithstanding the bureaucratized nature of the usual school system which integrates organizationally various levels of educational goals and strategies, the education profession clung to the view that it was akin to the individualistic professions. But teachers are mass employed like nurses, social workers, and engineers. The percentage of self-employed engineers amounts to 3% of the profession as compared to 60% for physicians and 59% for attorneys. The level of remuneration for these other professions is much higher than for engineers.[9]

Interestingly, engineers in the private sector are not implacably opposed to unionism *per se;* many of them believe that unions could give them more say in their work. Their opposition centers on their strongly felt perception that a union would gain excessive power at the expense of the engineer's professional status. Engineers recognize that improvements are needed in pension benefits, salary, and employment security, but they believe that loss of professional image and individualism is not worth the price of unionization.

Employers have not been slow to grasp this perception and have designed their personnel policies to accommodate the engineer's mind set. Creating a work environment that reinforces the engineer's frontier spirit is the best insurance against unionization initiatives.

During the period of 1968–1980, however, it appeared that the situation might change when the defense outlooks, rising unemployment, and high inflation wreaked havoc in the technology-based industries. The sense of security fostered by the full-employment days of 1965–1968 was now replaced by feelings of isolation and helplessness, and engineers in some quarters were more receptive to the idea of unionism. As engineers felt isolated from the corporate decision-making process which affected their lives and their work became more routine, or at least appeared to be so, they began to reevaluate the question of unionism.

Mindful of the engineers' traditional distrust of unions, employers aggressively emphasized the negative aspects of unionization. By arguing that a union would create a barrier between management and engineers and contending that engineers would lose professional status, employers were frequently successful in defeating union organizing drives.

In the 18 representation elections conducted under the National Labor Relations Board between 1968 and 1980, involving private-sector bargaining units of over 100 engineers, the engineers' unions lost 13 times.[10]

These results are not surprising since employers devised strategies that were in tune with the engineers' latent feelings. In fact, where an election resulted in a bargaining agency being certified, the employer took an extremely militant position at the bargaining table. The National Engineers Professional Association (NEPA) was decertified in 1976 at the General Dynamics, Convair Division when it could not negotiate a collective agreement. The union, unable to enforce its demands with the prospect of a successful strike, was powerless to conduct effective negotiations. The membership would not commit itself to direct action and this unwillingness to act aggressively irretrievably weakened its bargaining posture. The Convair Division had the ability to maintain operations, despite a strike and the engineers knew this strength.

Nor is an evocation of professionalism the only argument employers use.

In the high technology industries where the Japanese have shown a remarkable ability to develop quality products, employers are quick to point out that an adversarial work environment will help foreign competitors. However, perhaps the most potent argument lies in the working conditions which employers create for their engineers (and other employees, for that matter). If they enhance job security and working conditions that are conducive to creative work, it will be exceedingly difficult to convince engineers that a bargaining agent is needed.

Correlatively, the argument that engineers are managers, and are not covered by the protective provisions of the National Labor Relations Act (NLRA), has been given added support by the United States Supreme Court's decision in *Yeshiva University v. Roth* (1980). In that case the Supreme Court decided that professors at Yeshiva University, a private institution, were managerial employees and hence outside the ambit of the NLRA. The court defined managerial employees as those persons who formulate and effectuate management policy by expressing and making operative the decisions of the employer, and those persons who have discretion in the performance of their duties independent of their employer's established policy. While the Court went on to explain why managerial status is conferred upon persons in executive-type positions, it also pointedly noted that professional status by itself is not routinely equated with managerial status.

However, since professors at Yeshiva University were found to be part of the decision and implementing process, the Court reasoned that they were managerial employees. If this decision is broadly construed by employing institutions other than colleges and universities, it could confuse and impede any attempts to organize engineers. Employers can simply argue that engineers function in the same type of collegial decision-making environment as professors. However, employers will have to demonstrate by convincing organizational detail that engineers are truly managers. Incidental managerial functions will not be persuasive and the burden of proof will devolve upon the employers. This means that to be successful and consistent with the criteria in the *Yeshiva* decision, employers will have to organize work so that engineers are substantially performing managerial duties. Effectuating such outcomes will require demanding organizational analysis.

35.7. COLLECTIVE BARGAINING PROCESS

In spite of these actual and potential impediments, it is likely that union sentiment among engineers will increase in the late 1980s and early 1990s. As previously indicated, the changing composition of the workforce and the rapid changes in technology-based industries will create pressures that will force engineers to reassess their condition. While the organized portion of the labor force has experienced setbacks in recent years, including the use of bankruptcy proceedings by such firms as Continental Airlines, Johns Manville Corporation, and Bildisco in order to abrogate existing labor contracts, the pendulum will swing as the economy improves. Settlements will not be as high as the 1960s, but employees will be provided a greater degree of employment protection. Of course, this does not imply that guaranteed lifetime employment will be the norm, only that employers will undertake additional efforts to retrain and help relocate employees whose jobs cannot be saved.

The 1983 AT&T Agreement is a case in point. The wage increases for the

Communication Workers of America (CWU) amounted to 16.4% over a 3-year contract period, but the contract charted new ground in job security. It is not a milestone contract, but it reflects an awareness of competitive realities. Employees whose positions are affected by the introduction of new technology will be provided with retraining at the company's expense. This benefit will not accommodate the needs of employees nearing retirement age, but it will lessen the unemployment effect for younger employees whose careers might otherwise be disrupted.

If engineers believe that their salaries and job statuses are out of line with comparable professionals, or that unionized technicians are making disproportionate gains, they might well consider unionization as a reasonable response. In addition, if the career path to higher managerial levels is blocked because of the growing number of qualified applicants, a substitute mechanism for the coveted benefits might be sought. As the evidence indicates, it will not be easy to organize engineering unions, especially when the professional societies eschew this form of organization, but if unionization is successful in some of the more prestigious firms, the focus of bargaining will have to be on matters that heretofore were not considered important by traditional union negotiators. Engineering unions will have to formulate bargaining demands that address such concerns as professional standards, ethical codes, education, performance evaluation, and professionalism, while at the same time try to reconcile their differences with the societies. If they are able to hold their memberships together and convince them that militancy is the *sine qua non* of unity, they might be able to weather the storm.

Employers will argue that engineers are not blue-collar workers, but managerial employees led astray by outside union bosses, and if employers can further discern the sources of discontent by the use of survey questionnaires, they might be able to improve substantially the engineers' task environment. This might be sufficient to defeat the unionization campaigns, particularly if the unionization drive is in the beginning stage. It will not be a pleasant encounter but labor-management relations are seldom a symbiotic process.

If a union is certified to represent an engineering bargaining unit, the structure of the employment relationship will change significantly. Both sides will develop short- and long-term strategies to maximize their positions. Management will be forced to modify its methods of supervision, while the employee organization learns the importance of militance. Whatever naiveté either side had when it formulated its bargaining position will soon be dissipated at the first bargaining session. The parties will not disagree on the necessity of maintaining professional standards; they will disagree on the methods and costs of achieving them. Inventiveness and aggressiveness are premium qualities in this power relationship. Employers will fight hard to retain broad managerial rights while employees will fight hard to eliminate or minimize any threat to their job security or professional status. It has chess game qualities which should appeal to the engineers' analytical mind. It requires the transforming type of leadership that shapes the parties' collective relationship well into the future. A poorly negotiated contract cannot be undone during the next round of negotiations. It could take 5–10 years to reverse the damage. If either the union or the employer conducts collective bargaining without thinking through the strategic requirements of their relationship, they will be disadvantaged. Neither the threat of a strike nor a lockout will do any good in these circumstances. A handicapped position cannot be promptly corrected by direct-action threats. It requires in the first instance that the negotiators develop a thoughtful strategic plan of action.

Seasoned negotiators have learned to structure their bargaining tactics so as to secure the best gains possible, while at the same time, subtly changing the attitudes and perception of their opposition. By way of illustration, an employee group not interested in a particular fringe benefit agrees to appoint two of its representatives to a joint labor-management study committee. The idea at first seems harmless enough since there is no commitment to concede this benefit. But as time unfolds, the benefit seems less threatening. If the union was forcefully pushing another fringe benefit which was low on its priority objectives, it might withdraw this benefit in exchange for the benefit being studied by the joint labor-management committee. The benefit does not appear unreasonable now that it is better understood, and the union, at least in appearance, has withdrawn an item that raised tempers at the bargaining table. The employer has been persuaded that it won a major victory while the union obtained the fringe benefit it really wanted. An opposite result can be obtained if the employer's approach is structured analogously.

In preparing for bargaining, data regarding the employer's workforce in terms of sex, age, seniority, shift, time, and position classification, is needed in order to avoid adverse employment impacts. The Courts will overrule contract provisions which are discriminatory. Pay and benefit entitlements must be carefully examined within the context of the industry's standard and the rates and benefits provided by competitors. Internal economic data such as overtime costs, discretionary benefit plans, and earning levels are significant measures since the employer would prefer to negotiate benefits that are more widely distributed. Engineers would prefer, conversely, agreements that are consonant with the profession. By definition, they would not like to be classified as part of the employer's workforce. Different proposals must be financially calculated under various bargaining outcomes to ascertain the probable cost of an agreement. It is important for both parties to make rational choices that are realistically affordable. In most labor contracts where salary, wage, and benefit provisions are in effect, some or all of the following benefits are normally accorded covered employees: life insurance, hospital and surgical insurance, major medical insurance, disability insurance, dental care, retirement, severance pay, deferred compensatory increases, cost-of-living adjustments, and incentive plans. Usually, in determining whether these benefits are justified, the parties will consider the firm's economic status and the going rate among comparable firms in the relevant geographical area. Assessing the true significance of the data is often itself a matter of controversy.

Grievance procedures are also of critical importance to unions. A grievance arbitration process allows the union to challenge unilateral management actions that appear at variance with the meaning and application of a contract provision. The procedures set forth in a hierarchical appeals format the persons responsible for considering the grievance and the specific time limits at each step before a determination is made. If the aggrieved employee's claim is denied at the final stage of the appeals process, he or she has the contractual option to appeal it further to a third-party neutral arbitrator. In many cases where an employee feels that discipline has been assessed unjustly, the grievance arbitration process provides the forum for determining whether the suspension or discharge was justly imposed. In a nonunion setting, the discipline will be routinely enforced unless the person affected is legitimately protected by one of the employment discrimination laws.

In Japan, for example, grievances are minimized or mooted early because employers recognize the importance of managerial accessibility. An open-door policy is followed which permits aggrieved employees to take a real or perceived problem directly to higher-level management. This reduces the employee's disquiet, since the problem is dealt with promptly and by a managerial decision maker who has the full authority to resolve the dispute. Whether the employment setting is union or nonunion, however, the key to a good working relationship is a system of open and direct communications.

The negotiated labor agreement now governs the employment relationship and it must be administered in accordance with the intended objectives of the negotiators. The employer as a rule will provide an orientation seminar to its management personnel to acquaint them with the details and purported meaning of each contract provision. The union will provide the same type of training to its constituency. This is an important task which must be done properly, otherwise one side or the other will take unfair advantage of the situation. If they are unable to agree on the appropriate interpretation and/or application of a contested provision, it can be appealed to arbitration. The arbitrator's decision is binding on the parties. Moreover, the employer will have to make policy decisions regarding its political relationship with the union. Should it cooperate with the union on matters of equal concern or should it take independent action? Should it try to build a pragmatic cooperative relationship with the union, or should it remain distant and combative? These are important policy questions which affect the foundation and structure of the collective relationship. If the employer is intent on defeating the union in a decertification election, it will take a hardnosed, uncooperative stance. If it succeeds in this endeavor, it will be union-free, but if it loses and loses significantly, it will be faced with militant organization.

It is too early to determine with any degree of accuracy the future configuration of unionization in the engineering profession. If the past can be used as a precedent, and no unusual changes occur, then the engineer's traditional frontier spirit will preclude any meaningful union growth. If engineers in large organizations perceive their roles as part of a mass-employed entity, they might be persuaded to seek union representation. This will be more evident if they experience career blockage and feel that their mobility is limited by their age, or tenure, or by managerial indifference to, or exploitation of technological change as it affects engineering jobs. Since employers will not acquiesce in unionization efforts, the prospect is for significant conflict, the outcome of which may be affected by careful consideration of the factors previously enumerated.

35.8. EMPLOYMENT AT WILL

Traditionally, the law has held that an employer may terminate an employee for any reason, or at times, for no reason, as long as no law or contract has been violated. For many employees, including nonunionized engineers at all levels, instant dismissal is an ever present workplace reality. A notable example occurred when Lee Iacocca, who was president of Ford Motor Company, was fired by Chairman Henry Ford II on July 13, 1978, but the chance of instant job loss is a reality for lower ranks as well, perhaps indeed much more so. The rationale of the employment at will doctrine has a long historical genesis and was always the inherent prerogative of the employer, but it was judicially refined and extended to meet the needs of an industrial society.

Under the law of contracts, the parties to an agreement were held to have accepted certain legal obligations that included the right to quit a job or be terminated at will by the employer. It was a reciprocal, although independent, decisional process that devolved equally upon both the employer and employee. The courts took the position that the parties entered into the contractual employment relationship with full knowledge of the unilateral freedoms permitted. Thus, contrary to the Implied In Law Theory that developed earlier in England and enjoined employers from firing employees where an indefinite hiring relationship existed, the judicial emphasis in industrializing 19th century America significantly limited the employers' duties to their employees. In the absence of a fixed-term employment contract, the employee could freely terminate an employee at will. The longer job tenure that was normally associated with the cyclical vagaries of agriculture were not present in the mass production environment. Replacement employees were now always available. In the 1890s, less than half of all American employees were wage and salary workers. By 1970 the figure had risen to 90%. It is estimated that about 75% of the employees in the workforce today are subject to instant dismissal, but this inherent management prerogative is not absolute.

Since passage of the National Labor Relations Act in 1935, employees involved in union organizing campaigns were protected from employer interference. The ambit of this protection has been thoughtfully articulated and extended by several thousand National Labor Relations Board decisions. The equal employment opportunity laws passed in the 1960s and 1970s extended protection to other segments of the working population and laws such as the Occupational Safety and Health Act (1971) extended employment protection to employees summarily dismissed for refusing to work in unsafe or unhealthful working conditions.

In recent years the Public Policy Exception Doctrine has been frequently used to protect employees from unjust discharge where the employee acted in the public interest, as previously noted in our discussion of whistle-blowing. Similarly, it would contravene public policy, if an employer fired an employee for serving on a jury. Courts have ruled that firing an employee for jury service would thwart the public interest. It would also violate public policy if an employee were fired for providing information to law enforcement authorities about the possible criminal conduct of another employee.

A derivative doctrine, the good faith and fair dealing exception, provides that irrespective of what an employer says or does to make employment at will a policy, the employer must deal with the employee fairly and in good faith. California, Massachusetts, and Montana, have enacted good faith and fair dealing legislation. In other states the courts have held that when an employer hires a person for an indefinite period and agrees that termination will occur only for good or just cause, the agreement is enforceable. The Supreme Court of Massachusetts in *Fortune v. National Cash Register Company* ruled that an implied covenant of good faith and fair dealing existed where an employee with 24 years of service was discharged so that the employer could avoid paying commissions.

Nevertheless, even under the most liberal interpretation of the employment at will doctrine, the recognized exceptions apply only to a small fraction of the roughly three million employees fired yearly. For those who are members of unions, the grievance arbitration provision of a collective bargaining agreement require just cause for employment termination. In a nonunion setting and particularly for engineers, it would be wise policy for employers to provide standards and codes that protect employees from capricious dismissal. Work standards that are properly drafted enable employers to terminate employees pursuant to definable dismissal criteria and this insures consistency and fairness.

As a rule of thumb, employers should not promulgate rules that threaten dismissal for every type of workplace infraction or contingency. Discipline should be corrective and commensurate with the nature of the offense. Employers should conduct regular and systematic evaluations. Troublesome or incompetent employees should be identified early in the employment relationship to avoid acrimony, hardships, and litigation in the future. Counseling should be an integral part of the employee-management process and reports should be circulated periodically to supervisors and managers.

Many employers have found it beneficial to set up internal grievance procedures and some companies have established complaint and dispute resolution systems that incorporate binding arbitration. Grievance procedures would follow a hierarchical time delimited appeals process.

Step 1. Complaint to immediate supervisor
Step 2. Appeal to next level of management
Step 3. Appeal to vice-president of Personnel, or other equivalent senior officer

Addressing complaints and encouraging employees to use the complaint system not only enhances morale, but it also eliminates or reduces needless conflict. Employers are not barred from terminating employees because of bona fide business reasons such as retrenchment, reorganization, mergers, and so on but peremptory, whimsical dismissals should not reflect abuse of managerial discretion.

An increasing political and legal concern with the workings of employment at will doctrine has brought the issue to the forefront of employment practice and it is likely to remain there. In partial response, there have been books on how to avoid trouble,[11] notably by making disclaimers of employer responsibility to employees as legally airtight as possible. Yet employers should have learned from long experience, dating back at least to the demise of the doctrine of *caveat emptor,* that disclaimers may not stand the test of time. Court decisions eroding employment at will have now been handed down in 30 states, and unjust-dismissal legislation is being considered in several states. Thus just cause for dismissal may be a widespread legal requirement before long.[12]

Finally, whatever the legal ramifications and their current process of change, telling an employee at time of hiring or later that the employer in effect owes him or her nothing, casts a chill on the relationship which is bound to call for a response in kind whenever opportunity offers. It is thus clear that whether in relation to unions or general conditions of employment, the engineering industries face considerable problems which are not made easier by the business difficulties that many of them face. An awareness of the issues is thus clearly essential for engineering managers.

NOTES

1. G.W. Latta, "Union Organization Among Engineers: A Current Assessment," *Industrial and Labor Relations Review* (October 1981): 31–33.

2. E. B. Hoffman, *Unionization of Professional Societies,* (New York: The Conference Board Inc., 1976), pp. 4–6.

3. D. Chamon, "Beyond Unions," *Business Week* (July 8, 1985): 73.

4. "Local 368 United Federation of Engineers, International Union of Electrical, Radio, and Machine Workers, AFL-CIO v. Western Electric Company," *Federal Supplement* **359** (St. Paul: West Publishing Company, 1973), pp. 651–652.

5. "General Dynamics Corporation Decision of the National Labor Relations Board," *Labor Relations Reference Manual* **87** (Washington, DC: Bureau of National Affairs, 1975), pp. 1717–1726.

6. A.P. Wilson v. Sealtest Food Division of Kraftco Corporation, 501 F.2d 84 (CA–4, 1974).

7. G.S. Roukis, "Protecting Workers' Civil Rights: Equality in the Workplace," *Labor Law Journal* (January 1975): 11.

8. *Ibid.,* 12.

9. Latta, *op.cit.,* 31.

10. *Ibid.*

11. J. Barbash and J.B. Kauff, eds., *Unjust Dismissal 1983: Litigating, Settling and Avoiding Claims,* (New York: Practicing Law Institute 1983).

12. "Beyond Unions," *op.cit.,* pp. 73, 77.

MANAGING
YOUR CAREER

36

SELF-ASSESSMENT
FOR THE ENGINEERING
MANAGER

Marc A. Dorio

MARC A. DORIO is vice-president of McCooe & Associates, a management consulting firm located in Ridgewood, New Jersey. Mr. Dorio has had significant experience in human resources, consulting with major corporations both here and abroad. His areas of expertise include management development and training and career development. Mr. Dorio holds master degrees in both Divinity and Theology, which he received from St. Bernards College in Rochester, New York and a Master of Science Degree in Industrial Organizational Psychology from Stevens Institute of Technology in Hoboken, New Jersey. He has written articles on a variety of management topics. Mr. Dorio is also Professor of Human Resources in the Business department of Bloomfield College in Bloomfield, New Jersey where he teaches supervision, personnel management, and training and development.

36.1. INTRODUCTION

Engineering is a demanding profession requiring of its practitioners a unique combination of exactitude, discipline, and creativity, as well as a curious mix of both technical and human relations skills. It is in the present order of things an art as much as it is a science. This unique mixture of diverse and, at times, contradictory requirements is a constant challenge and demands of the professional engineer a continuing need for self assessment (s.a.) in order to improve growth, personally, professionally, and as a manager.

Furthermore, engineers of today are faced with the quickening pace of technological development which places demands on them in order to avoid professional obsolescence. At the same time the loner image of the past is gone. More and more the engineer is required to exercise technical skills, managing others in a team effort on a variety of projects. Therefore, not only is technology necessary, but the engineer must possess a keen awareness of interpersonal skills.

This continuing need for s.a., if dealt with effectively, can be the most valuable instrument in the engineer's toolbox. To a large extent, in today's world career success depends upon the ability to identify needs, solve problems, and react quickly. This same principle must be applied to personal self-development.

If you are fortunate enough to work for a company either here or abroad which has a career development and s.a. program, consider yourself fortunate. If, however, this is not your situation, then you must take responsibility yourself. The s.a. process proposed in this Chapter may be just what you need.

36.2. THE PERSONAL SELF-ASSESSMENT PROCESS

This process is a systematic step-by-step do-it-yourself self-assessment program utilizing a project management approach. Its goal is to enable you to identify needs for self-improvement and to take concrete action in the form of an action plan. The flow chart in Figure 36.1 outlines the steps of this program. It begins with the assumption that each one of us every day is constantly bombarded with feedback. This feedback comes in many shapes and forms and is often given to us whether we like it or not. Our children, our spouse, friends, bosses and coworkers are an infinite source to us and provide us with the data which makes the feedback process go. As well as these external sources of feedback, we also have the internal mechanism of the so-called "little voice" inside us which is also our critic and tells us things

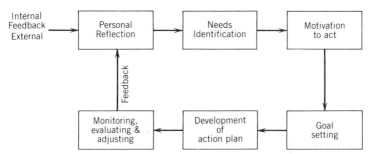

FIGURE 36.1.   Six-step self-assessment process.

like whether the presentation was successful, whether we like or dislike something, or how we are coming across in a given situation. Feedback both internal and external is constantly with us and it can be for us the starting point for s.a.

Step 1: Personal Reflection.   Feedback in and of itself is meaningless without personal reflection. If we do not take time to hear and understand what is being said to us and about us, feedback will have no effect. If we do not respond to its clues, it is as if someone is speaking to us while we are absorbed in another task. If, however, we take time to listen and to focus our attention on the feedback in our environment, then and only then can it become the first step in a s.a. program. Taking time to reflect is what s.a. is all about. As we look at the various aspects of our life; our personal relationship, our professional development, our managerial skills and our career plans, we must be open to the feedback in these areas and listen with interest which helps us to understand what our environment says to us about us. It is through this personal reflection that we begin to develop needs and to do something about them. We might also add that during this personal reflection, we might also discover our strengths too. Consequently, from the very beginning, the s.a. process is positive and dynamic.

Step 2: Needs Identification.   Personal reflection on the feedback we receive is not sufficient. We must internally convert this data and make it specific enough so that we clearly define a need or area of improvement. For a working definition a need is fundamentally a lack of something in our every day life. We are confronted constantly with a variety of needs, the most basic include food, security, money, love, and friendship. The needs you will develop as you go through this s.a. may be more sophisticated than these or may have a direct relationship to these more basic needs.

In the s.a. process, a second step therefore, is to clearly define the needs that we see in each of the areas we are evaluating. As you go through this process, you may find that a number of needs can be combined or that they are very similar in nature. The important thing is to clearly state the need or the area of improvement so that you can deal with it in a concrete and rational manner.

Step 3: Motivation to Act   Once we have articulated needs clearly, we must be willing to do something about satisfying them. It is well for me to say about myself that I need to be more patient with my subordinates, but it is a significant step to be able to say "I will do something about it."

Motivation, then, is the power that drives us to act. If it is lacking, the best of intentions will go unfulfilled and we are left with the knowledge that we should be doing something and will do it some time. The s.a. process requires that we do not procrastinate. If you are willing to evaluate yourself, that in itself is a positive step, and therefore, if you have come this far, you are a person who has the motivation necessary to take the next step.

Step 4: Goal Setting.   High achievers and successful people tend to be goal-oriented. The s.a. process helps you identify where you are. The goal-setting step enables you to clearly chart your course and say to yourself where you want to go. We feel it is important for goals to be realistic, achievable and, clearly "speak" to the need. Often our goals are not achievable. If, for example, your goal is to break 70 in your golf game when you are currently a 25 handicap, the goal in and of itself is admirable and good, but how realistic is it for you? In the s.a. process, it is important to achieve what we call "psychological success." In Figure 36.2 we have outlined a simplified model of psychological success. Basically, it says that when we achieve our performance goals, we feel successful, our self-esteem rises and our effort increases. If my goal is realistic, for example, if it is to compete in a three-mile race and after much effort I am able to do so, I feel good about myself and probably will intensify my efforts and shift my goal to a five-mile target. If, after a considerable amount of effort, I do not achieve my initial goal, chances are I might give up or find it very difficult to continue. There-

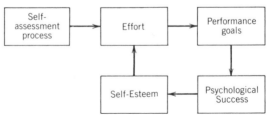

FIGURE 36.2.   Psychological success.

fore, the setting of goals is critical to self-improvement and is really the key to the success of the s.a. process.

Step 5: Development of Action Plan.   In this part of the process, it is important to indicate specific objectives and activities that you will perform to achieve the goal. For example, to say that my goal is to run a three-mile race and can only be achieved if many small steps are taken—exercises, running smaller distances, scheduling my workouts, and so on. Goals often times are set, and we wonder why nothing has happened as we look back months later on these goals. Much the same as New Year's resolutions which appear to be so definite on December 31, and seem to be so unattainable three months later. The action plan, as outlined in Figure 36.3, should contain the activities, objectives and a timetable which will help us achieve the goals. The first thing, however, that you will be asked to do in developing an action plan is to actually specify your first step. How will you begin, what will you do first. Chances are if you take the first step, your plan will be put into action and has a good chance of being implemented.

Step 6: Monitoring, Evaluating and Adjusting.   Once the action plan has been set into motion, you will need to monitor your progress on a periodic basis. Much as you would do on a major project, you will have milestones that will clearly give you feedback concerning the implementation of your plan. At the same time, you should be evaluating whether your plan is realistic and is achieving what you want to achieve, for example, if in my action plan I say I will run five days a week in order to achieve my goal of competing in a three-mile race, I may find that because of my travel schedule for a current project, I am unable to do this five days a week. Also, I may find that running that much is having a bad effect on me and maybe I am overdoing it. Based on this kind of data, I can evaluate and adjust my plan and say that I will now run four days a week. The important thing is that along the way, we are trying to insure our success in self-improvement. All of us tend to want quick results and we tend not to achieve them. We get discouraged. Again, it is important for us to be successful in this process. This step in the process is important because the feedback from it and our personal reflection as to our progress helps us to further define and refine our needs so that a s.a. and self-improvement program becomes a part of our daily life.

36.3.   LOOKING AT YOU—THE PERSON

Who are you? What are your interests, values, and personality characteristics? These questions seem so simple, yet isn't it strange that the person we

Areas for Improvement:_____

_____

_____

Specific Needs:_____

_____

_____

Goals:_____

_____

_____

My first step will be:_____

_____

Objectives, Activities and Time Schedules:_____
1.
_____

2.
_____

3.
_____

4.
_____

5.
_____

How will I evaluate my progress?_____

_____

FIGURE 36.3.   Self-assessment action plan.

often know least is our own self. Yet success demands self-knowledge. In the exercise below, you will have the opportunity to assess yourself and then put into practice the self-assessment process we have just outlined. Let us now begin to look at you, the person.

Interests. Based on your own personal feedback as received via your past education and work history, try to identify the areas of activities in which you have the greatest interest. Examples of such interests are listed here. Select the five that are most important to you.

Analyzing	Investigating	Producing
Communicating	Leading	Promoting
Controlling	Liaising	Record keeping
Coordinating	Managing	Researching
Counseling	Marketing	Selling
Creating	Maintaining	Servicing
Directing	Negotiating	Systematizing
Evaluating	Persuading	Training
Forecasting	Planning	Writing
Interacting with others	Problem solving	Other

If you had only six months left to work before retirement, how would you like to spend that time at work?

Values. Everyone has their own list of priorities as to which aspects of life are most important. From the following list select the values that are most important in your life:

Self	Respect	Compassion	Security
Family	Community	Equality	Power
Travel	Aesthetics	Tolerance	Success
Religion	Productivity	Conformity	Change
Honesty	Leisure	Individuality	Trust
Loyalty	Competition	Political	Environmental
"Fellow man"	Customers	Profit	Stability

Personality. If you were asked to describe yourself, what would you say? What do other people say about you? What is the picture you present to the world? There are a number of ways in which we receive feedback concerning our personality. Using the list below, please circle the words you feel best describe you. We suggest that you give this list to someone who knows you and have the other person do the same. In both cases, please choose from the words you have selected the five that best describe you.

Aggressive	Literate	Punctual	Hard
Cautious	Stern	Stable	Just
Cynical	Shrewd	Modest	Knowledgeable
Deliberate	Shy	Mature	Limited
Hurried	Amiable	Humorous	Motivated
Imperfect	Creative	Poised	Natural
Impatient	Competitive	Resourceful	Organized
Impulsive	Devious	Honest	Polite
Incisive	Energetic	Brash	Resolute
Lazy	Forceful	Dignified	Remarkable
Moody	Considerate	Vigorous	Stylish
Merry	Careful	Flippant	Versatile
Nervous	Calculating	Grave	

36.4. REFLECTION

Your answers to these few exercises can be most revealing. Take a few moments and review your answers. What do they say to you about your interests, values and personality? Do you see in your answers current strengths and areas to be enhanced? Do you also see areas for improvement? For each of these areas, develop your needs and write them down.

Personal Needs Identification

Interests:

_____

_____

_____

Values:

_____

_____

_____

Personality:

_____

_____

_____

36.5.   PERSONAL GOAL SETTING

For each of the areas where you have identified needs, write a goal which speaks to the need, that is, it is your target for improvement, your expected accomplishment. This is what you will strive to do to improve yourself.

Interests:

_____

_____

_____

Values:

_____

_____

_____

Personality:

_____

_____

_____

36.6.   PERSONAL ACTION PLAN

Based on your goals, what will be your action plan for improving yourself as a person? The action plan should have the following components using the steps outlined in Figure 36.3.

1.   List the specific activities and the dates of completion for these activities
2.   Write down the first step you will take in each of the areas
3.   The method of evaluating your progress

36.7.   LOOKING AT YOU: THE PROFESSIONAL

Career success demands that we not only seek to improve ourselves personally, but also professionally. The following exercises in Figure 36.4 will take you through a professional assessment focusing on self-development activities and your accomplishments. Please review the following list below. For each item, please indicate whether this is regular activities, an occasional activity, or one you have not done at all. Use as your timeframe the past year.

		Yes	No
1.	Did you volunteer to take a formal training program to further your progress?	_____	_____
2.	Did you read any management books?	_____	_____
3.	Did you take a self-study course or listen to cassettes to improve your skill?	_____	_____
4.	Did you increase your participation in company activities?	_____	_____
5.	Did you volunteer for any special assignments or projects not normally within your job?	_____	_____
6.	Were you involved in community activities, civic associations, or church groups?	_____	_____
7.	Did you observe others who you consider to be successful to find their secrets of success?	_____	_____
8.	Did you learn more about other functions and activities in the company outside your own job?	_____	_____
9.	Did you keep up with the state of the art in your specialty?	_____	_____
10.	Did you write an article, give a talk, or take some other step to improve your communication skills?	_____	_____
11.	Did you show a willingness to expose yourself to new experiences?	_____	_____
12.	Did you find out more about your business or industry?	_____	_____
13.	Did you take an active part in a professional association?	_____	_____
14.	Did you take a course at an outside university or organization?	_____	_____
15.	Do you devote more time to learning various project management methods?	_____	_____
16.	Did you devote time to learning about various styles of management?	_____	_____
17.	Did you subscribe to and read one or more professional journals in your area of specialty?	_____	_____

FIGURE 36.4.   Self-development checklist.

36.8.   ACCOMPLISHMENTS

Looking back on the past five years, please list your professional accomplishments. Where possible, be specific and quantify.

Objectives, Activities and Time Schedules:_____

_____

_____

_____

_____

Personal Reflection.   What do the answers to these exercises reveal to you about your professional growth? Were you surprised by your accomplishments? Did you find that in the self-development checklist you did very few of these on a regular basis? What do you see as you step back and look at yourself as a professional? Based on these exercises, what are one or two professional needs that come to mind? Please write them down.

Your Professional Needs Identification

_____

_____

_____

Your Professional Goals   What would you like to do to improve yourself professionally? What accomplishments would you like to achieve in the next year? Write down your goals.

_____

_____

36.9.   YOUR PROFESSIONAL SELF-IMPROVEMENT ACTION PLAN

As you did with your personal improvement plan, please develop an action plan for professional improvement. It may be somewhat easier to measure your progress in the professional area, but again it is essential that you adjust your plan so that it is realistic and achieves the results you desire.

36.10.   LOOKING AT YOU—THE MANAGER

Your managerial skills are essential to your career success. As you know, you bring to your project technical expertise, but this alone is not enough. How you work with your superiors, subordinates and peers is important not only for your professional growth, but for your job satisfaction. Using the exercises below, take a few moments to evaluate your managerial expertise. Using the following list, please rank yourself on a scale of 1–5 for each item; 1 being poor, 5 being excellent. The list is as follows:

MANAGERIAL SKILLS

Supervisory ability	1	2	3	4	5
Decision making	1	2	3	4	5
Planning	1	2	3	4	5
Organizing	1	2	3	4	5
Problem solving	1	2	3	4	5
Conflict resolution	1	2	3	4	5
Delegating	1	2	3	4	5
Coaching	1	2	3	4	5

Subordinates.   What would you like your subordinates to say about you? How would you like them to evaluate you as a manager?

_____

_____

How do you think they now evaluate you as a manager?

_____

_____

Superiors.   If your boss was called as a reference to you, what would he or she say about you? What would you like him or her to say about you?

_____

_____

Peers.   What would your co-workers say about you? How would they evaluate your managerial skills?

_____

_____

Personal Reflection

What do the answers to these questions reveal about your managerial abilities? How would you measure up? What areas of improvement come to mind? What are your strengths? Using this and other feedback you may have received in the past, determine for yourself your managerial needs.
Your Managerial Needs

_____

_____

_____

Your Managerial Goals

_____

_____

_____

Your Managerial Action Plan

All of the exercises thus far have helped you look at yourself as a person, professional, and manager. The following exercises are designed to help you look at yourself in your career. You cannot leave your career development to others, but must take charge of it yourself. The first step is to know where you are now and the second is to decide where you want to go.

Where Are You Now?   Please draw a line in the box that will depict for you your career growth. Put an *X* to indicate where you are now in your career.

```
┌─────────────────────────────────────────────────────┐
│                                                       │
│                                                       │
│                                                       │
│                                                       │
│                                                       │
└─────────────────────────────────────────────────────┘
```

Where do you want to be in your career five years from now?
If a historian were to write your biography, what would you like that person to say about your career accomplishments?

Personal Reflection

Are you satisfied with your career progress? Is your career on a rising trend, flat or turning down? Why? Are your career goals clearly stated? As in any good project planned, do you have a definite picture of the completed task? How will you know when you have achieved your goals? Reflecting on your career, what are the critical needs you see at the present time?

Your Managerial Needs:

_____

_____

_____

Your Managerial Goals:

_____

_____

_____

Your Managerial Action Plan:

_____

_____

_____

36.11. CONCLUSION

Self-assessment is a life-long process. The s.a. process you have just completed is not reality, it is a life-long program. The process was intended to teach a life skill. This was not meant to be a one-time exercise whose results are noted and then forgotten. If you as an engineer, are serious about continually growing personally, professionally, and as a manager with a career, this process should become a part of your daily life. You may find that this process can be valuable not only to yourself, but to those close to you and therefore, can become an integral part of who you are and will enable you to reap the rewards of a richer life. Your personal development should not be left up to chance or dependent upon others. You are the one responsible for your life and in a very real sense, just as you build the earth as an engineer, you have the opportunity to build yourself and create your reality. You, therefore, should begin to hold yourself accountable for your future. Begin today. Take your first step. In short, do it! It is your most important engineering project.

37

KEEPING YOUR CAREER
ON A SUCCESS TRAJECTORY

Leonard J. Smith

37.1. ESTABLISHING YOUR PERSONAL CAREER GOALS

Like the vast majority of engineers, you are interested in achieving success in your career. This success involves determining the career for which you are best suited, preparing for this career, embarking on it, and then achieving and maintaining a success trajectory. This successful career achievement is not easily accomplished. For some engineers a lifetime will pass without ever finding the "right" career. Some, despite having found the career of their lives, will be unable to achieve success.

Your efforts for a successful career achievement must begin with the establishment of a personal career goal. You must decide for yourself what career you would like to pursue. It may be the one in which you are presently employed or it could be another career. This sets the direction for your efforts.

37.1.1. Factors Affecting Personal Careers

The establishing of a personal career goal involves your consideration of a number of factors. All of these are complex and have differing values. What may be important to someone else may have little value to you and vice versa. The importance of these factors to you also change during your life-

time. Thus the consideration of these factors depend on circumstances existing at the time your personal career goal is established or when you are
reexamining your success trajectory.

Your Interests.   In establishing your career goal it is advisable to ensure
that it satisfies your various interests. A study of engineering careers in terms
of types of interests involved (See Table 37.1) identifies a large number of
differing career interests. A review of this list will assist you in determining
your own interests. When these are identified you can evaluate possible
career choices which will enable you to satisfy these interests.

Embarking on a career which does not enable you to fulfill your interests
may cause you to be unhappy and dissatisfied with your work. This feeling
could interfere with your ability to achieve success. On the other hand you
might achieve success but at a personal sacrifice—doing work which has little or no interest to you and is not personally satisfying.

Table 37.1.   Interest Areas of Engineers and Scientists

1. Ideas-planning, concepts, processes, advertising sales promotion, methods or industrial engineering.
2. Things-products, equipment, packages, materials
3. Numbers/figures-mathematics, statistics, calculations
4. People-management, sales, purchasing, customer services
5. Creative, innovative and imaginative—design, new products and services.
6. Diversity versus repetitive activities
7. Complexity versus simplicity of activities, sophisticated vs basic.
8. Indoor versus outdoor
9. Clerical-recordkeeping, detail-oriented
10. Analytical-inductive and deductive reasoning
11. Problem solving and decision making—trouble shooting
12. Specialist versus generalist
13. Literary-writing reports and papers, proposal preparation, advertising copy-
    writing, preparation of instruction manuals.
14. Graphic-artistic, photography, illustration, drafting
15. Research versus applications
16. Legalistic-patents, liability
17. Financial-accounting, budgets and estimating
18. Manual versus mental
19. Conservative versus speculative—"play it safe" versus gambler
20. Academic versus practical—theory versus practice
21. Sociological—improving society
22. Environmental—protecting environment, hazardous or toxic waste handling
23. Linguistic—translator, interpreter

Your Opportunities.   A second factor to be considered is the opportunities available to you. These opportunities can be within your present organization. As an employee of an organization, you can observe career opportunities within that organization that might be attractive to you. These may include present or projected future vacancies or needs which you want to fill. They may be the result of company growth, normal progression plans, anticipated organizational changes, or new fields of endeavor being undertaken.

On the other hand, you may be unable to see desirable opportunities within your present organization. It may be necessary for you to seek opportunities elsewhere. This could include those in other companies as well as different industries or fields of endeavor other than engineering.

It makes little sense for you to consider a career in which there are few, if any, opportunities. By the time you have prepared to embark on your career you might find no openings. You might be attracted to a career with fewer opportunities, on the premise that there also would be little competition for available openings. However, the odds of your achieving success would not be any greater than in your selecting a career with many opportunities and many competitors. The more opportunities, the greater your chances of achieving a success trajectory.

Your Individual Needs.   A third factor is the influence of your individual needs on your career choice. There are a variety of different needs which will affect your decision in establishing your personal career goal. They have differing weights in your decision making but each has an impact on your career selection.

The first of these needs is personal. It involves what you seek or desire to obtain from your work. These include psychological and economic benefits to you. An adaptation of Maslow's Hierarchy of Needs identifies ten basic personal career or job needs.[1]

1.   Adequate Compensation & Benefits
2.   Security—permanence of employment and personal
3.   Safety, pleasant working environment
4.   Positive and informed leadership
5.   Being kept informed
6.   Sense of belonging—team spirit—pride
7.   Opportunity for growth and advancement
8.   Appreciation for efforts—professional recognition
9.   Right to be heard—to contribute—to publish

10.   Interesting and challenging assignments—constructive achievements

Once you have identified and established a priority for your personal needs, you are in a position to determine which career will enable you to fulfill these needs. Compromises may be necessary in your career choice but you should be aware of the compromises you make to judge your career success.

A second group of needs which will influence your career choice are those related to your family. The latter may consist of your parents, spouse, children, and other dependents you may include as part of your family. You may find yourself limited in your career selection in order to satisfy or to fulfill your family obligations. These obligations may overshadow your personal needs. In any event, they will affect your career choice. They may serve either as positive or negative influences on establishing your career goal.

A third group of needs are those related to your professionalism or desire for professional recognition. Careers differ in their professional importance. Some careers enhance your professional status while others do not. The degree to which you are influenced by professional considerations can be an important factor in your career goal.

Related to professional needs is a fourth group known as challenges. These challenges may motivate you to perform at a higher level than when functioning in a nonchallenging position. They may provide you with greater satisfaction than any other factor. There are careers which provide you with, or lend themselves, to challenges. There are others in which there is an absence of challenges.

A fifth group of needs involve your health and physical condition. If you are a normal healthy individual with all your faculties intact, there may be no impediment to your career selection. However, if you have any health problems, such as allergies or physical disabilities, these need to be part of your considerations. For instance, if you can't stand heights, certain kinds of field engineering may be out of bounds for you.

Finally, your career choice may be restricted geographically. In spite of increased population mobility, this is still an important consideration. It may be determined by family ties, sociocultural or related preferences, such as between cities or metropolitan or rural areas, different climates, or by the career needs of your spouse or prospective spouse who may have other geographic restrictions as well. A further important factor is the need to travel frequently, or be uprooted fairly often as career needs dictate. The latter is a significant part of engineering employment, especially in large construction projects or large conglomerate firms with multiple divisions.

Your Ambitions.   A final factor affecting your personal career is your ambition. This is also known as your self-motivation or drive. The greater your ambitions the greater your needs for a career in which you can grow with reasonable speed. Careers which rely heavily on length of service for progress and advancement are not careers for ambitious individuals. On the other hand, careers in which advancement is based on individual initiative and performance are not for individuals who have little or no self-motivation. If you are ambitious you will have difficulty working in a career in which you cannot exercise your drive and energy.

37.1.2   Types of Career Goals

Your decision regarding your personal career goal may not be a simple one. There are several different types of careers which an engineer can select. Some involve engineering, but others do not. However, to select a career in an entirely different field could wipe out years of study and experience. Ideally, you should primarily consider engineering-related careers. If you should reach the conclusion that these do not appeal to you as goals, then, and only then, should you look elsewhere.

Professional Growth and Development.   The most frequently selected career goals for engineers are those which involve professional growth and development. These positions are oriented toward higher-level engineering. They include research in new engineering technology, as well as assignments concerned with specialized and advanced design. In most instances, they are careers which require additional study, knowledge, and effort above and beyond that of the ordinary engineer.

These careers may be found with your present employer. This is particularly true in engineering-oriented companies with diversified and sophisticated products or services. Single product lines with established technology and limited objectives in the area of technological change offer few opportunities for professional growth. In these situations you may find it necessary to relocate with another employer to realize this type of career goal.

Engineering Managerial Careers.   The majority of engineers with growth potential are faced with the decision of moving out of an engineering career into one involving management. Most employers of engineers have a progression ladder which provides for growth into management. These organizations establish both a progression plan and a compensation plan which encourages movement from engineering to management.

To be successful in an engineering management career, you need to have knowledge and abilities which are different than those of an engineer. Success as an engineer does not by itself make for your success as an engineering manager. In fact, all too frequently successful engineers by accepting promotion to management diminish their potential for continued success.

However, you might find engineering management to be a stimulating and rewarding career. It utilizes your engineering knowledge and experience. In addition, it is people oriented, and involves the planning, supervising, and controlling of activities performed by others.

Operational Careers.   In addition to engineering management careers, you have a choice of managerial careers outside of engineering. You need not limit your goals to those careers which are directly related to engineering. You can consider a number of career opportunities in operations. These include manufacturing, purchasing, planning, customer service, and sales.

Each of these operational careers will permit you to utilize your engineering knowledge, experience, and skills. Instead of being an engineer as such, your engineering becomes the foundation for your career work performance.

As with movement into an engineering management career, selecting an operational career should be based on your abilities to successfully perform the work and to meet the requirements of these careers. You need to learn more about the careers to determine the demands they will make on you. You also have to analyze the opportunities each would provide.

Most frequently these careers will be found with your present employer. However, it is not uncommon to have opportunities for operational careers make available to you by other organizations. These may be offered on the basis that your engineering knowledge, experience, and potentials in operational activities will enhance your success in these careers.

You may find many opportunities to become familiar with operational careers and determine your interests and potential satisfaction during the performance of your present work. You should avail yourself of opportunities to visit your employer's manufacturing facilities. You should observe operations and meet with manufacturing personnel including managers. You should offer to visit customers both with the sales people and on service calls. When possible you should meet with suppliers' personnel to discuss materials and equipment. These exposures may serve to stimulate or dull your interests in these types of careers.

Entrepreneurial Careers.   Many engineers have opted for careers as entrepreneurs. These have been individuals who desire to function on their own. They include those who want to become rich as well as those who want to control their own destiny. Some have selected entrepreneurial careers because they have encountered obstacles and barriers to pursuing their professional careers while working for others.

As an entrepreneur, you are your own boss. Instead of others reaping the rewards of your engineering efforts, you reap the success. You should be aware that most entrepreneurial careers also involve managerial abilities. Being in business for yourself means that you have to manage that business. If you join forces with others to form a business, you have become an owner. This ownership involves only the financial aspects of the business. The work you perform in the business determines your career. If your work continues to be engineering, you are in an engineering career only with an employer in which you also have a financial interest.

Should you determine that you prefer a career in research, you could become your own R&D organization. This would permit you to engage in those research projects that appeal to you. You could undertake those assignments which you find interesting, challenging, or personally rewarding. However, as soon as you employ others to assist you, you have embarked on a managerial career. Instead of being a professional engineer, you may find yourself a businessperson.

You also could abandon engineering and enter an entirely different field as an entrepreneur. Although you may not be utilizing your professional knowledge and skills, you may find these to be of value in your own business. Usually this career goal is selected when you have an unexpected opportunity to go into business. A strong factor in the selection is the fact that your engineering career may not be as fulfilling as you desire.

Consulting Careers.   Closely related to entrepreneurial careers are consulting careers. These are careers which enable you to utilize your professional knowledge and skills in a variety of situations. Instead of routine and repetitive types of assignments, you could find yourself performing many diverse types of work. You could be called on to provide advice and guidance, solve problems, give expert testimony, or handle unusual situations. However, you should be aware that entering a consulting career may involve travel. It has periods of great demand and of little or no demand.

If you are to function as an independent consultant, you will need to develop and maintain a reputation that will attract clients. This could be as a specialist in one aspect of engineering or as a generalist. On the other hand if you join a consulting firm, you become an employee who may be assigned to projects selected by your superiors.

Academic Careers.   Your personal career goal may be in academia rather than business or industry. You may prefer to teach rather than perform engineering tasks. This type of career would enable you to assist in the education and training of future engineers. You might also consider teaching a subject such as mathematics rather than engineering.

On a college or university level, an academic career could include engineering research. The research could be an adjunct to your teaching or it could be a separate career. Grants could be obtained to finance research projects you might desire to undertake. In addition, there could be greater opportunities for you to publish articles, papers, and books. Note, however, that an earned doctorate in engineering or another appropriate subject is an absolute requirement for those who aspire to a career in the better institutions, as is a significant record of publications.

Frequently, it could be possible for you to combine an academic career with a consulting career. Affiliation with an academic institution could provide you with opportunities to undertake consulting assignments. In reality you could consider serving on a faculty as the means by which you support yourself when you embark on a consulting career.

Writing Careers.	If you enjoy writing and have the talent to write, you might select a writing career. Many engineers utilize their knowledge and experience as writers. You could write for a technical magazine or journal. You could serve as a technical editor for a publication. You could be a writer of textbooks. These careers also could be your extracurricular activities. They could provide you with professional recognition, and they could lead to career opportunities otherwise not available to you.

Nonengineering Careers.	You always have a choice of leaving engineering and embarking on a new career. This could be a new or different field or type of work. There are literally thousands of careers available to you. The decision to search outside of engineering should be reached only when you have found an engineering career unsatisfying or unfulfilling. Other engineere have successfully embarked on second careers. Some have waited until retirement to do so.

The Value of Strategic Overview.	To keep your career on a success trajectory it is necessary for you to develop a strategic overview. A strategic overview is an approach to understanding and effectively dealing with the numerous factors and events that you will encounter in your efforts to achieve success. It involves a study of what you can expect and the steps you may be required to take in your career goal efforts.

One of the first benefits of a strategic overview is that it permits you to ensure that you have set the proper direction for your career goal. This is accomplished by your viewing your career from where you have been to where you want to end.

Second, it enables you to determine the various competitive situations you will face in the course of moving toward your career goals. An identification of these potential situations will permit you to plan in advance how to deal with them when and if they arise. You might be able to avoid some of these competitive situations by knowing that they might arise in the future.

A third value of the strategic overview is that it permits you to strengthen your personal capabilities to meet your career requirements. As you review the projected path of your career, you will uncover the demands that will be made on you. Your shortcomings or weaknesses that require strengthening will become evident.

The strategic overview also enables to identify potential short- and long-term opportunities that could arise in the future. You could then be in a better position to take advantage of these opportunities. This includes being more sensitive and more aware of them when they arise. You also should be able to determine the best approach to take. Conversely, you will be in a position to recognize hazards and obstacles to your career success. Early recognition of these problems will enable you to take judicious action to avoid, minimize or overcome these hazards and obstacles.

The fifth value of the strategic overview is the fact that you are able to do your career planning with forethought. Not only can you determine the steps you should take to achieve success, but you can establish the timing of these steps. You also can develop your alternate steps to deal with the competitive situations, opportunities, hazards, and obstacles that you could encounter.

The final value of this effort is that it minimizes the subjectivity of your career goal establishment. As you conduct your strategic overview you see your career path in an objective light. You will view all aspects of the road to your career success. You are not unduly influenced by any one factor, positive or negative.

37.3.   BUILDING YOUR SUCCESS MODEL—ROLE MODELING

It will be extremely helpful in your efforts to achieve your career goal if you can build a success model to emulate. Role modeling permits you to select one or more individuals whose pattern of success you can follow. Building your role model will provide you with standards for comparison. In addition it permits you to study the factors that made for their success.

37.3.1.   Determining the Ideal Model for Success

Your ideal role model should be an individual whose successful career you would most like to duplicate. It could be a member of your family, someone you know, or an individual about whom you have heard or read. Preferably your role model should be someone about whom you can obtain personal information. In particular, you will need to learn the traits, characteristics, education, and experience that contributed to their success. In the process you will need to clarify for yourself the reasons for selecting the particular success model. This information will be helpful in your success trajectory.

37.3.2.   Networking as a Method of Building a Success Model

You might find it difficult to locate any one individual to serve as a role model. A technique that has proven helpful in this effort is networking. This is a process of getting others to refer you to, or to recommend individuals who might be your role model. Networking enables you to screen potential success models and identifying those patterns of success which appeal to you. In the process of screening, you will establish the criteria for your ideal success model.

37.3.3.   Identifying Personal Assets That Match Your Role Model

Once you have determined who will be your success model, you should undertake the task of identifying which of your personal assets match the model's. This will assist you in establishing a similarity. If there is little similarity, it might be desirable for you to abandon the role model and seek another whose assets are more closely related to your own. The more your personal assets match your role model's, the greater your chances of achieving similar success.

37.3.4.   Action Plans for Achieving Other Aspects of the Success Model

In addition to identifying matching personal assets you will uncover assets of the role model which you lack. To achieve your desired successful career goal, you may have to acquire or develop these assets. This will necessitate that you prepare action plans which will enable you to develop these as your personal assets. The action plans should set forth the activities and efforts you will undertake and the timing of these steps to build yourself in the image of your success role model.

37.4.   CAREER PLANNING MILESTONING

If you have completed all the preliminary reviews and analyses, you should be in a position to engage in an exercise known as career planning milestoning. This is the function of organizing your plans to achieve success in your career goal. It establishes the efforts or steps you plan to take in order to keep on a success trajectory. These steps may include the acquisition of: additional knowledge, higher degrees, professional recognition, personality changes, financial resources, and experience.

The milestoning of these steps is a technique by which you identify their sequence and determine the length of time required to complete the activity being planned. You use it to identify the relationship between the various career planning steps. Some of your plans can be undertaken simultaneously; others can overlap, while still others need to be scheduled end-to-end. You also will need to establish the priority of the steps. That is, the identification of those plans which are of major importance to your career success. The sequence of these major steps form your critical path in the milestoning.

37.5. CREATIVE ALTERNATIVES

No career planning is complete without the establishment of career alternatives. As you reduce your career plans to writing you may realize that many things can interfere with your achievement of these plans. For each plan or step to be undertaken, there should be one or more alternatives to effectively overcome any "what ifs." The more your creativity, the more likely you are to uncover alternative methods for achieving your career success. For example, original research and the publication of your findings might offset your need to acquire an advanced degree. Both would provide you with professional recognition. Another example would be personal research to substitute for work experience in a special field.

In addition to alternatives in the steps or efforts to be undertaken, you might want to consider alternative career goals. These may be necessary in the event that changes take place to make your personal career goal unattainable. Some of the changes which could affect you are personal—your changing values as the years pass or changes in your family status. Others are organizational, technological, regulatory, political, and economic. These changes also may affect your ability to keep your career on a success trajectory even when you have achieved success.

Creative career alternatives might include secondary careers. Sometimes an avocation can be turned into a vocation. A secondary career is one you can undertake while pursuing your primary career goal. It may be related to your career goal or it could be a previously considered career but one which was subordinated by your final choice.

You always should be prepared for unanticipated opportunities. These are opportunistic situations which can arise in the course of your career goal trajectory. They may be considered "once in a lifetime" opportunities. They are unplanned alternatives. They can be more fulfilling and provide more satisfaction than your career goal. On the other hand, they may serve to distract you from your success career trajectory. It should be noted that frequently these unanticipated opportunities are created by the previously mentioned changes.

37.6. TACTICAL CAREER DECISION MAKING

The selection of your personal career goal and the planning to keep on a success trajectory involve choices or decisions. Your ability to effectively make these decisions will be enhanced if you develop your own decision-making formula. This requires that you determine which factors need to be considered in arriving at your final decision. After identifying these factors, you need to assign weights or values to them. All of this must be in terms of your thoughts on both your personal career goal and your career success.

The various alternatives or choices which you consider can be evaluated in terms of your decision-making formula. That which achieves the highest value should be your most appropriate career decision. It is difficult to determine whether or not your career goal is the best one for you. But, under the circumstances, if and when you consider the various factors and their value to you, you should arrive at a worthwhile career goal. If you are already on your success trajectory, the decision making will enable you to continue your progress to the successful attainment of your personal career goal.

NOTES

1. A.H. Maslow, *Motivation and Personality* (New York: Harper & Row, 1970).

38

COMMUNICATIONS AS
A MANAGEMENT TOOL

George Black

GEORGE BLACK is Chairman of the Board of the BRM International Division of Bozell & Jacobs, one of the world's largest advertising and public relations agencies. A specialist in communications, Mr. Black received his B.A. in Literature and Education from Brooklyn College, then completed special studies in Materials Engineering at Johns Hopkins University. His work experience includes being an inspection foreman in detail manufacturing, supervisor of employee training programs, writer of technical manuals and instruction bulletins, technical editor, and director of advertising and public relations for industrial firms. Mr. Black is an accomplished lecturer, writer for the trade and business press, and author of several books including the basic textbook, *Sales Engineering: An Emerging Profession* (Gulf, 1979, Second Edition).

38.1.   INTRODUCTION

The secret of good engineering management lies in the ability to shorten the path from conception to performance by enlisting the talents of others in behalf of understood objectives. It doesn't lie in working yourself to death because no one else can do it as well or as fast as you can. It doesn't lie in creating human machines that carry out your bidding to the letter. Efficient engineering management depends on efficient communications, because it

741

is only through efficient communications that you can get things done, through other people, the way you want them done.

That's about as close to a meaningful definition of communications as you'll ever find. To put it more formally: Communications is the transfer of meaning through planned input into a prepared environment. In many ways it is like cultivating a garden. You have to know what it is you are planting, prepare the soil to receive it, do the actual planting, protect against the environment, and keep checking to be sure nothing goes wrong.

Another analogy that clarifies the meaning of communications is the relationship between a quarterback and a pass receiver on a football team. No matter how true the pass, it has no value if it's not caught. No matter how skillful the maneuvering of the receiver, your team can't gain ground on a pass that's not completed. Between the passer and the receiver, there are a whole host of obstacles, each one of which must be taken into consideration by the team on the offense. What must never be lost is the purpose behind the throwing of the ball, or as in the previous analogy, the purpose behind the planting of the seed. Purposeful understanding is a significant part of our definition of communications.

38.2. WHY IS COMMUNICATIONS A PROBLEM?

One of the reasons communications is a problem is our educational system—particularly the educational orientation of engineers. You've been taught that two and two, when added together, always equals four, and that if you push the right buttons on the computer and feed the facts into the formula, the answer must be right. The engineer lives in a world of facts or truths.

That may be fine when dealing with the inanimate. But when it comes to relationships between people, facts are few and far between, if they exist at all. Your boss is not my boss. Your desk not my desk. Your job is not my job. This is true even if we share the same boss, the same desk, and the same job.

What is mine is not yours, even if they are the same. This is so because your eyes, ears, sense of taste, sense of smell, physical feelings, and emotions are all different. These differences change what you see, hear, taste, smell, touch, or interact with in any way.

Let's look at it another way. I'm sure your parents and teachers and clergymen have taught you that the major difference between humans and animals is that humans are rational beings. If you believe that, you might just as well be walking around in the belief that the world is flat. A person is a reactive rather than rational being. Reason plays a minor role in initial decision making. If you have been trained to base your strategies and presentations on the rationality of the reader or listener, you're in trouble.

Your reason is the fine tuner on a television set that is used after the channel has been selected. It helps to justify your decision and make you comfortable with the reactive decision you've already made. The mood you bring to the table affects the taste of the food. Your mental set affects what you hear or read or see. As a general rule it can be said that reason will not displace emotion. Usually it takes another emotion to do that.

That's why engineers who have proved the customer wrong end up winning the argument but losing the customer. That's why managers who pressure their subordinates to do their bidding end up having it done incorrectly or inefficiently. One of the most important objectives you can achieve as an engineer is to get your colleagues and contacts to put their emotional clutches in neutral. That's the only way to get them to forget their prejudices, listen, and be truly receptive.

To restate this idea: as an engineer, you have been educated along Aristotelian concepts of logic. This means that you tend to believe:

1. *A is A.* Consistency is critical.
2. *A cannot be both B and non-B.* Something cannot be and not be at the same time.
3. *A must be either B or non-B.* A fact is singular. It must be true or it is false.

But there is another approach known as "General Semantics" which emphasizes a non-Aristotelian logic. It says:

1. *A is not A, except at a particular time.* It changes constantly with time. You are not the same in the morning as you are at night, the same before an argument as after, or the same when worried as when confident.
2. *A may be both A&B.* A boss and a subordinate, a teacher and a student, or a father and a son.
3. *A may not be rigidly classified.* It might be "true" in one context and "false" in another. If you say "I am lying," the statement is true if it is false and false if it is true.

Einstein once said: "As far as the laws of mathematics refer to reality, they are not certain: as far as they are certain, they do not refer to reality." Effective communications requires open-mindedness and being aware of meaning changes due to a variety of experiences, needs, desires, fears, and so on.

38.3.   WITH WHOM DO WE COMMUNICATE?

We must recognize and accept that communications is a two-way street and that effective communications must take into account the emotional and mind set of both the sender and the receiver. Now let us ask ourselves with whom we communicate or interact everyday. This is a fundamental and significant part of communications workshops which I have given or attended, the reasons for which should be obvious. The people we interact with affect us in subtle and not so subtle ways, affecting our own actions and reactions: the things we do and say, the things we see and hear. It is important for you to be aware of this as a communicator and as a manager. The communications pattern between you and a colleague who is in the midst of a divorce, or has just had a death in the family, or has won a million dollars in the lottery, will and should vary.

An interesting exercise is to make a list for yourself of the daily influences on your life. These might include such things as:

1.   The environment into which you awaken in the morning. Is your bed, your room, your home, cheerful and relaxing or somber and tension-building? Does it help you start your day with a smile or a frown?

2.   Is your wife, roommate, or first human contact loving, pleasant or irritating?

3.   Do you eat breakfast at home or in a diner? Pleasant company, pleasant surroundings? Enjoyable or boring?

4.   Is your car a source of pleasure or a source of trouble? Did you choose your vehicle or was it a company requirement?

5.   The trip to work. What does the trip to work do to your frame of mind? Your emotional situation? Is it a pleasant country drive or bumper to bumper crawling? Are you taking a bouncing bus or a crowded train to work?

6.   Is your work area physically pleasant? Is it adequate, cramped, or annoying? Is it too warm or too cold?

7.   Are support personnel, such as the telephone operator, secretary, and clerks, pleasant, helpful, and competent?

8.   Do you enjoy working with your colleagues and other people on the job? Do they enjoy working with you?

9.   What kind of relationship do you have with your boss? Is it open or closed, formal or friendly? What he or she says and how it is said can completely change you and your day.

10.   How is your physical health? How is your financial health?

We could go on and on with this list of people and things you communicate with, and how they affect the communications pattern, but these things should be obvious:

1. Our environment affects and changes us.
2. Our environment is made up of people and objects.
3. Our needs and our desires affect and change us.
4. Our fears and other emotions affect and change us.
5. Our attitudes and our motivations affect and change us.

To communicate efficiently it is extremely important that we be aware of all of the factors which affect the sending or receiving of messages. Who we are is one of the most significant factors to consider.

People are multiple, complex identities. The same "man" is an employee in one situation and a boss in another. He can be a father, son, uncle, neighbor, husband, stockholder, taxpayer, tenant, landlord, spectator, team player, soldier, peacemaker, and on and on. To communicate effectively with this one man, it is important to know those aspects which will affect what he hears as you speak, and to realize that the way you send or receive messages is dramatically affected by who you are at the time you send or receive.

38.4. HOW DO WE COMMUNICATE?

It has often been said that we communicate in more ways than we know, to more receivers than we ever dream about, and that these multiple ways and receivers affect the efficiency of our communications. Let's take a close-up look at the highway of communications in terms of sending and receiving messages. (See Figure 38.1.)

We send them by our appearance, clothes, voices, actions, written as well as spoken words, touch, and through a variety of intangibles such as our positions of responsibility, our reputations, and our attitudes.

We receive messages through all of our senses—sight, sound, taste, touch, and smell. The receipt of these messages is conditioned by many things in the environment such as available light, noise, distractions—plus our experience, memory, knowledge, and emotions. This myriad of sending and receiving mechanisms in which we are involved every waking hour can be conveniently divided into one or two "languages." They are either verbal, that is they use words to help transfer meaning. Or they are nonverbal—the transfer of meaning comes about without the use of words.

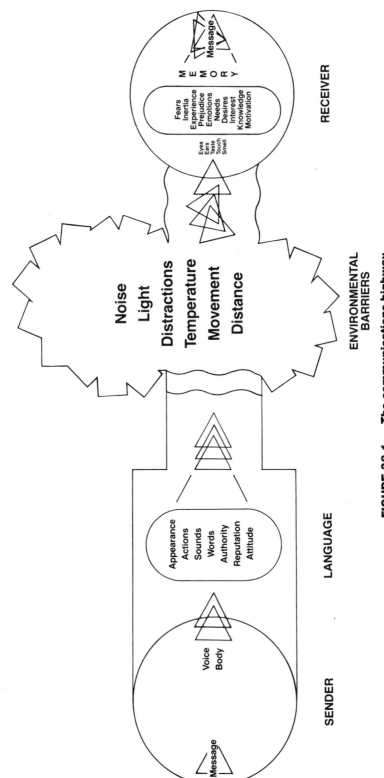

FIGURE 38.1. The communications highway.

38.5. NONVERBAL COMMUNICATIONS

Most engineers tend to play down the value of nonverbal communications. Yet there is ample evidence to indicate that a large proportion of emotions is communicated without the use of words.

And since emotions directly impact on the ability to receive messages accurately, this is an area that deserves serious consideration. In examining the list of the ways in which we communicate, look at the ways which are nonverbal in nature: Appearance, clothes, voice, actions, touch, attitude, reputation, and so on; the kind of person you are, way you smile, shake hands, walk, tone of your voice, way you look, feelings you inspire—all of these open or close communication paths. Many studies have been made and books written on the effect of "body language" as it relates to communications. There are even commercial parlor games designed to make this point and improve your ability to communicate nonverbally.

In one of these games called "Body Talk," the statement is made that as you play the game you are learning to "communicate with your mouth tied behind your back." Body language has been extensively discussed in popular literature as a human semaphore system capable of expressing an extraordinary range of emotions which significantly affects how messages are received.

Effective communications requires that you be aware of the signals you are sending (authority, appearance, voice, reputation, and so on) and how these signals may be received by the complex beings toward which they are directed. It is also essential to consider how the receivers may be affecting the way you project your message. Does your communications pattern reflect your reaction to the receiver? Are you humble? Bossy? Sarcastic? Do you show annoyance? Respect? Consideration? The interchange between sender and receiver affects both and consequently affects the message that gets through.

38.6. VERBAL COMMUNICATIONS

It is obvious that the intricacies of nonverbal language set up all kinds of problems or interferences, and these will be discussed in greater detail when we talk about barriers. But one thing is certain. If nonverbal language is a source of problems, words are even bigger culprits. The reason is that words are arbitrary symbols which have no meanings in themselves, but rather have multiple meanings, or meanings which differ from time to time, place to place, and person to person.

Words are symbols which were established arbitrarily at one time or another in our history—symbols whose meaning may have changed or disappeared with time. In many cases today's meanings are exactly opposite of the original ones. There are more than 600,000 words in our English dictionary. Only 2000–3000 of them are in daily use and these have 14,000 dictionary meanings. It's no wonder there's a problem!

Let's take a look at the various ways in which you use words to help you communicate. There are two major approaches with a number of subheads:

38.6.1. The Spoken Word

You can send a message face-to-face, one-on-one, to a small group, or a large group. You can send it by telephone, radio, or television. You can send it in an interchange situation by means of an extemporaneous presentation, or through canned speech. It can stand alone or be reinforced by graphic images or other symbols. You can be on the receiving end of the spoken word in all of the variations possible with the sending of messages. You can be receiving it one-on-one, as part of a small group, or as part of a large audience. You can be receiving it in person or through the telephone or electronic media.

38.6.2. The Written Word

You can send messages through the written word by means of letters or informal memoranda; and you can do this in your own handwriting or through a mechanical device such as a typewriter. In addition to letters, you can communicate through the written word with activity reports, technical reports, analyses, recommendations, and so on. These methods of communications can be accompanied by illustrations, photographs, or other nonverbal language approaches.

The same holds true with respect to receipt of the written word. You can be the receiver of the letter, memorandum, criticism, report, recommendation, and so on. You can be the reader of the catalogs, advertisements, or bulletins. Verbal communication comes in through the eyes or ears, but it comes in the form of words, and this is the source of trouble.

38.7. THE PROBLEM WITH WORDS

Words change their meaning with time and people. If we say something is "cool" today, there is no relationship to temperature. If we say that an idea is "hot," we are not the least bit concerned with thermal qualities. With the advent of CBs, we have a whole new language. I challenge the uninitiated to make any sense out of the verbal groupings now beamed along the airways by citizen band radios. (See Figure 38.2.)

At best, words are compromises. That's why it is so essential to be sure that what you have said has not only been heard, but is understood by your listener. How often have you heard the expressions, "my words fell on deaf

JARGON

Bear................	State Highway Patrol
Bear In The Air.............	Spy in The Sky
Bear Report..............	Where Are They?
Bear Taking Pictures	Radar
Bears Wall to Wall	Many Bears
Beat The Bushes	Find The Bears
Blow The Doors Off	Pass
Boulevard	Interstate Highway
Bounce-Around................	Return Trip
Break..............	Let Me On The Channel
Bubble Gum Machine ..	Police Emergency Light
Chicken Choker.............	A Poultry Truck
Chicken Coops	Weigh Stations
Clean....................	No Bears Ahead
Convoy	Organized Bear Hunt
Cotton Pickers	Anybody Else
County Mounty.................	Local Bear
A Dead Pedal	A Slow Moving Vehicle
Do It To It	Put The Hammer Down

FIGURE 38.2.   **Words are arbitrary symbols whose meanings change with time and events.**

ears," "it was like talking to a wall," or "he didn't hear a word I said." The reason these expressions are close to the truth is that the listeners heard words rather than meaning. The words couldn't jump the barriers.

38.8.   BARRIERS TO COMMUNICATIONS

Since communication involves a sender and a receiver and since we know from experience that it's not easy to transfer meaning from our head to the heads of our listeners, it will be worthwhile to explore the significant barriers that stand in the way of achieving purposeful understanding and examine ways to overcome them.

1.   *Inertia: Resistance to Change.* We are all creatures of habit doing what comes naturally—doing what we have been doing. This goes for the clothes we wear, games we play, cars we drive, and entertainment we seek. It is easier to do that which we have done before than to try something new. Change comes hard. The ruts of automatic behavior run deep. Intellectually, your colleagues or subordinates may want to change their methods, follow your instructions, and do things your way. But resistance to change is often stronger than the intellectual appreciation.

Remember we said that humans are reactive beings. The tendency is to react negatively to a suggestion that anything be done in a new way. What does this mean to you as a manager? It means you must concentrate on making the new way familiar or making the change feel like no change at all. You must think in terms of building the desire in the mind of the listener so that the path you outlined will feel right. To put it another way, in order to overcome inertia you must supply the grease that changes the high friction surface into a slide.

2.   *Prejudice: Emotional Bias.* Whether we like to admit it or not, we are all subject to prejudice. A prejudiced person is merely one who tends to prejudge or reach conclusions prior to hearing the whole story. A prejudiced person bases decisions on what is known rather than on the current evidence being received by his or her eyes and ears. We tend to develop an emotional bias, generally a negative one, in relation to people who are different. A person's color, religion, speech pattern, monetary position, appearance, accent, and so on, all elicit a reaction from us—a reaction based on previous experience, fears, and desires. Specific words affect us the same way. *Sex. Abortion. Communist. Radical. Women's Lib.* This emotional bias affects how we hear what is being said and how we say what is on our mind. It affects how we react to what we read and how other people react to what we

write. Emotional bias or prejudice interferes with our ability to get across an idea or understand an idea that is being presented.

3. *Ignorance: Insufficient Knowledge.* We live in a complicated world— one that gets more complicated each day. This is frequently referred to as an age of specialists, because the body of knowledge is so extensive that it's becoming increasingly difficult for an educated person to be broadly knowledgeable. Computer literacy is a requirement of this new age. The slide rule is ancient history. Chips are in and vacuum tubes have disappeared. Outer space is part of the new environment. Miniaturization is accepted with hardly an exclamation.

Ignorance takes its toll in communications because the people we deal with may not know or understand the words or symbols we use. Because they are "educated" they may hesitate to tell us they don't understand. This ignorance leads to fear, inefficient action, or no action at all. Recognition of this information gap is a key to efficient management. Without talking down to associates or employees or up to bosses, we must be sure they understand what we are saying and why what we are saying is significant to them. We must get feedback to be sure the message is getting through.

4. *Seeing.* Whether we send our messages through our voices, bodies, or printed materials, sight plays a major role in how it is received. Impaired eyesight can change the meaning of the message or prevent it from being received altogether. If you are showing someone a diagram and explaining its meaning, make sure you are being understood. Since you know the diagram, the lines on the paper have much greater meaning for you. They look larger and they make sense. But the person to whom you are talking doesn't have the same brain image you have. The lines will be less distinguishable and their meaning will be cloudy. It's the difference between looking at a roadmap covering the town in which you live and looking at a similar map covering a strange city.

Type size is also an important factor. You may be pointing to catalog references with type so small that only young eyes see it clearly. The person to whom it is being shown may not want to admit the difficulty, and particularly if the subject is not well understood, the person may have trouble following or understanding what you are saying.

The problem with "seeing" goes even deeper. Anticipation, expectation, prejudice, desire, and many other barriers get in the way of sight—even for those with 20/20 vision. (Figure 38.3.) People tend to see what they want to see, what their mind set tells them is there, even when their reason tells them otherwise. That's one of the reasons proofreading is best done by fresh eyes rather than by the person who did the writing. (Figure 38.3, 38.4, 38.5.)

5. *Speaking.* The way you talk conditions to some extent how well you

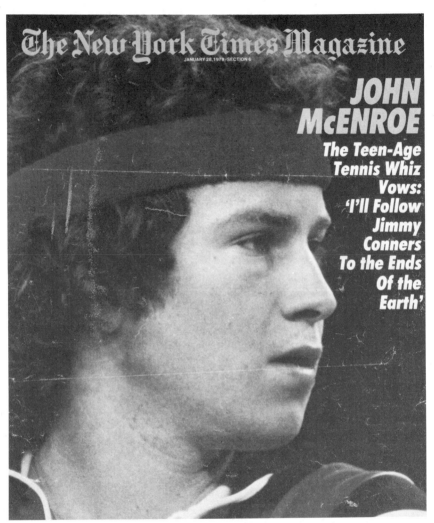

FIGURE 38.3.   Proofreading is difficult because we tend to see what we expect to see rather than what is there. (New York Times Magazine)

are heard. The sound of your voice, speed at which you speak, voice pattern, distance from the listener all affect how well you are heard. Be on the alert for those who are hard of hearing but whose egos won't permit the wearing of hearing aids. Be alert to background noises and interferences which may make it difficult for your listener to hear what you are saying. Watch for those distracting influences, momentary interruptions, or just plain boredom.

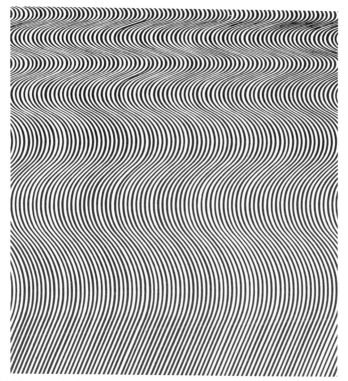

FIGURE 38.4. **Are you sure these lines aren't moving? Seeing can be very deceiving. These lines are not moving.**

With some variation these hints will work equally well whether you are speaking one-on-one in person or via phone, or whether you are addressing a small or large group. Interchange is a bit more difficult from the lecture platform but you can compensate by using more effective visual aids.

The way you speak has a direct impact on how well you communicate. Any number of courses are available to you at local schools or at franchised commercial programs; and there are dozens of exercise books at the library. They are all helpful, but here are a few hints to point you in the right direction.

1. Have something to say and a reason for saying it.

2. Think in terms of your receiver or audience. Why is your message important to them?

3. Plan your presentation. The sequence of thoughts is significant because it affects how the thought is received and what is retained.

4. Relax. Tension breeds tension. Be conversational. Don't make speeches.

5. Be enthusiastic. It's contagious.

6. Speak clearly. Make sure you can be heard.

7. Listen as you speak. Listen to extraneous noises which may distract your listener to signs of inattentiveness or boredom.

8. Vary your voice pattern and speed. Avoid the monotone.

9. Don't monopolize. Let the listener get a word in. Encourage the asking of questions so that you can get the feedback you need to let you know you are getting through.

10. Reinforce your message with visual aids—photographs, diagrams, samples—whenever possible.

6. *Listening.* Listening is a lot more difficult than speaking. In spite of the fact that effective listening is essential to living, it is not taught in the schools. When it is mentioned at all, it is identified with restriction: shut up and listen. The sickening fact is that 75% of what we hear and think we learn or understand we will forget completely in 24 hours.

FIGURE 38.5.   It's hard to believe this card is selling "Better Communication" . . .

Another important thing to remember is that people can listen approximately four times faster than you would normally speak. While you're talking they can be thinking of a luncheon they are going to, a dinner, a night out on the town, a football game or what have you. For effective communications, it is critical that you tie up this free listening time. You can do this by using visual aids such as photographs and diagrams and samples, or by asking questions to get the feedback you need to be sure you are understood.

Research studies indicate that we spend as much time listening as we do speaking, maybe more. If we combine it with reading and seeing, the other receiving skills, its importance is even greater.

These guidelines should help improve your listening skills:

1. Know why you are listening.
2. Get a handle on your own prejudice with respect to the speaker and the subject.
3. Suspend judgment. Listen to understand what is said not to refute.
4. Concentrate. If your mind wanders off track, bring it back.
5. Look the speaker in the eye. Watch for nonverbal language that interprets meaning.
6. Ask questions if thoughts are not clear. Don't refute or try to embarass.
7. Ask questions if convenient. But make notes of ideas only, not direct quotes or supportive data.
8. Restate your understanding so that the speaker knows that you heard not only what was said, but what was meant.

7. *Reading.* It has been estimated that engineers read a million words a week, with no more proficiency than they did on entering secondary school. Reading training tends to stop very early in our educational system. In recent years there has been an increased emphasis on improving reading speed and a more recent movement aimed at improving comprehension. Numerous programmed instruction kits are available as well as many textbooks, commercial courses, and in-plant training programs.

Here are a few hints to start you in the right direction:

1. Read with a purpose. Are you seeking information or passing the time? Are you looking for signposts or studying in depth?
2. Read selectively, with your purpose in mind. Scan for subject clues and then intensify.
3. Underline key material that matches your purpose.

4. Check your tools. Your eyes, glasses, posture, available light, chair, and so on. Make sure they are helping, not hindering your reading.

5. Try not to vocalize or lip read. Don't slow yourself down by mouthing the words. Let the message go directly from the printed page to your eyes to your brain.

8. *Writing.* It has long been the view of engineering educators and others that the engineer's first step toward the executive suite may not be so much interdisciplinary education but some very basic training in one very demanding discipline—writing. A growing number of states are enacting "plain language" laws which require contracts or agreements to be written so as to be understandable by the average citizen. If you want your letters or memoranda, reports or recommendations to be read and reacted to favorably, you have to sharpen up those writing skills. A detailed exposition of the subject appears in Chapter 39, but here are a few hints that may be helpful:

1. Have a reason for writing—a clear purpose.

2. Focus on the reader in terms of position, education, interest, need, and so on.

3. Plan your piece, whether it be a simple memo or a book. Structure it to state your purpose, tie up the reader's interest, and present the facts clearly.

4. Avoid big or unusual words which may not be understood. Avoid complex sentences.

5. Be direct. Aim your message at your target. Write with a specific person in mind, not a vague "someone". Get the reader involved by talking directly.

6. Summarize. Itemize. List.

7. If action is required, state it clearly. Restate it at the end.

8. Re-read what you've written. Is its purpose clear? Is its message uncomplicated? Is action to be taken spelled out?

39.9.   GENERAL GUIDELINES FOR IMPROVING COMMUNICATION SKILLS

Many books have been written on the subject of developing more effective communication skills, and there are literally hundreds of different correspondence, continuing education, and other courses to help you practice what they preach. It is not within the scope of this chapter to provide definitive workshop material, but I'd like to emphasize three specific concepts which may work wonders for you.

1. *Slow Down.* The vast majority of people on the sending or receiving end of the communications pass are in much too much of a hurry. They rattle off their instructions or their messages or they listen with one ear. Slow down. Before you write a memo or a letter, before you pick up the phone to make a call, before you call your secretary in to dictate or to give her a message, before you open your mouth to talk, slow down and think. And think again. Ask yourself some meaningful questions. Start with what it is that you want to accomplish—not what you want to say, but what you want to accomplish. Only then should you start to talk or to write. Make yourself a little outline to guide what you say or write. Take the time to determine why you are going to talk or write.

2. *Don't Assume.* The world and the people in it are too complicated for you to take action based on what you think they already know or based on the assumption that they understand what you are asking them to do. Get feedback. Don't ask "Do you understand?" The answer will generally be "yes" although the facts may be in reverse. Ask the simple questions that tell you, without insulting your listener, that they have received the message you sent.

3. *Delay Judgment.* This tends to come with slowing down, but not automatically. Listen or read to learn what is being said. Make sure you understand the other person's point of view before you start building arguments to knock it down. Learn to listen or read without judging. Absorb and understand what you hear. Then judge.

If you concentrate on these three simple steps: slow down, don't assume, and delay judgment, you are bound to become a more efficient communicator. Wasted words and wasted motions mean wasted time. We have all sat through meetings and heard the chairman of the group say "We're going around in circles. This is getting us nowhere." Even when we haven't heard it said, we've known it. And this is just as true on a one-on-one as it is in a group. Studies indicate that 40–50% of all letters written need never to have been written at all. They tend to be explanations or interpretations of letters previously sent or of past conversations.

Listen to your colleagues or to yourself the next time you get into a discussion of a plant or an office problem. You will readily see how much time is being lost because there is no real communications. Efficient communications can save a lot of time because you don't have to chew your cud twice. You don't have to right wrongs or have the jobs done over again. If you follow these thoughts and remember the barriers which stand in the way of effective communications, you begin to realize how important it is to package yourself and your ideas for maximum acceptance. A significant key is attitude. Work at making your attitude one that encourages rather than discourages those with whom you are trying to communicate. Be interested in what you have to say and what they have to say. Show it. Be enthusiastic about it. Interest and enthusiasm are contagious. They rub off. Get your listener or reader involved and be involved yourself. Communications takes place when people are not just passively sending or receiving but are interacting.

Look for the "hot buttons." Look for motivation angles. What's in it for the listener or reader? Why should they do as you ask? If you concentrate on their needs and their interests, your communications is bound to be more effective.

Effective communications is essential to sound engineering management. You can take a giant step in the right direction if you remember that human beings are reactive before they are rational. To move a meaningful message from your head to another's you must be concerned with how you are sending it, who is receiving it, and all of the barriers that lie in between.

39

WRITING WITH CLARITY, PRECISION, AND ECONOMY

John Louis DiGaetani

JOHN LOUIS DiGAETANI has a Ph.D. from the University of Wisconsin and is an assistant professor of English at Hofstra University. He has specialized in teaching business and technical writing. In addition to his work at Hofstra, he has also taught writing at the Harvard Business School, Long Island University, and the University of New Orleans. He is the author of several books, most notably *Writing Out Loud: A Self-Help Guide to Clear Business Writing* and *Writing for Action: A Guide for the Health Care Professional,* and he edited *The Handbook of Executive Communication.* Dr. DiGaetani's essays have appeared in *The Wall Street Journal, Business Horizons, Computer World, International Management,* and *The Journal of Business Communication.*

39.1.   INTRODUCTION

Clarity, precision, and economy are the major characteristics of good technical writing, and a manager in engineering should write with all these qualities. Clarity, precision, and economy sound so obvious and easy to produce, but actually they are very difficult to achieve because most writing for technical fields is vague, confusing, jargon-ridden, and wordy. Such writing does not communicate a good image of the person whose name is attached to the document, nor to the company he or she represents.

But how can one improve one's writing? How can engineering managers insure that their writing is clear, precise, and economical? The best long range solution for writing problems is to read more. Generally, writing and reading problems go together. The person who does not read much is generally the person who writes poorly. Reading in your field will give you more awareness of your field, but it will also give you an instinct for writing in that field. General reading, for example of *The Wall Street Journal* every day, will also improve your writing. But the more immediate way to improve your writing as a manager is to revise carefully, using the nine guidelines for better writing that follow.

39.2. GUIDELINES

39.2.1. Gear Your Writing to Your Audience

This is something you do automatically when you speak, and if you can do it as smoothly in your writing you can have more control over your reader's response. Often in business writing you will not know precisely who your audience is, but you will know the type of person or persons you are writing to, and you have some knowledge of the characteristics of that type. Capitalize on that knowledge. Regardless of the message, you would probably write to a superior differently from the way you write to a colleague. And a letter to a stockholder should not sound like a letter to a creditor. The master of style knows how to apply a knowledge of what others expect, need, and want in order to tailor a message for a particular audience.

Thus, if you know the person you are writing for, that knowledge can help make your writing successful. For example, if you are writing for a manager who loathes jargon and complex engineering terminology, you would be wise to eliminate them. But, alas, there are also managers who love jargon. They are probably naive, but some executives want documents sprinkled with jargon since they feel that jargon indicates that the writing is really professional. If you have to write for such a naive soul, then shovel in the engineering terminology. Such people, however, are in the minority these days when an avoidance of jargon is generally preferred. For example, examine the following two paragraphs to a prospective client. Which is more appropriate for the audience?

Version A

Your request for information about purchasing silicon chips has been turned over to me. I have enclosed our current catalogue. This catalogue should answer all your questions.

Version B

I am pleased that you are interested in our company's silicon chips. I have enclosed our current catalogue, which gives specifications and prices. If you have any questions about any of these products, please call me at (800) 865-1509.

The writer of Version B sounds like a person who wants some business. The pleasant tone is clearly better for attracting business than the abrupt, matter-of-fact tone in Version A. A sensitivity to the audience creates the difference in these two versions.

39.2.2. Be Assertive and Clear

Assert what you want as clearly as possible, and you will find that you can often get it. Remember that when most people read your writing, they will ask themselves: "What does this person want from me?" So try to answer that question as quickly as possible. However, being assertive is different from being aggressive, which can work in some speaking situations but can be very overbearing and abrasive in writing. There will also be times, of course, when you want to withhold information until the end of a letter or memo for a particular reason. If your message will be distressing and you want to soften the blow, putting the bad news at the end may help. But this approach should be the exception rather than the general rule in your professional writing.

More often than not, your reader wants to know your main point so a decision can be reached about responding. If your reader can't figure out what you want rather quickly, he or she will have less time and energy to do what you want—and your reader may become annoyed as well. Which of the following two paragraphs is more clear and assertive?

Version A

I am writing to recommend that the #12 couplings for the bridge be built with tempered steel rather than carbon steel because of the climate conditions the couplings will have to withstand.

Version B

Tempered steel has been tested for rain, snow, and sleet conditions in addition to shrinkage due to heat and dry air, and has generally been found to be better able to endure such weather conditions than carbon steel.

Version A is very clear, precise, and assertive about the writer's position on the choice of material for the #12 couplings. Most business readers would much prefer Version A because it is clear, assertive, and as a result it communicates more successfully. Version B is a report on some tests and does not convey what the writer wants to see done.

39.2.3.   Be Brief

Notice the differences between the following two passages.

Version A

It is also essential that the interior wall surface of the reactor be maintained in a dry condition and that some means be provided for drying continually the peripheral interior wall surface of the reactor during operation of the device in order to avoid accumulation of moisture and liquid matter about the peripheral interior surface area of the wall of the reactor.

Version B

The interior wall of the reactor must be kept dry at all times.

Version A is no more accurate or technical than Version B, but it wastes a lot more words to say the same thing. Be sure to edit your writing carefully to avoid the dreary wordiness of Version A. Remember that something should not be "very unique," but only "unique," not "a new innovation," but just "an innovation," not "repeatedly the same response," just "a repetition." Redundancy and wordiness add flatulence and boredom to your writing.

Remember as well that few of your readers actually want to read your letter or report, so take as little of their time as possible. Executive time is especially valuable, so without being curt or rude come to the point. As a corollary to this rule, try to use short paragraphs except in formal reports. Long paragraphs, while sometimes unavoidable, create a psychological barrier to your writing. The visual effect of a long paragraph strikes most readers as a formidable reading chore. If the same material can be broken down into shorter paragraphs, it is likely to get a better reception.

39.2.4.   Avoid Jargon When You Can

Certainly, technical language is sometimes necessary in technical reports, but often not only technicians will be reading that report, so avoid jargon and technical terminology when you can. In most professional writing for business, the special use of unusual words should be avoided in favor of simple, direct, and specific words. Remember, your task as a manager in engineering is not to dazzle your reader with your repertory of technical terms, but to communicate. This is best accomplished if you can use words that you know others will understand.

There are several distinct types of jargon, all to be avoided if possible. Technical jargon attempts to use technical terms whenever possible to impress the reader with one's education and professional expertise—for example: "galvanize the personnel reports," "synchronize the time schedules," "digitize the new motors," and so on. Jargon can also use bureaucratic and legalese terms like "as stated heretofore," "network instrumentation diffusion," and "analysis structure facilitation." Administrative jargon is generally from the administration of a particular company and tries to use the "in" terms of that company: "please X-38 those new employees," "will J.R. see the new N.T. foreman this afternoon," "Crompton likes the nuke reports cued by the T.R. Committee before he reads them," and so on. Often jargon is related to a particular industry and becomes excessively used by people within that industry: "Did that student matriculate with Dr. Jones?" "The job market for Mech. Eng. looks hot this year." "Structural Engineering II demands trig and stats." And there is the jargon generated by the latest trend-setting company or industry, and that jargon is picked up quickly by people trying to appear most trendy. These days, second only to the armed forces which have long generated their own language, the computer industry provides the major source for this category of jargon because of its current prestige in engineering circles. Thus this kind of jargoneer might refer to: "data-base matrix structure," "network interface feedback," "modem byts," "floppy disk programmer," or "software packaging."

Often the person who uses jargon most is the naive soul who feels he is impressing people. And he is right some of the time, for there are naive people around and some of them are very impressed with jargon. These people generally believe that the sign of a really educated man is that no one can understand what he is saying. But the more sophisticated and knowledgeable are unimpressed with jargon and find the whole attempt rather pathetic. Jargon in fact often generates suspicion since many people connect jargon with attempts to obfuscate or purposely confuse an audience.

It must also be admitted, however, that some technical journals prefer jargon and long, windy sentences (usually in the passive voice) because they feel that is the style their readers expect in a professional journal. So, if you are writing for one of these journals, you may have to use jargon to get an article accepted. But for most situations and most audiences, jargon is to be avoided since it interferes with immediate and clear communication.

Notice the differences between the following two paragraphs.

Version A

Salaries due you for your vacation leave necessitate submission of a vacation pay voucher (form # 1372 XD) before Wednesday of the week preceding your vacation period. Said voucher must be received by this office, reviewed, and validated before said funds can be approved and released."

Version B

Please send me a vacation pay voucher (form # 1372 XD) by Wednesday of the week before your vacation begins.

The writer of Version A is obviously a lover of jargon who wants to impress the reader with all the big words he or she knows and prove how educated and professional he or she is, but the reader just wants to understand what the memo is trying to say. Clearly, Version B is more lucid because of its use of a simple vocabulary when that is all the situation calls for.

Since engineers are often guilty of jargon where clear, simple vocabulary is possible, another example is in order.

Version A

The specification data that were forwarded to me were herewith examined by me despite their poor programming and tardy input into this office. The data were due by Dec. 10 yet the information was not received in this office until Jan. 9, which is almost a full month beyond the date due. Please restructure this material so that it can be examined and audited as per our request.

Version B

The specification data you sent us were confusing and arrived here late. Please revise these figures, put them in some order, and get them to me within a week so they can be examined properly.

Version B is easier to understand because of its clear writing style and avoidance of jargon. Version A reads like an especially weak rough draft which, after much revision, finally became Version B. Most engineers would prefer to read Version B because of its clarity.

39.2.5. Avoid Clichés

Clichés are overused, weary, unimaginative expressions like "talking straight from the shoulder," "keeping an open mind," "time is money," "sound as the dollar," "pretty as a picture," and "seeing is believing." Clichés indicate a lack of sincerity because they are so worn out and so frequently used that they have become meaningless. If you always write in clichés, people will suspect that you think in clichés as well. Revise carefully to eliminate clichés. If you end all your memos with "thank you for your cooperation," no one will find much sincerity at the end of your memos. If you begin all your memos with "in response to your memo of. . . . ," you will brand yourself a clichéd writer.

Examine the following two sentences, both from a letter responding to a customer's written complaint.

Version A

In response to your letter of March 14, please be advised to examine your contract with us as per our agreement.

Version B

The complaint you mentioned in your last letter can be answered if you would examine your contract with us.

Notice the human tone in Version B, versus the dreary, mechanical clichés in Version A. Which sentence better represents an engineering firm? Most customers would surely prefer Version B.

39.2.6.   Use the Conversational Test

A way of producing the clear, simple, unhackneyed diction we have been suggesting is a method of revision called the conversational test. As you revise, ask yourself if you would ever *say* to your reader what you are *writing*. If you were talking to a business colleague, would you ever say, "In response to your memo of 11/18/85?" If you did, he would probably laugh. If you were speaking to a customer, would you ever say, "Enclosed please find your order for three silicon chips?" If you did, your customer would surely think you were weird.

But the conversational test, while a valid method for revising business writing, is not a mandate for slang. If you were speaking to that colleague, you would maintain a certain formality unless you were close friends. If you were speaking to that customer, you might be talking to a total stranger. Certainly, you would not use slang. But using the conversational test will eliminate the mechanical sounds of business jargon, clichés, and excessive examples of the passive voice. These are the things that can make your business writing so embarrassing. Once they disappear, your own humanity will add color and interest to your writing.

Thus, if you were speaking to a colleague, you would probably say, "I am writing about the question you raised in your last memo." If you were speaking to that customer, you would probably say, "Here are the three silicon chips you ordered." The conversational test will help you to write as naturally and persuasively as you speak.

You might imagine a regional sales manager calling up one of his district managers to say, "We don't have the final figures for the quarter yet, but word of mouth has it that they're bad. Inventories are too high again, and we might be cutting out some dealerships soon. Just wanted to let you know. I'll send you the figures as soon as I can, and I'll be calling a meeting for the district managers and the whole sales force."

Too often, such a person's follow-up memo will read:

Version A

Re our telephone conversation of June 14, final sales totals for the quarter ending in June are enclosed herewith. A planning conference for all sales personnel will be scheduled for the near future and these figures will be discussed. It is hoped that all district managers will be aware that the figures are such that reductions in the total number of retail units may be indicated. Thank you for your cooperation.

Why does the man who seems so direct and clear on the phone make himself sound pompous, mechanical, and stilted in his writing (not to mention the familiar redundancy "enclosed herewith")? First, he is probably insecure about his writing skills. He doesn't trust his own use of language enough to let himself write naturally, as he speaks. And he thinks that somehow jargon, wordy expressions, the passive voice, and puffy sentences will make him appear better educated or more polished than he fears he really is. Second, he may be under the impression that business writing is supposed to be stuffy, roundabout, and impersonal, since so many of the memos he gets read that way. Finally, he may be timid about putting certain information—in this case, bad news—too bluntly, especially in writing.

But chances are your command of English is plenty good enough if you just write things down as naturally as you say them. Trying to puff things up to give added heft to your everyday thoughts is rarely necessary in business and only makes you sound like a blowhard or a computer printout. Such writing impresses only the naive. Lots of business people adopt such a stilted style when writing, but the truly savvy are not impressed.

Direct, clear writing which approximates the way reasonable people speak will never embarrass you. Trying to hide bad news in a fog of wordiness just doesn't work. A reader of the memo printed above, once he or she manages to figure out its content, is going to be annoyed and maybe amused. If you're the writer, you might as well state your message simply and clearly. Spare your reader the added burdens of trying to puzzle out a meaning, whether pleasant or otherwise. And the conversational test will help you here.

Using the conversational test, that memo could be revised to the following new version:

Version B

Please attend a sales conference on Wednesday, July 16, in the conference room starting at 10. A.M. We will be discussing the June sales figures, which are attached, and their implications for our current strategy. The figures might indicate the need to reduce inventories so we will discuss this option at the meeting. All district sales managers and salespeople should plan to attend.

Most readers will find Version B, revised by using the conversational test, much easier to understand. The new version also makes the writer sound like a reasonable, responsible manager rather than a pompous, stilted bureaucrat. But one warning: the conversational test will work best if your parents did not speak a foreign language or a dialect of English at home. Hearing standard American English at home gives most people an instinct for the language that foreigners or dialect speakers may lack. If you did not learn standard American English at home, you will have to be more careful and learn to revise according to formal rules of grammar and usage. For most business people, however, the key to better professional writing is as close as their mouths. Using the conversational test will vastly improve the quality of their business writing. It will make their writing as interesting and human as they are.

Lastly, as in most business situations, your own natural humanity is a valuable asset. Most people don't like working for someone who comes across as not really human; and if your memos as a manager sound inhuman, their unintended message may subvert your own efforts to create a better rapport with your subordinates, colleagues, and superiors.

39.2.7.   Favor the Active Voice

Active voice: Ralph designed a new turbine engine.
Passive voice: A new turbine engine was designed by Ralph.

In general, use the active voice rather than the passive voice. Most English sentences are in the active voice, so most people find the active voice easier to understand. Given that reality, you will communicate more easily if you use the active voice. There are some exceptions to this rule, but not many.

Also, most people don't use the passive voice in conversation, so avoid it in your professional writing. There are times when a passive voice construction is useful—for example, when the direct object is more important than the subject of a particular sentence, for variety, or for purposes of tact—but these are rare occurrences. Notice the differences in verb voice in the following two paragraphs.

Version A

Your letter of July 20 was forwarded to me. The insurance claim which was filed by you was received by me on August 15. This claim was investigated and approved, and your check for $530 was mailed out today under separate cover.

Version B

We have reviewed and approved the insurance claim you filed on July 20. We have issued a check for $530 to you for this claim. The check has been mailed separately and will reach you soon.

The major difference between these two paragraphs is verb voice. All the verbs in Version A are in the passive voice while those in Version B are in the active voice. Most readers much prefer Version B because the active voice does succeed in making that version sound more human and more immediately understandable.

39.2.8.   Adopt a Positive Outlook

Emphasize *can* and *will,* rather than *cannot* and *will not.* For example, if you can't get the widgets out until the end of the month, you can tell a customer to expect the widgets at the end of the month rather than that you "can't get them out" till then. It is better to emphasize the positive.

There are three major advantages to emphasizing the positive in your managerial writing. The first is that it usually reflects well on you, suggesting that you are a "can do" person. Emphasizing the positive also generates good PR for the company you represent. And finally, such a positive approach gives your reader good news, which makes for happier customers.

39.2.9.   Be Tactful

Avoid antagonizing people in your professional communication; and when you can, avoid blaming people, even if they deserve it. You do not benefit from making enemies, and they might get even later. Tact is that wonderful and rare ability to be sensitive to the way your reader hears what you are saying and to communicate without creating suspicious or hostile reactions.

To be promoted in a large corporation often depends on a network of friends, but the tactless person usually has a network of enemies he does not even know he has. But how can writing be tactful? There are several methods. The first is certainly word choice. You can say that a person is "incompetent" or you can say the person "needs more training." If you call a person "incompetent," you have created an enemy. Being aware of the connotations of words, their shades of meaning, is a way of avoiding the words with negative connotations, and remaining tactful yet still honest. Other ways of being tactful include using the passive voice to avoid blame and using conditional words ("should," "would," or "might") to be evasive when that is to your advantage as a manager.

For example, look at the following two memos.

Version A

All employees must attend our quarterly meeting this Monday, August 10, from 4 to 5 P.M. in Conference Room C.

Version B

Please attend our department's quarterly meeting this Monday, August 10, from 4 to 5 P.M. in Conference Room C.

Version A would certainly offend many people because of its lack of tact, while Version B works better. Why treat people as children to be ordered around, when they are adults who will resent such treatment? The tactful manager will get much more cooperation from his or her staff.

39.3.   A CONCLUDING NOTE

A manager of an engineering firm has special communication problems. He or she usually has to communicate with both technical and nontechnical people. By using the above guidelines, your writing can be a professional document that is a credit to your department, company, and yourself. Clear, precise, economical writing communicates more effectively in a business setting. Memos, letters, proposals and reports, when written with these nine guidelines in mind, can make communication easier to understand and more successful.

40

EFFECTIVE LISTENING

Donald A. Christman

DONALD A. CHRISTMAN has over 15 years of experience as a training, marketing, and business development consultant. He has done a wide variety of specialized
projects for major industrial corporations and financial institutions. Mr. Christman
has also written a number of books and articles on marketing, business development
and human resource subjects. He is Senior Vice President with responsibility for
marketing, product development, and client services with the Performance Services
Group, Inc., a consulting firm in New York City.

40.1. OUR POINT OF DEPARTURE

This chapter explains how the engineering manager can (and indeed must)
use the skills of effective listening in dealing with both subordinates and superiors within the organization. We will undertake to show that the process
of listening is a critical determinant of functional efficacy in the organization, and that the qualitative level at which listening takes place affects individual and collective performance to a degree not widely apprehended by
engineering management people.

Nearly everyone can be, and should be, a better listener. Although the
subject has been widely addressed by human resources professionals, it has
been usually presented in the somewhat limiting context of sales training.

While highly relevant to that frame of reference, the subject transcends such attempts, whether intentional or unintentional to compartmentalize it. As a universal and indispensable ingredient in human interaction, we will all benefit from a more rigorous and systematic effort to understand its dynamics than the subject usually receives. Let us then begin with a working definition which will serve our purposes in this chapter. *Listening is a purposeful information gathering and processing activity which takes place at several levels of awareness and in a variety of interpersonal formats.*

40.2.   LISTENING AS AN ACTIVE VERSUS A PASSIVE PROCESS

The reason why so many people are poor listeners is because we become imbued early in life with the attitude that listening is a passive activity. From that fundamental misconception springs such pernicious progeny as the notion that listening is, therefore, of merely peripheral importance—a sterile interlude between what one has said last and what one intends to say next. A life-long habit thus cultivated, albeit unconsciously, is therefore rarely perceived as the gross impediment to efficacy and performance that it indisputably is. These habits of thinking, moreover, are tenacious and require vigorous remedial intervention and a high level of personal motivation to overcome. The effort to do so, however, will richly reward the committed individual through visibly better job performance at lower expenditure of energy, and across the entire spectrum of life experience. A personal decision to improve one's listening ability should be preceded by understanding and accepting the fundamental proposition that effective listening is always an active diagnostic process employing specific skills and techniques. It is not a passive, random, tune-in/tune-out activity. Forearmed with that knowledge, progress becomes possible and significantly enhanced listening abilities predictable.

40.3.   THE HEAVY PRICE OF POOR LISTENING

To describe the myriad ways in which poor listening habits create and perpetuate confusion, foster and exacerbate misunderstanding, impair efficiency, cripple productivity and cause endless mayhem and mischief in human and institutional affairs, would require more space than this chapter could provide and more patience on the part of the reader than we could reasonably demand. Our purpose will be better served by reminding the reader of an amusing old parlor game. A designated person verbally conveys a simple piece of information to another and asks that person to pass the information verbatim to the person seated next to him. This goes on until the information is passed in turn to perhaps eight or ten people. When the last person in the chain is asked to reveal what he or she has been told, the massive distortions and deviations in content and meaning will nearly always shock and amaze. Why does this occur? The answer is inescapably clear: poor listening skills are passed along with compound interest in the coinage of confusion from one link in the human chain to the next.

The engineering manager should consider the importance of effective listening in relation to both personal and organizational contexts. The engineering professional functions in a matrix involving organizational efficiency and personal performance. Both will be improved as the quality of listening becomes better.

40.3.1. Personal Effects of Poor Listening

Inefficient listening on the part of the engineering manager will inevitably create or intensify problems in a number of areas which bear directly upon his or her personal performance and productivity:

Impaired ability to interpret accurately and implement verbally communicated directives from higher management

Impaired ability to benefit maximally from staff input received in one-on-one dialogue or in staff meetings

Increased likelihood of misunderstanding and conflict with peers

Decreased sensitivity in interpersonal dialogue with subordinates—less awareness of actual or potential problems concerning attitude and morale

Functional estrangement from the organization resulting from lower level of awareness and understanding

40.3.2. Organizational Effects of Poor Listening

Poor listening habits on the part of the engineering manager will have a ripple effect at both intradepartment and intercompany levels.

Feedback problems resulting from instructions and directives—inefficient supervision of project development

Lost time through lapses in communication with subordinates

Compromised awareness of project and personnel problems as they arise due to poor feedback

Impairment of interdepartmental coordination due to lack of understanding or misunderstanding

Company-wide productivity penalty caused by the effect on work flow of performance and morale problems induced by defective listening

40.4.   GOOD LISTENING—WHAT DOES IT MEAN?

The many problems and performance impediments caused by poor listening become abundantly clear to anyone who takes the trouble to think the subject through. Until the subject is called to our attention, our awareness of it as a major problem in human communication probably remains at best at a subliminal level. But once the problem in all its productivity crippling potential has been presented to us, it should not be necessary to belabor it or make its obvious seriousness an exercise in redundancy. The question is, what, if anything can be done about it? First of all, a word of reassurance. The problem, even in notoriously poor listeners, is remediable. The malady is curable, but it does require a high level of commitment on the part of anyone who now finds himself a victim of poor listening habits acquired early in life and reinforced over a period of many years. We suggest that the problem be seen in its three fundamental causative aspects:

1.   Habit
2.   Inclination
3.   Lack of tools and techniques

Let's take them one at a time. Habit is a harsh taskmaster. Once enmeshed in an undesirable habit pattern of any kind, a genuine sense of commitment on the part of the individual is the first necessity. All else follows from that, and nothing does if it isn't there. There is a nuance in perspective here however, which is of critical importance to our prospects for success. It is difficult to the point of an impossibility to break an ingrained habit in a purely negative mode of attack. The individual who says to himself or herself, "this is a bad habit pattern, and I must therefore fight to overcome it" is nearly always doomed to a long, frustrating, and ultimately unrewarding battle. Much more successful is the positive attack mode which says in effect, "I have a habit pattern which is compromising my effectiveness, but the acquisition of certain positive behaviors will transform it." The focus makes all the difference. In the first instance, the individual focuses on the strictures of the habit and in the second instance he or she focuses on the remedial behaviors which need to be acquired. The first approach relies purely on an act of will. It is like stretching a rubber band. As soon as the force is removed, the rubber resumes its original shape. The second approach builds in new behavioral ingredients which alter the configuration of the problem, therefore setting in motion its metamorphosis into a non-problem.

The habit of poor listening is therefore our starting point. This is where we make our commitment to improve. But we make it by deciding to acquire the ingredients of behavior which will neutralize the problem, rather than to fight the problem itself.

The next aspect of poor listening is inclination. The simple truth is that we have been impelled toward the acquisition and reinforcement of poor listening habits through our natural inclination to do so. Ego is at the root of this problem. We occupy the center of our own universe. We are the most important person we know. Unfortunately, we are often more interested in talking than in listening. The activity of listening is therefore relegated to second-rate status in our discourse with others and becomes a mere interlude between our own conversational gambits, a pause dictated by social amenity, or the need to catch our breath. The absence of an active, diagnostic listening activity on our part destroys the possibility of efficient communication. When we do speak therefore, we talk *at* the person, not *to* him or her. If the other party to the dialogue shares this proclivity, the confusion quotient is correspondingly greater. In fact the very word dialogue becomes a misnomer. What you have is two monologues in time-contiguous orbits.

The line of attack against this aspect of the problem is simply to recognize it, admit its presence, and resolve to neutralize it. It is helpful to realize that our long-term ego rewards will be greater through acquiring and demonstrating a capacity to listen actively, creatively, and diagnostically. The results we achieve in better performance and better relationships will speak directly to our ego needs, thus reinforcing the acquired effective behavior.

So far in this discussion we have considered the role of *habit* and *inclination* with respect to listening problems. We have seen that poor listening habits are usually acquired early in life as a consequence of the mistaken perception that listening is a passive activity. These habits are then reinforced through inertia. There is a failure to recognize how dysfunctional and counterproductive they are, with the resulting absence of motivation to address the problem. We have also seen that natural inclination plays a role in poor listening. Human egocentricity is the culprit here. We tend to focus more on what we want to say rather than what the other person is saying. In its extreme form, the requirement to listen at all is accepted only grudgingly, and we spend the time in which we should be actively listening in planning what we are going to say next. When two people share this predilection, the only possible result is a dialogue of the deaf.

We now approach the third leg of the triad in terms of causation—a lack of knowledge of the tools and techniques with which to attack the problem. Effective listening is a skill or, more precisely, a set of related skills. If this were not true, this chapter would serve little purpose other than to provide a forum for lamentation on an intractable problem. Fortunately we can do better than that.

Good listening habits can be achieved through the systematic application of a few simple techniques—or skills as we prefer to call them. With adequate practice and reinforcement they can be grafted as new behaviors and integrated naturally with your personal style. Once you have made them a part of you, you will use them automatically. We think you will be surprised and pleased at the benefits you will experience—a significantly enhanced ability to listen, organize, and retain information with less expenditure of energy. This chapter will acquaint you with the principles of good listening and provide some examples of how the skills are used. Simply reading this material of course will not build a capability for more effective listening. You must make a commitment to yourself to work with these skills, to practice them until they become thoroughly internalized. With sufficient practice and day-by-day reinforcement, you will notice that the skills become self-activating—you will use them automatically without special effort. It's really no different from learning to play golf or tennis. Step one is learning *about* the techniques and step two is making them your own through practice and repetition. Now let's take a look at some simple, but efficient skills in good listening.

40.5.   PRIMARY AND SECONDARY POINTS

If you analyze verbal communication you will see that the significant and relevant portions of any person's conversation can be organized on the basis of its primary and secondary points. The primary point is the main idea which the person is trying to communicate. The secondary points (there may be only one or there may be several) relate to the primary point. Secondary points elaborate and expand the primary point by adding relevant information.

Example:

"Jim, this project schedule you gave me looks very tough. There's a lot of basic design work involved here, and it looks to me like there will be problems with the configuration we were talking about. I don't see how we can work our way through this, particularly with the heat and stress specifications we're stuck with, in only six weeks. I think we need to discuss the whole situation and rethink the time factor."

In the preceding example, the primary point was:

This project schedule is tough.

The secondary points were:

a lot of basic design work
problems with the configuration
heat and stress specifications
need to discuss and rethink

Do you see how the concept of primary and secondary points organize the content of whatever was said? Make it a rule whenever you are engaged in dialogue to scan mentally the verbal material so as to identify the primary point. Then go on and identify the secondary points which further elaborate the primary point and try to position them in a hierarchy of relative importance. You will find this to be a very useful mental exercise. It will organize the content of verbally transmitted material in a rational manner and put it in a framework of structure and meaning which will be easy to retain and retrieve whenever necessary. Let's look at another example.

"George, I've looked at the parameters we have to address in this design. Concerning the vibration and shock control aspect, I think we need a lot more information about the environmental factors, and the damping characteristics of the material under high G loads. Do you have any suggestions?"

Read the above example again and identify the primary point. Then go on and identify the relevant secondary points. Try to arrange the secondary points in a descending order of importance.

You probably identified "vibration and shock control" as the primary point. You then probably arranged the secondary points in approximately the order shown below.

environmental factors
damping characteristics under high G loads
suggestions?

Work with this concept of primary and secondary points. As you do, you should find that you are organizing verbally transmitted material much more efficiently and that you are better able to retain it and to retrieve it in concise summary form whenever you wish. When you use this technique, you are making the listening process active instead of passive. You are scanning the data for organization and meaning. You are listening *diagnostically.*

40.6.   KEY WORDS

What we call key words are really the facilitators of good listening techniques. If we may draw an analogy from the age of the personal computer, key words can be thought of as the entry software to our program for effective listening. Key words are memory joggers—"stand-out" words which help you to remember primary and secondary points. Key words are usually "action words," and of course if they are truly "key," they must connect with primary and secondary points in a logical way. You should remember that there are no absolute rules in selecting key words. Key words are simply a concept which helps you to zero in on the substantive elements of what is said. They should function as effective memory aids for *you.* Let's look at an example:

"Mr. Johnson, the microcircuitry on this design is giving us real problems. I'm afraid we may have a can of worms here. The people in the lab tell us they can't tell if these chips are wrong for the design application, whether they are inherently defective or whether we are seeing an unusual quality control problem. And, with the Air Force project evaluation team due next week, it looks like we're really under the gun."

In the preceding example, the key words were "can of worms" and "under the gun." Note that both of these are colloquial "stand-out" words which fix attention on the problems articulated by the engineer. They are good memory joggers because they are colorful phrases which will help the listener to remember the critical issues raised by the speaker. By picking out key words of this kind, the listener will find it easier to assimilate the material he has heard. Such key words facilitate both the reception and later retrieval of verbal material. As mentioned before, key words can thus be viewed as the "entry software" for your personal enhanced listening skills program. Let's consider one more example:

"Bill, I've heard that the modifications on this new light alloy engine are seriously behind schedule. If that's true, then it's going to be a case of locking horns with production because we can't live with any slippage of the new model introduction date. The competition would have a field day with that."

Read the above example again and pick out the key words. If you identified "locking horns" and "field day" as helpful key words for identifying and retaining the central message of the statement, you were correct. Can you see how keeping a lookout for memorable key words can contribute to better listening and better retention?

40.7. SCREENING OUT

Another important and effective technique in building better listening capability is the process of systematically screening out and discarding nonessential or superfluous material. Much human discourse is larded with a great deal of "excess baggage" in the form of verbal material which digresses from the subject being discussed. Such material makes no contribution to the quality of communication, damages clarity of language, impairs precision of meaning, and simply wastes time. One of the best listening skills you can cultivate is to be alert to the nonessential and extraneous in verbal communication, screen it out, and discard it. If you will make this a cardinal rule, you will be amazed at how much more effectively you can listen to what people are *really* saying. Shorn of all the verbal underbrush, it becomes a relatively simple matter to listen, organize, diagnose, and retain.

The three commonest types of "screen out and discard" verbiage are:

Irrelevancy
Bias
Redundancy

The following example will illustrate how irrelevancy, bias, and redundancy can overload and obscure what would otherwise be a clear and easily understandable message:

"I know that this project is running behind schedule, Steve, and I'm as concerned as you are. First of all, we're still waiting for the test results. Have you ever tried to talk to those characters over there? Acme Testing, I mean. Not one of those turkeys will ever give you a straight answer. Am I right? Remember that song and dance on that last project? Sam—what was his name, the guy who handled that for us? He's gone now. Do you know what happened to him, by the way? Well, anyway, I've always said you can't depend on that outfit to do anything on time. I've always said that they wouldn't recognize a target date if one walked up on two legs and bit them. In any case, I called them again this morning. I talked to Fred Black. Do you know him? He's not a bad guy, actually. At least you can talk to him. Anyway, Fred says he'll have the data over here by next Monday morning. So I guess we'll have to live with that."

What you just read was, of course, an extreme example of irrelevancy/bias/redundancy overload. When all of the superfluous verbal fat was discarded, the substantive content was reduced to two sentences:

1. We're still waiting for the test results.
2. Fred says he'll have the data over here by next Monday morning.

The above example, overdrawn to make the point, certainly demonstrates the value of rigorous attention to screening out and discarding verbal material which does not contribute to clarity or content.

40.8.   DISTRACTIONS

Another common roadblock to good listening are distractions of various kinds. The most frequent types of distractions experienced in verbal discourse are:

Emotion
Peculiarities of speech (accents, impediments, etc.)
Background noise

Many people drift into the habit of focusing on such things to the detriment of their listening ability. Emotion, whether positive or negative, tends to rivet the attention and when this occurs discernment and comprehension suffer. Likewise, language peculiarities have a tendency to subvert the attention. The listener idly focuses on the delivery system instead of the message, with disastrous results in terms of what is absorbed, understood, and retained. Background noise, such as traffic sounds, machinery, ringing telephones, and so on, also thrust their blatant but vacuous messages in front of us, undercutting our reception and comprehension of important verbal content.

There is no magic cure for these problems. But the challenge is clear. If listening ability is to improve, the mischievous role of distractions must be understood and acknowledged. A concerted effort must be made to screen out such material and focus the attention on the substantive elements of the verbal message. The analogy to learning golf or tennis again suggests itself. Read the principles by all means, but then must come the practice and reinforcement which alone can build and establish the desired behavior.

There are also a couple of facets to the subject of listening which do not involve the verbal message itself. It will be worthwhile to touch upon them briefly because they definitely are part of the picture. The kind of active, diagnostic listening advocated here often depends in part upon nonverbal messages. To the extent that these can be interpreted correctly, they often contribute to meaning by clarifying what might otherwise remain ambiguous.

We know that communication among human beings takes place on a number of levels. As human language evolved, the spoken word supplemented other modalities and became the focal point of the communication process. The verbal component was elevated to its present position of primacy and we began to relegate to an unconscious level other modes of communication. We have coined the term "infraverbal" to refer to these other modes. The Latin prefix *infra* means "below" or operating at a sublevel of awareness. The "body language," which has been the subject of a number of books in recent years, is certainly an example of infraverbal communication. Visual contact and eye behavior is another. Some verbal expressions inadvertently convey a reverse meaning on an infraverbal level, and these might be included in our definition. While it is beyond the scope of this Chapter to deal with infraverbal behaviors in detail, their significance in human interaction should be realized. The engineering manager with sharply honed listening skills, in addition to using the techniques explained in this chapter, will also have cultivated a heightened awareness and an acuteness of perception with respect to the infraverbal modes of communication. People do not always say what they mean or mean what they say. Skilled listeners will always keep their eyes, as well as their ears, trained upon the person or persons with whom they are engaged in verbal discourse. The effective listener will strive to integrate the words which he or she hears with the gestures and infraverbal behaviors which they see. His or her mind will be constantly at work trying to process the data and extract from it a distillate meaning.

Figure 40.1 depicts in visual terms the listening process as it should be practiced. The skills and techniques described in this Chapter, as well as infraverbal awareness, should be seen as complementary components of the posture which the really effective listener maintains in any and all verbal dialogue with superiors, peers, and subordinates.

As this chapter has attempted to show, effective listening does not occur as a happy accident. Listening must be learned. The skills and techniques to improve listening are there and ready to be put into practice by the engineering manager who recognizes the need and is motivated to act and follow

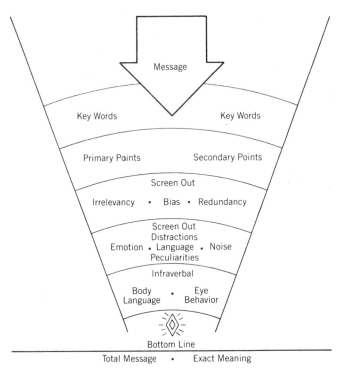

FIGURE 40.1.   The listening process.

through. The requisite behaviors can be built with remarkable rapidity if the commitment is there. To those who manifest that commitment to excellence, the rewards will follow naturally. Not the least of these will be the perception and reality of greater personal effectiveness. And it is not unreasonable to suppose that these will have their due impact upon a manager's future career trajectory.

41

SUCCESSFUL MEETINGS
AND PRESENTATIONS

Kenneth H. Recknagel

KENNETH H. RECKNAGEL, President, Leadership Development Associates (LDA) in Westwood, New Jersey, is a specialist in management and organization development and communications. He was co-founder of LDA and author and designer of its principal public program for management, The Leadership Workshop and The Participation Management Workshop. He also served as senior conference director and program developer for American Management Association's Executive Effectiveness Course; he is a consultant to management in a broad range of industries and public agencies, specializing in organization planning, managerial counseling, team development, and communication analysis and improvement.

Dr. Recknagel has B.S. and M.S. degrees in Industrial Sociology and Social Psychology from the University of Wisconsin and a doctorate in Intergroup Relations from Columbia University.

41.1. ASSESSING THE NEEDS

Trying to reach managers by phone in our own or other organizations is often a difficult and frustrating experience. Eight out of ten times the only result is a graciously apologetic or brisk, matter-of-fact message from the secretary—"No, I'm sorry, he's in a meeting." Repetitions of that experience leave you feeling that managers must spend 90% of their work day in meetings.

Managers who are caught up in that kind of schedule face serious consequences in terms of management of time—completing their own personal responsibilities and assignments and supervising and communicating with subordinates.

The opposite extreme also exists—managers initiating and conducting few if any group meetings with their subordinates. When working with executives and supervisors from a wide variety of companies and industries, it is surprising how frequently one encounters reports from subordinates indicating a serious need for more active communications and team-building efforts by departmental managers.

"I've worked for this boss for two years and she's only had a couple staff meetings during that entire period." Or, "We tried staff meetings for a few months, but they went so badly everyone was relieved when our boss discontinued them." "We never get together as a group—we all go our separate ways—do our own jobs, but there's no sense of teamwork or overall departmental direction."

So we do find a wide spectrum of management meeting practice from too often or too long to seldom if ever.

Managers need to learn to achieve that practical balance between overcommunication (group-think), and undercommunication (isolation). The manager, by title, inherits the responsibility to build and maintain a communications system inside and outside the organization; to develop an awareness, a sensitivity as to when there is a need to communicate face-to-face with members of his or her team, with upper management, interfacing departments, and customers or vendors.

The rule of thumb is "Call a meeting when *group interaction* will accomplish your goals." Conversely, this suggests that you select another communication medium if group interaction is not required. This general guideline is useful, but does have certain pitfalls. For example, the manager may be highly task-oriented and focussed largely on *work-production goals,* but be insensitive to more subtle *people needs* and *goals.* Engineering curricula often lack human relations courses or slight the topic and this, combined with prior personality traits, may leave an engineering manager ill-prepared to respond to the motivation, morale, and group affiliation needs of the people reporting to him or her. Group interaction becomes valuable when the manager's goals are both task- and people-oriented.

On this premise group meetings are customarily held:

To inform (present facts, needs, goals)

Plan strategies

Solve problems

Coordinate efforts

Motivate

Evaluate performance/results

Secure feedback reaction

Build a team

In summary, calling a meeting should not be a casual, off-hand decision. The need should be clearly identified, purpose defined, and plan for the meeting comprehensively outlined. "It must be Tuesday, we're having a staff meeting," captures the unthinking, routine nature of many manager's practices. Habit in such cases replaces productive purpose, and this is the downfall of many group gatherings. Habit, rather than rational management, is the downfall of many managers.

When you add up the hourly rate of the combined salaries of a department, then to call a meeting of that group simply as a formality or tradition is certainly not cost effective. The impact on productivity in the hours before and immediately following the meeting is an additional cost. And how about all those things that need to be handled and weren't, simply because, "Sorry, he's unavailable—he's tied up at a meeting?"

Be sure that in calling a meeting of the department, the manager is seeking *interaction* of the *group*. Otherwise a memo, telephone call, individual or subgroup conference is a less costly and more effective alternative.

41.2.   PREPARATION

Having determined that a group meeting (e.g. departmental) is justified, a precondition of success is that the manager and participants are prepared. Nonproductive meetings are the nemesis of managers and attendees. Without adequate preparation the meeting is likely to disintegrate into an exchange of hearsay, wallowing in self-centered opinion tugs-of-war, indecision, frustration, boredom, and passivity. A sure way to project a negative, ineffective image as a manager is to run poor staff meetings.

Here are some questions that need answering prior to the meeting. The questions are simple: The answers may need to be considerably more profound.

1.   Have I clearly defined the purpose in my own mind?

2.   How can I best communicate this purpose beforehand to those attending?

3.   How can I translate this purpose into a workable, productive agenda?

4. When should the meeting be scheduled and for how long?
5. Who should attend?
6. How should the agenda items be handled?
7. What should be my own role and behavior at the meeting?
8. What expectations do I have of other participants' roles and behavior?

Although these questions follow a sequential logic, they are actually interlocking and fluid as to order. The answer as to who should attend may well depend on how you feel the various agenda items should be handled. When you schedule the meeting or when it is held, may have a lot to do with who is there, what indirect message you want to send to participants and others, and what kind of perception you want to have others gain from the meeting.

For example, if you want to encourage a positive interfacing and team relationship between subordinates working at two different geographical locations, you can schedule staff meetings on a rotating basis between the two sites or you may bring everybody to a neutral, centrally located third site. On the other hand if you desire to reinforce the authority of your own position, have everyone come to your office, your home ground. This sends the message that authority is centralized here, in this office, in this position.

The convening time of the meeting also communicates the intent and process that you consciously or perhaps subconsciously desire. Scheduling a meeting at four o'clock on a Friday afternoon sends one message: brevity, efficiency, discipline, (or, alas, in some organizations, a chance for work maniacs to indulge their proclivities). Hold the meeting at an attractive resort on a Friday afternoon, followed by dinner with spouses and the expectations will be quite different. Managers often arrange staff meetings at the beginning or end of the week to enable the department to review the immediate past and plan for the upcoming period. (Friday relief/Monday anticipation).

Managers often synchronize the timing of their staff meeting with the corresponding upper-level meeting which they attend. This provides for a flow of information from top down with a minimum of delay and dissipation of "managements' message." A participative organization would reverse the sequence to encourage a bottom's-up process.

The length of time set aside for the meeting, either peer or subordinate group, if of a routine information- giving or sharing nature, should not exceed one hour and can often be shorter. Listening requires *concentration* of effort. The attention span of many is considerably shorter than the demands we often wish or need to make to accomplish our purposes. Compensation for this fact should be reflected in the scheduling of meetings, lengths of agenda, and presentation approach.

Often shorter sessions, held more frequently, are more productive. If your purpose is to present information, announce a decision, highlight a problem or concern, or pose a question which requires a thoughtful, delayed response, keep it brief. Work at projecting a crisp, organized, concise image, executed with a minimum of treading of water—Fifteen to twenty minutes, in and out with the facts presented, and the follow-up plan provided by a written outline so that everyone understands what's expected in response. In many instances, this kind of communication could be better accomplished without a meeting—presented instead by written report, memo, or request, if confidentiality is not a major concern.

Another important preparation step is the development and dissemination of the meeting agenda. All of us have probably experienced the meeting that opens with the question—"Well, what do you want to talk about?"— usually followed by dead silence. Or equally disastrous, the predictable talker takes over and that becomes the agenda.

A well-organized and prepared manager and participant develops an agenda a few days before the scheduled time of the meeting. Agenda items are obtained from participants, in addition to those provided by the convenor. Requests for reports, proposals, or other inputs by participants can be included. Whenever possible, the agenda should be distributed well before the meeting, even if such requests are not made. A typical agenda might be as follows:

October 10, 1985

MEMO TO: Engineering Department Section Heads and Supervisors
FROM: Chief Engineer
RE: Staff Meeting, 9:00 A.M., Thursday, October 21st
Conference Room A

AGENDA
1. Management announcements (Chief Engineer)
2. Status Report - Quarterly Performance against annual goal (Chief Engineer) *All* please be prepared to present your section's or group's performance against goals status report with hand-outs rating positive-negative variances and plans for addressing these variances. (10 minutes per person)
3. Summary presentation and discussion of results of recent departmental morale survey. (Personnel Manager and Chief Engineer) You all have copies of the questionnaire and your group results. Please be prepared with questions which are of particular concern to you regarding overall departmental morale and suggestions as to how we could address the priority improvements. I am

particularly looking for your suggestions regarding ways of increasing *non-supervisory employees participation* in the department or a plan to resolve problems contributing to poor morale.
COFFEE BREAK—10 minutes
4.   Reactions and suggestions regarding the company's cost reduction program. I have your written input and will highlight the department's response. Two recommendations are of particular interest. Section Manager B and Supervisor A will highlight their recommendations and rationale for your consideration. (see memo, September 30th) (15 minutes)
We will seek to reach consensus as to their implementation department-wide.
ADJOURNMENT: 11:15 A.M.

The sample agenda communicates where, when, and why the meeting is being scheduled and how long it is likely to last. The purposes of some agenda items are clarified to focus the preparation and direction of thinking of participants. Assignments have been given to specific individuals with approximate time deadlines in order to secure the responsibility and contribution of participants. As is noted, supporting documents were distributed in advance to permit study and preparation of reasoned responses.

This agenda communicates not only what will be conducted, but communicates the chief engineer's plan or expectation of the group process which will most appropriately address each issue. This group process ranges broadly in style and approach. It demonstrates that the manager has a plan. It reflects the intention of the manager to play an overall coordinating and moderating role, to make specific informational inputs or presentations, shift the focus to other members of the staff, and to expect and spark discussions and participation. An outside resource is also used, in this case the Personnel Director. Variety, tempo, and involvement are all ingredients in the plan. Presentation of facts, opinions, ideas, exploration and decision making are activities programmed into the process.

In addition to preparation of purpose, agenda, schedule, and participants, the manager has also to consider the physical setting of the meeting. Staff meetings are customarily held in a conference room made available by the organization. The room should provide a comfortable setting with table and ample seating space around it. Facilities for visual aids—chart pad, blackboard, corkboard, or screen—should be available as needed. There should also be arrangements for privacy and freedom from interruptions. Pads and pencils are provided as well. If practical, the manager should seat himself or herself at the foot of the table—leaving the visual aids area free for presentation.

41.3. CONDUCTING THE MEETING

One can of course do a good job of planning for a staff or other meeting, giving detailed attention to the form, but falter in dealing with the substance. Presentations by the manager set the norm in style and effectiveness for the group. If they are handled professionally, subordinates will generally reflect similar performance. Bosses can be inspiring or demoralizing role models.

When making presentations the following guidelines are helpful:

1. Clearly identify the purpose of your presentation and what you're expecting from the audience.
2. Have an outline of major headings on a chart pad prepared in advance.
3. As you are talking, print or write in key highlights in the outline.
4. Facing the audience, stand to the left of your visual aids to maximize the audience's field of vision.
5. Maintain eye contact with your audience, select a person to focus on while you are making a point. Then move to others. Communicate with your eyes, body position, and verbal references with as many people in the "audience" as possible. Remember, you convened the meeting because you felt that *interaction* was vital for the subject. Making a presentation requires interaction, if only nonverbal.
6. If time permits, ask a question, make a reference which encourages members of the group to express their agreement, seek clarification, register an opinion, or add a pertinent fact.
7. Practice variation of pitch, tempo, gestures, and body movement to provide emphasis and motivate the listener.
8. Avoid any distracting habits you have developed—dangling coins in your pocket, the "ah's," or "O.K.?" as fillers. Ask one of your colleagues (*not* a superior or subordinate!) to give you a reminder high sign when you slip back into the habits.
9. Organize the content so that it flows logically, but is not redundant. Avoid excessive detail.
10. Provide an oral summary—in the middle and at the end of your presentation, which highlights the issue and asks for a response. If appropriate, written handouts should be held to the end or at the completion of particular sections of your presentation.

When making a particularly significant or sensitive presentation, seeking out a resource person for a dry-run practice and coaching session can be helpful and confidence-building. While fulfilling a manager's role and responsibilities in other phases of the meeting his or her behavior must demonstrate flexibility, sensitivity, and maintenance of sense of direction and purpose. Leadership style in this context may range from a highly directive and closely controlled approach to a broad, participative approach involving active interaction among all members of the staff.

The leadership style should be consistent with the immediate and longer-range needs and goals that are concerned with both tasks and people. The approach needs to shift at the meeting from directive to consultative or participative, depending on the objective of the particular agenda item being addressed. If you want to inform; tell and sell. If you want others' ideas, feedback, or recommendations; ask, listen, and summarize what you have heard. If you are seeking full participation and consensus in reaching a mutual decision, move out of the subject content and focus on the group member maintenance and participation process. These objectives, functions, and requirements may be formally summarized as follows:

Task Functions

1.   Initiating: Proposing tasks or goals; defining a group problem; suggesting a procedure or ideas for solving a problem.

2.   Seeking information or opinions: Requesting facts; seeking relevant information about group concern; asking for expression of opinion; requesting a statement of estimate; soliciting expressions of value; seeking suggestions and ideas.

3.   Giving information or opinion: Offering facts; providing relevant information about group concern; stating a belief about a matter before the group; giving suggestions and ideas.

4.   Clarifying and elaborating: Interpreting ideas or suggestions; clearing up confusions; defining terms; indicating alternatives and issues before the group; building on others' ideas.

5.   Summarizing: Pulling together related ideas; restating suggestions after the group has discussed them; offering a decision or conclusion for the group to accept or reject.

6.   Consensus testing: Asking to see if group is nearing a decision; sending up trial balloon to test possible conclusion.

Maintenance Functions

Those functions which contribute to effective interaction and maintain group involvement include:

1.   Harmonizing: Attempting to reconcile disagreements; reducing tension; getting people to explore differences.

2.   Gate keeping: Helping to keep communication channels open; facilitating the participation of others; suggesting procedures that permit sharing remarks.

3.   Encouraging: Being friendly, warm, and responsive to others; indicating by facial expression or remark the acceptance of others; contributions.

4.   Compromising: When ideas or status is involved in a conflict, offering a compromise which yields status; admitting error, modifying in the interest of group cohesion or growth.

5.   Standard setting and testing: Testing whether group is satisfied with its procedures or suggesting new procedures, pointing out explicit or implicit norms which have been set to make them available for testing.

Every group needs both *task* and *maintenance* attention, and an adequate balance should be built and maintained between the two. It is first of all the manager's role responsibility to strike this balance through his or her own leadership style and approach.

In a directive-style group, major responsibility for organizing and conducting the effort characteristically rests with the boss or the chairman; *a majority of these functions are executed by him or her.* Delegation of authority is usually limited to minor routine assignments: "Joe, would you please watch the time for us." "Mary, please make sure we cover all the points." "Bill, please take notes." An able and aware manager will also be attempting to maintain participation, iron out differences, initiate or seek compromises, and observe group process and group maintenance-type functions. In addition the leader may desire to, or unintentionally, become involved in the actual content of the issues being discussed.

The leader is like a circus master watching the action in all three rings, but at the same time performing an act inside one of them. These many-sided responsibilities are often rather difficult to meet. Consequently, some needs of the group are often overlooked or left wanting.

Since "getting the job done" usually takes first priority in an authoritarian group with an authoritarian leader, *task needs take precedence. Maintenance needs and roles are often ignored or inadequately fulfilled.* Conflicts that are obvious to an observer may continue unabated and unperceived, or their consequences minimized, while the manager remains involved in the content of the task. Task-oriented members may also consider interpersonal relationship problems, or communication or participation issues as mere "personality clashes"—something to be avoided. When such problems do surface, they are not handled in public (before and with the group).

In some groups, members consider an authoritarian style to be supportive of their own needs and compatible with their values and standards as to "what is right" and "how businessmen should behave." In such instances maintenance concerns may be neglected with little negative effect. However, this is more the exception than the rule. Wherever people are interacting, there usually are differences in points of view. Conflicts of personal need and interest arise with some members dominating and others feeling they are being dominated, some talking and some not listening—all these opposing forces contribute to group maintenance problems. If these conflicts are left unresolved, they hinder the effectiveness of the decision-making.

Thus while an authoritarian, task-oriented group will be inclined to do the job, the quality and quantity of performance against time and effort expended often may be poor. The group process—"the *how*-we-go-about-what-we-do," that is, participation, member morale, or group understanding may block cooperation and creativity in dealing with the assigned job.

In the participative, democratic-oriented group, *task and maintenance leadership functions are distributed or informally assumed by several members of the group including the leader.* The manager may work with the rest of the group in deciding how the problems should be tackled, in what order, and through what procedures. Members in the group may, without solicitation, propose a specific role that the chairman assume, such as "time keeper and recorder," or "coordinator" so that everybody has a chance to express an individual opinion. As is implied by these examples, the role and authority of the leader of a democratically oriented group often cannot be defined by the leader on his or her own. The democratic principle, governing by the consent of the people, often comes into play even in the business setting.

By spontaneously assuming task and maintenance functions, members are able to secure additional or alternative ways of influencing others and expressing themselves. They gain self-esteem and the respect of others not only by contributing a creative solution for the problem at hand, but by displaying a particular talent for bringing others into a discussion and quieting down the overparticipator. Through expert listening one may be able to summarize effectively the important issues that have been raised and left unresolved. Competition for leadership often is siphoned off through this process of sharing task-maintenance functions.

The sharing of influence and functions helps to build and maintain positive group morale. All, or a majority of members, feel an incentive to continue to belong to the group. Through this association important personal satisfactions are achieved. "By being a member of this department, I feel important. It is an important department in the company. There are some sharp thinkers in this department. They challenge and stimulate my thinking, making it possible for me to come up with a better product." Members want to remain in a group like this where they have an opportunity to contribute and to be a part of the action.

In summary, *the participative group process,* and democratic leadership style *increases the potential for raising and maintaining high group morale, for inciting personal commitment, and providing opportunities for members to stretch their capacities to learn and develop.*

On the minus side, task productivity may sometimes suffer. Participative-style groups sometimes spend much time and effort engaged in "nontask" issues. These issues often include expressing feelings about what just happened in the department or at the staff meeting, encouraging some members to voice their points of view (perhaps even when they do not feel they have anything significant to give), discussing how to proceed in handling a particular task, or giving feedback to an "insensitive" leader or member.

It is often observed that groups moving from a highly controlled situation to a participative process, may become so fascinated, at least initially, with examining their feelings, giving feedback, and analyzing the process that they divert a great deal of attention and energy from the "real job they're being paid for."

As stated earlier *an adequate balance of task and maintenance must be found in order for the team to keep the gates open for full use of its resources.* This balance is not usually reached or maintained by happenstance. Trusting relationships must be established in order for openness and freedom to become the accepted standard. This means that in the process risk taking and defensiveness are reduced. To achieve this atmosphere of trust and balance of perspective, the manager and members must acquire group-maintenance skills. Such skills enable users to provide required remedial action.

While attention was given earlier to staff-meeting situations requiring the manager's leadership, there are other types of group interaction settings where your guidance and direction is needed, for example: *the task force team.* Special problems, study assignments, projects often set up on a matrix organization basis have some significant differences of history and needs as compared with the manager's regular direct-line staff meetings. Here the *team leadership* approach is particularly needed.

A task force or team must produce creative and practical plans and ideas to justify its existence. The right combination of group resources, utilized effectively, can result in a level of achievement greater than the effort of one individual or of several individuals working alone. This increase in the total effect that is greater than the sum of the effects taken independently is called *synergism.* Synergism is not automatic in the group process; it depends on the degree that group members are able to freely expose their knowledge and skills to one another and build constructively on their group potential.

For a team to build on its potential and stretch its output, team members need to develop an understanding and a "gut awareness" of their "here-and-now: interaction." If it is to BECOME —that is, develop into an effective, self-actualizing team—where individual members' and the total group's potential are more fully realized, members must KNOW what's going on in the group. You and they need to be able to identify and understand how group environment and group behavior are helping and hindering the synergistic process.

Some questions you need to address personally and within the group are:

How is group structure supporting or hindering our participation?

What are the strengths and weaknesses in the style and approach of our leadership?

In appraising individual member's attitudes and behavior, what personal improvements or skills are needed in order to release untapped resources within the group?

What kind of a group are we today?

What do we need to BECOME for the problems and challenges of tomorrow?

41.4.   CHOOSING THE BEST LINE-UP

To achieve synergism it is necessary to secure members who *complement* one another in talents, personality, experience, and motivation. Often the composition of the group is *too homogeneous.* People of similar experience and sentiments may come together and only agree with each other. They may unanimously point outside their group, placing the blame for their problems on others. When team participants find themselves not expressing ideas or reactions because everything has already been said by others in the group, it is likely that the membership is too alike—too homogeneous. Time is wasted in the repetition of facts and experiences. The challenge of differences and the stretch necessary for creativity and new approaches is sorely missing.

This sameness of personality type, participation style, or motivation may inhibit group effectiveness. A group may have too many "win-lose" personalities. Their overwhelming self-concern with competition blocks achievement and group progress. At the other extreme a group may have too many shy, nonparticipating types. Their lack of talent in communicating ideas, inspiring change, or challenging the status quo likewise blocks progress.

Therefore, in formulating a task force, care should be taken to *formulate a mixture of skills and resources:* outgoing innovators, precise theoreticians, group-process and problem-solving specialists, as well as some who are more knowledgeable regarding issues. Members with different experience or from different disciplines are brought together to allow for the exploration of a variety of alternative routes to a solution or decision.

Rotation and circulation of membership should also be considered wherever a group functions as a permanent or semipermanent body. New people coming into a stable group bring a refreshing new slant. They often question group plans and group standards, test the practicality of proposals, challenge complacency, and increase productivity.

41.5. WHO'S THE CAPTAIN?

Usually groups need an assigned or group-selected chairman or moderator during the early phase of formation. In business this leader is often the department head, the authority figure to whom some or all of the members have a full-time responsibility. However, there are also times when someone else in the group could equally, or more effectively, perform this moderator function.

For example, the nominal head could give the job to a subordinate who would benefit from the job-enrichment opportunity. This subordinate, who may be less personally involved in the history and content of the issues than the department head, may be more objective in implementing and coordinating the problem-solving process.

The following questions about group leadership should be critically considered:

Who is most appropriate to serve as chairman coordinator of the group? Should he or she be appointed by the organization or selected by the group?

What other special skills or knowledge will be required of the chairman in addition to his or her competencies in problem solving and conference leadership?

Will it be better if the moderator is not personally involved in the content of the task so that he or she can concentrate on helping the group move along rather than on selling his or her own ideas?

Would a change of leadership from department head to a group-selected or organization-appointed chairman be appropriate?

What leadership style would most encourage maximum productivity and/or critical development in group growth at this point in the life of the team?

Each group has a "personality" and operating style of its own. The personality of the group is determined by: its standards, resources, leadership and membership style, problem-solving approach, communication atmosphere, participation quality and quantity, influence pattern, and degree of satisfaction among its leaders and members. A single given type of group is not necessarily good or bad in itself. Rather, in order to judge whether a given group arrangement is good or bad or whether it is right or wrong, it is necessary to identify the group's "personality" and review the record of its results. This in turn must be evaluated against the company's need and purpose in creating the group. The relevant factors might be examined as follows in a typical situation:

Desired Results　　The top management of a company decided that a task force should be created for the official purpose of providing the company with freewheeling, creative ideas and radical alternatives regarding long-range company objectives and manpower policies and plans. Its recommendations would serve as a base for organization growth and development planning.

Group Membership Judged Appropriate　　Management felt that members of this planning group should be highly specialized technical experts from a variety of disciplines, with considerable knowledge about the industry as well as the company. Such employees would have strong, but well-tested opinions, be highly motivated to make responsible contributions, and reasonably disciplined to get the task accomplished in a responsible way.

Therefore, members selected had considerable tenure with the company and were well known for their technical expertise and their dependability and maturity. An executive who had a reputation for being a realistic taskmaster and expert in organization was chosen to head the task force. Leaders and members alike were believed to be results-centered, "no-nonsense," strong-willed people with definite opinions about what was needed for the organization and what its objectives should be.

Group's Interaction Style and Resulting Performance　　Because of the styles and backgrounds of its leader and members, as they continued to work together, they emerged into a highly organized, task-oriented, authoritarian-style group. They were, as expected, highly task-motivated and had strong opinions. However, because of their long tenure with the company and their rigid individual opinions, they were unwilling or incapable of being innovative. Their viewpoints were conflicting, and they took hard-line competitive positions.

Their leader, faced with an impasse with divergent points of view—and concerned about getting the job finished on time, forced a compromise solution which was neither creative nor satisfying to any of the group. The plan was reluctantly presented to top management by the task-force leader with other members of the group that were present. In the discussion following the presentation, it immediately became apparent that little or no commitment or consensus existed within the group regarding its recommendation. Top management rejected the proposal both for the lack of creativity and the absence of agreement.

The error in judgment on the part of management as to the group membership appropriate for the actual task requirements, resulted in a failure to achieve the desired results. The influence and relationship pattern among group members was in conflict with the conditions required for achievement of the company's purpose. In another situation this same group could be highly effective in developing a specific short-range plan or dealing with a series of more tangible problems. However, held to its original assignment, either the group would need to change its operating style, communication climate, and leadership dependency, or disband. The company would then be faced with the necessity of recruiting a new group which would be more appropriate in personality with the requirements of its purpose.

This latter option, recruiting a new group, is often impractical. It is generally more profitable for both the committee and the company to attempt to change and grow. If they can change, the group's potential is greatly increased. Members find motivational rewards of self-achievement and challenge.

It is important then, to consider further group-development concepts which can contribute to group change and growth. Heading your own group in staff meetings and special task-force assignments will require your maintaining a delicate balance of your attention between both technical and personal needs and goals. In these situations, as in management in general, getting results through people can best be accomplished by integrating those two objectives into an overall leadership approach, rather than trying to grope your way down that ambiguous middle road of compromise or trade-off. Flexibility, sensitivity and a broad repertoire of leadership skills need to be developed and utilized. This is the challenge and artistry of group management.

42

MANAGING YOUR TIME

Harold Lazarus

HAROLD LAZARUS, professor of Management and former dean, School of Business, Hofstra University, won trophies for teaching excellence there and at New York University's Graduate School of Business Administration. Professor Lazarus lectures in the United States and abroad on time management, leadership, management by objectives, communication, problem solving, decision making, and human relations. He is a consultant to corporations and government and was president of the Eastern Academy of Management and the Middle Atlantic Association of Colleges of Business Administration. Harold Lazarus serves or served on the Boards of Directors of Diplomat Electronics Corp., Ideal Toy Corp., Bond Clothing, North American Management Council, Continental Plastics, Superior Surgical Manufacturing Co., Ideal International, Interstate Molding, Crown Recreation, Rust Warehousing, and Alabe Products. He earned a Ph.D. in Management and an M.S. in Marketing from Columbia University.

42.1. INTRODUCTION

As engineering managers you are probably inundated with reading matter, meetings, and telephone calls and, most important of all, your project or task has time constraints and deadlines. The truth is that you have little control over time, you cannot really save it or manage it. You can however, utilize it appropriately. Do you? If you are like many engineering executives, the answer is probably "no." William Penn could have been observing engineering administrators when he said: "Time is what we want most, but what, alas, we use worst."

42.2. PURPOSE

This chapter is designed to help engineering managers make the best use of time. It will provide you with a set of simple, practical techniques which can be put to use immediately to get the most important things done promptly.

As an engineering administrator you need to review and organize the way in which you manage yourself and your time. The objectives of this chapter therefore are to present nuts-and-bolts systems and ideas for working smarter, not necessarily harder, and for making the best use of time and effort both on and off the job.

As an engineering executive you are probably barraged by interruptions, when you really need quiet time to think about such problems as, for example, environmental pollution. At this very moment, however, your desk is itself probably polluted with mounds of paper. How can you reduce the paperwork that plagues you? This chapter is designed to help you do that. It briefly lays out an arsenal of tested methods for eliminating nagging time wasters you may currently face. The chapter puts you in touch with approaches and options that will help you reach your goals faster and with fewer frustrations.

42.3. WHO SHOULD READ THIS CHAPTER?

This chapter is intended for those engineering managers who feel under time pressure. It is designed for engineers who like to be productive, yet want recreation and relaxation with friends, family, and hobbies. In addition it is for engineers who need more time to listen, think, plan, and enjoy their work.

The chapter was written with engineering executives in mind, for example, those managers of electronic and electrical engineers, mechanical engineers, civil engineers, chemical engineers, ceramic engineers, metallurgical engineers, and mining engineers.

The techniques discussed here will be useful for the supervisor of a construction site and the city engineer; those who work for manufacturing firms, utilities, government agencies, engineering consulting firms; those involved with technical sales; and those who estimate the time and cost of engineering projects. This chapter is also written for engineers who do not yet supervise other engineers and manage engineering functions but plan to do so in the future. A reader who is in this position recognizes that a time-wise engineer has a competitive edge in attempting to move up the organizational ladder.

42.4. NAGGING TIME WASTERS

The January–February 1984 issue of *Management Focus* reports on the results of a survey in which 120 personnel directors of medium-sized and large corporations estimated that the average American employee probably wastes as much as 4 months out of every year on the job, that is, more than 11 hours of paid worktime each week or 16 full workweeks each year. When asked what percentage of a typical American employee's paid workday they estimated is spent not working, these personnel directors estimated that on average it is approximately 32%.

Let us turn from employees in general to employees in engineering groups and list several of the most common ways in which engineering administrators waste time. How, for example, is time wasted in the largest branch of engineering? Electrical and electronic engineering managers have told this writer that poor procedures often waste time in the manufacture of electronic equipment, aircraft, business machines, and scientific instruments.

Electronic engineers who design, develop, test, and supervise the manufacture of computers, radar, and communications equipment report that their supervisors often "try to do it all themselves," that is, do not delegate sufficiently. Similarly, electrical engineers who design, develop, test, and supervise the manufacture of power-generating and transmission equipment used by electrical utilities describe their managers as "trying to do too much at once, so that little gets done really well."

A mechanical engineer who designs jet engines complains that his boss wastes time because of an inability to say "no." Although "the boss complains about interruptions, he welcomes visitors because they make him feel important," even though he needs time to plan without interruption.

A civil engineer who has helped design and construct factories, power plants, and airports in several countries reports that "in every one of them, confused responsibility has led to duplication of effort."

The manager of a chemical engineering department reports that her own fears and insecurities probably lead to the excessive testing of chemical products, overpreparing for meetings, and slow reading of engineering literature from the American Chemical Society and the American Institute of

Chemical Engineers. To help solve one of these problems she now has a young chemical engineer summarizing recent information for the firm and members of the department.

A ceramic engineer working with nonmetallic, inorganic materials used in energy conversion describes his supervisor as a "person who overcommunicates." He involves too many subordinates and gives them nonessential background information.

A metallurgical engineer working on jet engines that require metals which can withstand extreme heat reports to a "workaholic" supervisor who makes unnecessary work for himself.

What time wasters have you experienced in your engineering department? Unclear objectives? It is useful to clarify objectives, update them, and then keep the clear, updated objectives in mind.

Does lack of priorities waste time in your engineering firm? It is more productive to do a few important things than many unimportant ones.

Is time wasted in your engineering firm because of insufficient motivation, long telephone conversations, and excessive socializing? Is some of your own time wasted with junk mail? Should your secretary or assistant toss out the junk mail and classify the rest of it in terms of importance?

Do you tend to blame external factors for your time being wasted or do you accept some of the blame yourself? For example, how much of your time is spent attempting to impress others and/or attempting to win sympathy for workaholic martyrdom? Bad habits practiced for years are hard to break. With determination, it can, however, be done. Try substituting a good habit for a bad one. Make a resolution publicly, for example, to a friend, and reward yourself for approximations of good behavior, for keeping yourself on target, and for exercising self-discipline.

42.5.   RELEASE FROM TIME PRESSURES

How do the most productive people overcome time pressures and avoid working still harder and still longer? They have the same number of hours in the day that we do. Some, like Peter Drucker, save precious minutes, hours, and days by rejecting low-priority tasks. Drucker simply says "no" when this is appropriate. He has, therefore, rejected this writer's invitations to accept an honorary degree and to lecture at professional society meetings. When is it appropriate for you to say "no?"

A simple time log might help you determine when you should say "no." If you are caught in a time squeeze, it can help you analyze its causes in order to take corrective action. The secretary to a vice-president for engineering kept time logs for two days which compared his time plans with his actual performance. He learned that keeping time logs could help him to predict and reduce the length of meetings. Time logs also taught him that he must avoid unnecessary, self-generated tasks and that his open-door policy invited too many interruptions. As a result, his door is now open for part of the working day, not all of it.

For a couple of days you might ask your secretary to prepare simple, daily time logs containing three columns each. The left column headed "Time Plan," that is, your planned activities for the day should include your estimate of the time each activity is likely to take. He or she can then record your "Actual Performance" in the middle column, thus permitting you to compare it with your predictions and learn how long things actually take. The right-hand column can be headed "Learned." Recorded there is the information you acquired from the comparison, that is, did you make unnecessary work for yourself? As stated earlier, returns from the few most important projects are far greater than from numerous low-priority tasks. An engineering vice-president's two time logs are presented in Figures 42.1 and 42.2.

You will probably find, as would the electrical engineering managers mentioned earlier, that trying to do too many things at once means that none gets done well. Further analysis of your time logs may produce examples of overcommunication. As in the case of the ceramic engineering supervisor, you may involve too many people in the decision process and/or present unnecessary background information. Like the manager of the mechanical engineering department, you may welcome interruptions in an attempt both to win love and to feel important. The time log can help you to relearn the importance of clarifying and updating objectives, setting priorities, and then keeping these objectives and priorities in mind.

In addition to using time logs, ask objective, experienced engineering managers over whom you have no power to suggest ways in which you could better utilize your time. One recommendation might be that some meetings should be held standing up. Meetings rarely last long when all the parties are on their feet.

42.6.   TOOLS FOR TIME MANAGEMENT

A mechanical engineer who supervises a maintenance operation changed her behavior once she realized that there actually is enough time for the most important activities "if it's really important, I'll *make* time for it," she explains. In order to get better returns on her investment of time, she now

Time Plan for March 28	Actual Performance	Learned
9 a.m. arrive in the office and plan the day	8:55 arrived in the office and distributed notes written the night before	That's good; I'm starting to leave and arrive early.
9:15–10:30 Dictation	9:15–10:00 dictation and planned the day	Didn't really have to replan the day since I did it last night; dictation took less time than I expected.
10:30–12 Begin monthly report on operations	10–11:40 phone: Luke Moore Noble Adkinson Troncone Macleod Fisher signed my outgoing mail Ed stopped in for "a moment" again 11:40–12 worked on monthly report	Some of these conversations were unnecessary and too long; I let people impose, waste my time and then get angry, frustrated, tired, and not motivated to get back to work; let them know at the start that my time is limited and that conversation has to be short.

12–1:30 take staff to lunch	12–1:40 staff luncheon	That was fun and good for morale
1:30–5 complete monthly report on operations	1:45–2 dictated letter to Carton; spoke with Wicksel about department meetings and personnel matters	I need practice in predicting the length of time things will take
	2–2:50 discussed forecasts with Wicksel	
	2:55 stepped out of office with Ralph	
	3–3:45 worked on monthly report	
	3:45–3:50 men's room	
	3:50–5 worked on monthly report with Charlie	
5–5.45 meet with Kannellopoulous and Ganz to discuss research project	5–6:15 discussed research project	
6:00 leave for home	6:15–9:20 completed monthly report over dinner that was eaten at my desk; called Kannellopoulous and Ganz	Leaving work after 9 p.m. is workaholism; it's not fair to the family and to me.

FIGURE 42.1. Time log for a vice-president of engineering.

Time Plan for March 29	Actual Performance	Learned
9–10:00 dictation	9–9:15 discussed typing of monthly report with Eileen 9:15–10 dictation	
10–11 corrections of monthly report	10–11:30 meeting with Chasin and Roth on monthly report	Meeting should have taken 60 minutes, not 90 minutes.
11–12 review budget allocations and proposed schedule of vacations	11:30–11:45 spoke with Duffy about union negotiations and staff salaries 11:45–12 dictated notes to Barnes and Hammer	
12–1 lunch	12–12:25 lunch in office with Weiler; learned costs are up 9% 12:25–12:30 talked with Meyer about his career 12:30–12:45 examined room 201A with Smith and agreed upon repairs 12:45–12:55 signed letters	Only a workaholic takes 25 minutes for lunch while working; next time take a walk after lunch with Weiler. I should have delegated this; avoid unnecessary, self-generated tasks.

Plan	Activity	Comment
1–3:45 prepare forecasts so I can reserve the weekend for recreation	1:00–2:30 phone conversations with Hyer, Gutman, Meehan, Dragonetti, about CRASH program and need for office space.	Don't shoot from the lip and don't try to win everyone's love
	2:30–2:35 dictated letter of reference	
	2:35–4 prepared part of forecasts	
4–5 Dreyer's farewell party	4–4:10 Green stopped in	Open door invites interruptions; close it much of the time
	4:10–4:15 discussed Green's data with Selinger	
	4:15–5:30 farewell	Late for Dreyer's party
6 Promised I'd leave for home from Dreyer's farewell party	5:30–6:15 returned to office to complete 2 and 3 of forecasts	I did not review budget allocations and proposed vacation schedules; will spend Sunday evening on them. When will I learn to reserve weekends for family, friends and fun?
	6:15 left for home	

FIGURE 42.2. Time log for a vice-president of engineering.

uses prioritized daily "do lists." In them, highest priority activities are given the rank of A. Lower priority items are ranked B. Items that can be postponed are given C ranks. She also finds it useful to subnumber each activity. Then she tackles the activity designated A_1 before those designated A_2 and A_3 and certainly before those marked C_1 and C_2. In the author's case this chapter was written when it appeared as an item designated A_1 on his own "do lists." It had not been completed earlier because then it was ranked C_2 and therefore was postponed. As examples, two of the author's prioritized "do lists" appear as Figures 42.3 and 42.4.

The mechanical engineer mentioned earlier uses the famous C drawer for memos and letters with low priority that probably never need to be answered. Should any one of these then require response, that is, become an A or B item, the engineer pulls it out of the C drawer and responds. When the C drawer becomes filled, its contents are moved further away, until the "statute of limitations" permits the contents to be tossed out.

A chief industrial engineer was dissatisfied with the frequency, and productivity of meetings in his manufacturing company, so he recommended several changes. First, participants should know the purpose of meetings in advance. If there is no purpose, there should be no meeting. Second, agendas should be thoroughly planned and received in advance so people can prepare for meetings. Items on the agenda should be classified: (1) for action, (2) for discussion, and (3) for information. The chief industrial engineer also suggested the setting of beginning and ending times for all items on

A_1 Leave home early to avoid heavy morning traffic
A_2 Eat breakfast in the student cafeteria to learn something about the student body
A_3 Meet Brother Duffy, the department head
A_4 Interview student leaders
A_5 Talk with Dean O'Donnell
A_6 Lunch with professors Collins, Duffy, Berger, Schwartz, Galt, and Anderson
A_7 Evaluate Professor Collins' course in personnel management
A_8 Meet with Dr. Breslin, vice-president for Academic Affairs
A_9 While the experiences are still fresh, review and correct my notes on the evaluation of the Management Department.

B_1 Prepare the outline for my report to Dr. Dietemann, vice-president for Academic Affairs
B_2 Dictate my report, that is, the first draft
B_3 Return to my office to read two personnel files

C_1 Prepare outline for "Managing Your Time"
C_2 Dictate "Managing Your Time"
C_3 Sign and send memos and letters dictated earlier
C_4 Dictate answers to new mail

FIGURE 42.3.   Writer's prioritized "do list" for March 27.

A_1 Dictate "Managing Your Time"
A_2 Read Diplomat Electronics financials
A_3 Meet Carol for dinner at Marriott at 5:30

B_1 Invite Walter and Lola, Doug and Jill for dinner
B_2 Read G.B. 303 papers
B_3 Return phone calls
B_4 Sign my dictation
B_5 Answer new mail

C_1 Take George's call from Jacksonville
C_2 Send birthday cards to Cherie and Jean
C_3 Collect tax information for Ben
C_4 Were YPO brochures mailed?
C_5 Check on noise from rear of brown car
C_6 Haircut
C_7 Buy wild flower seeds
C_8 Should Phi Beta Kappa meet March 10 or 24 next year?
C_9 Who should be the Phi Beta Kappa speaker next March?
C_{10} The pianist?
C_{11} Call Mike about his contract

FIGURE 42.4.   Writer's prioritized "do list" for March 30.

the agenda and, thus for the whole meeting. His biggest gripe was that meetings began late and then ended late.

You can improve the effectiveness of your meetings if you analyze your responses to the 10 questions presented in Figure 42.5.

The industrial engineering chief mentioned earlier employs several time-saving techniques to control the growing demands on his time. He manages by exception (delegating routine details) and rejects reverse delegation, that is, he avoids solving problems that subordinates should solve, and, thus avoids becoming his subordinates' subordinate.

This chief of industrial engineering limits his span of control and therefore avoids the time pressure of having too many people reporting to him. He tries to handle a piece of paper no more than once. When appropriate, he writes answers on the original letter or memo and sends it back quickly, keeping a copy, if necessary. His subordinate industrial engineers prefer speedy replies to slower, formal responses. He uses PERT networks (see Chapter 17) and even old Gantt charts for planning and control purposes, and realizes that it is often faster and cheaper to prevent than to solve problems. Finally, although he attempts to put into practice the idea that doing a satisfactory job is often enough, he is still too perfectionistic about some unimportant projects.

A chemical engineering supervisor who is responsible for water quality

1. Was the agenda received in time to permit participants to prepare adequately for the meeting? _____ yes; _____ no
2. Did the meeting begin on time? _____ yes; _____ no
3. Did all participants understand the objectives of the meeting? _____ yes; _____ no
4. To what extent were the objectives of the meeting achieved?
5. Why is that? _____
   _____
6. Did participants know the items that were most important? _____ yes; _____ no
7. Were appropriate amounts of time spent on these? _____ yes; _____ no
8. Were items grouped appropriately, e.g., (a) for action, (b) for discussion, (c) for information? _____ yes; _____ no
9. Did your last meeting end on time? _____ yes; _____ no
10. What did you learn from your analysis of the last meeting that would help improve the effectiveness of the next ones? _____

FIGURE 42.5.   An analysis of your last meeting.

research avoids self-defeating procrastination by treating high-priority, time-consuming projects as if they were salamis. Frequently cutting slices from the "salami," it gets smaller and less awesome. She also uses a briefing board as a tool for planning, communication and control. This simple, magnetic board is a device for self control—the best kind of control.

This chemical engineering supervisor jots ideas for things to do on 3 x 5 cards and puts them under magnets in her "To Do" section at the top left corner of her briefing board. When it is time to start working on these projects, she moves the 3 x 5 cards to the center top section headed "Doing." She also jots additional notes to herself on the back of each 3 x 5 card for use in this project and next time it, or a similar project, surfaces, for example, "talk with Russ before beginning the report for next quarter."

When a task is completed the chemical engineering supervisor moves the 3 x 5 card to the "Done" section of the briefing board, that is, to the top right third of the magnetic board. This gives her a feeling of accomplishment and a place to hold cards for possible use again in the future.

On the lower left section of the briefing board (marked "Problems") the chemical engineering supervisor places 3 x 5 cards with notes about problems she cannot solve on her own. To the right of each of those problems in the section marked "Help" she lists, on a 3 x 5 card with an arrow pointing left, the names of those who could help her solve each problem, for example, her boss. A blank version of her briefing board is shown in Figure 42.6.

The chemical engineering supervisor stops unproductive activities quickly, recognizing that it is difficult to manage others when we do not manage ourselves well. "How can I expect subordinates to be on time, if I keep them waiting? So, I set my watch two minutes fast in order to leave and

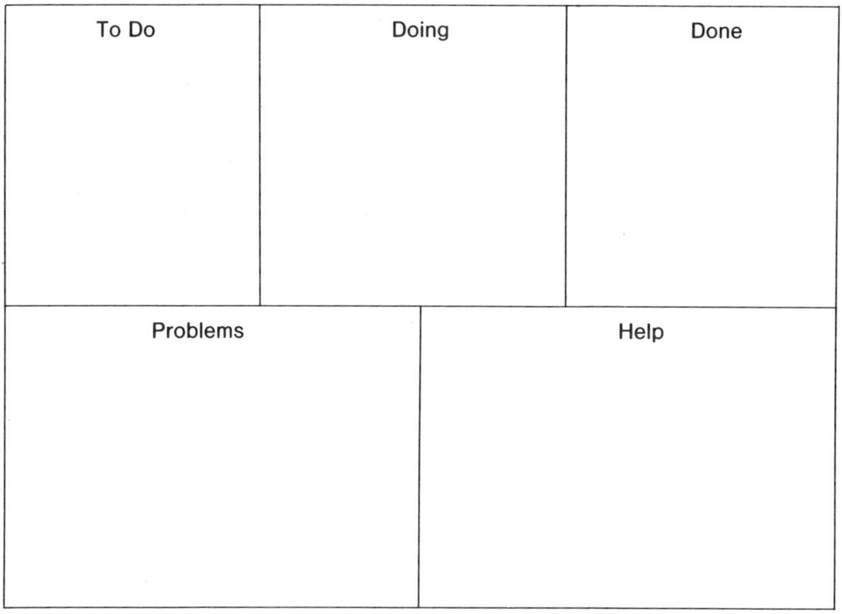

To Do	Doing	Done

Problems	Help

FIGURE 42.6.   A blank briefing board.

arrive a bit early. I tried setting it 10 minutes fast, but it didn't work for me. I couldn't fool myself that much," she laughed. "But when I set my watch a couple of minutes fast, I'd 'believe' the watch, leave early, and didn't have to drive like a maniac," she added.

Even though clients often keep him waiting, they rarely waste the time of an effective electrical engineer currently involved with technical sales. While waiting, he uses Scan Cards to plan high priority projects and his next calls. These Scan Cards contain important information about clients and potential clients and their needs on 3¼ x 3¼-inch color-coded cards. Thus, much of this engineer's "office" travels with him in four rows of slots, protected by a leather-colored, plastic 12½ x 9½ x ½-inch binder. The inside of his Scan-Card binder has four columns that this engineer has headed: "High-Priority Projects," "Today's Projects," "Standing Projects," and "Future Projects." In each of these four columns of slots are placed Scan Cards with names, addresses, phone numbers, product and service needs of clients and the status of their orders, if any. A simplified diagram of the inside of the Scan-Card binder is shown as Figure 42.7. (The Scan Cards and binder are manufactured by Executive Scan Card Systems of Columbus, Ohio.)

The electrical engineer mentioned in the last paragraph uses waiting time to catch up on his reading or to relax. He revises his life goals quarterly,

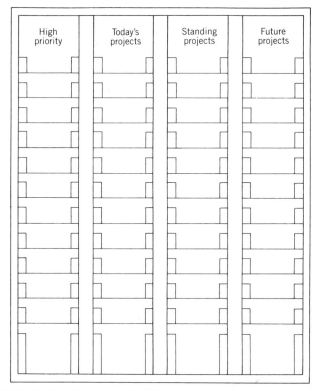

FIGURE 42.7.   Scan Cards help a sales engineer manage time.

keeps those goals in mind, and attempts to prevent problems, realizing that "an ounce of prevention . . . ." This engineer often asks himself, "What's the best use of my time right now?" Capitalizing on insights from the field of psychology, he rewards himself with time off and special treats for doing things that really matter, for meeting important deadlines, and for getting necessary paperwork done on time.

42.7.   EXCHANGING TIME-SAVING HINTS

Have you tried exchanging time-saving hints with engineering colleagues? How do they compensate for the fact that so many projects take longer than expected? How do they resist inappropriate interruptions? What can you learn from them about ways to respond when you are busy with an important task and are interrupted with the question "Do you have a minute?" How do effective engineers focus attention on high-priority tasks?

One effective senior engineer showed a group of assistant engineers and associate engineers two plastic, six-sided cubes that he kept on his desk. On each of the twelve sides of the cubes were time-management hints useful to him.

These hints are reproduced in Figure 42.8. Since the cubes shown were always on the senior engineer's desk, he frequently glanced at the hints. If

FIGURE 42.8. Time management hints on plastic desk cubes.

Time Wasters	Possible Causes	Possible Solutions
Not setting priorities	Lack of strong direction from officers	Setting time aside to determine priorities Taking a course in time management
Lengthy phone calls and apologies Discussion of number of calls before getting the party	Insufficient self discipline; Need to socialize	3-minute "hour" glass times all calls Set time blocks for telephone calls Audit business value and length of calls and ask quick questions
Interruptions and distractions	Cutting it short will anger speaker Submissive behavior Internal and external causes and inability to say "no"	Practice saying "no" Set time blocks for telephone calls
Working on small details, not important items	Moving too fast or without thinking	Stick to priorities
Duplication of effort	Too many people doing the same task	Organization chart Job description
Doing what one likes to do and letting unpopular work slide or pile-up	Poor leadership; lack of self-discipline	Concentrate on priorities
Being able to organize other's work but needing guidelines for self	Insufficient planning	Analysis of one's work habits Set goals
Marking time	Work does not flow evenly; there is slack time [when people have little to do]	Improve work flow so that quiet periods are less quiet and busy periods are less busy
Socializing	Is easier than working and a substitute for work one wants to avoid	Set deadlines and reward for meeting them
Visitors	They need information or attention and want to talk with you rather than a subordinate	Use telephone for brief conversations or instructions Well thought-out procedures Don't acknowledge presence of a visitor until you have time

FIGURE 42.9. Time survey for an engineering firm.

INDEX